W9-CPE-002

PRINCIPLES OF CMOS VLSI DESIGN

PRINCIPLES OF CMOS VLSI DESIGN

A Systems Perspective

Neil H. E. Weste
AT&T Bell Laboratories

Kamran Eshraghian
University of Adelaide

ADDISON-WESLEY PUBLISHING COMPANY
Reading, Massachusetts • Menlo Park, California
Don Mills, Ontario • Wokingham, England • Amsterdam
Sydney • Singapore • Tokyo • Mexico City
Bogotá • Santiago • San Juan

M. C. Varley, *Sponsoring Editor*

Hugh Crawford, *Manufacturing Supervisor*
Mary Crittendon, *Project Supervisor*
Richard Hannus, *Cover Designer*
Maureen Langer, *Text Designer*
Elydia P. Siegel, *Production Editor*

This book is in the Addison-Wesley **VLSI Systems Series**
Lynn Conway and Charles Seitz, *Consulting Editors*

Library of Congress Cataloging in Publication Data

Weste, Neil.
Principles of CMOS VLSI design.

Bibliography: p.
Includes index.
1. Integrated circuits—Very large scale integration
—Design and construction. 2. Metaloxide semiconductors,
Complementary. I. Eshraghian, Kamran. II. Title.
III. Title: Principles of C.M.O.S. V.L.S.I. design.
TK7874.W46 1985 621.381′73 84–16738
ISBN 0–201–08222–5

Reprinted with corrections June, 1988

KLMNOPQRST-HA-89

To Avril, Melissa, and Tamara
and Deidre, Michelle, and Kylie

The VLSI Systems Series

Lynn Conway *Consulting*
Charles Seitz *Editors*

VLSI Signal Processing: A Bit-Serial Approach	Peter Denyer and David Renshaw, 1985
The Design and Analysis of VLSI Circuits	Lance A. Glasser and Daniel W. Dobberpuhl, 1985
The VLSI Designer's Library	John Newkirk and Robert Mathews, 1983
Structured VLSI Design	Charles Seitz, 1986
Principles of CMOS VLSI Design: A Systems Perspective	Neil Weste and Kamran Eshraghian, 1985

FOREWORD

The subject of VLSI systems spans a broad range of disciplines, including semiconductor devices and processing, integrated electronic circuits, digital logic, design disciplines and tools for creating complex systems, and the architecture, algorithms, and applications of complete VLSI systems. The Addison-Wesley VLSI Systems Series is being organized as a set of textbooks and research references that present the best current work across this exciting and diverse field, with each book providing for its subject a perspective that ties it to related disciplines.

Principles of CMOS VLSI Design: A Systems Perspective by Neil Weste and Kamran Eshraghian provides both students and practicing system designers with a solid introduction to custom VLSI design in the complementary MOS (CMOS) technologies. The past several years have seen a rapid shift in the technology of choice for high-complexity digital microelectronics from nMOS to CMOS. This shift has occurred because CMOS offers high performance at low power, and scales extremely well to small feature size. In spite of its advantages, and its extensive use in semi-custom gate-arrays and custom commodity parts, CMOS has yet to be exploited to its full potential by the VLSI system design community. CMOS design and layout presents several intimidating complexities that Weste and Eshraghian have effectively put to rest by the way in which they have adapted hierarchical, structured design methods and layout abstraction to CMOS technology.

The presentation of a coherent design style together with many practical design examples allows this book to be used either as a text or as a reference. Those readers already skilled in VLSI design in the nMOS medium will find that this book yields almost instant access to knowledge about CMOS design and many starting points for CMOS VLSI system design. The book clarifies the similarities and differences among the CMOS process variants and their influence on system design. The symbolic layout approach developed in the book further helps the designer step back from the differential details of the process variants. We believe that designers, researchers, and design tool builders will all find this book to be a valuable reference as they jointly explore the frontiers of CMOS VLSI design.

Lynn Conway
Ann Arbor, Michigan

Chuck Seitz
Pasadena, California

PREFACE

Recently there has been an interest in expanding the set of people engaged in the design and specification of integrated circuits. This has occurred in two main thrusts. The text *Introduction to VLSI Systems* by Mead and Conway advocated what is now commonly called "structured hierarchical design," accompanied by a reduced and simplified geometric and electrical rule set. That text was based upon an nMOS depletion load technology. Design responsibility extended down to layout details. An alternative movement, largely supported by industry (as opposed to academia), has placed custom I.C. design capability at the logic level at the disposal of many system designers. This has largely been in the form of CMOS gate arrays and, more recently, CMOS standard cells.

This text has been written to assist those who wish to go beyond the standard cell and gate array approaches and realize fully custom designs that completely utilize the potential of the silicon surface.

The material in this book is divided into several parts. The first part deals with CMOS circuit design and CMOS processing technology. The second part deals with design issues and sub-system design. The last part is devoted to a rich set of examples of custom-designed CMOS circuits from which the reader may draw on the experience of other VLSI system designers.

A centralized theme in the book is the adoption of a symbolic layout approach to CMOS design. Most layout examples are given in this form with some mask level layouts for a typical bulk CMOS process. However, the symbolic designs are provided where necessary to provide layouts with some lifetime.

This text originated with a course that Weste taught at the University of North Carolina (Chapel Hill) and Duke University in the spring of 1982. An expanded course was taught by Eshraghian at Duke in the spring of 1983 and at the University of Adelaide in 1983/84. Dr. Kishor Trivedi taught the Duke course in the spring of 1984. A similar course was also taught at AT&T Bell Labs (Holmdel) in 1983.

The authors would like to acknowledge the support and help of many people during the preparation of the text. Bryan Ackland provided outstanding contributions to the outcome of the text, including key rewrites to Chapters 2 and 4. Kishor Trivedi was kind enough to debug a draft form of the text in his CMOS class. Steven H. F. Law, Gershon Kedem, Dave Ditzel, Don MacLennan, Malcolm Haskard, Alan Marriage, Marcus Paltridge, Jim Cherry, Richard Lyon, Mike Maul, Randy Katz, Jonathan Allan, and colleagues in the Computer Systems Research Laboratory of AT&T Bell Labs provided much needed comments and criticisms of the first draft. Jay Borris provided a great resource in the assembly of the first draft. The support of R. L. Andersson, S. C. Knauer, J. H. O'Neill, A. Huang, John W. Poulton, Henry Fuchs, Alan Paeth, R. H. Krambeck, and N. S. Vasanthavada is appreciated for their contributions to Chapter 9. Alex Dickinson, Charles Poirier, and Martin Levy provided support in the latter stages of the book.

Furthermore, the authors would like to acknowledge AT&T Bell Labs management, especially Bill Ninke, and staff for providing the experience, the atmosphere, and the resources without which this book would never have been completed. Additionally, The Microelectronics Centre of North Carolina and the associated universities, particularly Duke and UNC (Chapel Hill), provided the academic environment where work on this text was started. The University of Adelaide and Symbolics Inc. provided ongoing support for the work on the book.

Cambridge, Massachusetts N. W.
Adelaide, South Australia K. E.

ABOUT THE AUTHORS

Neil Weste is the Director of VLSI Systems at Symbolics Inc., in addition to holding a position as an Adjunct Professor in Computer Science at Duke University. Prior to joining Symbolics Inc., Weste spent six years at AT&T Bell Labs in Holmdel, New Jersey. He worked one year at the Microelectronics Center of North Carolina with teaching duties at Duke University and the University of North Carolina (Chapel Hill). Weste received his B.S., B.E., and Ph.D. from the University of Adelaide, South Australia.

Kamran Eshraghian is a senior lecturer in Electrical Engineering at the University of Adelaide, South Australia. In addition to CMOS VLSI Design, his research interests include Signal Processing. Eshraghian received his B.S., B.E., and Ph.D. from the University of Adelaide, South Australia. Eshraghian spent one year at the Microelectronics Center of North Carolina and Duke University. Prior to teaching, Eshraghian was with Philips Ltd. as an IC designer.

CONTENTS

2

MOS TRANSISTOR THEORY 31

3

CMOS PROCESSING TECHNOLOGY 63

PRINCIPLES OF CMOS VLSI DESIGN

INTRODUCTION TO CMOS TECHNOLOGY

PART 1

This Part orients the system designer toward CMOS technology. Chapter 1 gives a brief overview of CMOS circuit design. Chapter 2 deals with basic MOS transistor theory. Chapter 3 summarizes some CMOS processing technologies and also introduces typical geometric design rules. Chapter 4 introduces techniques to estimate performance of CMOS circuits. Chapter 5 covers at some depth the various alternatives available to the CMOS circuit designer.

INTRODUCTION TO CMOS CIRCUITS

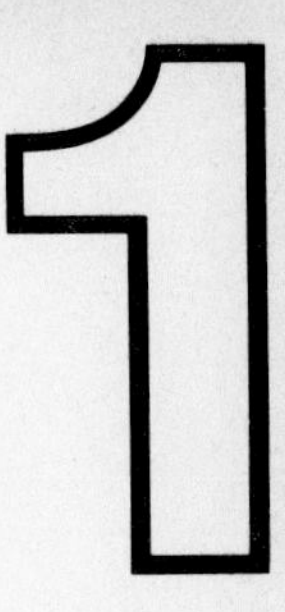

1.1 Introduction

Over the past few years, Complementary Metal Oxide Silicon (CMOS for short) technology has played an increasingly important role in the world integrated circuit industry. Not that CMOS technology is that new. In fact the basic principle behind the MOS field effect transistor was proposed by J. Lilienfeld as early as 1925 and a similar structure closely resembling a modern MOS transistor was proposed by O. Heil in 1935. Material problems foiled these early attempts. Experiments with early field effect transistors led to the invention of the bipolar transistor. The success of the latter device led to a decline in interest in the MOS transistor. MOS devices remained an oddity until the invention of the silicon planar process in the early 1960s. Material and quality control problems dogged the introduction of the MOS device into commercial uses until around 1967 [Cobb70]. Even then, single polarity p-type transistors or n-transistors were favored. The use of both polarity devices on the same substrate was initially used for very low power applications such as watches. As the processing technology required in the fabrication of CMOS circuits was more complex than that for single polarity transistors, CMOS was sparingly applied to general system designs. As nMOS production processes became more complicated, the additional complexity of the basic CMOS process decreased in importance. Additionally, system designers were being faced with very large chip sizes and power consumptions. For this, and other reasons that will become evident during the course of this book, CMOS technology has increased in its level of importance as a VLSI technology.

The purpose of this book is to provide the designers of hardware or software systems with an understanding of CMOS technology, circuit design, layout, and system design sufficient to feel confident with the technology. The text deals with the technology down to the layout level of detail, thereby providing a bridge from a circuit to a form that may be processed. At the present time, relatively automated design approaches can take a logic schematic and automatically convert these to a chip layout. However, these approaches do not really capitalize on the fundamental objects available in an IC — transistors. Hopefully, with texts such as this, software systems may be constructed that capture an expert's knowledge so that arbitrarily structured and architected silicon systems of enormous complexity can be built rapidly and accurately.

The book is divided into three main sections. Chapters 1–5 essentially provide a circuit view of CMOS IC design. In the first chapter, a rather idealized view of CMOS technology will be taken and some basic forms of logic and memory will be introduced. This is aimed at providing an unencumbered picture of the technology without delving into unnecessary detail. Chapter 2 deals at a greater depth with the operation of the MOS transistor and the DC operation of the CMOS inverter and a few other basic circuits of interest. The phenomenon commonly known as *latch-up* is also discussed. Some background to CMOS processing technology is presented in Chapter 3. The basic processes in current use are described along with some interesting process enhancements. Some representative design rules are also presented in this chapter. Chapter 4 treats the important subject of performance estimation and characterization of circuit operation. This covers speed and power dissipation. A section summarizing some first order scaling effects is also included. A summary of basic CMOS circuit forms is provided in Chapter 5. Various clocking schemes are discussed with the emphasis being placed on circuit design and layout. The second section comprises Chapters 6–8. These chapters present a *sub-system* view of CMOS design. Chapter 6 focusses on a range of current design methods, identifying where appropriate the factors in common with CMOS. A section on testing is also included. Symbolic layout techniques are discussed in Chapter 7, with particular emphasis placed on a design system implemented by one of the authors. This is included to give an idea of some of the components required in a custom CMOS design system. Chapter 8 is a rather hefty chapter on sub-system design, using the circuits discussed in Chapter 5. Discussion is commenced with a variety of adder designs. RAMs, ROMs, and PLAs are then covered. The final section is contained in Chapter 9. It consists of five examples of CMOS ICs of varying complexity. The purpose of these chapters is to illustrate the architectural decisions that lead to a full custom chip design. Where appropriate the specific relationships to CMOS technology are noted.

1.2 MOS transistors

An MOS (Metal-Oxide-Silicon) structure is created by superimposing several layers of conducting, insulating, and transistor forming materials. After a series of processing steps, a typical structure might consist of levels called diffusion, polysilicon, and metal that are separated by insulating layers. CMOS technology provides two types of transistors (also called *devices* in this text), an n-type transistor (nMOS) and a p-type transistor (pMOS). These are fabricated in silicon by using either *n*egatively doped silicon that is rich in electrons (negatively charged) or *p*ositively doped silicon that is rich in holes (the dual of electrons and positively charged). Typical physical

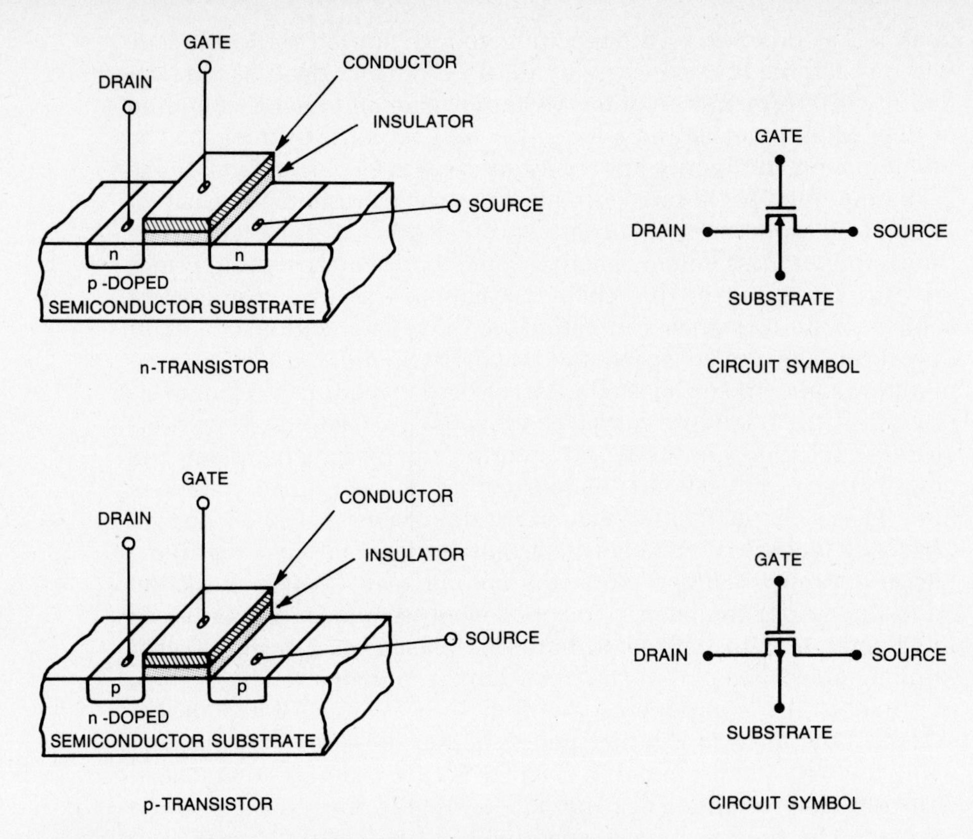

FIGURE 1.1. MOS transistor physical structures

structures for the two types of MOS transistors are shown in Fig. 1.1. For the n-transistor, the structure consists of a section of p-type silicon separating two diffused areas of n-type silicon. The area separating the n regions is capped with a sandwich consisting of an insulator and a conducting electrode called the GATE. Similarly, for the p-transistor the structure consists of a section of n-type silicon separating two p-type diffused areas. The p-transistor also has a gate electrode. For the purpose of introduction, we will assume that the transistors have two additional connections, which are designated the DRAIN and the SOURCE, these being formed by the n (p in the case of a p-device) diffused regions. The gate is a control input — it affects the flow of electrical current between the drain and source. In fact the drain and source may be viewed as two switched terminals. They are physically equivalent and the name assignment depends on the direction of current flow. For now, we will regard them as interchangeable.

1.3 MOS transistor switches

The gate controls the passage of current between the drain and source. Simplifying this to the extreme allows the MOS transistors to be viewed as simple on/off switches. In the following discussion, we will assume that a '1' is a high voltage that is normally set to 5 volts and called POWER or V_{DD}. The symbol '0' will be assumed to be a low voltage that is normally set to 0 volts and called GROUND or V_{SS}. The strength of the '1' and '0' signals can vary. The "strength" of a signal is measured by its ability to sink or source current. In general, the stronger a signal, the more current it can source or sink. Where the term output and input are used, the output will be the source of stronger '1's and '0's than the input. The power supplies (V_{DD} and V_{SS}) are the source of the strongest '1's and '0's.

The nMOS switch (N-SWITCH) is shown in Fig. 1.2a. The schematic representation is shown along with a switch representation. The gate has been labeled with signal s, the drain a, and the source b. In an N-SWITCH, the switch is closed or 'ON' if the drain and source are connected. This occurs when there is a '1' on the GATE. The switch is open or 'OFF' if the drain and source are disconnected. A '0' on the GATE ensures this condition. These conditions are summarized in Fig. 1.2b. An N-SWITCH is almost a perfect switch when a '0' is to be passed from an output to an input (say a to b).

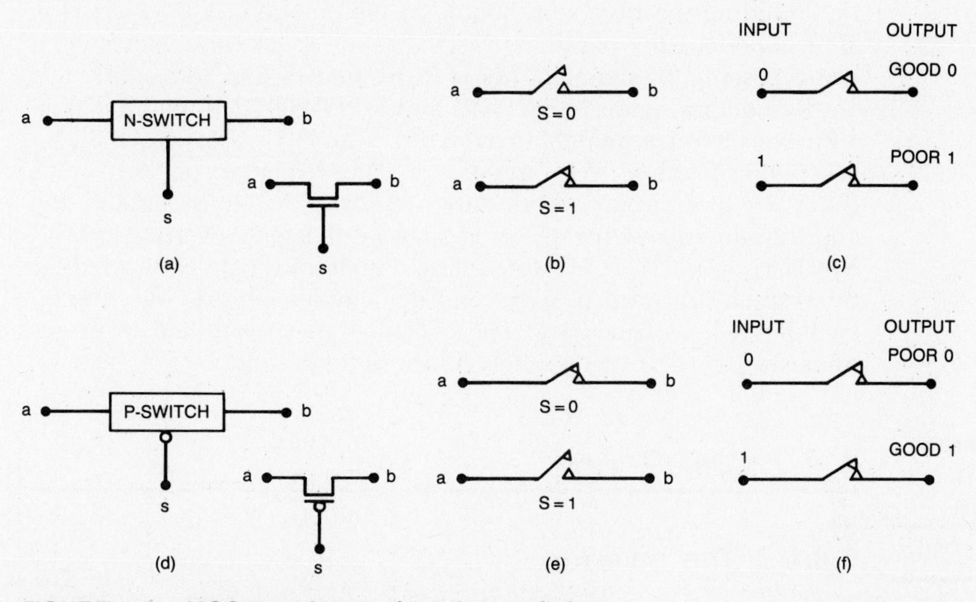

FIGURE 1.2. MOS transistors viewed as switches

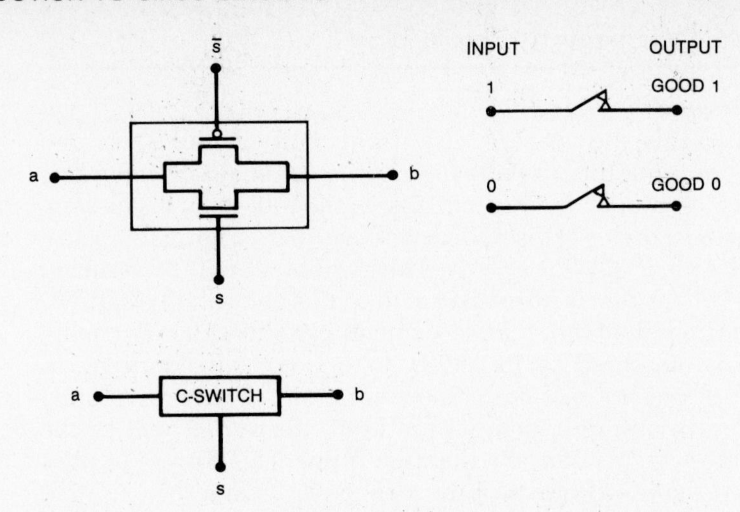

FIGURE 1.3. A complementary switch

However, the N-SWITCH is an imperfect switch when passing a '1'. In doing this, the '1' voltage level is reduced a little. (This is explained in Section 2.5.) These cases are shown in Fig. 1.2c. The pMOS switch (P-SWITCH) is shown in Fig. 1.2d. It has different properties than the N-SWITCH. The P-SWITCH is closed or 'ON' when there is a '0' on the gate. The switch is open or 'OFF' when there is a '1' on the gate. Fig. 1.2e depicts these conditions. Notice that the pMOS and nMOS switches are ON and OFF for complementary values of the gate signal. We denote this difference for a P-SWITCH by including the inversion bubble in the notation. A P-SWITCH is almost perfect for passing '1' signals but is an imperfect switch when passing '0' signals. This is illustrated in Fig. 1.2f.

By combining an N-SWITCH and a P-SWITCH in parallel (Fig. 1.3), we obtain a switch in which '0's and '1's are passed in an acceptable fashion. We term this a complementary switch or C-SWITCH. In a circuit where only a '0' or '1' has to be passed, the appropriate sub-switch (N or P) may be deleted, reverting to a P-SWITCH or N-SWITCH. Note that a double rail logic is implied for this switch (the control input and its complement are routed to all switches where necessary. The control signal is applied to the n-transistor and the complement to the p-transistor.).

TABLE 1.1. Inverter truth table

INPUT	OUTPUT
0	1
1	0

1.4 CMOS logic

1.4.1 The inverter

Table 1.1 outlines the necessary states to implement a logical inverter. If we examine this table we find that when there is a '0' on the

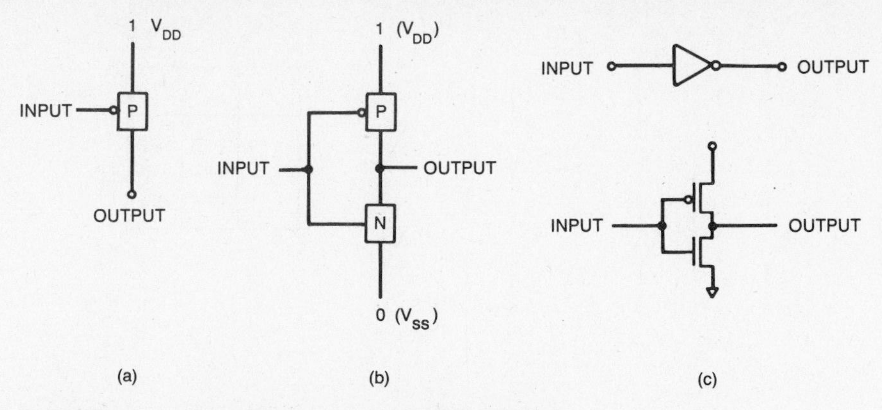

FIGURE 1.4. Construction of a CMOS inverter

input, there is a '1' at the output. This suggests a P-SWITCH connected from a '1' source (V_{DD}) to the output as shown in Fig. 1.4a. When there is a '1' on the input a '0' has to be connected to the output. This suggests the addition of an N-SWITCH between the output and a '0' source (V_{SS}). The completed circuit is shown in Fig. 1.4b. Note that as the lower switch only has to pass a '0' (the V_{SS} source of '0's is stronger than the output of the inverter), only an N-SWITCH is needed. By similar reasoning, the upper switch, which only has to pass a '1', needs only a P-SWITCH. The transistor schematic and schematic icon forms for this are shown in Fig. 1.4c. In general, a fully complementary CMOS gate always has an N-SWITCH (pull-down) array to connect the output to '0' (V_{SS}) and a P-SWITCH (pull-up) array to connect the output to '1' (V_{DD}).

1.4.2 Combinational logic

If two N-SWITCHES are placed in series, as shown in Fig. 1.5a on page 10, then the composite switch constructed by this action is closed (or ON) if both switches are closed (or ON) as illustrated in Fig. 1.5a.This yields an "AND" function. The corresponding structure for P-switches is shown in Fig. 1.5b. The composite switch is closed if both inputs are set to '0'.

When two N-SWITCHES are placed in parallel (Fig. 1.5c), the composite switch is closed if either switch is closed (either input is a '1'). Thus an 'OR' function is created. The switch shown in Fig. 1.5d is composed of two P-SWITCHES placed in parallel. In contrast to the previous case, if either input is a '0' the switch is closed.

By using combinations of these constructions, CMOS combinational gates may be constructed.

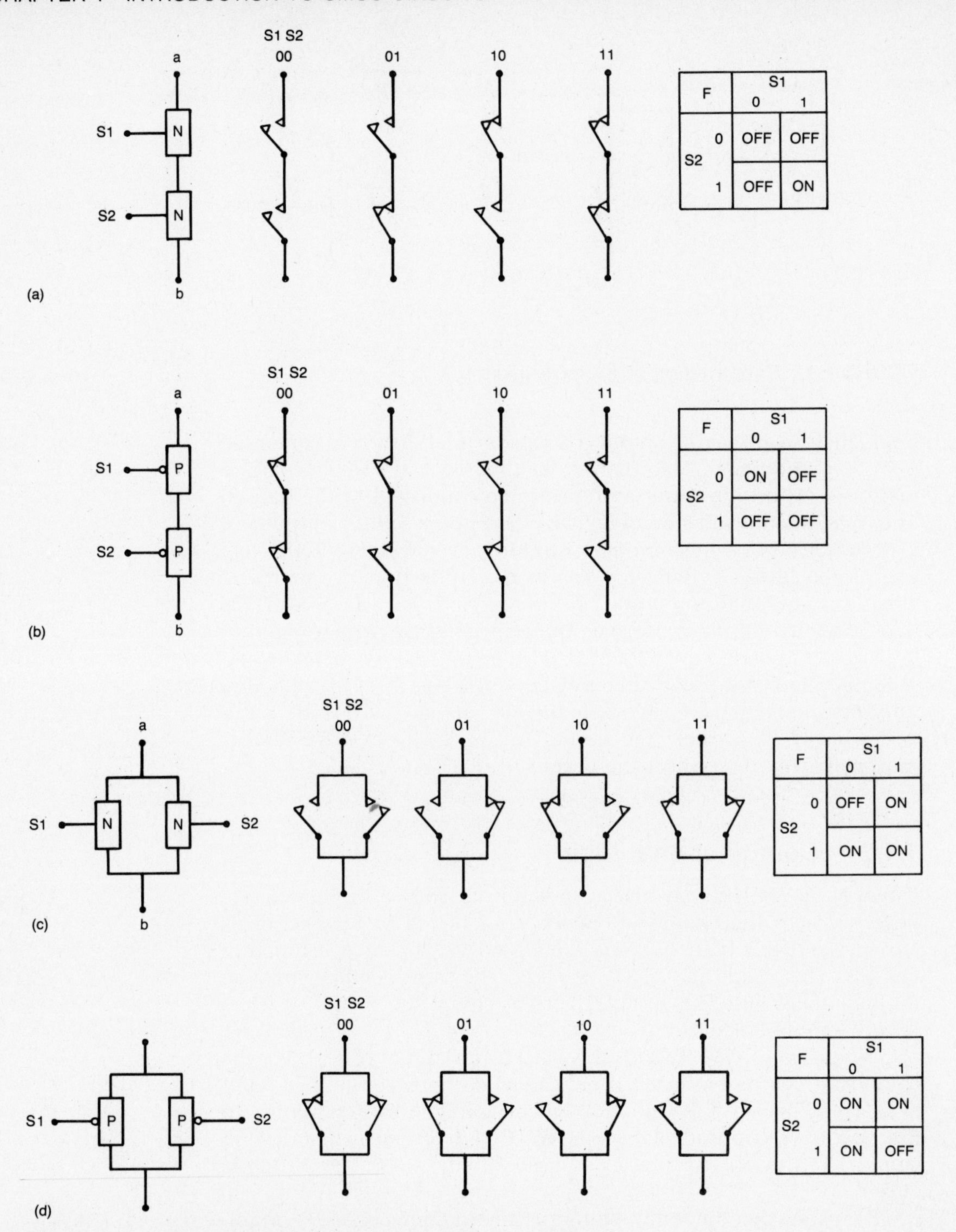

FIGURE 1.5. Series and parallel CMOS switch combinations

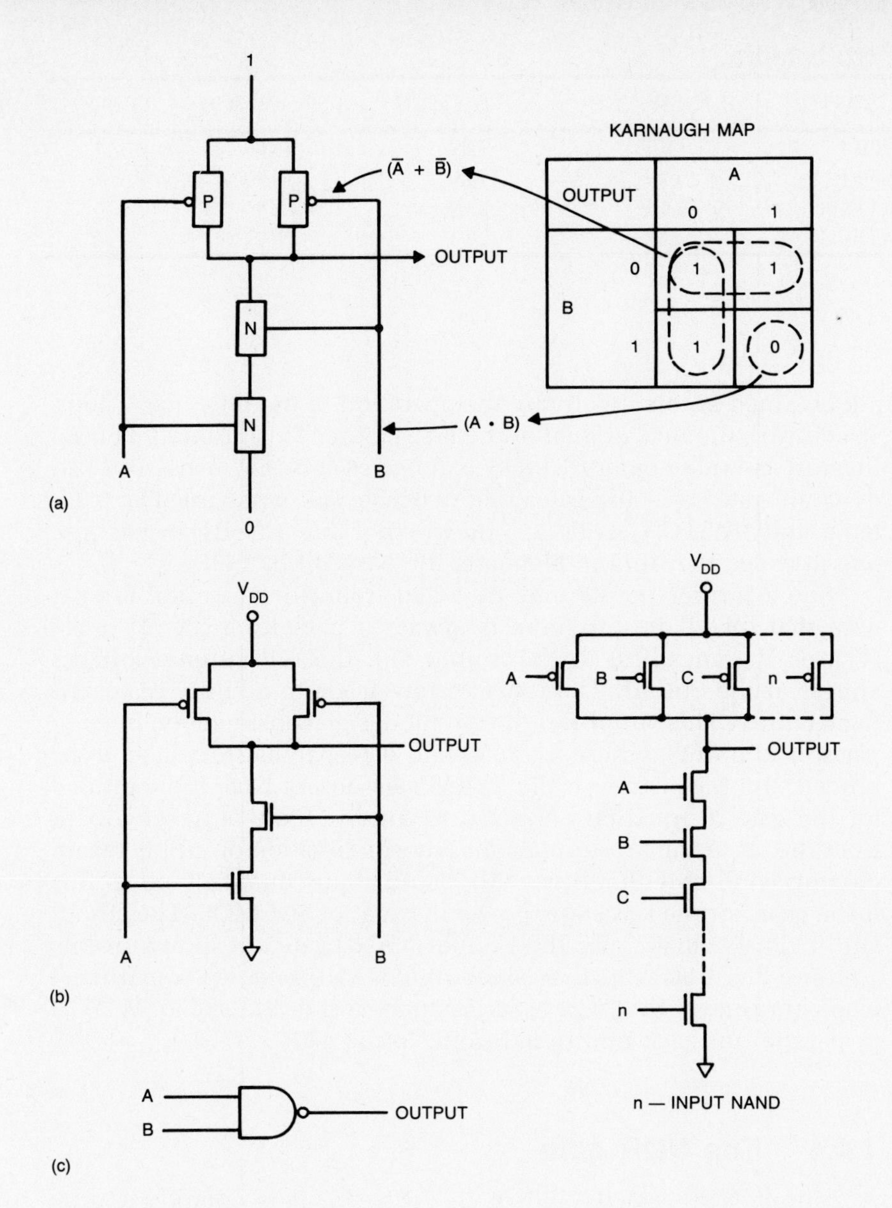

FIGURE 1.6. A CMOS NAND gate

1.4.3 The NAND gate

Fig. 1.6 outlines the construction of a 2-input NAND gate using the constructions introduced in Fig. 1.5a and Fig. 1.5d. These structures are derived by examining the Karnaugh map in Fig. 1.6a. The '0' term (pull-down to '0') dictates an AND structure (A.B). Grouping the '1's together results in a structure to perform $\overline{A} + \overline{B}$. This is realized by the parallel p OR structure. The complemented signals

TABLE 1.2. NAND gate truth table

A INPUT	B INPUT	A N-SWITCH	B N-SWITCH	A P-SWITCH	B P-SWITCH	OUTPUT
0	0	OFF	OFF	ON	ON	1
0	1	OFF	ON	ON	OFF	1
1	0	ON	OFF	ON	OFF	1
1	1	ON	ON	OFF	OFF	0

are obtained automatically by the operation of the p-device. The p-structure is the logical dual of the n-structure. This property is used in most complementary CMOS logic gates (but not necessarily in dynamic gates or static gates that dissipate static power). The truth table and SWITCH states are shown in Table 1.2. By inspection, one may see that this implements the NAND function.

Some further points may be noted from this example. Firstly, note that for all inputs there is always a path from the '1' or '0' (V_{DD} or V_{SS} supplies) to the output and the full supply voltages appear at the output. The latter feature leads to a "fully restored" logic family. This simplifies the circuit design considerably. In comparison to nMOS, where the load and driver transistors have to be ratioed, the transistors in the CMOS gate do not have to be ratioed for the gate to function correctly. Secondly, there is never a path from the '1' to the '0' supplies for any combination of inputs (again in contrast to nMOS). As we will learn in subsequent chapters, this is the basis for the low static power dissipation in CMOS. The circuit and logic schematics for the 2-input NAND gate are shown in Fig. 1.6b and Fig. 1.6c. Note that larger input NAND gates are constructed by placing one N-SWITCH in series on the n side and one P-SWITCH in parallel for each additional input to the gate.

1.4.4 The NOR gate

A 2-input NOR gate is shown in Fig. 1.7a. It is composed from sections introduced in Fig. 1.5b and Fig. 1.5c, according to the Karnaugh map. Note that the N and P switch combinations are the dual or complement of that for the NAND gate. The truth table is shown in Table 1.3. This implements a logical NOR operation. The corresponding schematics are shown in Fig. 1.7b and Fig. 1.7c. In comparison to the NAND gate, extra inputs are accommodated in the NOR structure by adding N-SWITCHES in parallel and P-SWITCHES in series with the corresponding switch structures.

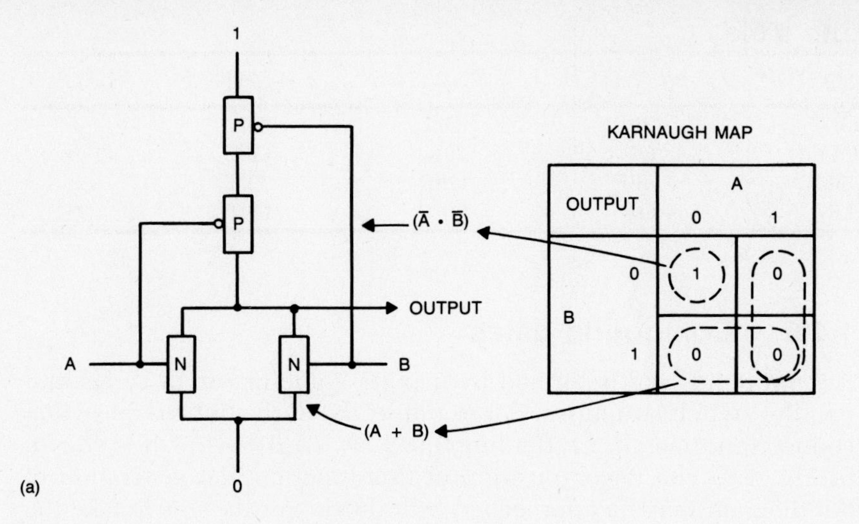

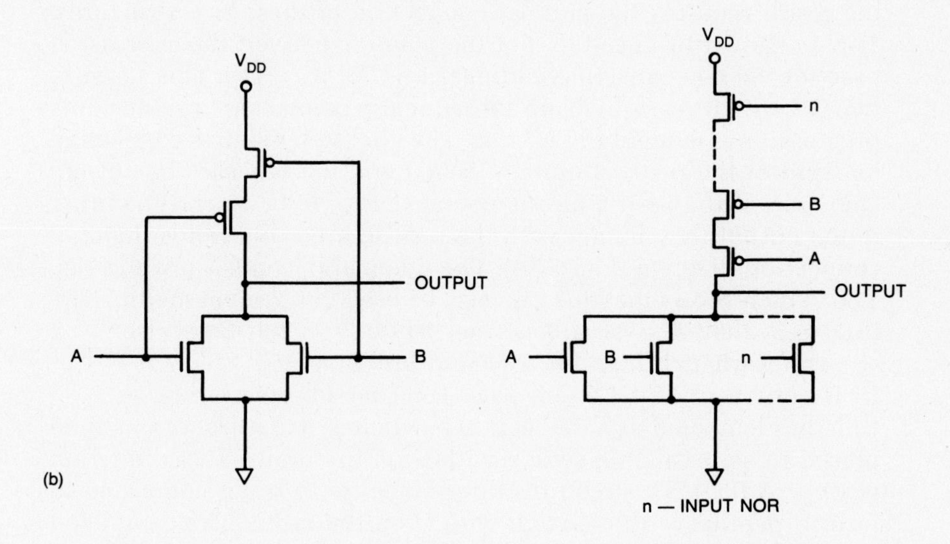

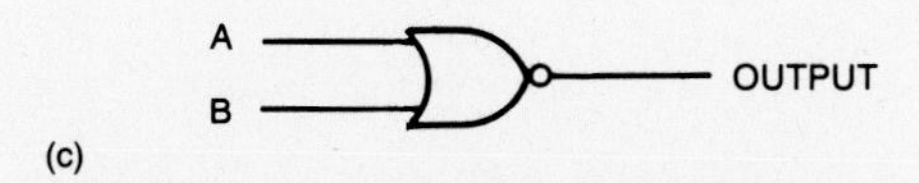

FIGURE 1.7. A CMOS NOR gate

TABLE 1.3. NOR gate truth table

A INPUT	B INPUT	A N-SWITCH	B N-SWITCH	A P-SWITCH	B P-SWITCH	OUTPUT
0	0	OFF	OFF	ON	ON	1
0	1	OFF	ON	ON	OFF	0
1	0	ON	OFF	OFF	ON	0
1	1	ON	ON	OFF	OFF	0

1.4.5 Compound gates

A compound gate is formed by using a combination of series and parallel switch structures. For example, the derivation of the switch connection diagram for the function $F = \overline{((A.B) + (C.D))}$ is shown in Fig. 1.8. The decomposition of this function and generation of the diagram may be approached as follows. For the n-side take the uninverted expression $((A.B) + (C.D.))$. The AND expressions $(A.B)$ and $(C.D)$ may be implemented by series connections of switches as shown in Fig. 1.8a. Now taking these as subentities and ORing the result requires the parallel connection of these two structures. This is shown in Fig. 1.8b. For the p-side we invert the expression used for the n-expansion, yielding $((\overline{A} + \overline{B}) . (\overline{C} + \overline{D}))$. This suggests two OR structures, which are subsequently connected in series. This progression is evident in Fig. 1.8c. The final step requires connecting one end of the p-structure to '1' (V_{DD}) and the other to the output. One side of the n-structure is connected to '0' (V_{SS}) and the other to the output in common with the p-structure. This yields the final connection diagram (Fig. 1.8d). The schematic icon is shown in Fig. 1.8e, which shows that this gate may be used in a 2-input multiplexer. If $C = \overline{B}$, then $F = \overline{A}$ if B is true, while $F = \overline{D}$ if B is false.

The Karnaugh map for a second function $F = \overline{((A + B + C) \cdot D)}$ is shown in Fig. 1.9a on page 16. The sub-function $(A + B + C)$ is implemented as three parallel switches. This structure is then placed in series with a switch with D on the input. The p-function is $(\overline{D} + \overline{A}.\overline{B}.\overline{C})$. This requires three switches in series connected in turn in parallel with a switch with D on the input. The completed gate is shown in Fig. 1.9b on page 16. In general, CMOS gates may be implemented by analyzing the relevant Karnaugh map for both n- and p-logic structures and subsequently generating the required series and parallel combinations of transistors.

1.4.6 Multiplexers

Complementary switches may be used to select between a number of inputs, thus forming a multiplexer function. Fig. 1.10a on page 17 shows a connection diagram for a 2-input multiplexer. As the

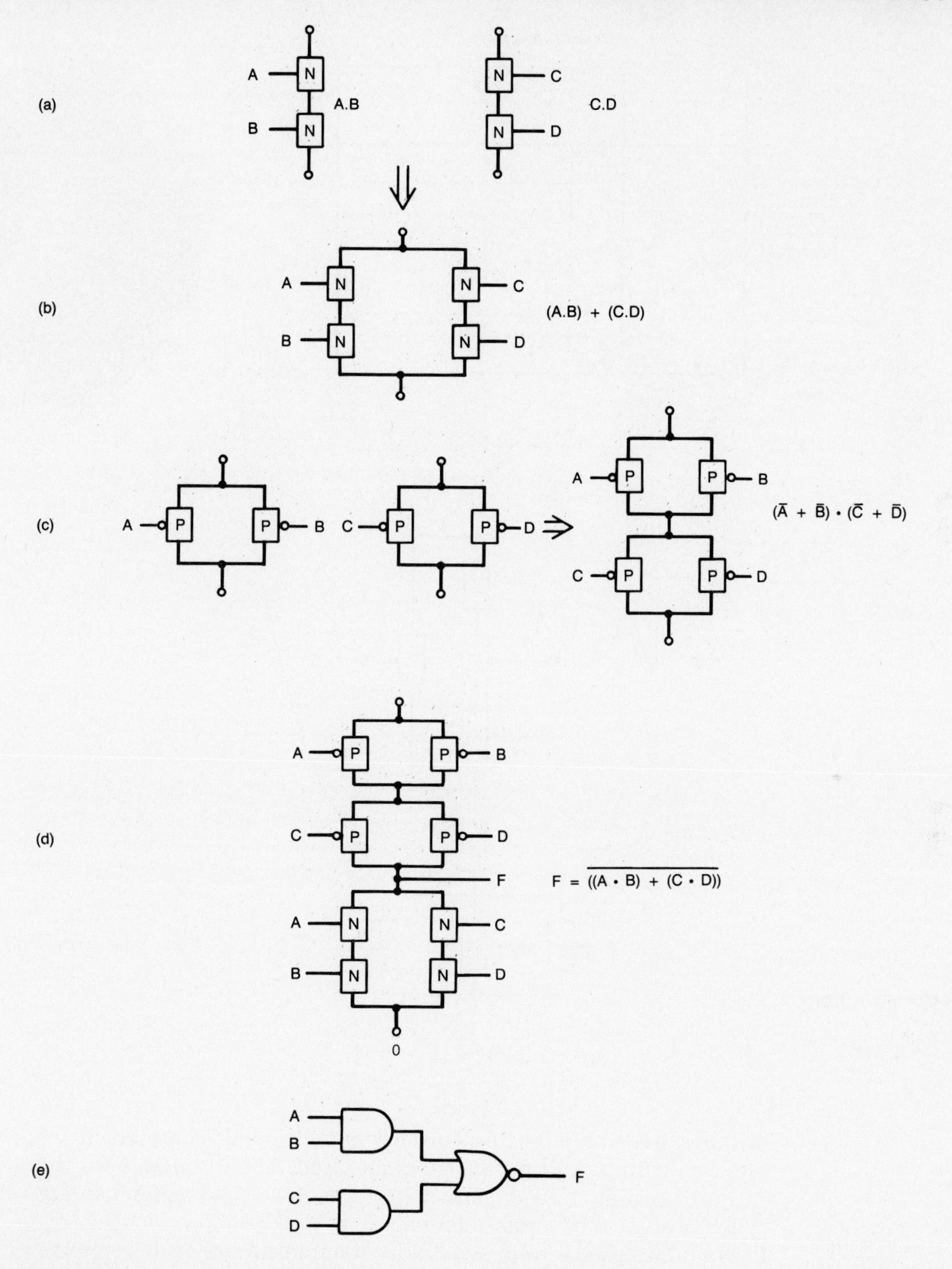

FIGURE 1.8. Construction of function $F = \overline{((A.B) + (C.D))}$

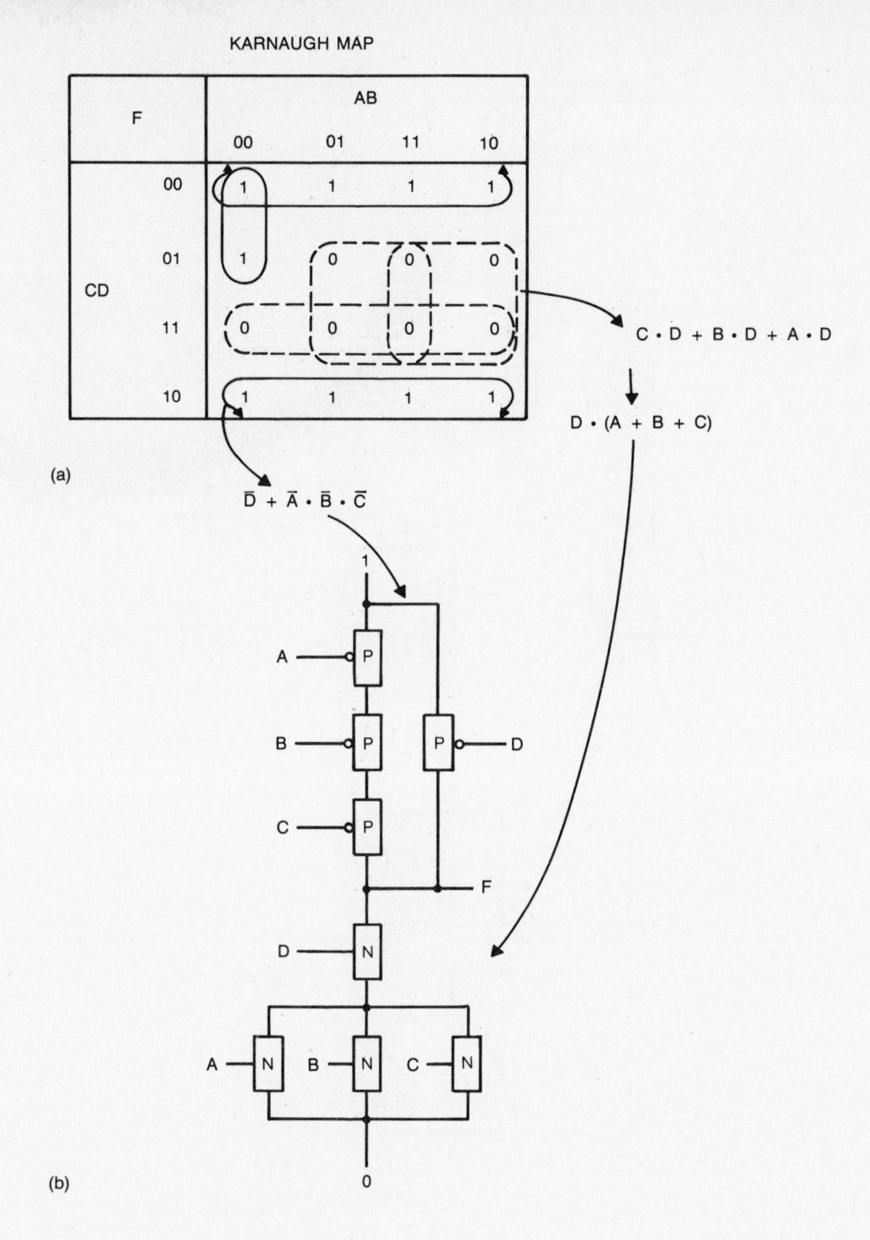

FIGURE 1.9. Construction of function $F = \overline{((A + B + C) \,.\, D)}$

switches have to pass ‘0’s and ‘1’s equally well, complementary switches with n- and p-transistors are used. The truth table for the structure in Fig. 1.10 is shown in Table 1.4. The complementary switch is also called a transmission gate or pass gate (complementary). A commonly used circuit symbol for the transmission gate is shown in Fig. 1.10b. The multiplexer connection in terms of this symbol and transistor symbols is shown in Fig. 1.10c.

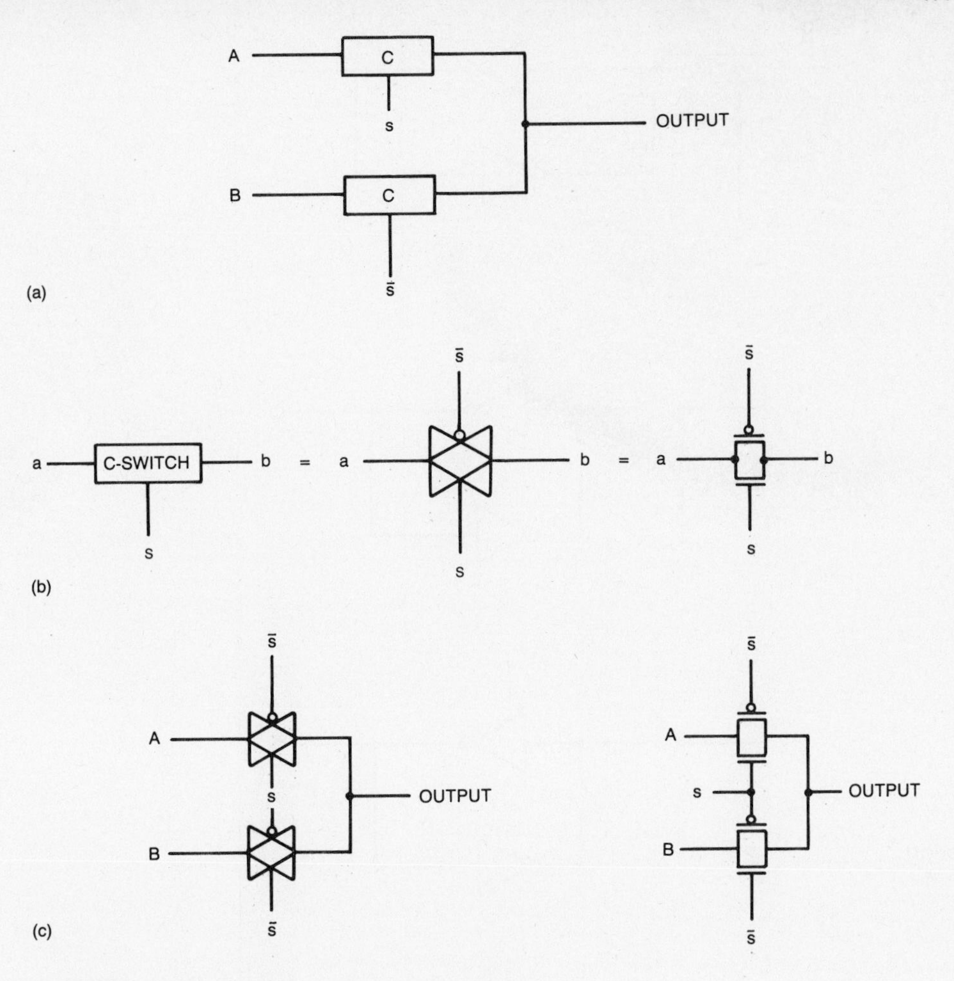

FIGURE 1.10. A CMOS 2-input multiplexer

1.4.7 Memory

We have now constructed a sufficient set of CMOS structures to enable a memory element to be constructed. A simple flip-flop using one 2-input multiplexer and two inverters is shown in Fig. 1.11a.

TABLE 1.4. Two input multiplexer truth table

A	B	S	$\overline{S}$	OUTPUT
X	0	0	1	0(B)
X	1	0	1	1(B)
0	X	1	0	0(A)
1	X	1	0	1(A)

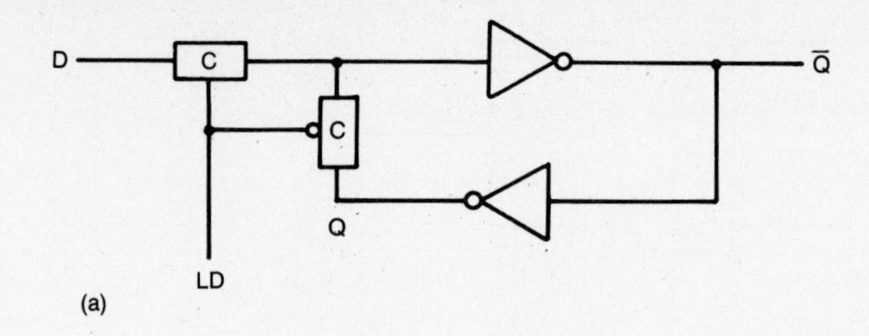

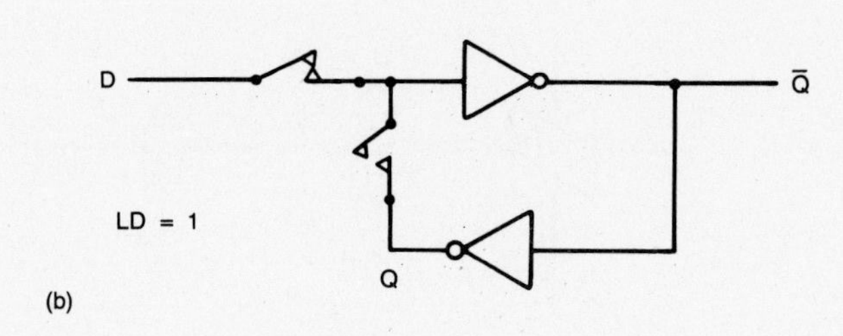

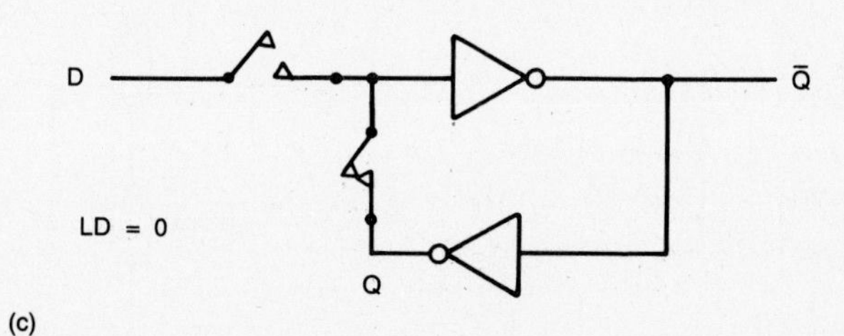

FIGURE 1.11. Connection of components for a simple CMOS flip-flop

When LD = '1', $\overline{Q}$ is set to $\overline{D}$ and Q is set to D (Fig. 1.11b). When LD is switched to '0', a feedback path around the inverter pair is established (Fig. 1.11c). This causes the current state of Q to be stored. While LD = '0' the input D is ignored.

1.5 Alternate circuit representations

In this section we will examine some alternate representations for the circuits developed so far. Generally, a design can be expressed in terms of behavioral, structural, and physical properties.

1.5.1 Behavioral representation

A behavioral representation describes how a particular design should respond to a given set of inputs. In Section 1.4 the behavior of a gate was defined in terms of its boolean function,

$$F = \overline{((A + B + C).D)}.$$

This is a technology independent behavioral specification at the logic level. No notion of how to implement this function is implied, nor is any speed performance implied. Higher levels of behavioral description are possible. For instance, an add operation may be summarized in a high level language by

```
sum = a + b
```

Here no method of addition is implied and the word length is assumed to be that of the machine. A further example of the behavior of the flip-flop designed previously is as follows:

```
IF(LD == 1)
THEN
   Q = D;
```

Note there may be some ambiguity associated with this style of behavioral representation. It could also represent a multiplexer without implying storage of state. Higher levels of behavioral specification can specify the types of registers involved in a design and the transfers that occur between them. Even less information about implementation is implied. At some stage it is possible to express behavior as an algorithm written in a high level language. The aim of most modern design systems is to convert some such specification into a system design in a minimum time and with maximum likelihood that the system will perform as desired.

1.5.2 Structural representation

A structural specification specifies how components are interconnected to perform a certain function (or achieve a designated behavior). We will use as an example of a complete structural description language, MODEL, a language conceived by Lattice Logic Ltd. [Latt82]. The specification for the inverter is

```
Part   inv  (in)-> out
   Nfet  out  in  vss
   Pfet  out  in  vdd
End
```

The first line declares a part called `inv` followed by a list of inputs, in this case `in`. The outputs appear on the other side of the symbol `->`, in this case `out`. Following this is a list of transistors with their type and connections in the form

```
Transistor-type   drain-conn   gate-conn   source-conn
     Nfet            out          in           vss
```

Thus the first statement describes an n-transistor with `drain` = `out`, `gate` = `in`, `source` = `vss`. The second statement describes a p-transistor with `drain` = `out`, `gate` = `in`, `source` = `vdd`.

The description for a 2-input NAND gate would be

```
Part nand2 (a,b) -> out
Signal i1
     Nfet i1 a vss
     Nfet out b i1
     Pfet out a vdd
     Pfet out b vdd
End
```

In this description the internal signal `i1` is declared by the keyword `Signal`. A diagram of this appears in Fig. 1.12a. It is worthwhile to compare this description with a possible behavioral description.

```
out = -(a&b)
```

or

```
out = (not (and a b ))
```

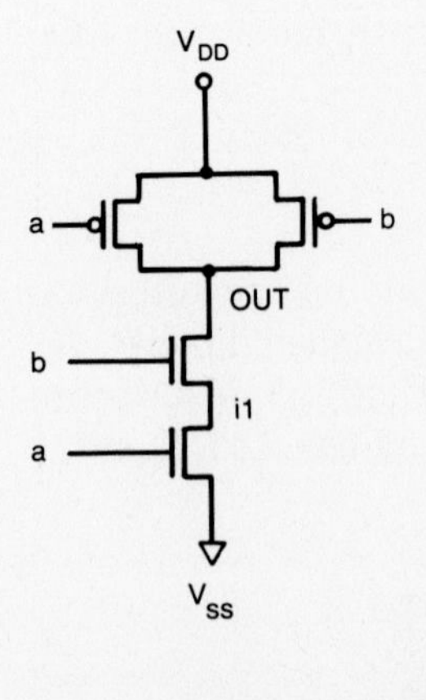

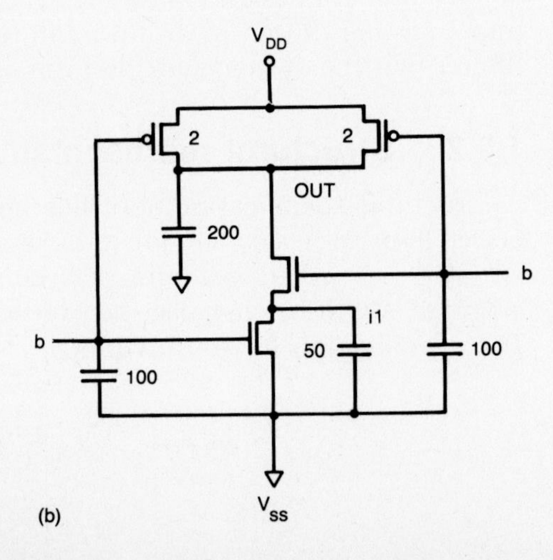

FIGURE 1.12. Graphical versions of structural descriptions for a CMOS NAND gate (schematics)

Note that we can infer all of the transistor connections from these code fragments. However, the intermediate node `i1` is "hidden." In the MODEL description, we may augment the purely structural description with some parameters such as capacitance and transistor sizing that will affect performance. Although notation for this can be added to the behavioral representation, rather baroque forms result and the simple elegance of the logical statement is lost. An expanded MODEL description might be

```
Part nand2 (a,b) -> out
Signal i1
     Nfet  i1 a vss
     Nfet out b i1
     Pfet out a vdd size = 2
     Pfet out b vdd size = 2
     Capacitance i1 50
     Capacitance  a 100
     Capacitance  b 100
     Capacitance out 200
End
```

Here the capacitance (in some units) has been specified. In addition, the p-transistor sizes have been modified according to some notional size parameter. This is shown in Fig. 1.12b. As we will learn in subsequent chapters, this type of information is crucial to the performance of CMOS circuits. In other words, the behavioral description ensures that the function may be correctly implemented but no reference is necessarily made to speed or other operational parameters. The structural description allows the specification of all components that affect performance. The circuit simulator SPICE [Nage75] uses a circuit description for the specification of transistor connectivity. The specification of the NAND gate might look like the following:

```
.SUBCKT NAND2 VDD VSS A B OUT
MN1 I1 A VSS VSS NFET W=8U L=4U AD=64P AS=64P
MN2 OUT B I1 VSS NFET W=8U L=4U AD=64P AS=64P
MP1 OUT A VDD VDD PFET W=16U L=4U AD=128P AS=128P
MP2 OUT B VDD VDD PFET W=16U L=4U AD=128P AS=128P
CA A VSS 50fF
CB B VSS 50fF
COUT OUT VSS 100fF
.ENDS
```

In this description the internal model in SPICE calculates the parasitic capacitances inherent in the actual device using the device dimensions specified. The capacitance statements in the above description add extra routing capacitance.

Defining a transmission gate in MODEL, we have

```
Part tg (a,c,cb) -> b
 Nfet a c b
 Pfet a cb b
End
```

We can now define the flip-flop (also called a D latch) as follows (a signal appended with "bar" is a complemented signal):

```
Part flipflop (in, ld, ldbar, q, qbar)
Signal a
      tg (in, ld, ldbar) -> a
      inv (a) -> qbar
      inv (qbar) -> q
      tg (q, ldbar, ld) -> a
End
```

Now we can use the flip-flop and other similarly constructed parts to hierarchically build larger and larger circuits.

A more familiar type of structural description is the schematic diagram shown in Fig. 1.13. Here we have the graphical hierarchy

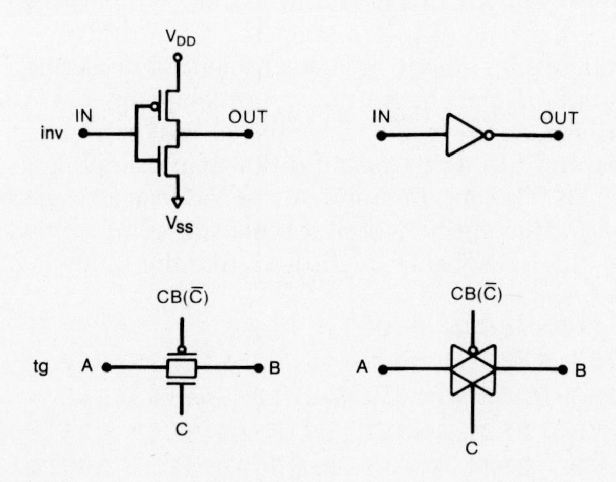

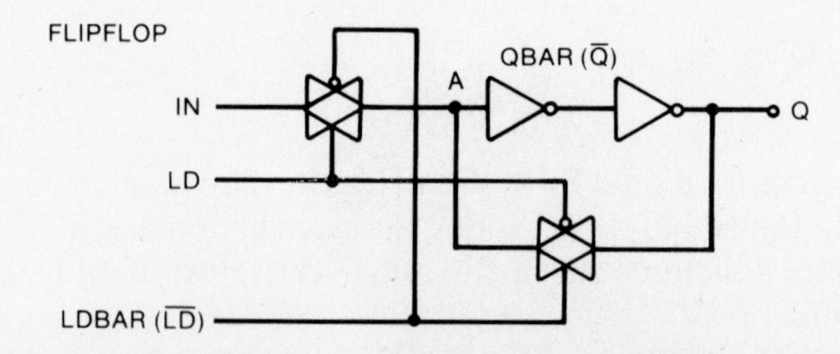

FIGURE 1.13. Schematic representation of CMOS flip-flop

of parts corresponding to the MODEL descriptions that we have developed. The specifications range from the circuit level (transistors), to the logic level (gates), to the functional block level (memory and collections of gates).

To all intents, both types of description are interchangeable, with preference for use dependent on the user. The schematic description is more immediately descriptive — "a picture is worth a thousand words." However, the language representation has some particular benefits, especially if the high level constructs such as looping, conditionals, and parameter passing are available. For instance, if we wanted to change the size of transistors in the inverter we might say

```
Part inv (in) [n] -> out
      Nfet out in vss    size = n
      Pfet out in vdd    size = 2*n
End
```

Here, n is a parameter passed to the inverter description to specify the size of the transistors. Emerging design systems show promise of dealing with language and graphical aspects of a design in a consistent fashion.

1.5.3 Physical representation

The physical specification for a circuit is used to define how the particular part has to be constructed to yield a specific structure and hence behavior. In an IC process, the lowest level of physical specification is the photo-mask information required by the various processing steps required on the fabrication process (see Chapter 3). At this stage, we will not dwell on these details but propose a simple model for the physical nature of a CMOS circuit, assuming that a program can translate our notation directly to the format needed for fabrication.

A typical physical representation for a transistor would consist of two rectangles representing the lithography required to fabricate the transistor. Precise "design rules" specify the size of each rectangle. In addition, for each different process, these rules change and the corresponding dimensions change — not necessarily linearly. Rather than try and remember these rules, we will use a single symbol to represent a transistor in a non-metric format. We will retain a form that reflects the physical nature of the transistor. The physical symbol for an n-transistor is shown in Fig. 1.14a and Plate 1. This mirrors the physical realization in which at least two process levels are overlaid. As we have seen, the gate connection is on one layer of the process and the source and drain on another layer. A similar

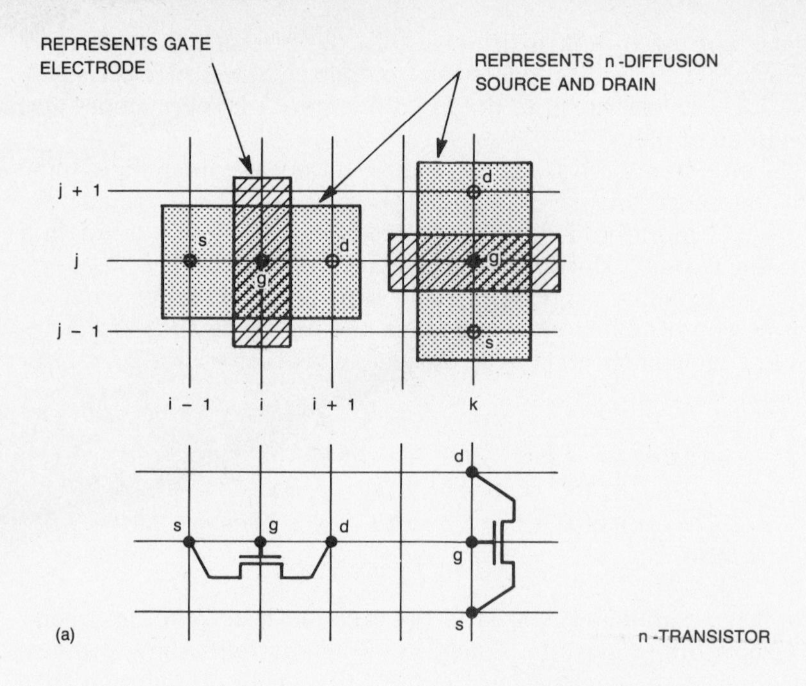

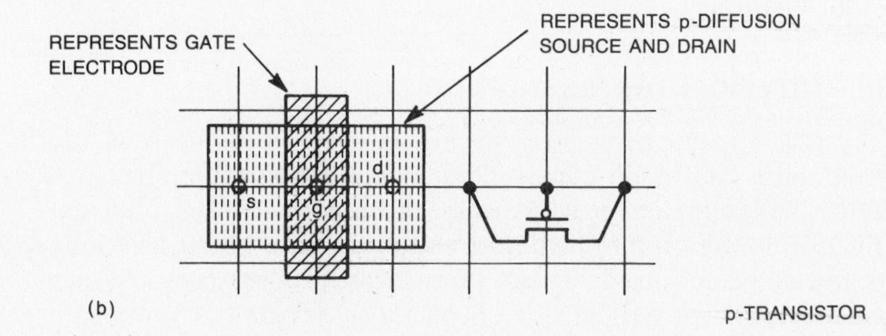

FIGURE 1.14. Physical symbols for transistors and simple circuits

symbol is used for the p-transistor, as shown in Fig. 1.14b and in color in Plate 1. Here, a "horizontal" transistor is shown. These symbols are overlaid on a grid. The transistor symbol occupies three grid points. The center grid point is the connection point for the gate of the transistor. The grid point to the right (or above) is the drain and the grid point to the left (or below) is the source. These two terminals are interchangeable. The schematic symbols are also shown for the n- and p-transistors in terms of grid connection points.

A symbolic layout for an inverter may be constructed using these symbols. It is substantially the same as the schematic but we have had to be careful about the layers in which connections have been made. We have "wires" on four layers. The interaction of these

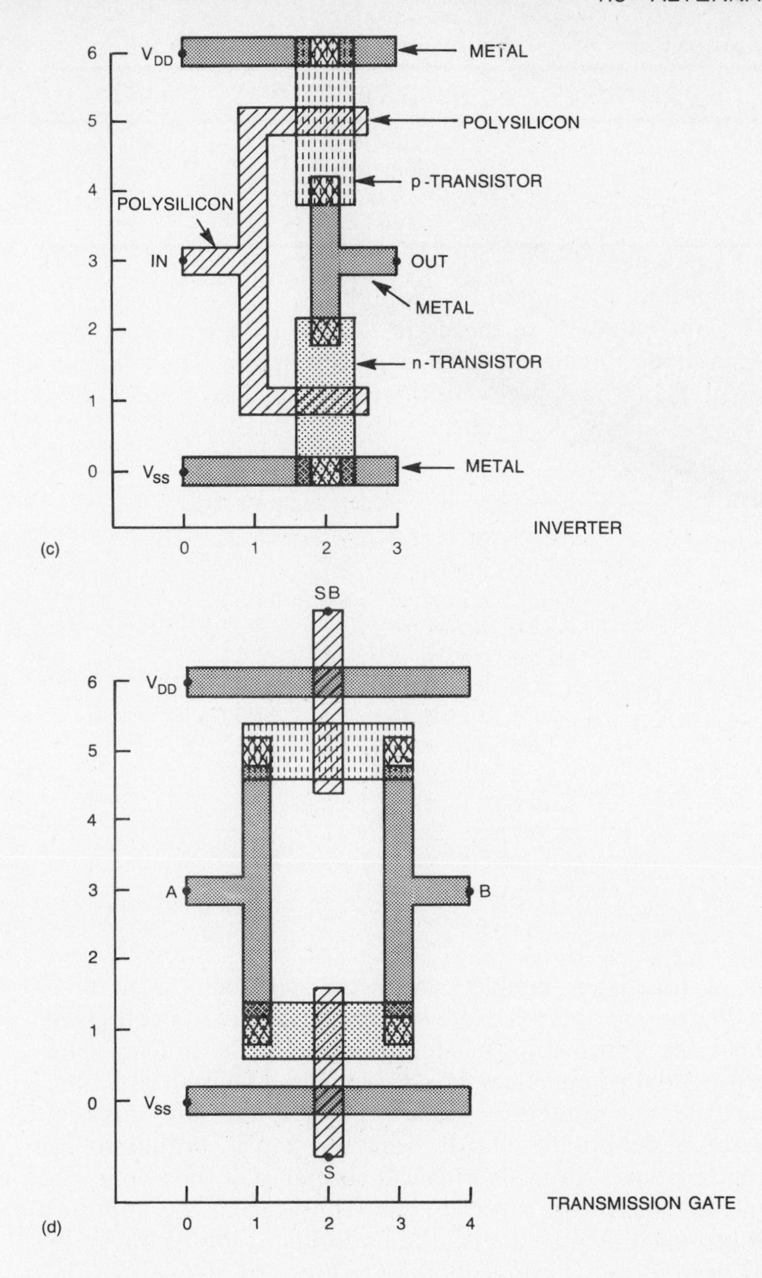

FIGURE 1.14. (Continued)

layers is summarized in Table 1.5. OK denotes that a connection may be made, while an X designates that a direct connection may not be made between the two layers. Any off-diagonal OK requires a "contact" (C) to connect the two layers.

A completed symbolic layout for the inverter is shown in Fig. 1.14c. The symbolic layout for a transmission gate is shown in

TABLE 1.5. Physical layer interactions

	n-DIFFUSION	p-DIFFUSION	POLYSILICON	ALUMINUM
n-diffusion	OK	X	Transistor	OK (C)
p-diffusion	X	OK	Transistor	OK (C)
polysilicon	Transistor	Transistor	OK	OK (C)
aluminum	OK (C)	OK (C)	OK (C)	OK

Fig. 1.14d. Color versions of these are found in Plate 1. This may also be expressed in the form of a language description. The following represents the transmission gate in the ICDL language (see Chapter 7):

```
begin tg
t1:     device n (2,1) or=east
t2:     device p (2,5) or=east
        wire alum (0,0) (4,0)
        wire alum (0,6) (4,6)
        wire poly (2,-1) (2,1)
        wire poly (2,7) (2,5)
        wire alum (1,1) (1,5)
        wire alum (3,1) (3,5)
        wire alum (0,3) (1,3)
        wire alum (3,3) (4,3)
        contact md (1,1)
        contact md (3,1)
        contact md (1,5)
        contact md (3,5)
end
```

It consists of transistor, contact, and wire statements with type qualifiers and grid coordinates. Note that although this is an abridged form of the mask information needed to fabricate the transmission gate, it is substantially larger than the corresponding MODEL structural description. Using a number of such structures, we can construct a physical sub-assembly that constitutes a flip-flop according to Fig. 1.15. Fig. 1.15a shows a simple physical abutment of the two transmission gate cells and two inverter cells. The V_{SS} and V_{DD} supplies have been arranged to feed across the bottom and top of the cells. A feedback line in metal connects the Q output to one side of the input multiplexer. In Fig. 1.15b, the internal circuit details of the cells are displayed in the form of schematic circuit symbols. Finally, Fig. 1.15c shows the symbolic layout representation, which is identical in topology to that shown in Fig. 1.15b. Plate 2 illustrates this example in color.

To a large extent, most CMOS IC design involves the steps illustrated in the preceding sections. Once a behavior is defined, the logic corresponding to that behavior is designed. That leads to

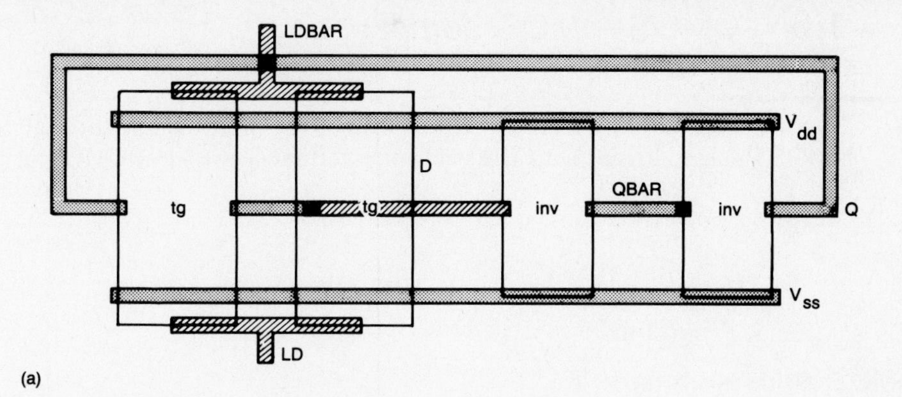

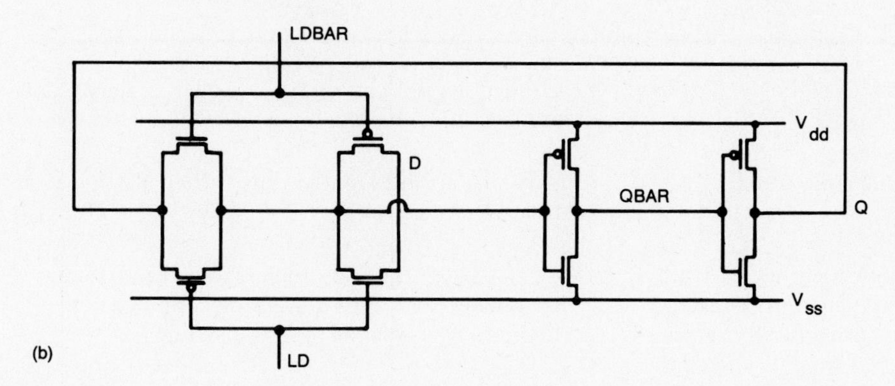

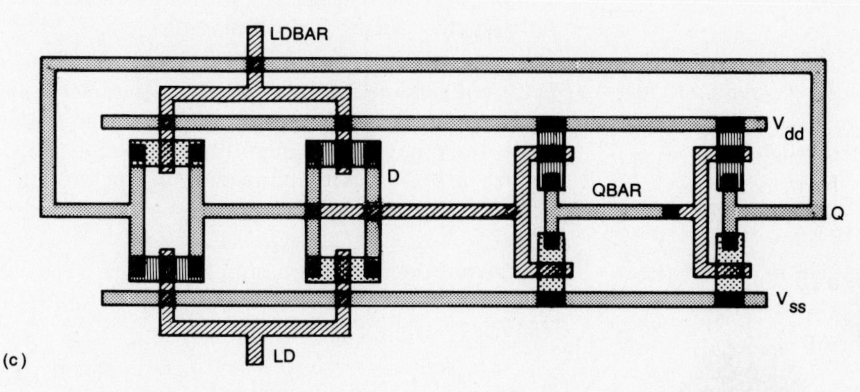

FIGURE 1.15. Physical construction of a CMOS flip-flop

a transistor circuit description. Finally, a layout may be designed for the particular logic function. Quite often, frequently used logic structures may be designed at the layout level and placed in a library. These library elements may then be assembled as illustrated in Fig. 1.15 to build more complex structures. An alternative to this approach is to compose primitive elements such as flip-flops from individual transistors. This yields more efficient layouts. Methods for designing in this manner will be treated in subsequent chapters.

1.6 CMOS–nMOS comparison

Many designers may have been introduced to VLSI design through nMOS design. In order to illustrate the salient features of CMOS, a

TABLE 1.6. CMOS-nMOS quick summary

CMOS	nMOS
(i) Logic Levels • Fully restored logic, i.e., output settles at V_{DD} or V_{SS} (GND).	• Output does not settle at V_{SS} (GND) — hence degraded noise margin.*
(ii) Transition Times • Rise and fall times are of the same order.	• Rise times are inherently slower than Fall times.*
(iii) Transmission Gates • Transmission gate passes both logic levels well. The output of transmission gate can be used to drive the input of other transmission gates.	• Pass transistor transfers logic '0' well but logic '1' is degraded. Pass transistor cannot drive the gate of a second pass transistor.
(iv) Power Dissipation • Almost zero static power dissipation. However power is dissipated during logic transition.	• With output of a given gate = '0' power is dissipated in the circuit in addition to power dissipated during logic transitions.*
(v) Precharging Characteristics • Both n-type and p-type devices are available for precharging a bus to V_{DD} and V_{SS}. Nodes can be charged fully to V_{DD} or alternatively to V_{SS} in a short time.	• With enhancement mode transistor the best one can do (with normal clocking) is to charge a bus to ($V_{DD} - V_t$). Generally use of bootstrapping or hot clocking is needed to precharge to V_{DD}.
(vi) Power Supply • Voltage required to switch a gate is a fixed percentage of V_{DD}.	• Somewhat dependent on supply voltage.
• Variable range 1.5 to 15 volts.	• Fixed.
(vii) Packing Density • Require 2N devices for N inputs for complementary static gates. Less for dynamic gates.	• Require (N + 1) devices for N inputs.
(viii) Pull-up to Pull-down Ratio • Load-to-driver device ratio is typically 1:1 or 2:1.	• Load-to-enhancement-driver ratio is typically 4:1 to optimize the logic '0' output level and minimize current consumption.
(ix) Layout • CMOS encourages regular layout styles.	• Depletion load and different driver transistor sizes inhibit layout regularity.

"quick summary" is presented in Table 1.6. However, it should be stressed that this is a broad overview and individual points may vary widely in importance. The comparisons are true of ratioed logic styles for nMOS and those items marked with an asterisk are not valid for ratioless nMOS logic design (dynamic circuits). The main points to note are that the output logic levels of CMOS are fully restored, a CMOS gate consumes no DC power when the output is at '1' or '0' level, and 2N devices are required for an N-input gate for a fully complementary gate.

1.7 Summary

This chapter introduced a simple model for an MOS transistor and developed logic that uses p-transistors and n-transistors. This led to a basic discussion of the various levels of representation of circuits and methods of composing these representations. The remainder of this book will elaborate on the material introduced in this chapter.

1.8 Exercises

1.1 Design a 4-input NAND gate using CMOS switch elements. Draw the full transistor circuit for the function.

1.2 $F = \overline{AB + BC + AC}$ implements a complemented carry function. Design a complementary CMOS gate to perform this function.

1.3 Design a 3-input OR gate. To what conclusions do you come?

1.4 A 4-input multiplexer structure is needed to multiplex four busses to a register in a microprocessor. Show two ways in which this may be implemented. Can you think of any reasons why one method is preferable to the others?

1.5 Using graph paper and colored pencils, complete a symbolic layout for the gates designed in Exercises 1.1, 1.2, and 1.3. What problems do you encounter?

1.6 Design and complete a symbolic layout for a CMOS memory element other than that shown in Fig. 1.11. Include waveform sequencing required for operation.

MOS TRANSISTOR THEORY

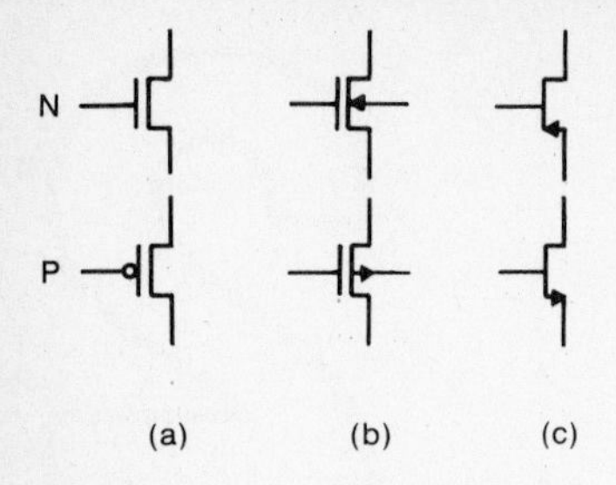

FIGURE 2.1. MOS transistor symbols

2.1 Introduction

In Chapter 1 the MOS transistor was introduced in terms of its operation as an ideal switch. In this chapter we will examine the characteristics of MOS transistors in more detail to lay the foundation for predicting the performance of the switches, which is less than ideal. Fig. 2.1 shows some of the symbols that are commonly used for MOS transistors. The symbols in Fig. 2.1a will be used where it is only necessary to indicate the switch logic necessary to build a function. If the substrate connection needs to be shown, the symbols in Fig. 2.1b will be used. Fig. 2.1c shows an example of the many symbols that may be encountered in the literature.

An MOS transistor is termed a *majority-carrier* device, in which the current in a conducting channel between the source and drain is *modulated* by a voltage applied to the gate. In an n-type MOS transistor (i.e., nMOS), the majority carriers are *electrons*. A positive voltage applied on the gate with respect to the substrate *enhances* the number of electrons in the channel (the region immediately under the gate) and hence increases the conductivity of the channel. For gate voltages less than a *threshold* value denoted by V_t, the channel is *cut-off*, thus causing a very low drain-to-source current. The operation of a p-type transistor (i.e., pMOS) is analogous to the nMOS transistor, with the exception that the majority carriers are *holes* and the voltages are negative with respect to the substrate.

The first parameter of interest that characterizes the switching behavior of an MOS device is the threshold voltage, V_t. This is defined as the voltage at which an MOS device begins to conduct ("turn on"). One can graph the relative conduction against the difference in gate-to-source voltage in terms of the source-to-drain current (I_{ds}) and the gate-to-source voltage (V_{gs}). These graphs for a fixed drain source voltage V_{ds} are shown in Fig. 2.2. It is possible to make n-devices that conduct when the gate voltage is equal to the source voltage, while others require a positive difference between gate and source voltage to bring about conduction (negative for p-devices). Those devices that are normally cut-off (i.e., nonconducting) with zero gate bias (gate voltage-source voltage) are further classed as *enhancement* mode devices, whereas those devices that conduct with zero gate bias are called *depletion* mode devices. The n-channel transistors and p-channel transistors are the duals of each other; that is, the voltage polarities required for correct operation are the opposite. The threshold voltages for n-channel and p-channel devices are denoted by V_{t_n} and V_{t_p} respectively.

In CMOS technologies both n-channel and p-channel transistors are fabricated on the same chip. Furthermore, most CMOS integrated circuits, at present, use transistors of the enhancement type.

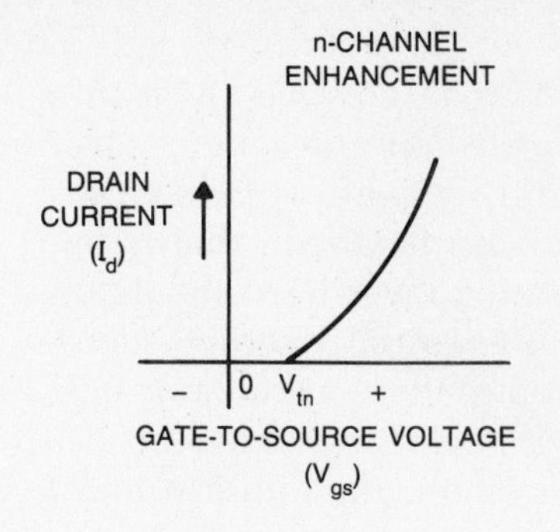

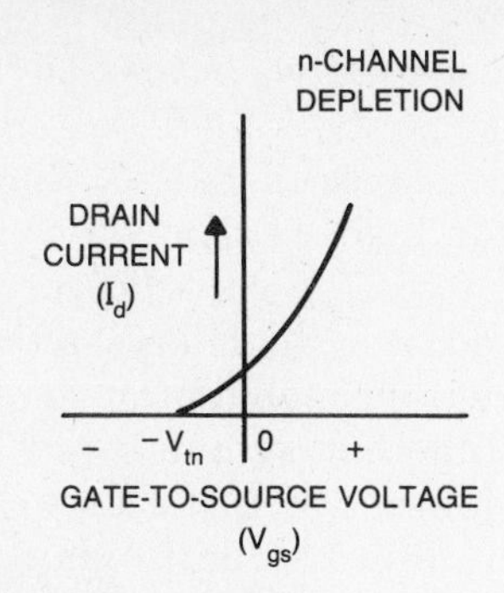

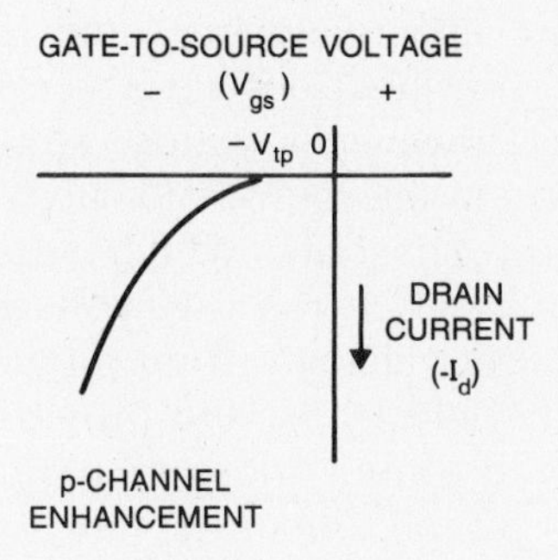

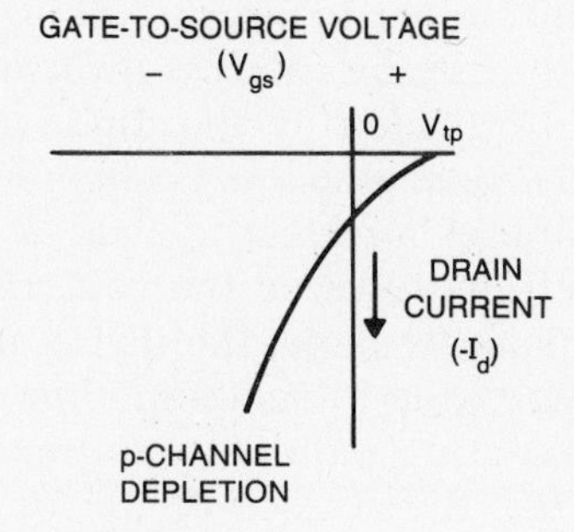

FIGURE 2.2. Conduction characteristics for enhancement and depletion mode transistors (assuming fixed V_{ds})

2.1.1 nMOS enhancement transistor

The structure for an n-channel enhancement type transistor shown in Fig. 2.3 consists of a moderately doped p-type silicon substrate into which two heavily doped n^+ regions, the *source* and *drain*,

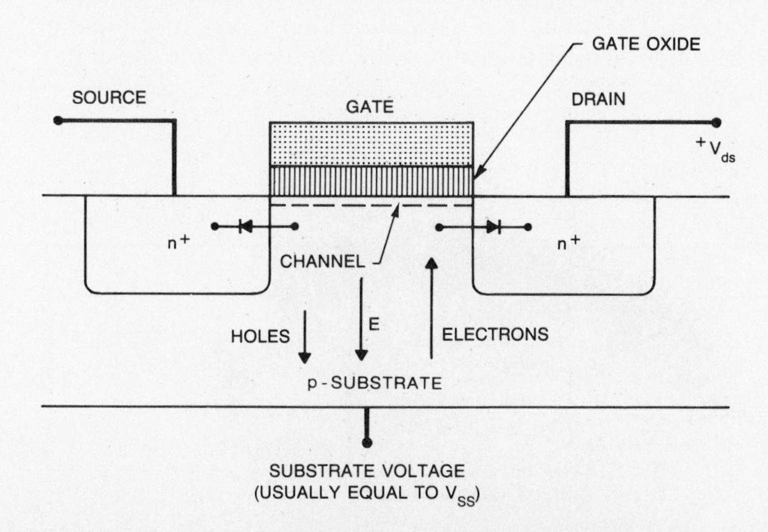

FIGURE 2.3. Physical structure of an nMOS transistor

are diffused. Between these two regions there is a narrow region of p-type substrate called the *channel*, which is covered by a thin insulating layer of silicon dioxide (SiO_2) called *gate oxide*. Over this oxide layer is a polycrystalline silicon (polysilicon) electrode, referred to as the *gate*. Polycrystalline silicon is silicon that is not composed of a single crystal. Since the oxide layer is an insulator, the current through the gate and channel is essentially zero. Because of the inherent symmetry of the structure, there is no physical distinction between the drain and source regions. Since SiO_2 has relatively low loss and high dielectric strength, the application of high gate fields is feasible.

In operation, a positive voltage is applied between the source and the drain (V_{ds}). With zero gate bias ($V_{gs} = 0$), no current flows from source to drain because they are effectively insulated from each other by the two reversed biased p-n junctions shown in Fig. 2.3 (indicated by the diode symbols). However, a voltage applied to the gate, which is positive with respect to the source and substrate, produces an electric field E across the substrate, which attracts electrons towards the gate and repels holes. If the gate voltage is sufficiently large, the region under the gate changes from p-type to n-type (due to accumulation of attracted electrons) and provides a conduction path between the source and drain. Under such a condition, the surface of the underlying p-type silicon is said to be *inverted*. The term *n-channel* is applied to the structure. This concept is further illustrated by Fig. 2.4, which shows the initial distribution of mobile positive holes in the silicon insulating layer before the application of a positive gate voltage and the final distribution after the application of gate voltage. As these ions drift toward the interface, they tend to induce more negative charge at the silicon surface beneath the gate, resulting in the formation of the inversion layer.

The difference between a p-n junction that exists in a bipolar transistor or diode (or between the source or drain and substrate)

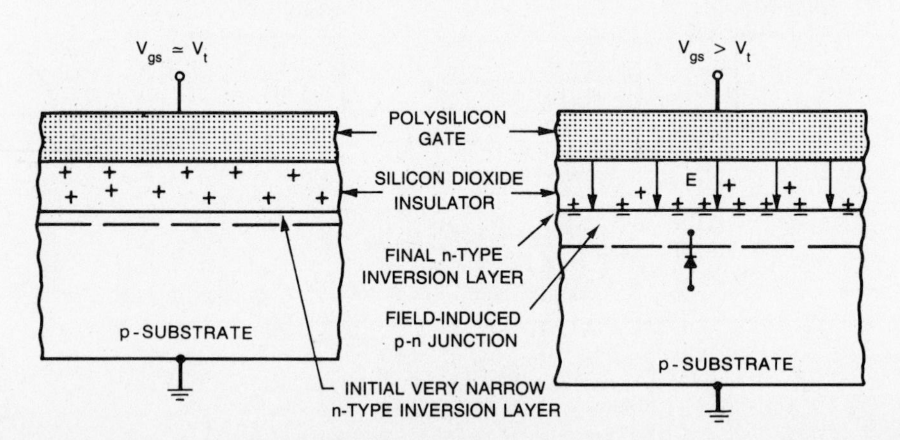

FIGURE 2.4. Creation of an inversion layer in an n-transistor

and the inversion layer-substrate junction is that in the p-n junction, the n-type conductivity is brought about by a metallurgical process; that is, the electrons are introduced into the semiconductor by the introduction of donor ions. In an inversion layer-substrate junction, the n-type layer is induced by the electric field E applied to the gate. Thus, this junction, instead of being a metallurgical junction, is a *field-induced* junction.

Electrically, an MOS device therefore acts as a voltage-controlled switch that conducts initially when the gate-to-source voltage, V_{gs}, is equal to the threshold voltage, V_t. When a voltage V_{ds} is applied between source and drain, with $V_{gs} = V_t$, the horizontal and vertical components of the electrical field due to the source-drain voltage and gate to substrate voltage interact, causing conduction to occur along the channel. The horizontal component of the electric field associated with the drain-to-source voltage (i.e., $V_{ds} > 0$), is responsible for sweeping the electrons from the channel towards the drain. As the voltage from drain-to-source is increased, the resistive drop along the channel begins to change the shape of the channel characteristic. This behavior is shown in Fig. 2.5. At the source end of the channel, the full gate voltage is effective in inverting the channel. However, at the drain end of the channel, only the difference between

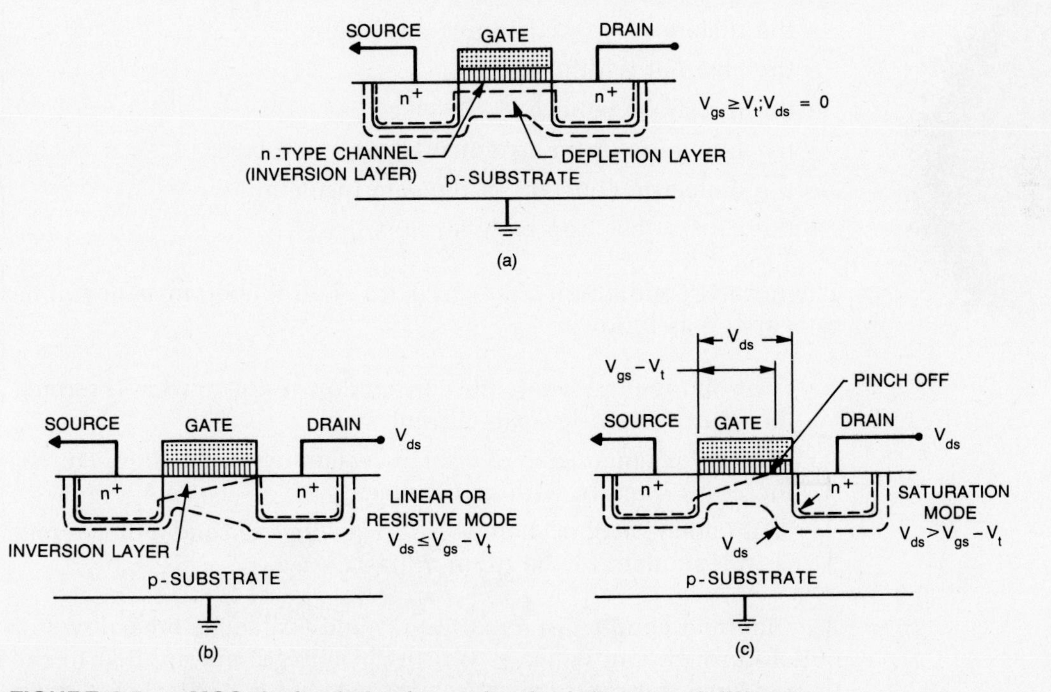

FIGURE 2.5. nMOS device behavior under the influence of different terminal voltages

the gate and the drain voltages is effective. When the effective gate voltage ($V_{gs} - V_t$) is greater than the drain voltage, the channel becomes deeper as V_{gs} is increased. This is termed the "linear," "resistive," or "unsaturated" region, where the channel current I_{ds} is a function both of gate and drain voltages. If $V_{ds} > V_{gs} - V_t$, then $V_{gd} < V_t$ (V_{gd} is the gate-to-drain voltage), and the channel becomes *pinched-off* — the channel no longer reaches the drain. This is illustrated in Fig. 2.5c. However, in this case, conduction is brought about by a drift mechanism of electrons under the influence of the positive drain voltage. As the electrons leave the channel, they are injected into the drain depletion region and are subsequently accelerated towards the drain. The voltage across the pinched-off channel tends to remain fixed at ($V_{gs} - V_t$). This condition is the "saturated" state in which the channel current is controlled by the gate voltage and is almost independent of drain voltage. It should be noted that a depletion region is devoid of mobile carriers and therefore is able to insulate the channel from the rest of the substrate. Thus no significant current passes through the substrate because, in effect, a reverse biased p-n junction is formed with the channel (Fig. 2.5c). For fixed drain-to-source voltage and fixed gate voltage, the factors that influence the level of drain current I_{ds} flowing between source and drain (for a given substrate resistivity) are:

- the distance between source and drain
- the channel width
- the threshold voltage V_t
- the thickness of the gate-insulating oxide layer
- the dielectric constant of the gate insulator
- the carrier (electron or hole) mobility μ.

The normal conduction characteristics of an MOS transistor can be categorized as follows:

- "Cut-off" region: where the current flow is due to what is termed the source-drain leakage current.
- "Linear" region: region of weak inversion where the drain current increases linearly with gate voltage.
- "Saturation" region: channel is strongly inverted and drain current is independent of the drain voltage.

An abnormal conduction condition called avalanche breakdown or punch-through can occur if very high voltages are applied to the drain. Under these circumstances, the gate has no control over drain current.

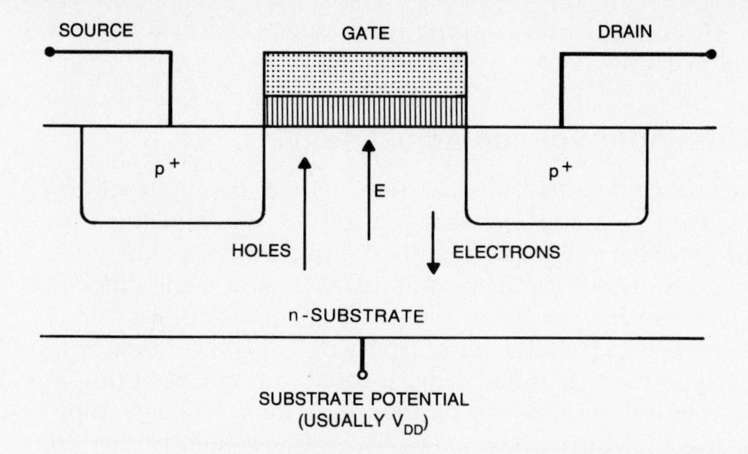

FIGURE 2.6. Physical structure of pMOS transistor

2.1.2 pMOS transistor

So far our discussions have been primarily directed towards nMOS. However, reversal of n-type and p-type regions yield a p-channel MOS transistor. This is illustrated by Fig. 2.6. Application of a negative gate voltage (w.r.t. source) draws holes into the region below the gate, resulting in the channel changing from n-type to p-type. Thus similar to nMOS, a conduction path is created between the source-to-drain. In this instance, however, conduction results from the movement of holes (vs. electrons) in the channel. A negative drain voltage sweeps holes from the source, through the channel to the drain.

2.1.3 Threshold voltage

The threshold voltage, V_t, for an MOS transistor can be defined as the voltage applied between the gate and source of an MOS device below which the drain-to-source current I_{ds} drops to zero. In general, the threshold voltage is a function of a number of parameters including the following:

- gate material
- gate insulation material
- gate insulator thickness
- channel doping
- impurities at silicon-insulator interface
- voltage between source and substrate V_{sb}.

In addition, the absolute value of the threshold voltage decreases with an increase in temperature. This variation is approximately

−4 mV/°C for high substrate doping level, and −2 mV/°C for low doping level [VaGr66].

2.1.4 Threshold voltage adjustment

It is often necessary to adjust the native (original) threshold voltage of a device. Two common techniques used for the adjustment of the threshold voltage entail varying the doping concentration at the silicon-insulator interface through ion implantation, or using different insulating material for the gate. In this latter approach, a layer of silicon nitride (Si_3N_4) (relative permittivity of 7.5) is combined with a layer of silicon dioxide (relative permittivity of 3.9), resulting in an effective relative permittivity of about 6, which is substantially larger than the dielectric constant of SiO_2. Consequently, for the same thickness as an insulating layer consisting only of silicon dioxide, the dual dielectric process will be electrically equivalent to a thinner layer of SiO_2. In order to prevent the surface of the silicon from inverting under the regions between transistors, the threshold voltage in these field regions is increased by heavily doped diffusions, implants of the silicon surface, or by making the oxide layer very thick. MOS transistors are self-isolating so long as the surface of the silicon may be inverted under the gate, but not in the regions between devices by normal circuit voltages.

2.1.5 Body effect

As we have seen so far, all devices comprising an MOS device are made on a common substrate. As a result, the substrate voltage of all devices is normally equal. (In some analog circuits this may not be true.) However, in arranging the devices to form gating functions it might be necessary to connect several devices in series as shown in Fig. 2.7 (for example the NAND gate shown in Fig. 1.6). This

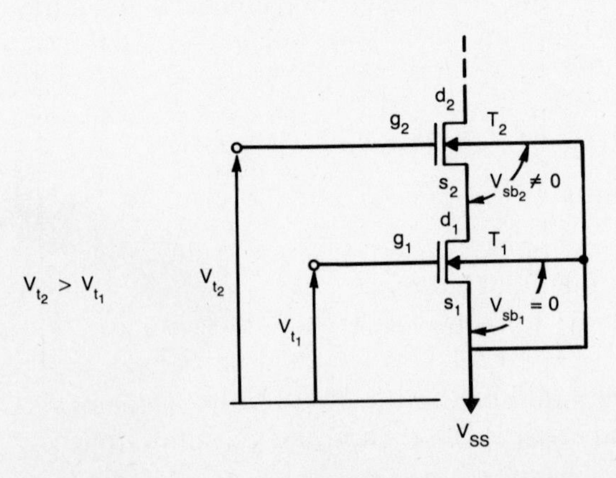

FIGURE 2.7. The effect of substrate bias on series connected transistors

may result in an increase in source-to-substrate voltage as we proceed vertically along the series chain ($V_{sb_1} = 0$, $V_{sb_2} \neq 0$).

Under normal conditions, that is, when $V_{gs} > V_t$, the depletion layer width remains constant and charge carriers are pulled into the channel from the source. However, as the substrate bias V_{sb} ($V_{source} - V_{substrate}$) is increased, the width of the channel-substrate depletion layer also increases, resulting in an increase in the density of the trapped carriers in the depletion layer. For charge neutrality to hold, the channel charge must decrease. The resultant effect is that the substrate voltage V_{sb} adds to the channel-substrate junction potential. This increases the gate-channel voltage drop. The overall effect is an increase in the threshold voltage V_t ($V_{t_2} > V_{t_1}$). The effective threshold voltage can be approximated by the following expression:

$$V_t = V_{t(0)} \pm \gamma (V_{sb})^{1/2}, \tag{2.1}$$

where $V_{t(0)}$ is the threshold voltage with V_{sb} equal to zero and γ is a constant which depends on substrate doping. The negative sign is used for pMOS. Typical values for γ lie in the range of 0.4 to 1.2. As we shall learn in Chapter 3, the type of CMOS process can have a large impact on this parameter for both n- and p-transistors. The threshold voltage effectively increases, leading to smaller device currents, which in turn leads to slower circuits.

2.2 MOS device design equations

As stated previously, MOS transistors have three regions of operation:

- cut-off region
- linear region
- saturation region.

The ideal (first order) equations [Cobb70][Sah64] describing the behavior of an nMOS device in the three regions are:

$$I_{ds} = \begin{cases} 0; & V_{gs} - V_t \leq 0 \quad \textit{cut-off} \quad (a) \\ \beta\left[(V_{gs} - V_t)V_{ds} - \dfrac{V_{ds}^2}{2}\right]; & 0 < V_{ds} < V_{gs} - V_t \quad \textit{linear} \quad (b) \\ \dfrac{\beta}{2}(V_{gs} - V_t)^2; & 0 < V_{gs} - V_t < V_{ds} \quad \textit{saturation} \quad (c), \end{cases} \tag{2.2}$$

where I_{ds} is the drain-to-source current, V_{gs} is the gate-to-source voltage, V_t is the device threshold, and β is the MOS transistor gain

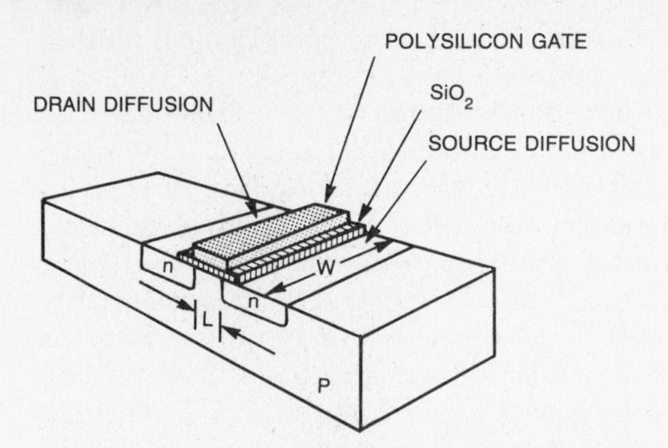

FIGURE 2.8. Geometric terms in the MOS device equation

factor. β is dependent on both the process parameters and the device geometry and is given by

$$\beta = \frac{\mu\varepsilon}{t_{ox}}\left(\frac{W}{L}\right), \tag{2.3}$$

where μ is the effective surface mobility of the electrons in the channel, ε is the permittivity of the gate insulator, t_{ox} is the thickness of the gate insulator, W is the width of the channel, and L is the length of the channel. The gain factor β thus consists of a process dependent factor $(\mu\varepsilon/t_{ox})$, which contains all the process terms that account for such factors as doping density and gate oxide thickness and a geometry dependent term (W/L), which depends on the actual layout of the device. The geometric terms in Eq. (2.3) are illustrated in Fig. 2.8 in relation to the physical MOS structure.

Typical values (for an n-device) are as follows: $\mu_n = 500\text{cm}^2/\text{V-sec}$, $\varepsilon = 4\varepsilon_0 = 4 \times 8.85 \times 10^{-14}$ F/cm, $t_{ox} = 500$ Å. Hence a typical n-device β would be

$$\frac{500 \times 4 \times 8.85 \times 10^{-14}}{.5 \times 10^{-5}}\frac{W}{L} = 35\frac{W}{L}\,\mu\text{A/V}^2.$$

The cut-off region described by Eq. (2.2a) is also referred to as the *subthreshold* region, where I_{ds} increases exponentially with V_{ds} and V_{gs}. Although the value of I_{ds} is very small ($I_{ds} \approx 0$), the finite value of I_{ds} may affect the performance of circuits dependent on dynamic charge storage such as memory cells.

The approximation describing I_{ds} in Eq. (2.2c) is derived under the assumption that the current in the channel saturates (i.e., is constant) and is independent of the applied drain voltage. In practice, the drain current in saturation increases slightly with increasing drain voltage. A more accurate model that takes this behavior into account is represented by the following equation:

$$I_{ds} = \frac{\beta}{2}((V_{gs} - V_t)^2(1 + \lambda V_{ds})), \tag{2.4}$$

where λ is an empirical *channel length modulation* factor having a value in the range $0.02V^{-1}$ to $0.04V^{-1}$.

The term $\mu(\varepsilon/t_{ox})$ is frequently referred to as the *process gain factor*, K_p. It is a common parameter used in the SPICE MOS model specification. For a typical process K_p is in the order of 10 to 30 μAV^{-2}. It is not unusual to expect a spread of around 10 to 20 percent for K_p for a given process, primarily due to variations in the starting material and variation in SiO_2 growth.

The mobility, μ, describes the ease with which carriers drift in the substrate material. It is defined by

$$\mu = \frac{\textit{average carrier drift velocity } (v)}{\textit{Electric Field } (E)}.$$

If the velocity (V) is given in cm/sec, and the electric field (E) in V/cm, the mobility has dimensions of $cm^2/V\text{-}sec$.

A more detailed expression that defines the threshold voltage is given by

$$V_t = V_{t(0)} + \gamma[\sqrt{V_{sb} + 2\phi_F} - \sqrt{2\phi_F}], \tag{2.5}$$

where ϕ_F is a constant, V_{sb} is the substrate bias, V_{t0} is threshold voltage for $V_{sb} = 0$, and γ is the constant that describes the substrate bias effect. It may be expressed as

$$\gamma = \left(\frac{t_{ox}}{\varepsilon_{ox}}\right)\sqrt{2q\varepsilon_{Si}N}, \tag{2.6}$$

in which q is the charge on an electron, ε_{ox} is the dielectric constant of the silicon dioxide, ε_{Si} is the dielectric constant of the silicon substrate, and N is the concentration density of the substrate. In common with K_p, λ, and $V_{t(0)}$, γ is also a SPICE model parameter.

With $V_{sb} >> \phi_F$, the expression reduces to the form introduced in Eq. (2.1). It should be noted that simplified equations that describe the behavior of an MOS device assume that carrier mobility is constant, do not take into account the variations in channel length due to the changes in drain-to-source voltage V_{ds}, and furthermore neglect leakage currents. For long channels, the influence of channel variation is of little consequence. However, as devices are scaled down, this variation should be taken into account. A reduction in channel length increases the (W/L) ratio, thereby increasing β as the drain voltage increases. Thus there is a finite output impedance in the saturated region.

The effective channel length is approximated by:

$$L_{eff} = L - \sqrt{2\varepsilon_o \frac{\varepsilon_{Si}}{qN}(V_{ds} - [V_{gs} - V_t])}. \tag{2.7}$$

The same MOS device equations also apply to the pMOS device.

The only difference in the result is the change in the sign associated with the voltages and drain current.

2.2.1 V-I characteristics

The voltage-current characteristics of the n- and p-transistors in the linear and saturated regions are represented in Fig. 2.9. Note that we use the absolute value of the voltages concerned to plot the characteristics on the same axes. The boundary between the linear and saturation regions corresponds to the condition $|V_{ds}| = |V_{gs} - V_t|$ and appears as a dashed line in Fig. 2.9.

The output resistance (i.e., channel resistance) in the linear region can be obtained by differentiating Eq. (2.2b) with respect to V_{ds}, which results in an output conductance of

$$\lim_{V_{ds} \to 0} \frac{dI}{dV_{ds}} \approx \beta(V_{gs} - V_t). \tag{2.8}$$

Upon rearrangement the channel resistance R_c is approximated by

$$R_{C(linear)} = \frac{1}{\beta(V_{gs} - V_t)} \tag{2.9}$$

which indicates that it is controlled by gate-to-source voltage. The relation defined by Eq. (2.9) is valid for gate-to-source voltages that maintain constant mobility in the channel. In contrast, in saturation [i.e., above $V_{ds} = (V_{gs} - V_t)$], the MOS device behaves like a current

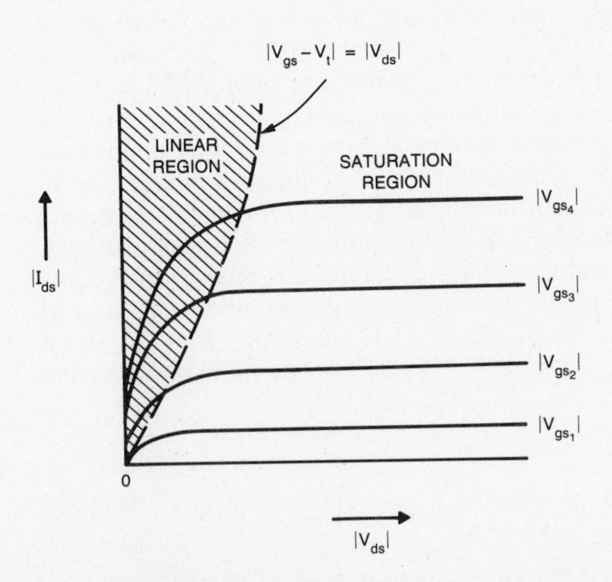

FIGURE 2.9. V-I characteristics for n- and p-transistors

source, the current being almost independent of V_{ds}. This may be verified from Eq. (2.2c) since

$$\frac{dI_{ds}}{dV_{ds}} = \frac{d\left(\frac{\beta}{2}(V_{gs} - V_t)^2\right)}{dV_{ds}} = 0. \tag{2.10}$$

The transconductance g_m expresses the relationship between output current I_{ds} and the input voltage V_{gs}, and is defined by

$$g_m = \left.\frac{\partial I_{ds}}{\partial V_{gs}}\right|_{V_{ds}=constant} \tag{2.11}$$

It is used to measure the gain of an MOS device. In the linear region, g_m is given by

$$g_{m\,(linear)} = \beta V_{ds}, \tag{2.12}$$

and in the saturation region by

$$g_{m\,(sat)} = \beta(V_{gs} - V_t). \tag{2.13}$$

For example, the value of transconductance for an n-type transistor in the linear region is

$$g_{m_n} = \left(\frac{\mu_n \varepsilon}{t_{ox}}\right)\left(\frac{W_n}{L_n}\right)V_{ds}. \tag{2.14}$$

Since transconductance must have a positive value, absolute values are used for voltages applied to p-type devices.

2.3 The complementary CMOS inverter — DC characteristics

A complementary CMOS inverter is realized by the series connection of a p- and n-device, as shown in Fig. 2.10. In order to derive the DC transfer characteristics for the inverter (output voltage V_O as a function of V_{in}), we start with Table 2.1, which outlines various regions of operation for the n- and p-transistors. In this table, V_{t_n} is the threshold voltage of the n-channel device, and V_{t_p} is the threshold voltage of the p-channel device. The objective is to find the variation in output voltage (V_O) for changes in the input voltage (V_{in}).

We commence with the graphical representation of the simple algebraic equations described by Eq. (2.2) for the two transistors

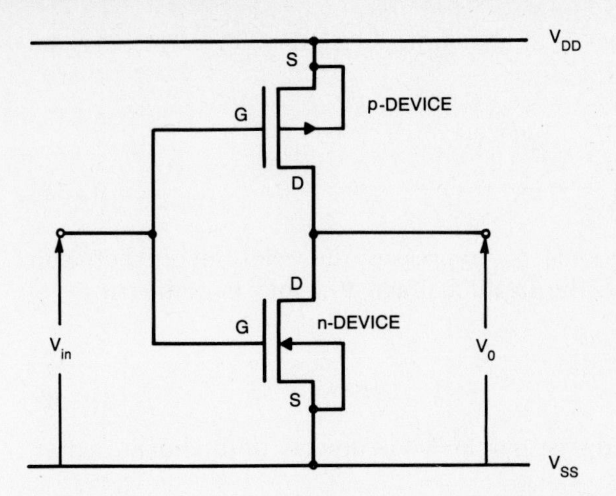

FIGURE 2.10. A CMOS inverter (with substrate connections)

shown in Fig. 2.11a [CaMi72]. The absolute value of the p-transistor drain current I_{ds} inverts this characteristic. This allows the V-I characteristics for the p-device to be reflected about the x-axis (Fig. 2.11b). This step is followed by taking the absolute value of the p-device V_{ds}, and superimposing the two characteristics yielding the resultant curves shown in Fig. 2.11c. The input/output transfer curve may now be determined by the points of common V_{gs} intersection in Fig. 2.11c. Thus, solving for $V_{in_n} = V_{in_p}$ *and* $I_{ds_n} = I_{ds_p}$ gives the desired transfer characteristics of a CMOS inverter as illustrated in Fig. 2.12. The switching point is typically designed to be 50 percent of the magnitude of the supply voltage: $\approx V_{DD}/2$. During transition, both transistors in the CMOS inverter are momentarily 'ON', resulting in a short pulse of current drawn from the power supply. This is shown by the dotted line in Fig. 2.12.

TABLE 2.1. Relations between voltages for the three regions of operation of a CMOS inverter

	CUTOFF	LINEAR	SATURATION
p-device	$V_{gs_p} > V_{t_p}$;	$V_{gs_p} < V_{t_p}$;	$V_{gs_p} < V_{t_p}$;
	$V_{in} > V_{t_p} + V_{DD}$	$V_{in} < V_{t_p} + V_{DD}$	$V_{in} < V_{t_p} + V_{DD}$
		$V_{gd_p} < V_{t_p}$;	$V_{gd_p} > V_{t_p}$;
		$V_{in} - V_O < V_{t_p}$	$V_{in} - V_O > V_{t_p}$
n-device	$V_{gs_n} < V_{t_n}$;	$V_{gs_n} > V_{t_n}$;	$V_{gs_n} > V_{t_n}$;
	$V_{in} < V_{t_n}$	$V_{in} > V_{t_n}$	$V_{in} > V_{t_n}$
		$V_{gd_n} > V_{t_n}$;	$V_{gd_n} < V_{t_n}$;
		$V_{in} - V_O > V_{t_n}$	$V_{in} - V_O < V_{t_n}$

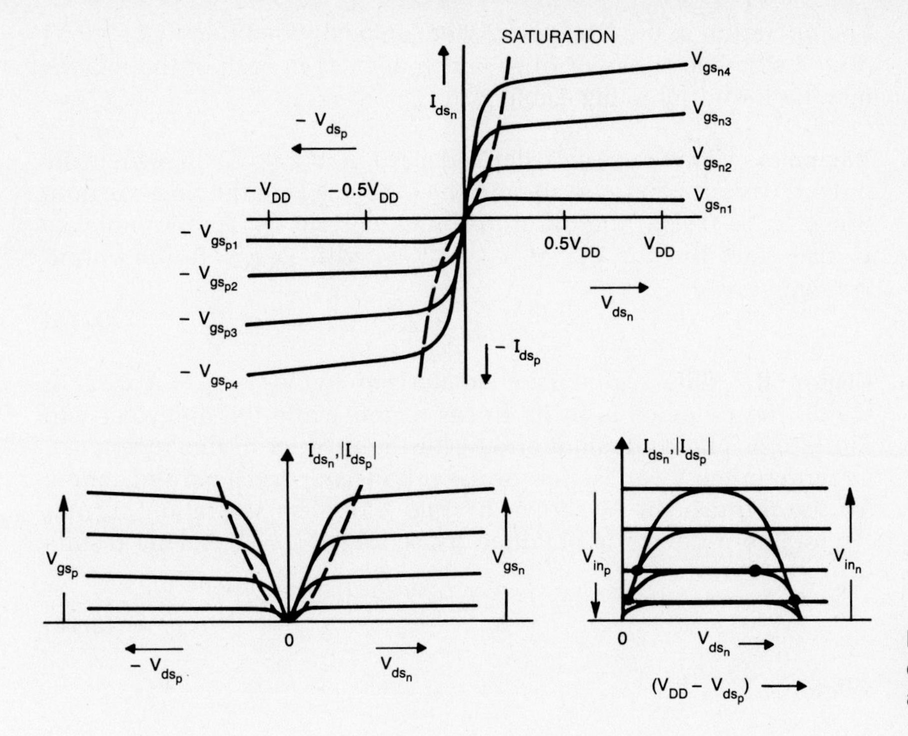

FIGURE 2.11. Graphical derivation of inverter characteristic (load line)

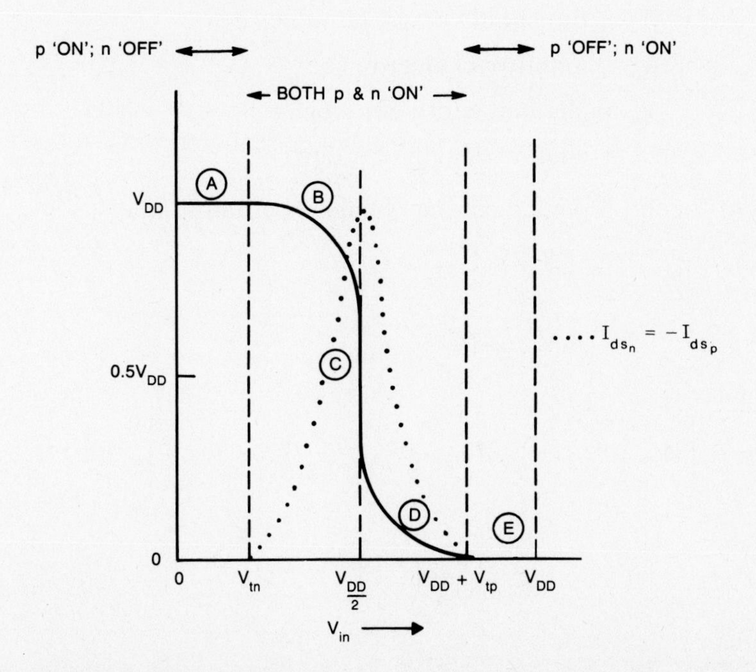

FIGURE 2.12. CMOS inverter DC transfer characteristic and operating regions

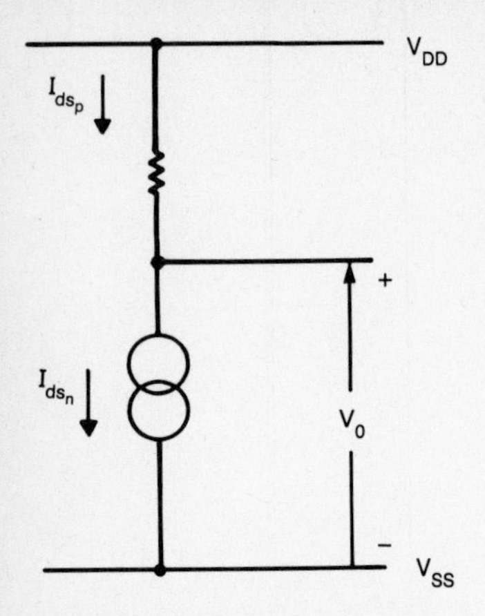

FIGURE 2.13. Equivalent circuit for region B of inverter operation

The operation of the CMOS inverter can be divided into five regions (Fig. 2.12). The behavior of n- and p-devices in each of the regions may be found by using Table 2.1.

Region A. This region is defined by $0 \leq V_{in} \leq V_{tn}$ in which the n-device is cut-off ($I_{ds_n} = 0$), and the p-device is in the linear region. Since $I_{ds_n} = -I_{ds_p}$, the drain-to-source current I_{ds_p} for the p-device is also zero. But for $V_{ds_p} = V_O - V_{DD}$, with $V_{ds_p} = 0$, the output voltage is

$$V_O = V_{DD}. \tag{2.15}$$

Region B. This region is characterized by $V_{t_n} \leq V_{in} < V_{DD}/2$ in which the p-device is in its linear region while the n-device is in saturation. The equivalent circuit for the inverter in this region can be represented by a resistor for the p-transistor and a current source for the n-transistor as shown by Fig. 2.13. The saturation current I_{ds_n} for the n-device is obtained by setting $V_{gs} = V_{in}$. This results in

$$I_{ds_n} = \beta_n \frac{[V_{in} - V_{t_n}]^2}{2}, \tag{2.16}$$

where

$$\beta_n = \frac{\mu_n \varepsilon}{t_{ox}} \left(\frac{W_n}{L_n} \right)$$

and

$$V_{t_n} = \text{threshold voltage of n-device}$$
$$\mu_n = \text{mobility of electrons}$$
$$W_n = \text{channel width of n-device}$$
$$L_n = \text{channel length of n-device.}$$

The current for the p-device can be obtained by noting that

$$V_{gs} = (V_{in} - V_{DD})$$

and

$$V_{ds} = (V_O - V_{DD})$$

and therefore

$$I_{ds_p} = -\beta_p \left[(V_{in} - V_{DD} - V_{t_p})(V_O - V_{DD}) - \frac{1}{2}(V_O - V_{DD})^2 \right], \tag{2.17}$$

where

$$\beta_p = \frac{\mu_p \varepsilon}{t_{ox}} \left(\frac{W_p}{L_p} \right)$$

and

$$V_{t_p} = \text{threshold voltage of p-device}$$
$$\mu_p = \text{mobility of holes}$$
$$W_p = \text{channel width of p-device}$$
$$L_p = \text{channel length of p-device.}$$

Substituting

$$I_{ds_p} = -I_{ds_n}, \tag{2.18}$$

the output voltage V_O can be expressed as

$$V_O = (V_{in} - V_{t_p}) + \left[(V_{in} - V_{t_p})^2 - 2\left(V_{in} - \frac{V_{DD}}{2} - V_{t_p}\right)V_{DD} - \frac{\beta_n}{\beta_p}(V_{in} - V_{t_n})^2\right]^{1/2}. \tag{2.19}$$

Region C. In this region both the n- and p-devices are in saturation. The saturation currents for the two devices are given by

$$I_{ds_p} = \frac{1}{2}\beta_p(V_{in} - V_{DD} - V_{t_p})^2$$

$$I_{ds_n} = \frac{1}{2}\beta_n(V_{in} - V_{t_n})^2$$

with

$$I_{ds_p} = -I_{ds_n}.$$

This yields

$$V_{in} = \frac{V_{DD} + V_{t_p} + V_{t_n}\sqrt{\dfrac{\beta_n}{\beta_p}}}{1 + \sqrt{\dfrac{\beta_n}{\beta_p}}}. \tag{2.20}$$

By setting

$$\beta_n = \beta_p \text{ and } V_{t_n} = -V_{t_p},$$

we obtain

$$V_{in} = \frac{V_{DD}}{2}, \tag{2.21}$$

which implies that region C exists for only one value of V_{in}. The possible values of V_O in this region can be deduced as follows:

$$\text{n-channel:} \quad V_{in} - V_O < V_{t_n}$$
$$V_O > V_{in} - V_{t_n}$$
$$\text{p-channel:} \quad V_{in} - V_O > V_{t_p}$$
$$V_O < V_{in} - V_{t_p}.$$

Combining the two inequalities results in

$$V_{in} - V_{t_n} < V_O < V_{in} - V_{t_p}. \tag{2.22}$$

This indicates that with $V_{in} = V_{DD}/2$, V_O varies within the range shown. Of course, we have assumed that an MOS device in saturation behaves like an ideal current source with drain-to-source current being independent of V_{ds}. In reality, as V_{ds} increases, I_{ds} also increases slightly, thus region C has a finite slope. The significant factor to be noted is that in region C we have two current sources in series, which is an "unstable" condition. Thus a small input voltage has a large effect at the output. This makes the output transition very steep, which contrasts with the equivalent nMOS inverter characteristic. The relation defined by Eq. (2.20) is particularly useful since it provides the basis for defining the *gate threshold* V_{inv}, which corresponds to the state where $V_O = V_{in}$.

Region D. This region is described by $V_{DD}/2 < V_{in} \leq V_{DD} - V_{t_p}$. The p-device is in saturation while the n-device is operating in its linear region. This condition is represented by the equivalent circuit shown in Fig. 2.14. The two currents may be written as

$$I_{ds_p} = -\frac{1}{2}\beta_p(V_{in} - V_{DD} - V_{t_p})^2$$

and

$$I_{ds_n} = \beta_n\left[(V_{in} - V_{t_n})V_O - \frac{1}{2}V_O^2\right]$$

with

$$I_{ds_p} = -I_{ds_n}.$$

The output voltage becomes

$$V_O = (V_{in} - V_{t_n}) - \left[(V_{in} - V_{t_n})^2 - \frac{\beta_p}{\beta_n}(V_{in} - V_{DD} - V_{t_p})^2\right]^{1/2} \tag{2.23}$$

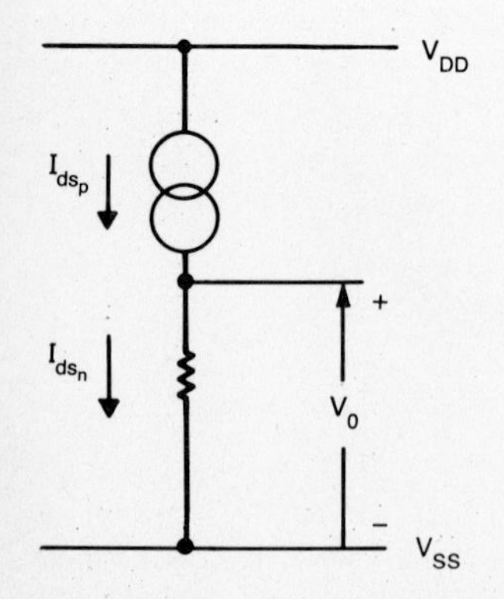

FIGURE 2.14. Equivalent circuit for region D of inverter operation

Region E. This region is defined by the input condition $V_{in} \geq V_{DD} - V_{t_p}$, in which the p-device is cut-off ($I_{ds_p} = 0$), and the n-device

TABLE 2.2. Summary of CMOS inverter operation*

REGION	CONDITION	p-DEVICE	n-DEVICE	OUTPUT
A	$0 \leq V_{in} \leq V_{t_n}$	linear	cut-off	$V_O = V_{DD}$
B	$V_{t_n} \leq V_{in} < \frac{V_{DD}}{2}$	linear	saturated	$^*V_O = (V_{in} + 1) + \sqrt{15 - 6V_{in}}$
C	$V_{in} = \frac{V_{DD}}{2}$	saturated	saturated	$V_O \neq f(V_{in})$
D	$\frac{V_{DD}}{2} < V_{in} \leq V_{DD} - V_{t_p}$	saturated	linear	$^*V_O = (V_{in} - 1) - \sqrt{6V_{in} - 15}$
E	$V_{in} \geq V_{DD} - V_{t_p}$	cut-off	linear	$V_O = 0$

* Parameters assumed $V_{DD} = +5$ volts; $V_{t_p} = -1$ volt; $V_{t_n} = +1$ volt; $\frac{\beta_n}{\beta_p} = 1$.

is in the linear mode. Here, $V_{gs_p} = V_{in} - V_{DD}$, which is more positive than V_{t_p}. The output in this region is

$$V_O = 0. \tag{2.24}$$

From the transfer curves of Fig. 2.12, it may be seen that the transition between the two states is very "steep." This characteristic is very desirable as the noise immunity is maximized. This is covered in more detail in Section 2.3.2. For convenience, the characteristics associated with the five regions are summarized in Table 2.2.

2.3.1 Influence of β_n/β_p ratio on transfer characteristic

In order to explore the variations of the transfer characteristic as a function of β_n/β_p, it is possible to plot the transfer curve for several values of β_n/β_p shown in Fig. 2.15. Here, we note that the gate threshold voltage V_{inv} defined by the state in which

$$V_{in} = V_O \tag{2.25}$$

is dependent on β_n/β_p. Thus, for a given process, if we want to change β_n/β_p, we need to change the channel dimensions, i.e., channel length L and channel width W. From Fig. 2.15 it can be seen that as the ratio β_n/β_p is decreased the transition region shifts from left to right; however, the output voltage transition remains sharp and hence the switching performance is not affected. This behavior should be contrasted with the nMOS inverter, where the transition gain depends critically on the β ratio of the load (pull-up) and driver (pull-down) transistors.

For the CMOS inverter a ratio of

$$\frac{\beta_n}{\beta_p} = 1 \tag{2.26}$$

may be desirable since it allows a capacitive load to charge and discharge in equal times by providing equal current source and sink capabilities. This will be expanded upon in Chapter 4.

Another factor that needs to be considered is the influence of temperature on the transfer characteristics [Cobb66]. As the temperature of an MOS device is increased, the effective carrier mobility μ in the channel decreases. This results in a decrease in β, which is related to temperature T by

$$\beta \propto T^{-3/2}. \tag{2.27}$$

Therefore

$$I_{ds} \propto T^{-3/2}. \tag{2.28}$$

Since the voltage transfer characteristics depend on the ratio β_n/β_p, and the mobility of both holes and electrons are similarly affected, this ratio is independent of temperature to a good approximation. Both V_{t_n} and V_{t_p} decrease slightly as temperature increases. This implies that as temperature increases, the extent of region A is reduced while the extent of region E increases. Thus the overall transfer characteristics of Fig. 2.15 shift to the left as temperature increases. Based on the figures given earlier, if the temperature rises by 50°C , the thresholds drop by 200mV each. This would cause a .4 V shift in the input threshold.

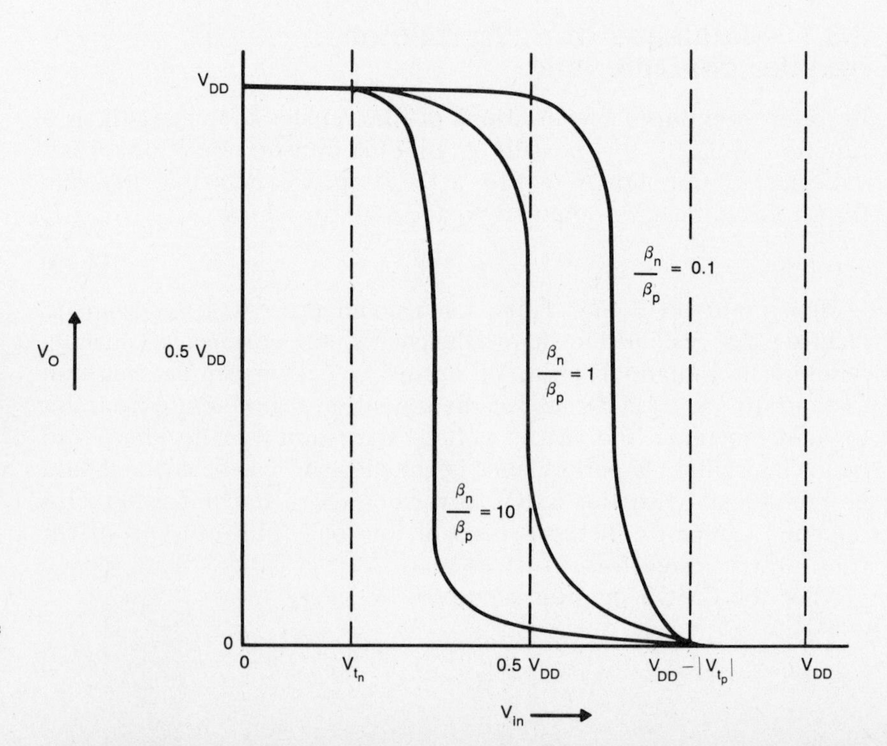

FIGURE 2.15. Influence of $\frac{\beta_n}{\beta_p}$ on inverter DC transfer characteristic

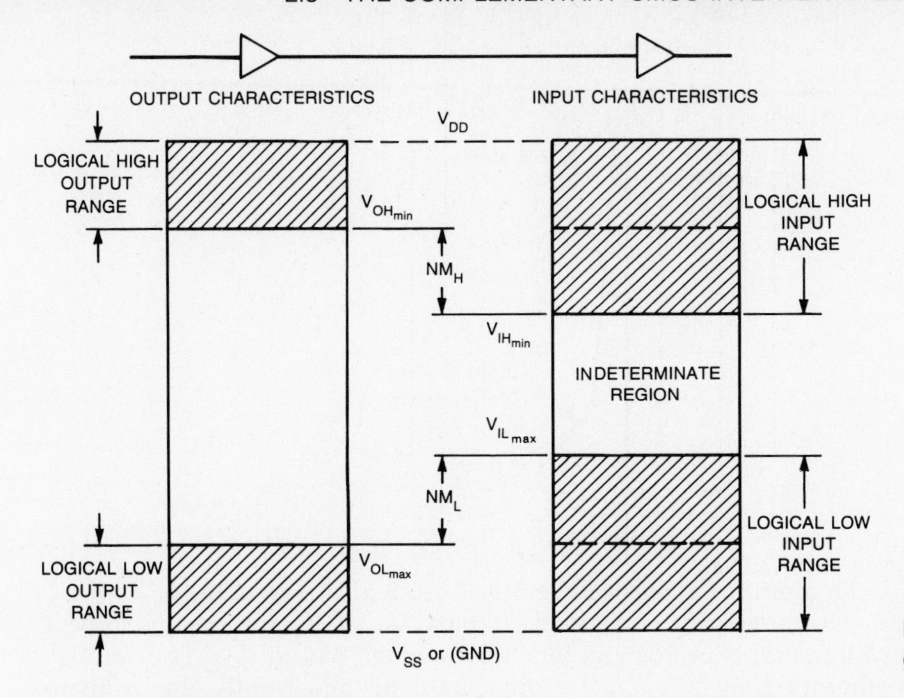

FIGURE 2.16. Noise margin definitions

2.3.2 Noise margin

Noise margin is a parameter closely related to the input-output voltage characteristics. This parameter permits one to determine the allowable noise voltage on the input of a gate so that the output will not be affected. The specification most commonly used to specify noise margin (or noise immunity) is in terms of two parameters — the *LOW* noise margin, NM_L, and the *HIGH* noise margin, NM_H. With reference to Fig. 2.16, NM_L is defined as the difference in magnitude between the maximum LOW output voltage of the driving gate and the maximum input LOW voltage recognized by the driven gate. Thus

$$NM_L = |V_{IL_{\max}} - V_{OL_{\max}}|. \tag{2.29}$$

The value NM_H is the difference in magnitude between the minimum HIGH output voltage of the driving gate and the minimum input HIGH voltage recognized by the receiving gate. Thus

$$NM_H = |V_{OH_{\min}} - V_{IH_{\min}}|, \tag{2.30}$$

where

$V_{IH_{\min}}$ = minimum HIGH input voltage

$V_{IL_{\max}}$ = maximum LOW input voltage

$V_{OH_{\min}}$ = minimum HIGH output voltage

$V_{OL_{\max}}$ = maximum LOW output voltage.

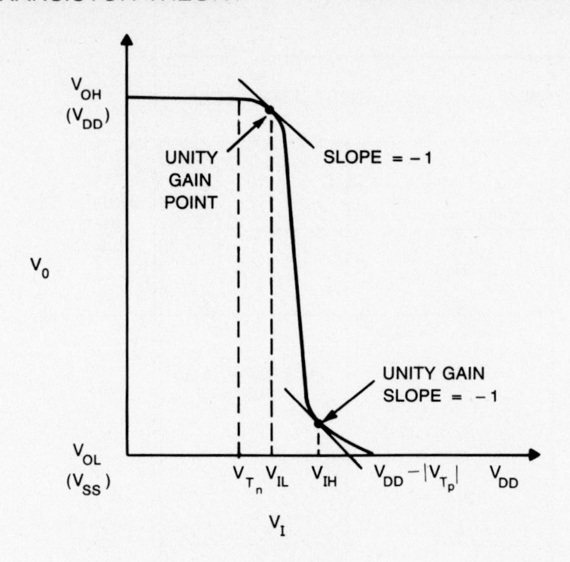

FIGURE 2.17. CMOS inverter noise margins

The definitions on page 51 are illustrated in Fig. 2.16.

Generally, it is desirable to have $V_{IH} = V_{IL}$ and for this to be a value that is midway in the "logic swing," V_{OL} to V_{OH}. This implies that the transfer characteristic should switch abruptly, that is, there should be high gain in the transition region. For the purposes of calculating noise margins, the transfer characteristic of the inverter and the definitions of voltage levels V_{IL}, V_{OL}, V_{IH}, V_{OH} are shown in Fig. 2.17. To determine V_{IL}, we note that the inverter is in region B of operation, where the p-device is in its linear region while the n-device is in saturation. The result of analyzing these quantities ($\beta_n = \beta_p$) (see Appendix A) are as follows:*

$$NM_H = \frac{3V_{DD} + 5|V_{tp}| - 3V_{tn}}{8} \tag{2.31}$$

and

$$NM_L = \frac{3V_{DD} - 3|V_{tp}| + 5V_{tn}}{8}. \tag{2.32}$$

In the case $V_{t_n} = |V_{t_p}| = 0.2\ V_{DD}$, we have

$$NM_L = NM_H = (3 + 1 - 0.6)\frac{V_{DD}}{8} = 0.425\ V_{DD}.$$

Note that if $|V_{t_p}| = V_{t_n}$, then NM_H and NH_L *increase* as threshold voltages are *increased*. Note that if either NM_L or NM_H for a gate are reduced ($\approx .1V_{DD}$), then the gate may be susceptible to switching noise that may be present on the inputs. This is the reason to keep track of noise margins. Quite often, noise margins are compromised

* Derivation by K. Trivedi, Duke University.

to improve speed. Circuit examples later in the book will illustrate this trade-off.

2.4 Alternate CMOS inverters

Fig. 2.18a on page 54 shows an inverter that uses a p-device pull-up that has its gate permanently grounded. This is roughly equivalent to the use of a depletion load in nMOS. This circuit is used in a variety of CMOS logic circuits. It has the disadvantage that it dissipates DC power when the n-transistor (pull-down) is turned on. Similar to the complementary inverter, a graphical solution to the transfer characteristic is shown in Fig. 2.18b on page 54 for various sized p-devices. This shows that the ratio of β_n/β_p affects the shape of the transfer characteristic and the V_{OL} of the inverter. To determine the ratio of the n-transistor size to the p-transistor size the circuit in Fig. 2.19 on page 54 will be used. This shows two cascaded pseudo-nMOS inverters. In order to cascade inverters without degradation of signal levels, the following condition should be met:

$$V_O = V_{in} = V_{inv},$$

where

$$V_{inv} = \text{the gate threshold voltage.}$$

For equal noise margins, the gate threshold voltage V_{inv} should be set to approximately $0.5V_{DD}$. At this operating point, the n-device (pull-down) is in saturation ($0 < V_{gs_n} - V_{t_n} < V_{ds_n}$), and the p-device (pull-up) is in the linear mode of operation ($0 < V_{ds_p} < V_{gs_p} - V_{t_p}$). From Eq. (2.2c), the n-device I_{ds} with $V_{gs_n} = V_{inv}$ is

$$I_{ds_n} = \frac{\beta_n}{2}(V_{inv} - V_{t_n})^2. \quad \textbf{(2.33)}$$

Similarly the p-device I_{ds} with $V_{gs_p} = -V_{DD}$ is

$$I_{ds_p} = \beta_p\left[(-V_{DD} - V_{t_p})V_{ds_p} - \frac{V_{ds_p}^2}{2}\right]. \quad \textbf{(2.34)}$$

Equating the two currents we obtain

$$\frac{\beta_n}{2}(V_{inv} - V_{t_n})^2 = \beta_p\left[(-V_{DD} - V_{t_p})V_{ds_p} - \frac{V_{ds_p}^2}{2}\right]. \quad \textbf{(2.35)}$$

Upon rearrangement,

$$V_{inv} = V_{t_n} + \left[2\frac{\beta_p}{\beta_n}\right]^{1/2}\left[(-V_{DD} - V_{t_p})V_{ds_p} - \frac{V_{ds_p}^2}{2}\right]^{1/2}. \quad \textbf{(2.36)}$$

FIGURE 2.18. Pseudo nMOS inverter and DC transfer characteristics

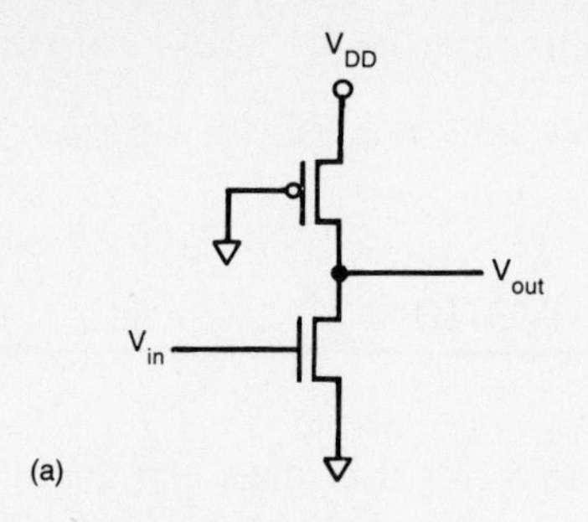

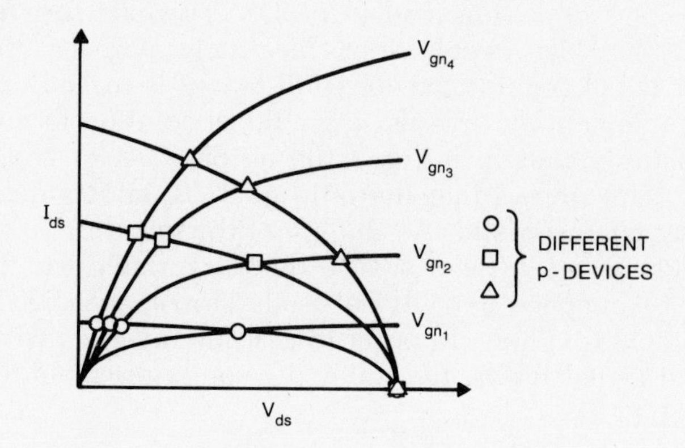

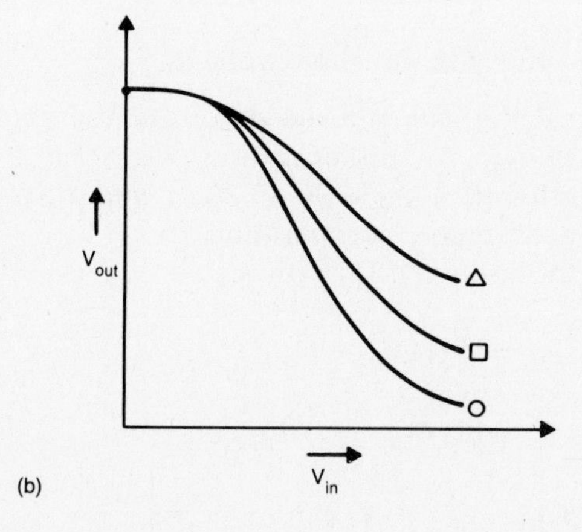

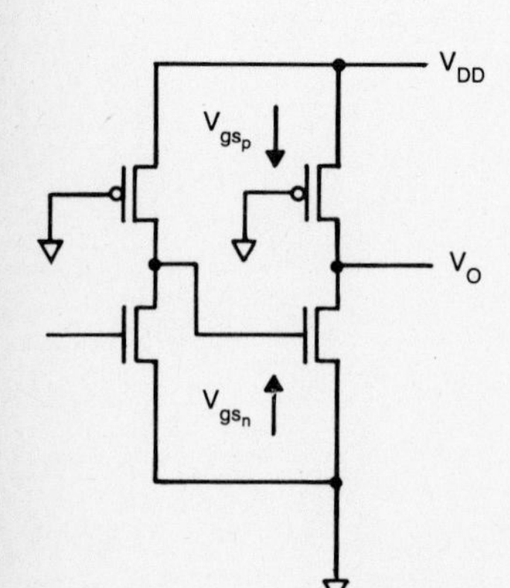

FIGURE 2.19. Cascaded pseudo nMOS inverters

With $V_{inv} = 0.5V_{DD}$, $V_{t_n} = |V_{t_p}| = 0.2V_{DD}$, $V_{DD} = 5$ volts, the following result is obtained for an adequate β ratio:

$$\frac{\beta_p}{\beta_n} = \frac{1}{6} \tag{2.37}$$

Recalling that the technology and geometry contributions to β, the ratio of widths of the n-device to the p-device should be approximately

3/1. For an inverter with slightly lower noise immunity a β_n/β_p of 4/1 may be used which parallels the popular nMOS ratio rule [MeCo80]. This inverter finds use in circuits where an "n-rich" circuit is required and the power dissipation can be tolerated. Typical uses include static ROMs and PLAs. Note that the circuit could use n-load devices and p-active pull-ups, if this was of advantage.

Another inverter of interest is the tri-state inverter shown in Fig. 2.20. When CL = '0', the output of the inverter is in a tri-state condition (the Z output is not driven by the A input). When CL = '1', the output Z is equal to $\overline{A}$. For the same sized n- and p-devices, this inverter is approximately half the speed of the inverter shown in Fig. 2.10. This inverter will be discussed in more detail in Chapter 5, as it forms the basis for various types of clocked logic, latches, multiplexers, and I/O structures.

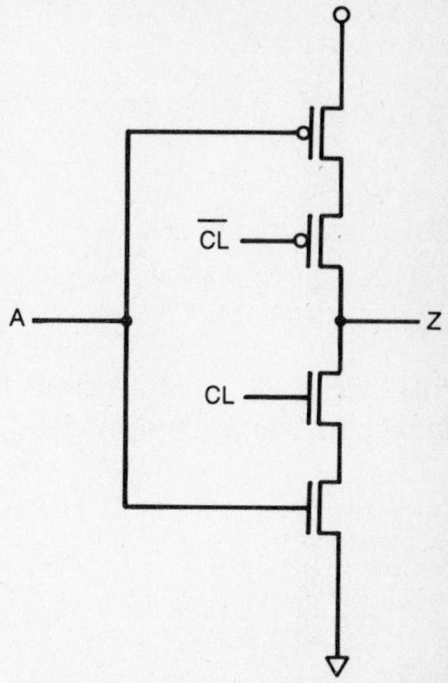

FIGURE 2.20. CMOS tri-state inverter

2.5 Transmission gate — DC characteristics

The transistor connection for a complementary switch or transmission gate is reviewed in Fig. 2.21. It consists of an n-channel transistor and a p-channel transistor with separate gate connections and common source and drain connections. The control signal ϕ is applied to the gate of the n-device, and its complement $\overline{\phi}$ is applied to the gate of the p-device. The operation of the transmission gate can be best explained by considering the characteristics of both the n-device and p-device as pass transistors individually. We will address this by treating the charging and discharging of a capacitor via a transmission gate.

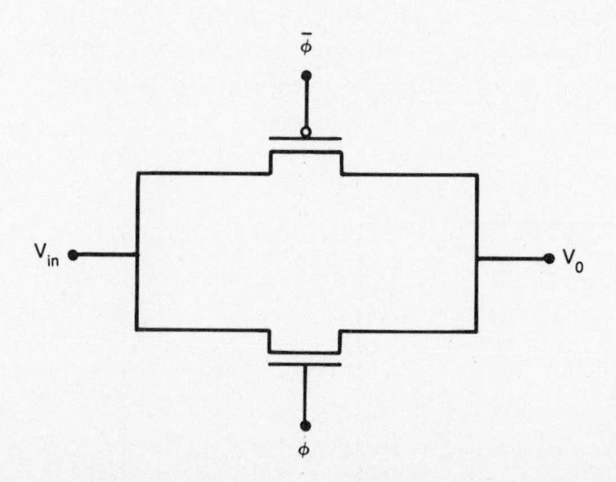

FIGURE 2.21. Transistor connection for CMOS transmission gate

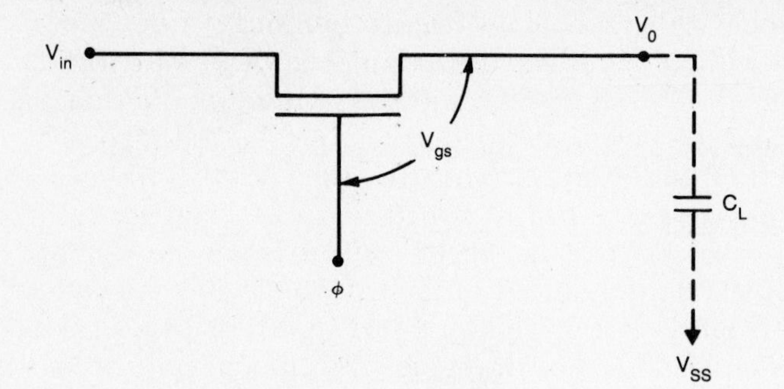

FIGURE 2.22. nMOS transistor in transmission gate

nMOS Pass Transistor. Referring to Fig. 2.22, the load capacitor C_L is initially discharged (i.e., V_O = 0). With ϕ = '0' (i.e., V_{gs} = 0 volts), I_{ds} = 0, then V_O = '0' irrespective of the state of the input V_{in}. When ϕ = '1', and V_{in} = '1', the pass transistor begins to conduct and charges the load capacitor towards V_{DD}, i.e., initially $V_{gs} = V_{DD}$. Since initially V_{in} is at a higher potential than V_O, the current flows through the device from left to right. As the output voltage approaches $(V_{DD} - V_{t_n})$, the n-device begins to turn off. Load capacitor, C_L, will remain charged when ϕ is changed back to '0'. Therefore the output voltage V_O remains at $(V_{DD} - V_{t_n})$. This implies that the transmission of logic '1' is degraded as it passes through the gate. With V_{in} = '0' and ϕ = '1', the pass transistor begins to conduct and discharge the load capacitor towards V_{SS}, i.e., $V_{gs} = V_{DD}$. Since initially V_{in} is at a lower potential than V_O, the current flows through the device from right to left. As the output voltage approaches V_{SS} (0 V), the n-device current diminishes. Thus the transmission of a logic '0' is not degraded.

pMOS Pass Transistor. Once again a similar approach can be taken in analyzing the operation of a pMOS pass transistor, as shown in Fig. 2.23. With ϕ = '1', V_{in} = '1', and V_O = '0', the load capacitor C_L remains uncharged. When ϕ = '0', current begins to flow and charges the load capacitor towards V_{DD}. However, when V_{in} = '0' and V_O = '1', the load capacitor discharges through the p-device

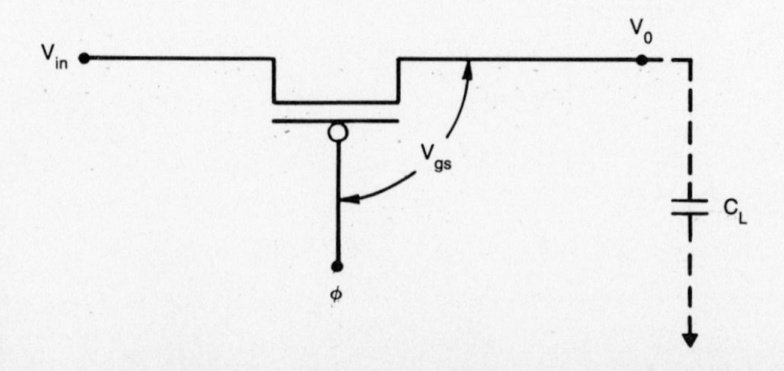

FIGURE 2.23. pMOS transistor in transmission gate

TABLE 2.3. Transmission characteristics of *n*-channel and *p*-channel pass transistors

DEVICE	TRANSMISSION OF '1'	TRANSMISSION OF '0'
n	poor	good
p	good	poor

until $V_O = |V_{tp}|$, at which point the transistor ceases conducting. Thus transmission of '0' is somewhat degraded through the p-device.

The resultant behavior of the n-device and p-device are shown in Table 2.3. By combining the two characteristics we can construct a transmission gate that can transmit both a logic '1' and logic '0' without degradation. As can be deduced from the discussion so far, the operation of the transmission gate requires both the true and the complement version of the control signal ϕ. The overall behavior can be expressed as:

$$\phi = \text{'0'};\begin{cases}\text{n-device} = \text{off};\\ \text{p-device} = \text{off};\\ V_{in} = \text{'0'},\ V_O = Z;\\ V_{in} = \text{'1'},\ V_O = Z,\end{cases} \tag{2.38}$$

where Z refers to a high impedance state and

$$\phi = \text{'1'};\begin{cases}\text{n-device} = \text{on};\\ \text{p-device} = \text{on};\\ V_{in} = \text{'0'},\ V_O = \text{'0'};\\ V_{in} = \text{'1'},\ V_O = \text{'1'}.\end{cases} \tag{2.39}$$

The corresponding output characteristic is shown in Fig. 2.24.

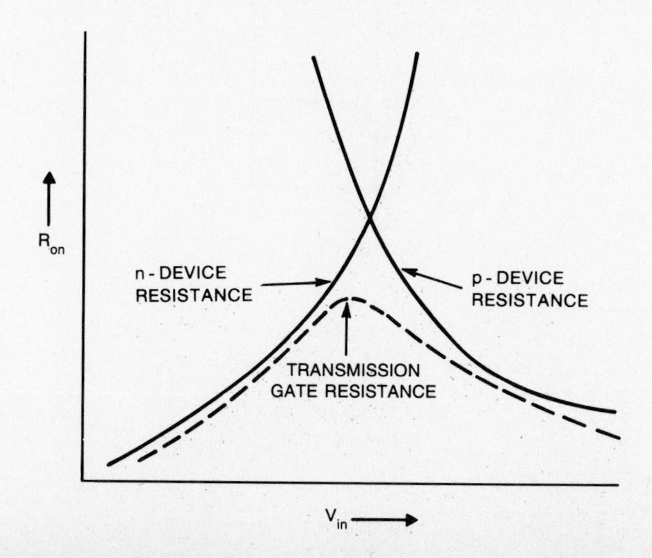

FIGURE 2.24. Transmission gate output characteristic

2.6 Latch-up

If every silver lining has a cloud, then the cloud that has plagued CMOS is a parasitic circuit effect called "latch-up." The result of this effect is the shorting of the V_{DD} and V_{SS} lines, usually resulting in chip self-destruction or at least system failure with the requirement to power down. This effect was a critical factor in the lack of acceptance of early CMOS processes but in current processes is controlled by process innovations and well understood circuit techniques. Before proceeding with this section the reader should read Chapter 3, to become familiar with process layer terminology.

The source of the latch-up effect [Estr80][EsDu82] is depicted in Fig. 2.25. The figure shows a cross-sectional view of a typical (p-well) CMOS process and, additionally, shows two parasitic bipolar transistors. Now, bipolar transistors have completely different characteristics than MOS transistors. There are two types of bipolar transistors: npn and pnp. The letters refer to the type of semiconductor material that comprises the three terminals of a bipolar transistor, which are called the collector, base, and emitter. In an npn transistor, the transistor passes current from collector to emitter if the collector is biased positive with respect to the emitter and the base is about

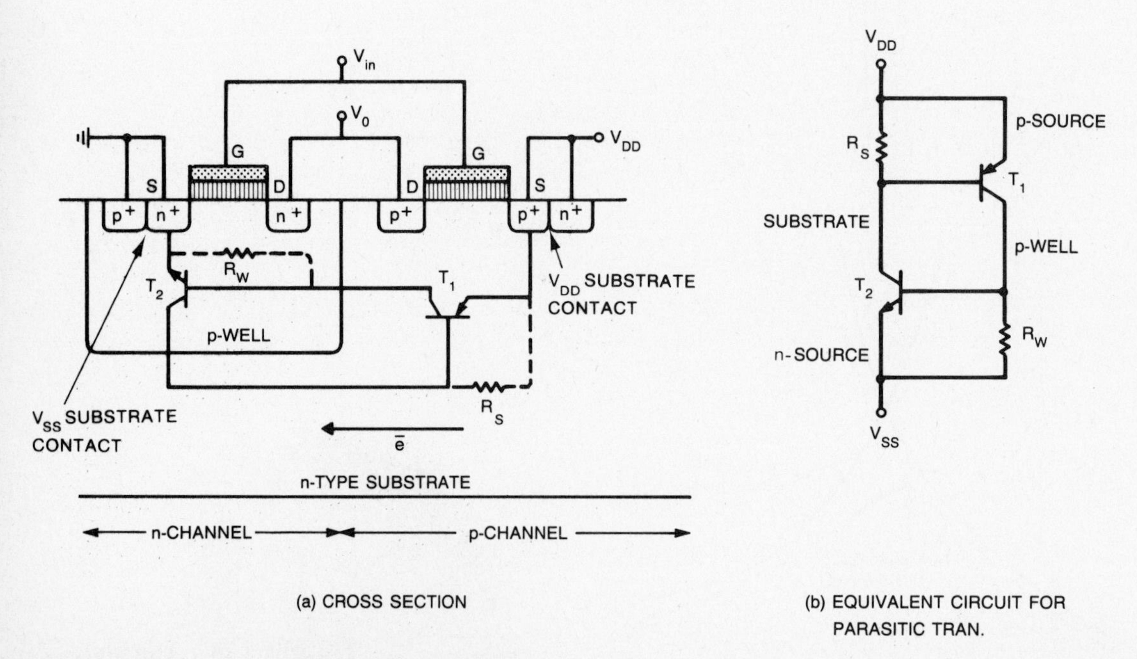

FIGURE 2.25. Source of latch-up in CMOS

.6 V more positive than the emitter. A pnp transistor has complementary characteristics. Although these descriptions are basic, they will serve to illustrate the source of latch-up. In the figure, we have indicated a pnp (T_1) and an npn (T_2) bipolar transistor, each of which has a gain factor associated with them. In addition, resistances R_S and R_W are critical in determining the latch-up susceptibility of a circuit. R_S is due to the n-substrate, while R_W is due to the p-well. The larger these resistances are, the more likely a structure is to latch-up. Fig. 2.25b shows the equivalent circuit of the latching circuit. If enough electron current is injected into the n-type substrate and collected by the p-well, the voltage drop across R_S will be of sufficient magnitude to switch the two bipolar transistors T_1 and T_2 into a low resistance mode. Thus, depending on the series resistance of the *parasitic path* (R_S), the current drawn from the power supply can be significant and may result in circuit failure. Alternatively, the flow of collected excess carriers through the well region may be sufficient to *forward bias* T_2. Once again, circuit failure may result. These effects are moderated by reducing the gain of the transistors and reducing the value of the parasitic resistors R_W and R_S. The latter action is achieved by using substrate contacts which effectively short out the resistors. These are shown in Fig. 2.25.

In most current processes the possibility of latch-up occurring in internal circuitry has been reduced to the point where a designer need not worry about the effect as long as *liberal* substrate contacts are used. The definition of "liberal" is usually acquired from designers who have completed "successful" designs on a given process. Actually, calculating the parasitics is possible, but the actual switching transients existent in the circuit have a great effect on any possible latch-up condition. Thus at this time it is quite difficult to "synthesize" the required number of substrate contacts. A few rules may be followed that reduce the possibility of internal latch-up to a very small likelihood.

- Every well must have a substrate contact of the appropriate type.
- Every substrate contact should be connected by metal directly to a supply pad (i.e., no diffusion or polysilicon underpasses in the supply rails).
- Place substrate contacts as close as possible to the source connection of transistors connected to the supply rails (i.e., V_{SS} n-devices, V_{DD} p-devices). This reduces the value of R_S and R_W. A very conservative rule would place one substrate contact for every supply connection.
- Place a substrate contact per 5–10 logic transistors.
- Lay out n- and p-transistors with packing of n-devices towards V_{SS} and packing of p-devices towards V_{DD} (see layout styles

later). Avoid "convoluted" structures that intertwine n- and p-devices in checkerboard styles.

The most likely place for latch-up to occur is in I/O structures where large currents flow, large parasitics may be present and abnormal circuit voltages may be encountered. In these structures two options can be taken. The first is to use proven I/O structures designed by experts who understand the process at a detailed level. Secondly, rules may be applied to the design of these structures that minimize the possibility of latch-up. Typical rules (p-well process) include:

- Physically separate the n- and p-driver transistors (i.e., with the bonding pad).
- Include p^+ guard rings around n-transistors connected to V_{SS}.
- Include n^+ guard rings around p-transistors connected to V_{DD}.
- Employ minimum area p-wells so that p-well photocurrent is minimized during transient pulses. In fact, in some n-well I/O designs, wells are entirely eliminated by using only n-devices in buffers.
- Source diffusion regions of the p-transistors should be placed so that they lie along equipotential lines when current flows between V_{DD} and the p-wells; that is, source fingers should be perpendicular to the dominant direction of current flow rather than parallel to it. This reduces the possibility of latch-up through the p-transistor source, due to an effect called "field aiding" [EsDu82].
- Shorting p-transistor source regions to the substrate and the n-transistor source regions to the p-well with metallization along their entire lengths will aid in preventing either of these diodes from becoming forward biased, and hence reduces the contribution to latch-up from these components.
- The p-well should be hard-wired (via p^+) to ground so that any injected charge is diverted to ground via a low resistance path. The p-well has a relatively high sheet resistance and is susceptible to charge injection.
- The spacing between the p-well p^+ and n-transistor source contact should be kept to a minimum. This allows minority carriers near the parasitic npn transistor emitter-base junction to be collected and reduces R_W.
- The separation between the substrate n^+ and p-transistor source contact should be minimized. This results in reduced minority carrier concentration near the pnp emitter-base junction.

2.7 Exercises

2.1 A process has a nominal K_p of $10\mu AV^{-2}$ and a thin-oxide thickness (t_{ox}) of 1000 $A°$. A batch of circuits have a t_{ox} of 800 $A°$. What would the nominal K_p be? What effect would this have on the drain current of n- and also p-transistors?

2.2 A transistor has a drawn W of 20μ and a drawn L of 5μ. During processing, the polysilicon is overetched by 1μ on all sides. The source and drain diffusions bloat by $.5\mu$. If $K_p = 15\mu AV^{-2}$, what is the β of the final transistor?

2.3 Calculate the $g_{m(sat)}$ of a transistor with $\beta = 40\ \mu AV^{-2}$, $\lambda = .03\ V^{-1}$, $V_t = 1.0\ V$, and $V_{gs} = 5\ V$, taking into account channel length modulation.

2.4 Calculate the input switching voltage for a 2-input NOR gate constructed of identical sized n- and p-transistors with one input held high and both inputs held high. How do the noise margins vary? What ramifications does this have for multiple input gates?

CMOS PROCESSING TECHNOLOGY

After a brief discussion of silicon semiconductor processing, this chapter will provide an overview of a variety of CMOS technologies that are currently in use in industry and a view of some current research directions. Following this, the geometric design rules for some representative processes will be summarized.

3.1 Silicon semiconductor technology: an overview

Silicon in its pure or *intrinsic* state is a semiconductor, having a bulk electrical resistance somewhere between that of a conductor and an insulator. The conductivity of silicon can be varied over several orders of magnitude by introducing *impurity* atoms into the silicon crystal lattice. These *dopants* may either supply free *electrons* or *holes*. Impurity elements that use electrons are referred to as *acceptors* since they accept some of the electrons already in the silicon, leaving vacancies or holes. Similarly, *donor* elements provide electrons. Silicon that contains a majority of donors is known as *n-type* and that which contains a majority of acceptors is known as *p-type*. When n-type and p-type materials are brought together, the region where the silicon changes from n-type to p-type is called a *junction*. By arranging junctions in certain physical structures and combining these with other physical structures, various semiconductor devices may be constructed. Over the years, silicon semiconductor processing has evolved sophisticated techniques for building these junctions and other structures having special properties.

3.1.1 Wafer processing

The basic raw material used in modern semiconductor plants is a *wafer* or disk of silicon, which varies from 75 mm to 150 mm in diameter and is less than 1 mm thick. Wafers are cut from ingots of single crystal silicon that have been pulled from a crucible melt of pure molten polycrystalline silicon. This is known as the "Czochralski" method (Fig. 3.1) and is currently the most common method for producing single crystal material. Controlled amounts of impurities are added to the melt to provide the crystal with the required electrical properties. The crystal *orientation* is determined by a *seed* crystal that is dipped into the melt to initiate single crystal growth. The melt is contained in a quartz crucible, which is surrounded by a graphite radiator. The graphite is heated by radio frequency induction and the temperature is maintained a few degrees above the melting point of silicon ($\approx$1425° C). The atmosphere above the melt is typically helium or argon.

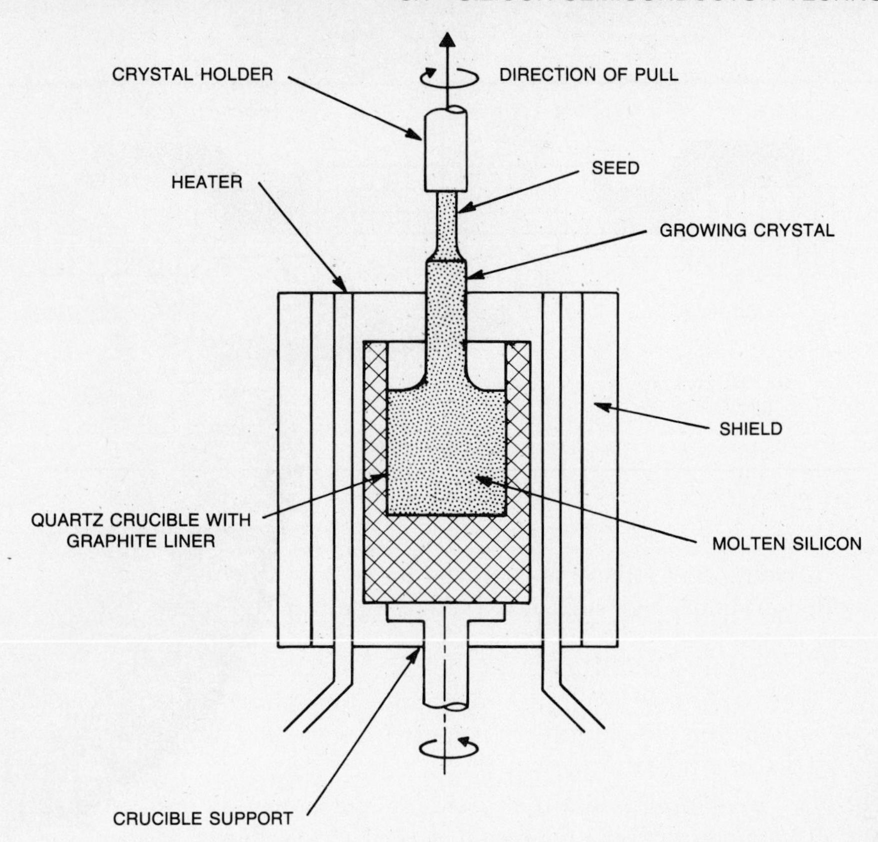

FIGURE 3.1. Czochralski process for manufacturing silicon ingots

After the seed is dipped into the melt, the seed is gradually withdrawn vertically from the melt while simultaneously being rotated. The molten polycrystalline silicon melts the tip of the seed and as it is withdrawn, refreezing occurs. As the melt freezes, it assumes the single crystal form of the seed. This process is continued until the melt is consumed. The diameter of the ingot is determined by the seed withdrawal rate and the seed rotation rate. Growth rates range from 30 to 180 mm/hour.

Slicing into wafers is usually carried out using internal cutting edge diamond blades. Wafers are usually between 0.25 mm and 1.0 mm thick, depending on their diameter. Following this operation, at least one face is polished to a flat, scratch-free mirror finish.

3.1.2 Oxidation

Many of the structures and manufacturing techniques used to make silicon integrated circuits rely on the properties of the oxide of silicon, namely, silicon dioxide (SiO_2). Therefore the reliable manufacture of SiO_2 is extremely important.

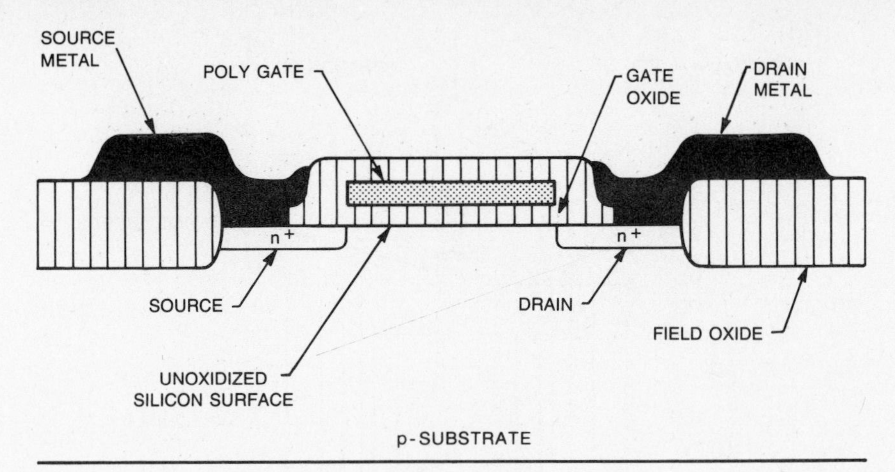

FIGURE 3.2. An nMOS transistor showing the growth of field oxide in both vertical directions

Oxidation of silicon is achieved by heating silicon wafers in an oxidizing atmosphere such as oxygen or water vapor. The two common approaches are:

- Wet oxidation: when the oxidizing atmosphere contains water vapor. The temperature is usually between 900° C and 1000° C. This is a rapid process.
- Dry oxidation: when the oxidizing atmosphere is pure oxygen. Temperatures are in the region of 1200° C, to achieve an acceptable growth rate.

The oxidation process consumes silicon. Since SiO_2 has approximately twice the volume of silicon, the SiO_2 layer grows almost equally in both vertical directions. This effect is shown in Fig. 3.2 for an n-channel MOS device in which the SiO_2 (field oxide) projects above and below the unoxidized silicon surface.

3.1.3 Selective diffusion

To create different types of silicon, containing different proportions of donor or acceptor impurities, further processing is required. As these areas are required to be precisely placed and sized, a means of ensuring this is required. The ability of SiO_2 to act as a barrier against doping impurities is a vital factor in this process called *selective diffusion*. The SiO_2 layer may be used as a *pattern mask*. Areas on the silicon wafer surface where there is an absence of SiO_2 allow dopant atoms to pass into the wafer, thus changing the characteristics of the silicon. Areas where SiO_2 overlays the silicon act as barriers to the dopant atoms. Thus selective diffusion entails:

- Opening *windows* in a layer of SiO_2 grown on the surface of the wafer.
- Removing SiO_2, but not Si, with a suitable etchant.
- Subjecting exposed Si to a dopant source.

The process used for selectively removing the oxide involves covering the surface of the oxide with an acid resistant coating, except where diffusion windows are needed. The SiO_2 is removed using an etching technique. The acid resistant coating is normally a photosensitive organic material called *photoresist* (PR), which can be polymerized by ultraviolet (UV) light. If the UV light is passed through a mask containing the desired pattern, the coating can be polymerized where the pattern is to appear. The unpolymerized areas may be removed with an organic solvent. Etching of exposed SiO_2 then may proceed. This process is illustrated in Fig. 3.3. In established processes using PRs in conjunction with UV light sources, diffraction around the edges of the mask patterns and alignment tolerances have limited line widths on the order of about 1.5 μm to 2 μm. However, during recent years, electron beam lithography (EBL) has emerged as a contender for pattern generation and imaging where line widths of the order of 0.5 μm with good definition are achievable. The main advantages of EBL pattern generation are:

- Patterns are derived directly from digital data.

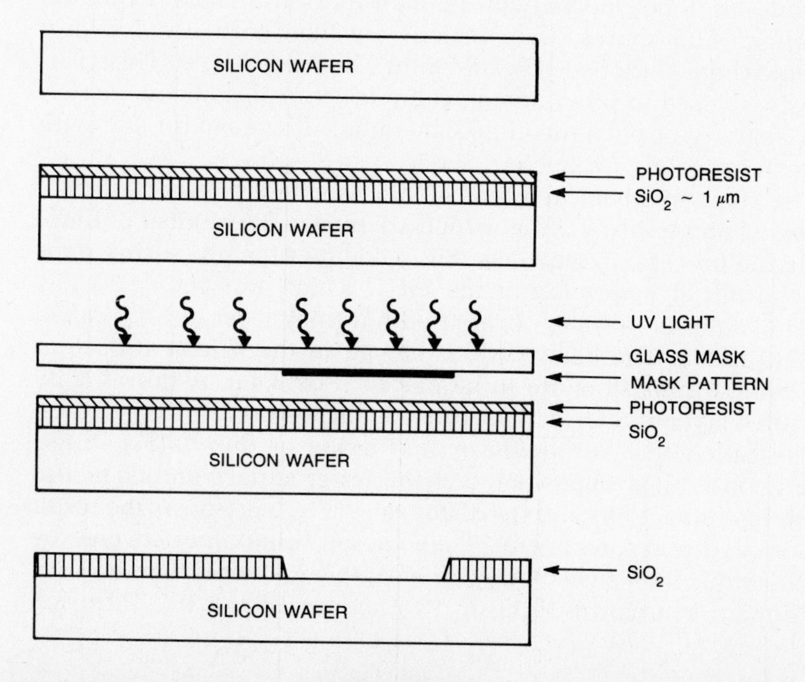

FIGURE 3.3. Simplified steps involved in the patterning of SiO_2

- There are no intermediate hardware images such as recticles or masks; that is, the process may be direct.
- Different patterns may be accommodated in different sections of the wafer without difficulty.
- Changes to patterns can be implemented quickly.

The main disadvantage that has precluded the use of this technique in commercial fabrication lines is the cost of equipment and the large amount of time required to access all points on the wafer.

3.1.4 The silicon gate process

So far we have touched on the single crystal form of silicon used in the manufacture of wafers and the oxide used in the manufacture and operation of circuits. Silicon may also be formed in an amorphous form (not having a carefully arranged lattice structure) commonly called *polycrystalline* silicon or *polysilicon*. This is used as an interconnect in silicon ICs and as the gate electrode on MOS transistors. The most significant aspect of using polysilicon as the gate electrode is its ability to be used as a further mask to allow precise definition of source and drain electrodes. This is achieved with minimum gate-to-source/drain overlap, which we will learn improves circuit performance. Polysilicon is formed when silicon is deposited on SiO_2 or other surfaces. In the case of an MOS transistor gate electrode, undoped polysilicon is deposited on the gate insulator. Polysilicon and source/drain regions are then normally doped at the same time. Undoped polysilicon has high resistivity. This characteristic is used to provide high value resistors in static memories. The resistivity of polysilicon may be reduced by combining it with a refractory metal (see Section 3.2.5).

The steps involved in a typical silicon gate process entail photomasking and oxide etching, which are repeated a number of times during the processing sequence. Fig. 3.4 shows the processing steps after the initial patterning of the SiO_2, which was shown in Fig. 3.3. The wafer is initially covered with a thick layer of SiO_2 called the *field oxide*. The field oxide is etched to the silicon surface in areas where transistors are to be placed (Fig. 3.4a). A thin, highly controlled layer of SiO_2 is then grown on the exposed silicon surface. This is called the *gate oxide* or *thin oxide* or *thinox* (Fig. 3.4b). Polysilicon is then deposited over the wafer surface and etched to form interconnections and transistor gates. Fig. 3.4c shows the result of an etched polysilicon gate. The exposed thinox (not covered by polysilicon) is then etched away. The complete wafer is then exposed to a dopant source, resulting in two actions (Fig. 3.4d). Diffusion

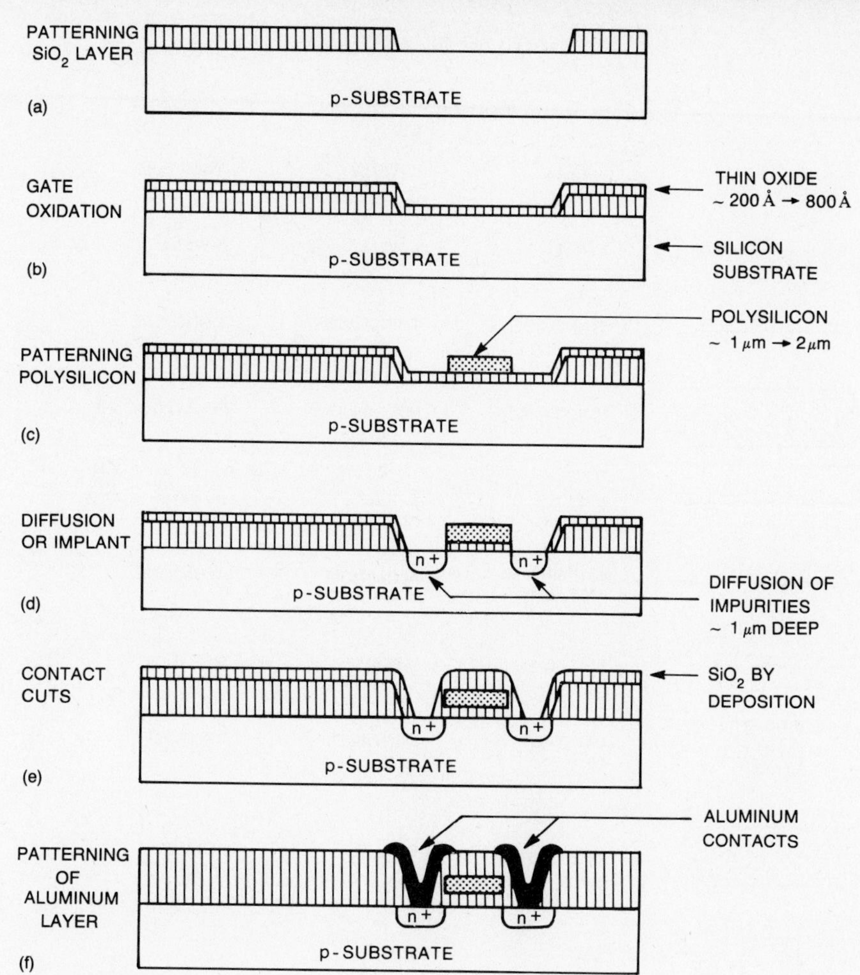

FIGURE 3.4. Fabrication steps for a silicon gate nMOS transistor

junctions are formed in the substrate and the polysilicon is doped with the particular type of dopant. This reduces the resistivity of the polysilicon. Note that the diffusion junctions form the drain and source of the MOS transistor. They are formed only in regions where the polysilicon gate does not shadow the underlying substrate. This is referred to as a *self-aligned* process because the source and drain do not extend under the gate. Finally, the complete structure is covered with SiO_2 and contact holes are etched to make contact with underlying layers (Fig. 3.4e). Aluminum or other metallic interconnect is evaporated and etched to complete the final connection of elements (Fig. 3.4f).

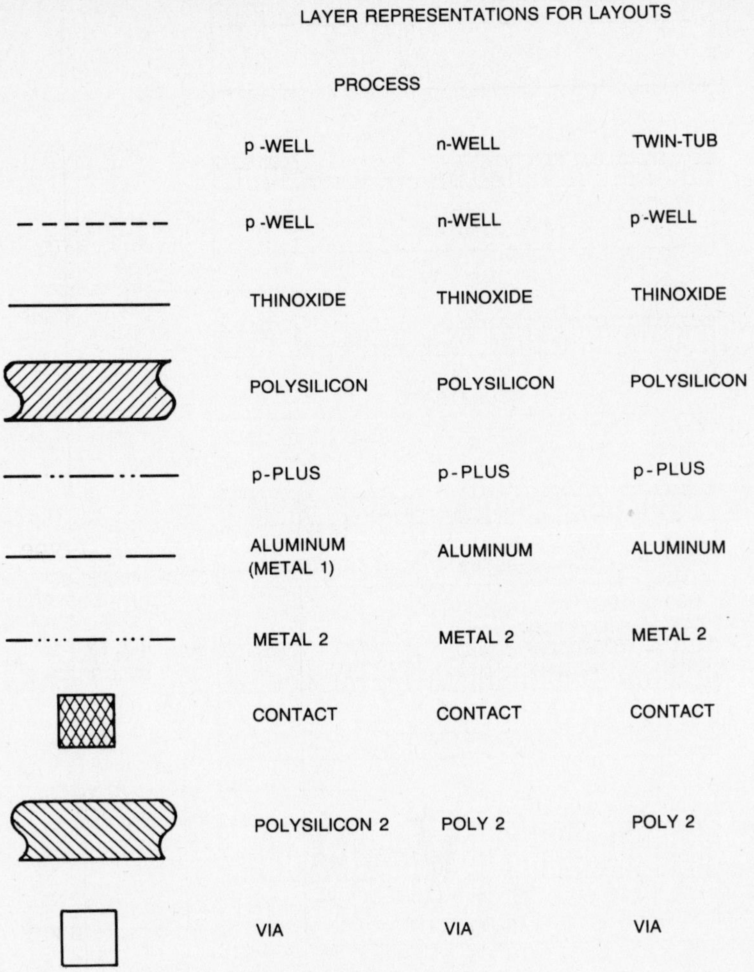

LAYER REPRESENTATIONS FOR LAYOUTS

PROCESS

p-WELL	n-WELL	TWIN-TUB
p-WELL	n-WELL	p-WELL
THINOXIDE	THINOXIDE	THINOXIDE
POLYSILICON	POLYSILICON	POLYSILICON
p-PLUS	p-PLUS	p-PLUS
ALUMINUM (METAL 1)	ALUMINUM	ALUMINUM
METAL 2	METAL 2	METAL 2
CONTACT	CONTACT	CONTACT
POLYSILICON 2	POLY 2	POLY 2
VIA	VIA	VIA

FIGURE 3.5. CMOS process cross-section and layout conventions

3.2 CMOS technologies

CMOS (Complementary Metal Oxide Silicon) technology is recognized as a leading contender for existing and future VLSI systems. CMOS provides an inherently low power static circuit technology that has the capability of providing a lower power-delay product than comparable design-rule nMOS or pMOS technologies. In this section we provide an overview of four dominant CMOS technologies, with a simplified treatment of the process steps. This is included primarily as a guide for better appreciation of the layout styles that are to follow.

The four dominant CMOS technologies are:

- p-well process
- n-well process
- twin-tub process
- silicon on insulator.

During the discussion of CMOS technologies, process cross-sections and layouts will be presented. Fig. 3.5 summarizes the drawing conventions.

3.2.1 The p-well process

A common approach to p-well CMOS fabrication has been to start with a moderately doped n-type substrate (wafer), create the p-type well for the n-channel devices, and build the p-channel transistor in the *native* n-substrate. Although the processing steps are somewhat complex and depend on the fabrication line, Fig. 3.6 on pages 72-73 illustrates the major steps involved in a typical p-well CMOS process. The mask that is used in each process step is shown in addition to a sample cross-section through an n-device and a p-device. Although we have shown a polysilicon gate process, it is of historical significance to note that CMOS was originally implemented with metal (aluminum) gates. This technology formed the basis for the majority of low power CMOS circuits implemented in the 1970s. The technology is robust and still in use in many areas.

As can be seen from Fig. 3.6, the mask levels are not organized by component function. Rather they reflect the processing steps.

- The first mask defines the p-well (or p-tub); n-channel transistors will be fabricated in this well. Field oxide (FOX) is etched away to allow a deep diffusion (Fig. 3.6a).
- The next mask is called the "thin oxide" or "thinox" mask, as it defines where areas of thin oxide are needed to implement transistor gates and allow implantation to form p- or n-type diffusions for transistor source/drain regions. The field oxide areas are etched to the silicon surface and then the thin oxide is grown on these areas (Fig. 3.6b). Other terms for this mask include *active area*, *island*, and *mesa*. In nMOS this would be the *diffusion* mask.
- Polysilicon gate definition is then completed. This involves covering the surface with polysilicon and then etching the required pattern (in this case an inverted "U"). As noted previously, the

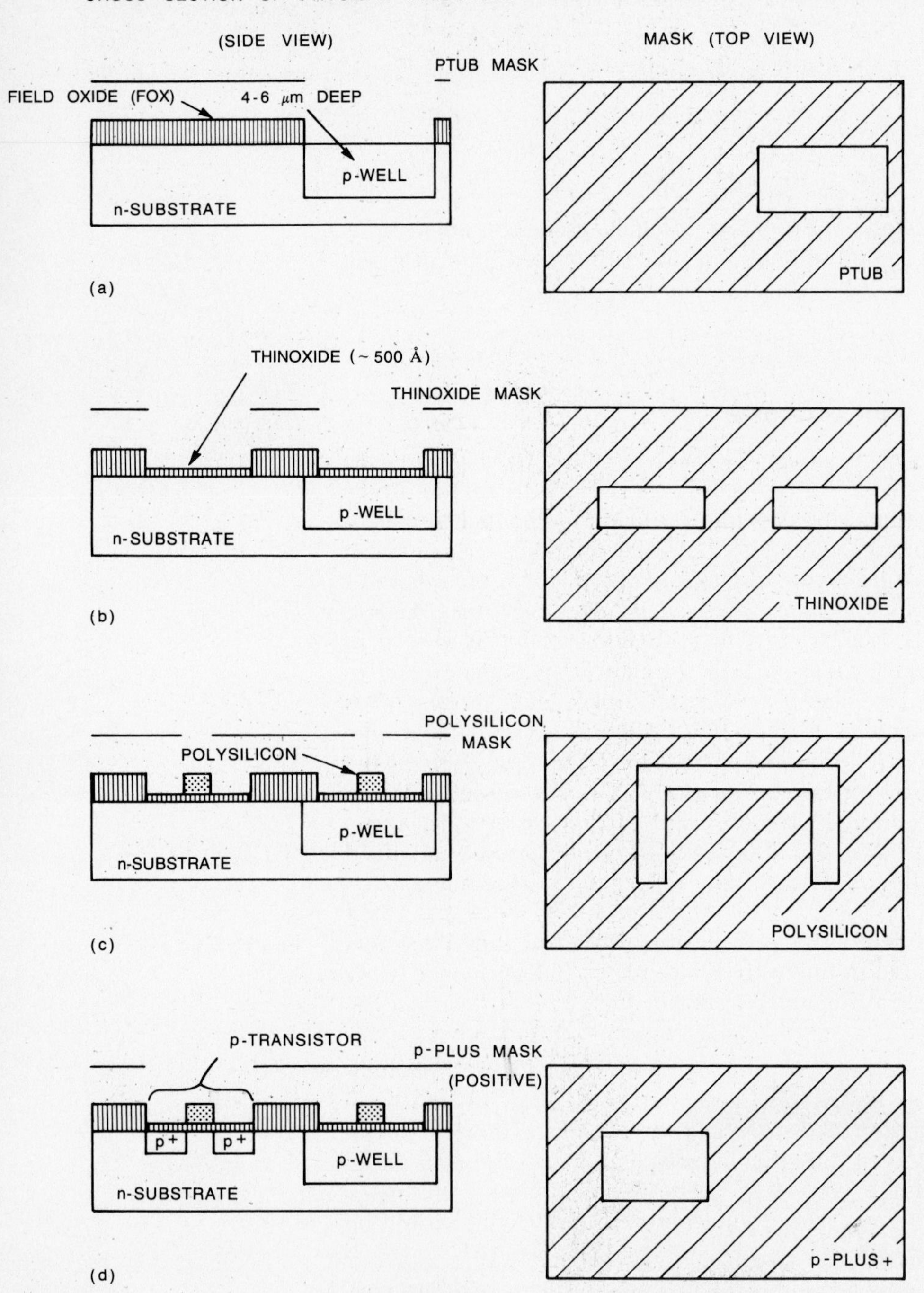

FIGURE 3.6. Typical p-well CMOS process steps with corresponding masks required

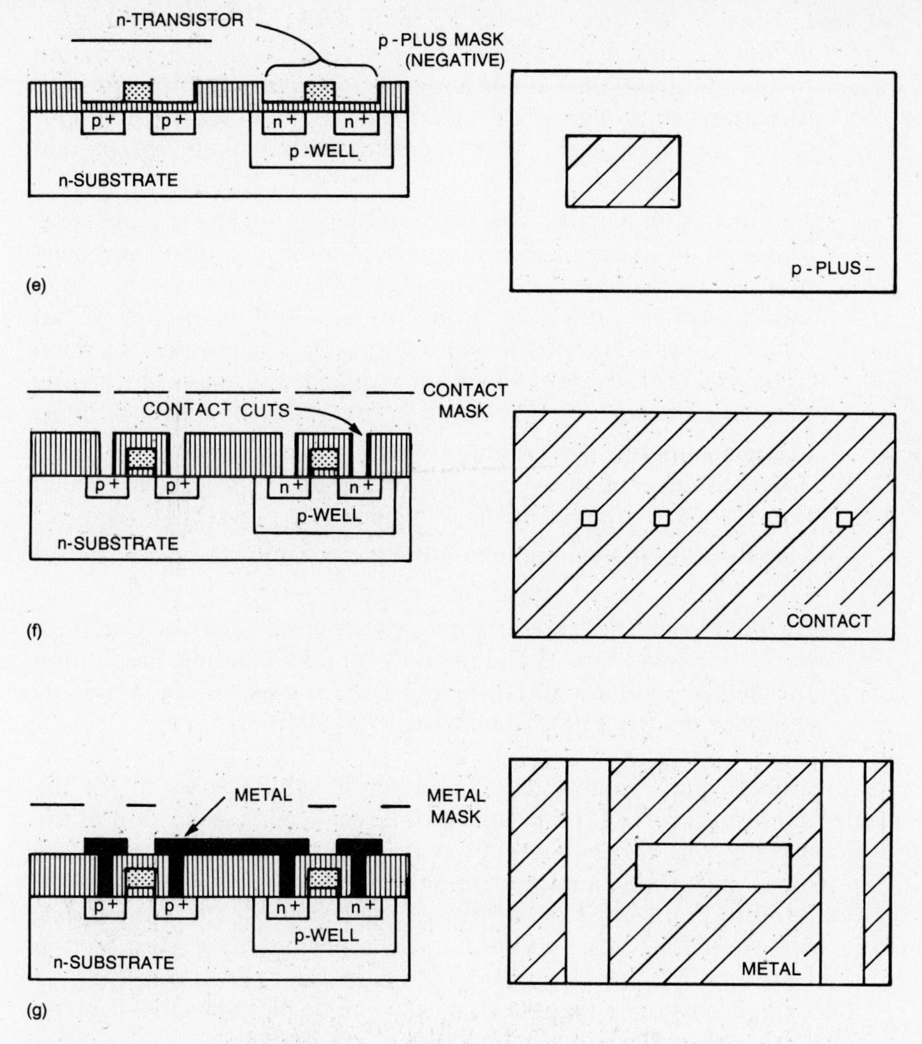

FIGURE 3.6. (Continued)

"poly" gate regions lead to "self-aligned" source-drain regions (Fig. 3.6c).

- A p-plus (p^+) mask is then used to indicate those thin-oxide areas (and polysilicon) that are to be implanted p^+. Hence a thin-oxide area exposed by the p-plus mask will become a p^+ diffusion area (Fig. 3.6d). If the p-plus area is in the n-substrate, then a p-channel transistor or p-type wire may be constructed. If the p-plus area is in the p-well (not shown), then an *ohmic* contact to the p-well may be constructed. An ohmic contact is one which is only resistive in nature and is not rectifying (as in the case of a diode). In other words, there is no junction (n-type and p-type silicon abutting). Current can flow in both directions in an ohmic contact. This type of mask is sometimes called the *select* mask as it *selects* those transistor regions that are to be p-type.

- The next step usually uses the complement of the p-plus mask, although an extra mask is normally not needed. The "absence" of a p-plus region over a thin-oxide area indicates that the area will be an n^+ diffusion or n-thinox. n-thinox in the p-well defines possible n-transistors and wires (Fig. 3.6e). An n^+ diffusion in the n-substrate allows an ohmic contact to be made. Following this step, the surface of the chip is covered with a layer of SiO_2.
- Contact cuts are then defined. This involves etching any SiO_2 down to the contacted surface (Fig. 3.6f). These allow metal (next step) to contact diffusion regions or polysilicon regions.
- Metallization is then applied to the surface and selectively etched (Fig. 3.6g).
- As a final step (not shown), the wafer is passivated and openings to the bond pads are etched to allow for wire bonding. Passivation protects the silicon surface against the ingress of contaminants that can modify circuit behavior in deleterious ways.

Additional steps might include threshold adjust steps to set the threshold voltages of the n- and p-devices. The cross-section of the finished p-well process is shown in Fig. 3.7c. The layout of the p-well CMOS transistors corresponding to this cross-section is illustrated in Fig. 3.7b. The corresponding schematic (for an inverter) is shown in Fig. 3.7a, while a more representative cross-section showing realistic topology is depicted in Fig. 3.7d. From Fig. 3.7 it is evident that the n-type substrate accommodates p-channel devices, while the p-well accommodates n-channel devices.

The p-well diffusion must be carried out with special care since p-well doping concentration and penetration depth affect the threshold voltages as well as the breakdown voltages of the n-channel devices. To achieve low threshold voltages (0.6V–1.0V), either deep well diffusion or high well resistivity is required. Deep junctions require larger spacings between the n-type and p-type transistors due to lateral diffusion, resulting in larger chip areas. High resistivity can accentuate latch-up problems (Section 2.6). In order to achieve narrow threshold voltage tolerances in a typical p-well process, the well concentration is made about one order of magnitude higher than the substrate doping density, thereby causing the body effect for n-channel devices to be higher than for p-channel transistors. In addition, due to this higher concentration, n-transistors suffer from excessive

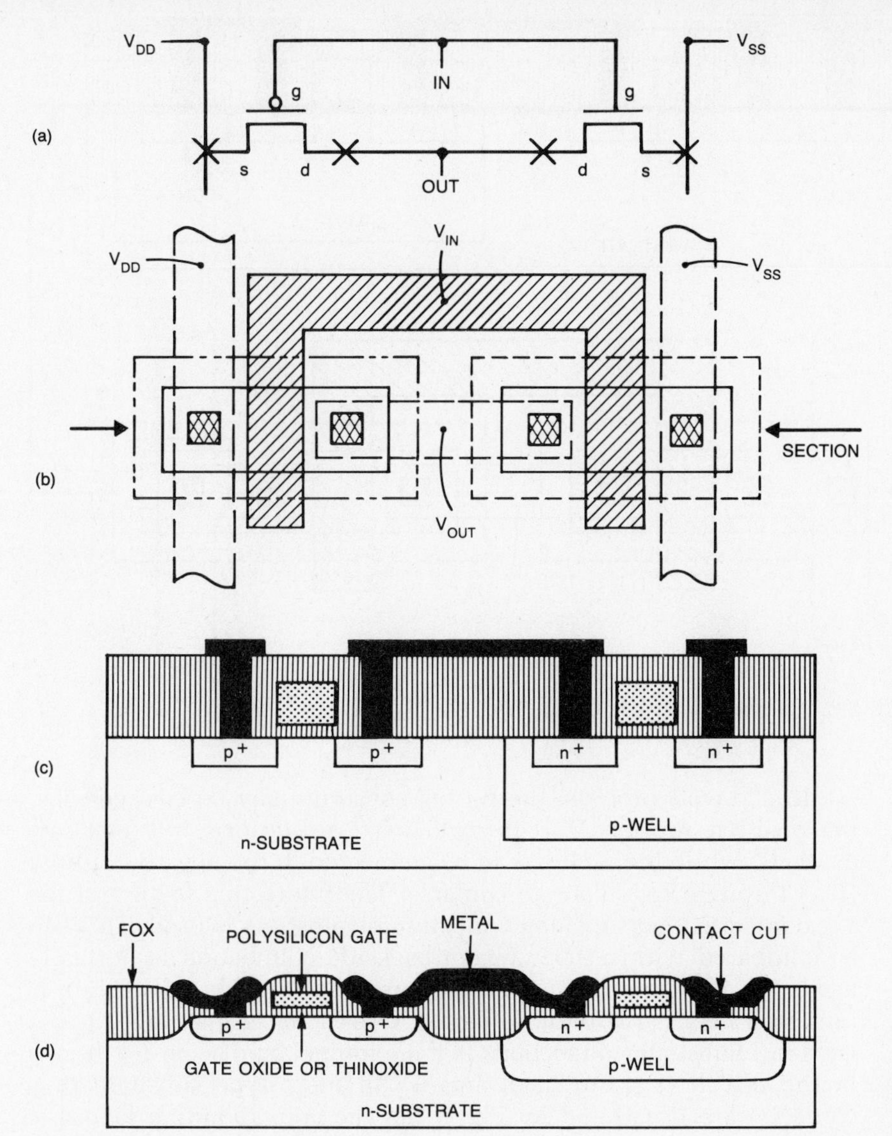

FIGURE 3.7. Layout and process cross-sections of transistors and inverter in p-well CMOS technology

source/drain to p-well capacitance. In general, the n-transistors are inferior to those that could be built on a native substrate (no well). Thus circuits involving n-transistors will tend to be slower than, say, for a typical nMOS depletion load process; degradation in circuit performance may be expected in some logic structures (see Chapter 4). Since the sheet resistance for a p-well is in the order of 1–10kΩ per square, as a measure against "latch-up," the well must be grounded in such a way as to minimize any voltage drop due to injected current in substrate that is collected by the p-well.

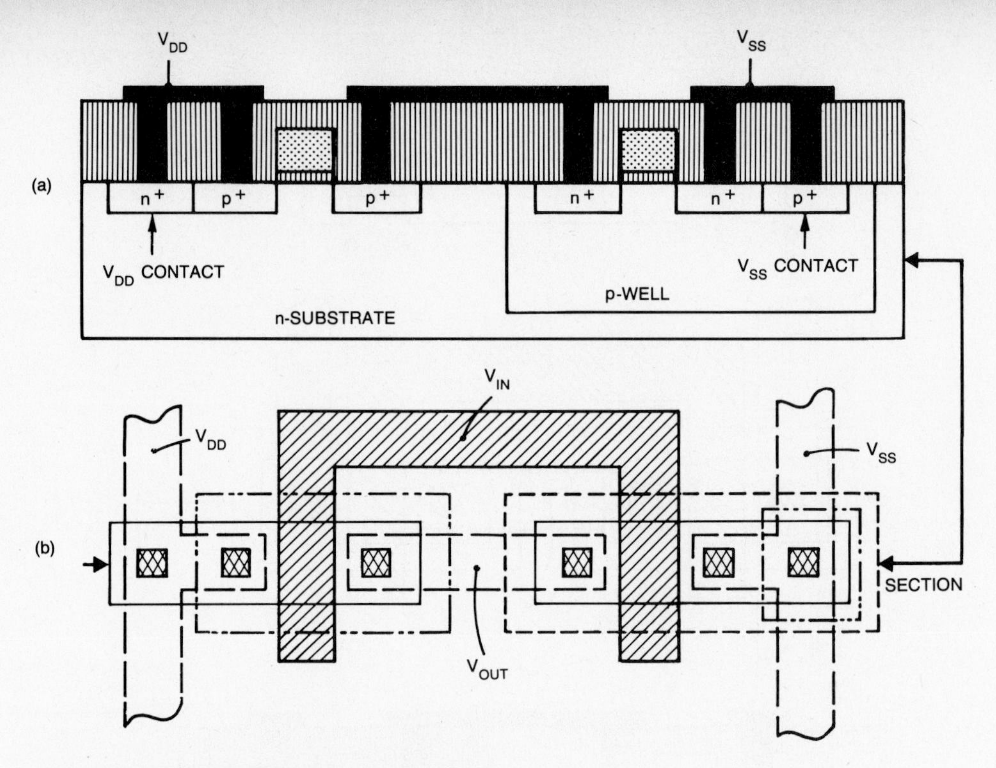

FIGURE 3.8. p-well substrate contacts

In a p-well process, the n-type substrate may be connected to the positive supply (V_{DD}) through what are termed V_{DD} substrate contacts, while the well has to be connected to the negative supply (V_{SS}) through V_{SS} substrate contacts. The interesting feature of the V_{SS} contact is that topside connection of substrate is used. This can be compared with nMOS, where backside connection is normally used. V_{DD} backside contact may be used but topside connection is preferred because it reduces parasitic resistances that could cause latch-up. Substrate connections that are formed by placing p^+ regions in the p-well (V_{SS} contacts) and n^+ in the n-type substrate (V_{DD} contacts) are illustrated by Fig. 3.8a. The corresponding layout is shown in Fig. 3.8b. Other terminology for these contacts include "well contacts" for the V_{SS} substrate connection or "body ties." We will use the term "substrate contact" for both V_{SS} and V_{DD} contacts, as this terminology can be commonly used for most bulk CMOS processes. It should be noted that these contacts are formed during the implants used for the p-channel and n-channel transistor formation.

In current fabrication processes the polysilicon is normally doped n^+. The p^+ doping phase reduces the poly doping such that the polysilicon inside the p-plus regions have a higher sheet resistance than the polysilicon outside the p-plus region. The extent of this

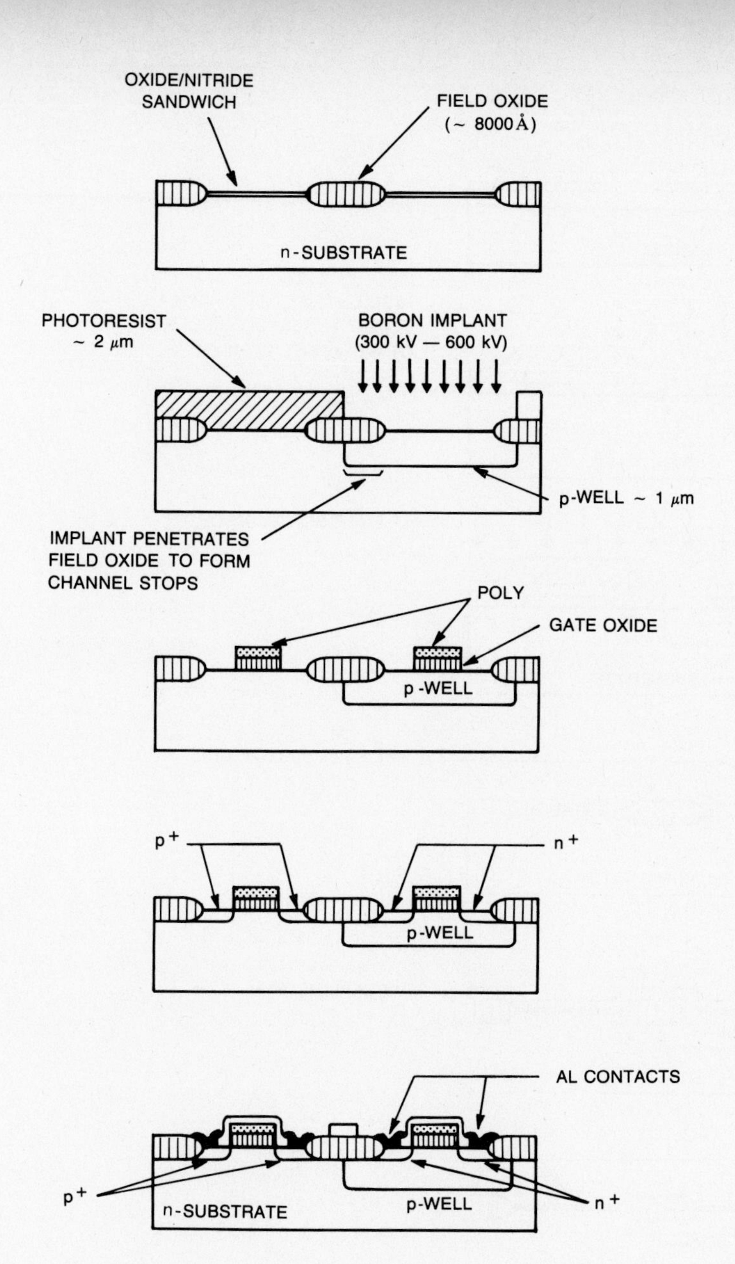

FIGURE 3.9. GE-Intersil's "retrograde p-well" process

reduction may influence the quality of metal-poly contacts within p-plus regions.

To meet the growing need for higher packing density, improvements in latch-up, and independent threshold adjustment, a number of improved p-well CMOS processes have emerged during recent years. We will examine two such processes in more detail: the "retrograde p-well CMOS" process developed by GE-Intersil, Inc. [Comb81], and the "CMOSC" process developed by Hewlett-Packard [HJLV83]. These are illustrated by Fig. 3.9 and Fig. 3.10, respectively.

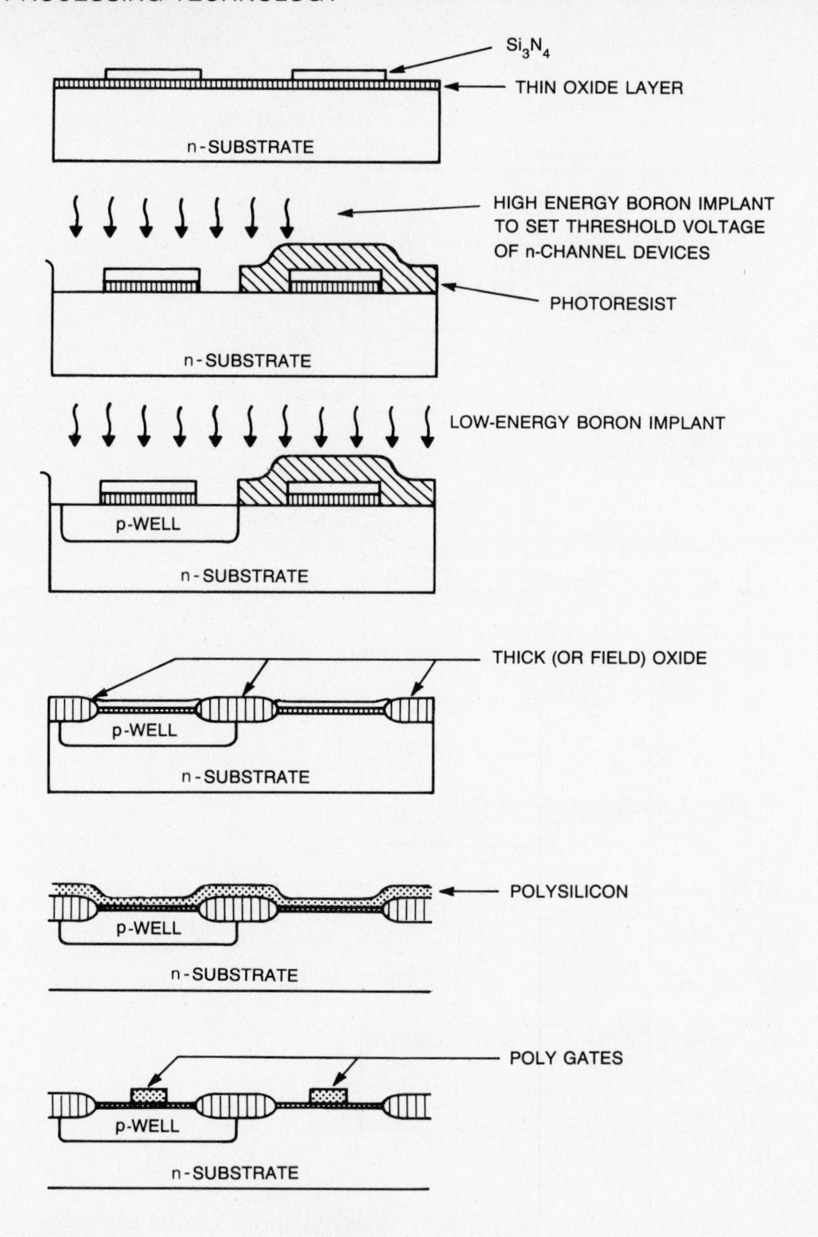

FIGURE 3.10. Hewlett-Packard's CMOSC process

In the retrograde p-well process the well is implanted with a high energy boron implant as opposed to a thermal diffusion process. As a result of this step and the fact that the implant is made after field oxide, the p-well impurities do not diffuse from their original implanted position, thus reducing the lateral diffusion of the well. This enables reductions in the spacing between p- and n-transistors. Further advantage of the retrograde process is that junction depth,

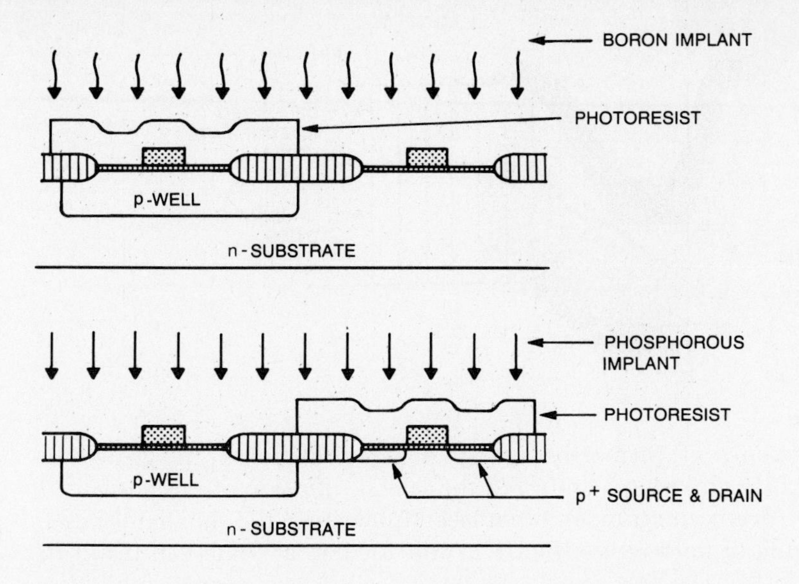

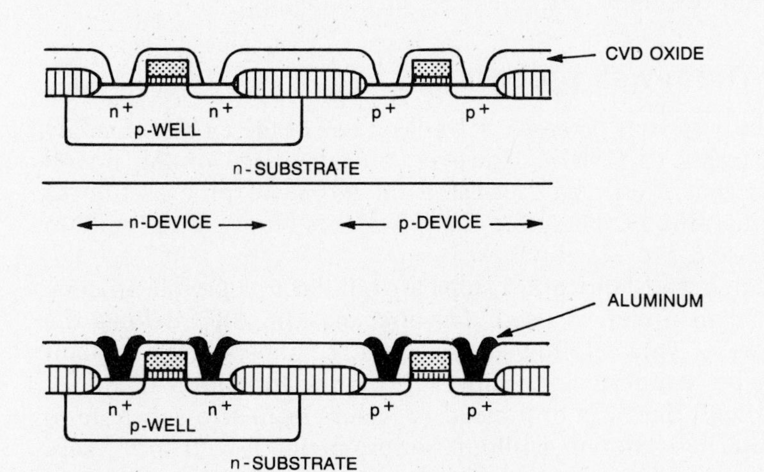

FIGURE 3.10. (Continued)

sheet resistance, and threshold voltage are independent, allowing separate adjustments to take place for optimizing the behavior of the CMOS devices.

A number of the process steps for the CMOSC process are shown in Fig. 3.10. A boron implant is used to define the p-transistors and a phosphorus implant is used to define the n-transistors. Improvements in CMOSC processes have resulted in an extremely low standby

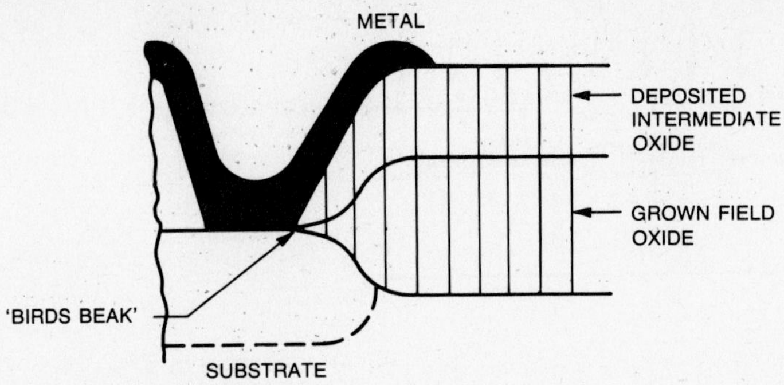

FIGURE 3.11. "Bird's beak" effect

leakage current, primarily through characterization of the growth of field oxide, improved control of the lateral diffusion of the implanted source/drain junction, and increased integrity of the gate oxide edge. Thinning of field oxide (Fig. 3.11) during contact etch can result in a nondestructive breakdown that increases leakage currents. Inhibiting this thinning effect, referred to as "bird's beak," also provides significant improvement in the leakage mechanism.

3.2.2 The n-well process

Until recently, p-well processes have been one of the most commonly available forms of CMOS. However, an advantage of the n-well process is that it can be fabricated on the same process line as conventional nMOS. Therefore this process is often "retrofitted" to existing nMOS processes [Ohz80].

Typical n-well fabrication steps are similar to a p-well process, except that an n-well is used. The first masking step defines the n-well regions. This is followed by a low-dose phosphorous implant driven in by a high-temperature step for the formation of the n-well. The well depth is optimized to ensure against p-substrate to p^+ diffusion breakdown, without compromising n-well to n^+ separation. The next steps are to define the devices and other diffusions, to grow field oxide, contact cuts, and metallization. An n-well mask is used to define n-well regions, as opposed to a p-well mask in a p-well process. An n-plus (n^+) mask may be used to define the n-channel transistors and V_{DD} contacts. Alternatively, we could use a p-plus mask to define the p-channel transistors, as the masks usually are the complement of each other.

Although there are a number of n-well CMOS processes becoming available, the n-well process developed at the University of California at Berkeley is chosen as a good illustration of the details of the fabrication steps. To illustrate this process, the process steps have been reproduced [GrLN83]. These are couched in a Process Input Description Language. The commands in this language are as follows:

- `SUBSTRATE <NAME> (*TYPE=[P,N] IMPURITY=[])`
  Specifies the substrate name, type, and impurity level.
- `OXIDE <NAME> THICKNESS = []`
  Specifies oxide layer and thickness.
- `DEPOSITION <NAME> (*) THICKNESS=[]`
  Specifies a layer and thickness of a deposited layer. `(*)` is followed by `TYPE=[] IMPURITY=[]` if it is silicon.
- `ETCH <NAME> DEPTH=[]`
  Specifies a material and an etch depth.
- `DOPE TYPE=[P,N] PEAK=[] DEPTH=[] DELTA=[] BLOCK=[]`
  Specifies parameters necessary to define a diffusion step.
- `MASK <RESIST NAME> <EXPOSED NAME> <MASK NAME> <POLARITY OF MASK>`
  Specifies a resist layer and associated information.

The complete process input file is as follows (with abbreviations):


```
1. LEVEL 1
2. SUBS SILICON TYPE=P IMPU=1e13
```

Initial oxidation

```
3. OXIDE OX1 THICK=0.1
```

n-well definition

```
 4. DEPO NTRD THICK=0.5
 5. DEPO RST THICK=0.5
 6. MASK RST DRST MNWL POSI
 7. ETCH DRST DEPTH=0.6
 8. ETCH NTRD DEPTH=0.6
 9. ETCH RST DEPTH=0.6
10. OXIDE OX2 THICK=0.5
11. ETCH NTRD DEPTH=0.6
12. DOPE TYPE=N PEAK=1.5e15 DEPTH=0.0 DELTA=1.5
    BLOCK=0.2
13. ETCH OX DEPTH=0.7
14. OXIDE OX3 THICK=0.1
```

All active area definition

```
15. DEPO NTRD THICK=0.5
16. DEPO RST THICK=0.5
17. MASK RST DRST MAA POSI
18. ETCH DRST DEPTH=0.6
19. ETCH NTRD DEPTH=0.6
20. ETCH RST DEPTH=0.6
```

Field dope for n-channel

```
21. DEPO RST THICK=1.0
22. MASK RST DRST MNWL POSI
23. ETCH DRST DEPTH=1.1
24. DOPE TYPE=P PEAK=1e21 DEPTH=0.05 DELTA=0.15
    BLOCK=0.2
25. ETCH RST DEPTH=1.1
26. OXIDE OX4 THICK=0.7
27. ETCH NTRD DEPTH=0.6
```

Threshold adjust dope

```
28. DOPE TYPE=P PEAK=1e20 DEPTH=0.0 DELTA=0.05
    BLOCK=0.2
```

Regrow gate oxide

```
29. ETCH OX DEPTH=0.1
30. OXIDE OX5 THICK=0.1
```

Poly gate definition

```
31. DEPO POLY THICK=0.30
32. DEPO RST THICK=0.5
33. MASK RST DRST MSI POSI
35. ETCH DRST DEPTH=0.6
35. ETCH POLY DEPTH=0.6
36. ETCH RST DEPTH=0.6
```

Arsenic dope for n-channel source and drain

```
37. DEPO RST THICK=1.0
38. MASK RST DRST MIIN POSI
39. ETCH DRST DEPTH=1.1
40. DOPE TYPE=N PEAK=1e22 DEPTH=0.0 DELTA=0.2
    BLOCK=0.2
41. ETCH RST DEPTH=1.1
```

Boron dope for p-channel source and drain

```
42. DEPO RST THICK=1.0
43. MASK RST DRST MIIN NEGA
44. ETCH DRST DEPTH=1.1
45. DOPE TYPE=P PEAK=1e22 DEPTH=0.0 DELTA=0.2
    BLOCK=0.2
46. ETCH RST DEPTH=1.1
```

LPCVD oxide (**L**iquid **P**hase **C**hemical **V**apor **D**eposition Oxide)

```
47. DEPO OX6 THICK=0.5
```

Contact definition

```
48. DEPO RST THICK=1.0
49. MASK RST DRST MCC NEGA
50. ETCH DRST DEPTH=1.1
51. ETCH OX DEPTH=1.1
52. ETCH RST DEPTH=1.1
```

Metallization

```
53. DEPO METL THICK=1.0
54. DEPO RST THICK=1.0
55. MASK RST DRST MME POSI
56. ETCH DRST DEPTH=1.1
57. ETCH METL DEPTH=1.1
58. ETCH RST DEPTH=1.1
```

Some of the abbreviations are as follows:

```
NTRD - Nitride
RST - Resist
METL - Metal (Aluminum)
NEGA - Negative
POSI - Positive
MNWL - N-well mask
MAA - Thin-oxide mask
MSI - Polysilicon mask
MIIN - Nplus mask
MCC - Contact mask
MME - Metal mask
```

Using the abbreviations and language definitions, the sequence in processing may be traced out. For instance, steps 31–36 deposit and etch the polysilicon layer. Step 31 deposits .3μ of polysilicon.

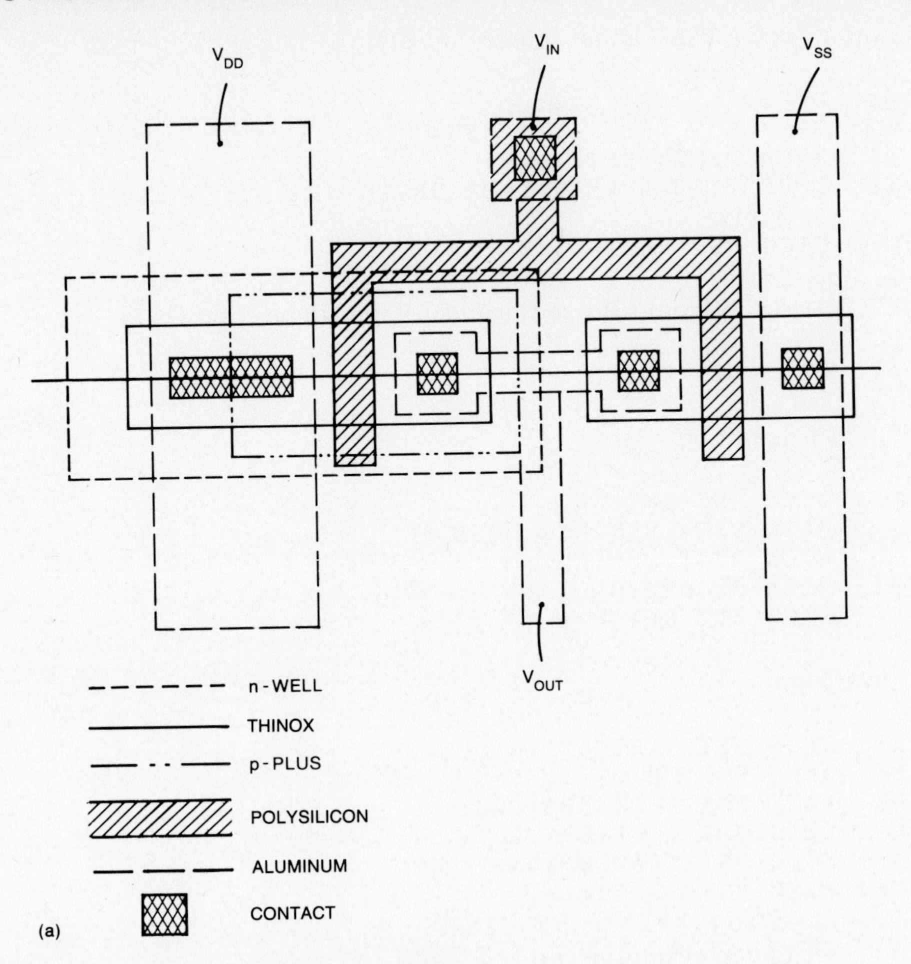

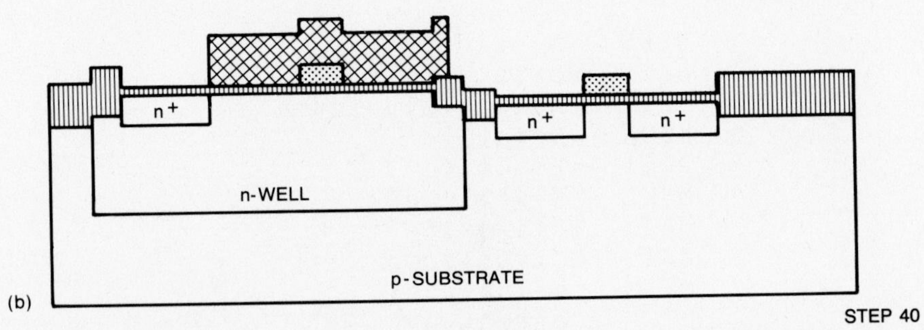

FIGURE 3.12. Berkeley n-well process snap-shots and layout for n-well inverter © IEEE 1983 ([GrLN83])

Step 32 deposits $.5\mu$ of resist called RST. Step 33 masks this resist with a positive polysilicon mask, calling the exposed resist DRST. Step 34 etches DRST to a depth of $.6\mu$. The exposed polysilicon is then etched to a depth of $.6\mu$ by step 35. Finally resist RST is etched away, leaving the final polysilicon pattern. Fig. 3.12 shows

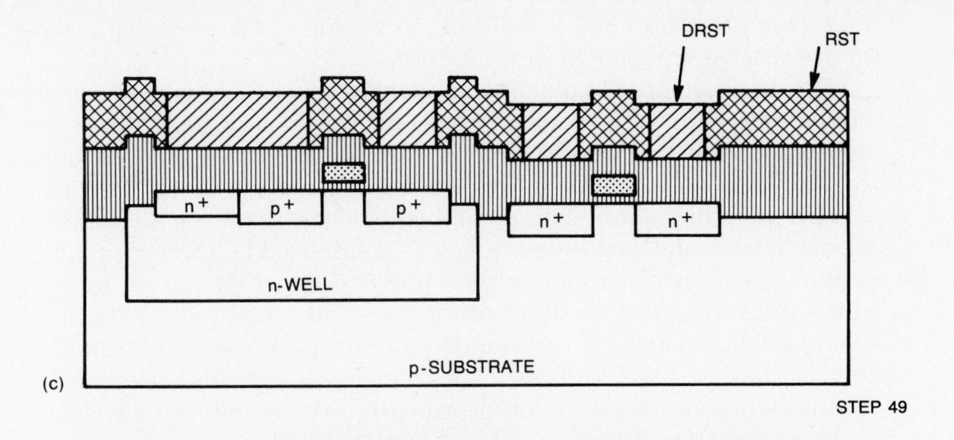

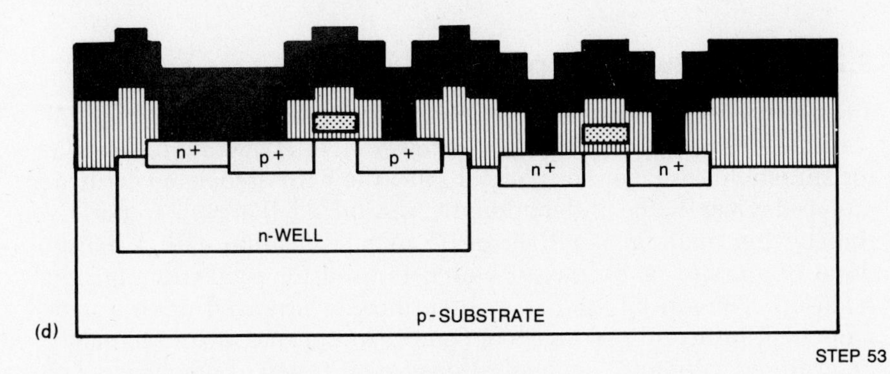

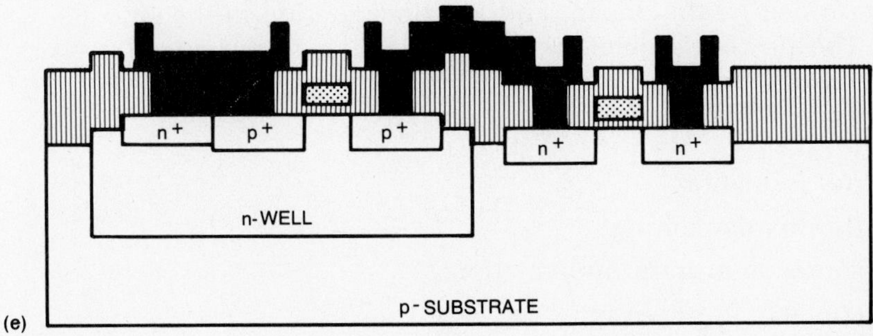

FIGURE 3.12. (Continued)

snap-shots of the cross-section of part of the layout for a number of steps during processing. The layout corresponding to the cross-section is shown in Fig. 3.12a. The cross-sections may be generated automatically from this process file using the SIMPL-1 program [GrLN83]. Fig. 3.12b shows the n^+ implant step. Fig. 3.12c shows the step required to define contact windows. Fig. 3.12d illustrates the wafer prior to metal definition, while Fig. 3.12e demonstrates the completed inverter in cross-section.

Due to differences in mobility of charge carriers the n-well process creates nonoptimum p-channel characteristics, such as high junction capacitance and high body effect (in the same manner that the p-well influences n-transistors). However, many emerging CMOS designs contain more n-channel than p-channel devices, so the overall effect of poor p-transistor performance may be minimized by careful circuit design. The n-well technology provides a distinct advantage here, where optimum device characteristics are only required for the n-channel transistors and not for the p-transistors. Thus n-channel devices may be used to form logic elements to provide speed and density, while p-transistors could primarily serve as pull-up devices. Fully n-type I/O circuits may be also used to advantage.

3.2.3 The twin-tub process

Twin-tub CMOS technology provides the basis for separate optimization of the p-type and n-type transistors, thus making it possible for threshold voltage, body effect, and the gain associated with n- and p-devices to be independently optimized [Parr80]. Generally the starting material is either an n^+ or p^+ substrate with a lightly doped *epitaxial* or *epi* layer, which is used for protection against latch-up. The aim of *epitaxy* (which means "arranged upon") is to grow high purity silicon layers of controlled thickness with accurately determined dopant concentrations distributed homogeneously throughout the layer. The electrical properties for this layer are determined by the dopant and its concentration in the silicon.

The process sequence, which is similar to the p-well process apart from the tub formation where both p-well and n-well are utilized, entails the following steps:

- tub formation
- thin oxide etching
- source and drain implantations
- contact cut definition
- metallization.

Fig. 3.13 illustrates the steps involved in the AT&T-Bell Laboratories twin-tub process. Since this process provides separately optimized wells, better performance n-transistors (lower capacitance, less body effect) may be constructed when compared with a conventional p-well process. Similarly, the p-transistors may be optimized. Note that the use of threshold adjust steps is included in this process. These masks are derived from the thinox and n-plus masks.

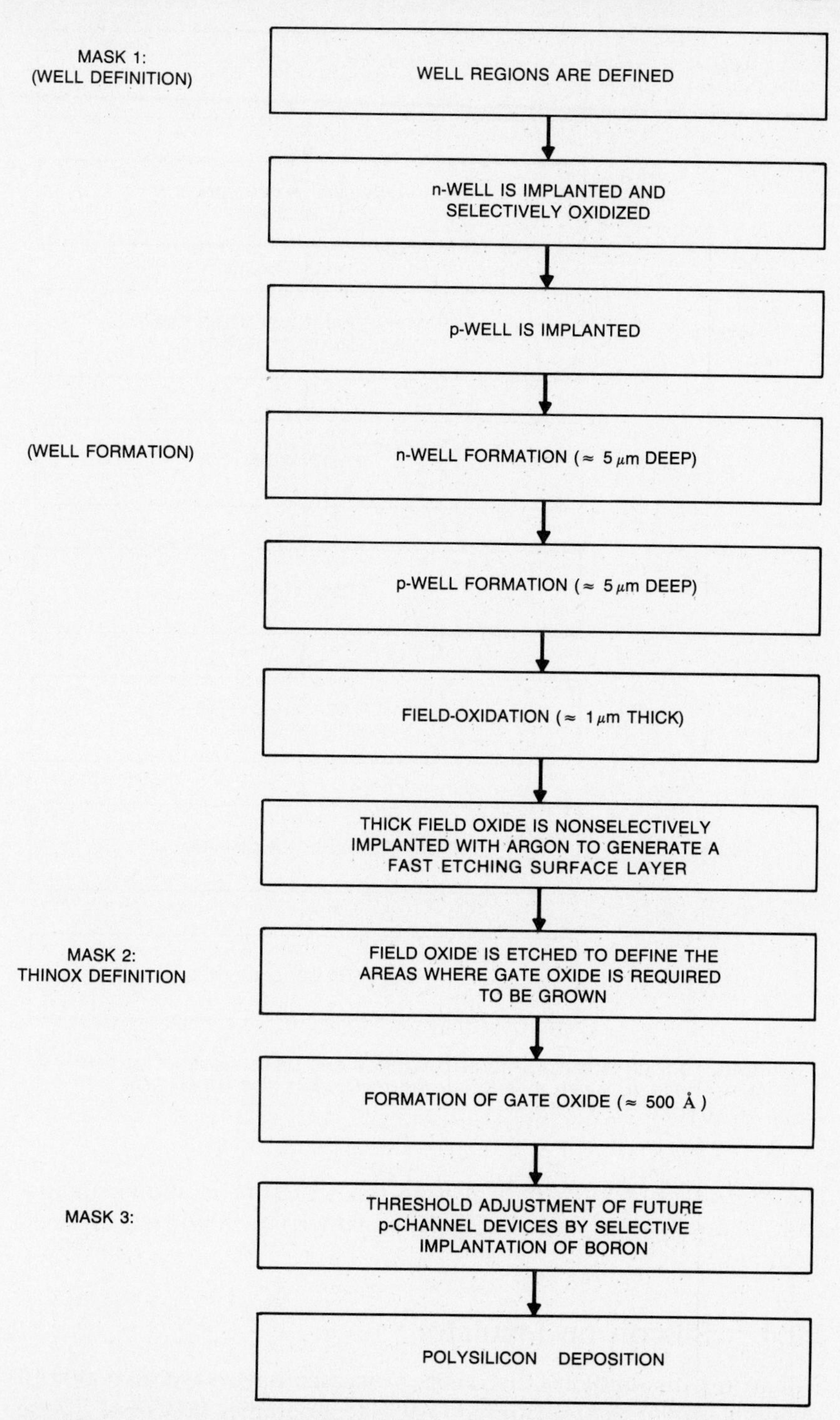
MASK 1:
(WELL DEFINITION)
WELL REGIONS ARE DEFINED
n-WELL IS IMPLANTED AND SELECTIVELY OXIDIZED
p-WELL IS IMPLANTED
(WELL FORMATION)
n-WELL FORMATION (≈ 5 μm DEEP)
p-WELL FORMATION (≈ 5 μm DEEP)
FIELD-OXIDATION (≈ 1 μm THICK)
THICK FIELD OXIDE IS NONSELECTIVELY IMPLANTED WITH ARGON TO GENERATE A FAST ETCHING SURFACE LAYER
MASK 2:
THINOX DEFINITION
FIELD OXIDE IS ETCHED TO DEFINE THE AREAS WHERE GATE OXIDE IS REQUIRED TO BE GROWN
FORMATION OF GATE OXIDE (≈ 500 Å)
MASK 3:
THRESHOLD ADJUSTMENT OF FUTURE p-CHANNEL DEVICES BY SELECTIVE IMPLANTATION OF BORON
POLYSILICON DEPOSITION

FIGURE 3.13. AT&T Bell Laboratories' twin-tub CMOS process steps

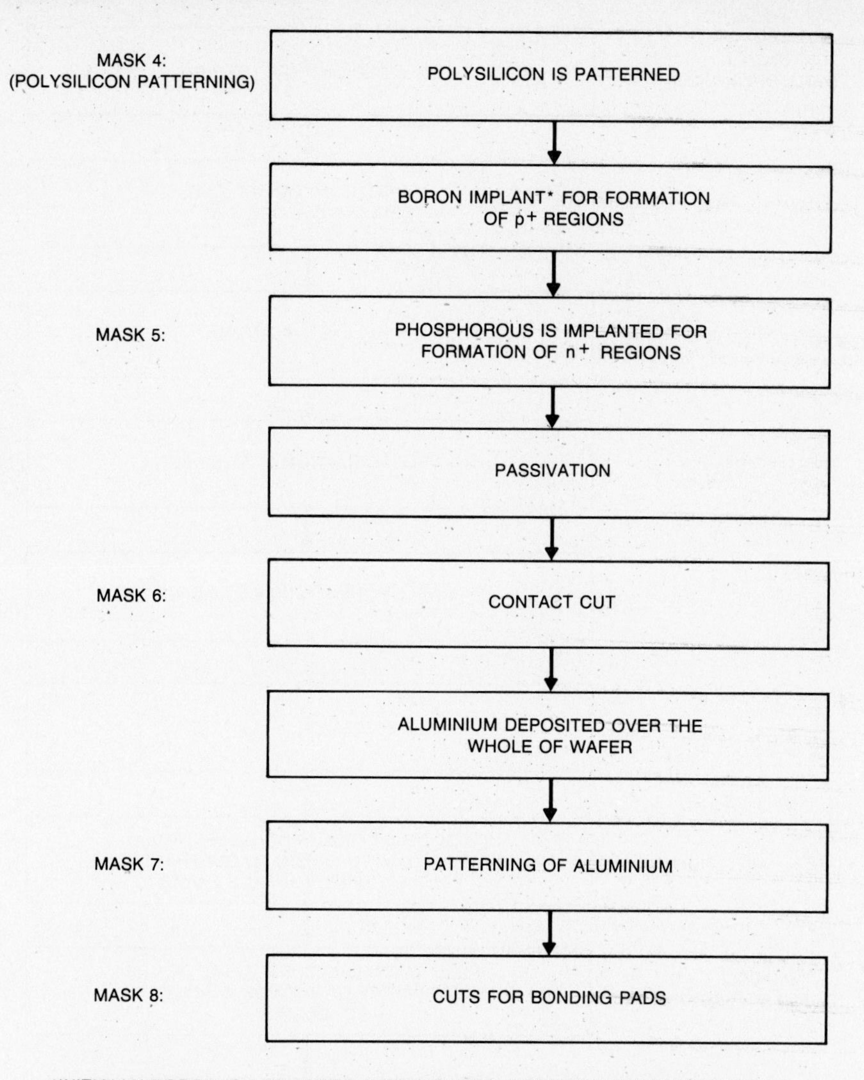
MASK 4:
(POLYSILICON PATTERNING)
POLYSILICON IS PATTERNED
BORON IMPLANT* FOR FORMATION OF p+ REGIONS
MASK 5:
PHOSPHOROUS IS IMPLANTED FOR FORMATION OF n+ REGIONS
PASSIVATION
MASK 6:
CONTACT CUT
ALUMINIUM DEPOSITED OVER THE WHOLE OF WAFER
MASK 7:
PATTERNING OF ALUMINIUM
MASK 8:
CUTS FOR BONDING PADS

*INITIALLY BORON IS IMPLANTED INTO SOURCE AND DRAIN REGIONS OF BOTH TRANSISTORS FOLLOWED BY PHOSPHOROUS IMPLANT FOR FORMATION OF n+ REGIONS

FIGURE 3.13. (Continued)

The cross-section of a typical twin-tub structure is shown in Fig. 3.14. The substrate contacts (both of which are required) are also included.

3.2.4 Silicon on insulator

Silicon on insulator (SOI) CMOS processes have several potential advantages over the traditional CMOS technologies [MaSi64]. These include higher density, no latch-up problems, and lower parasitic

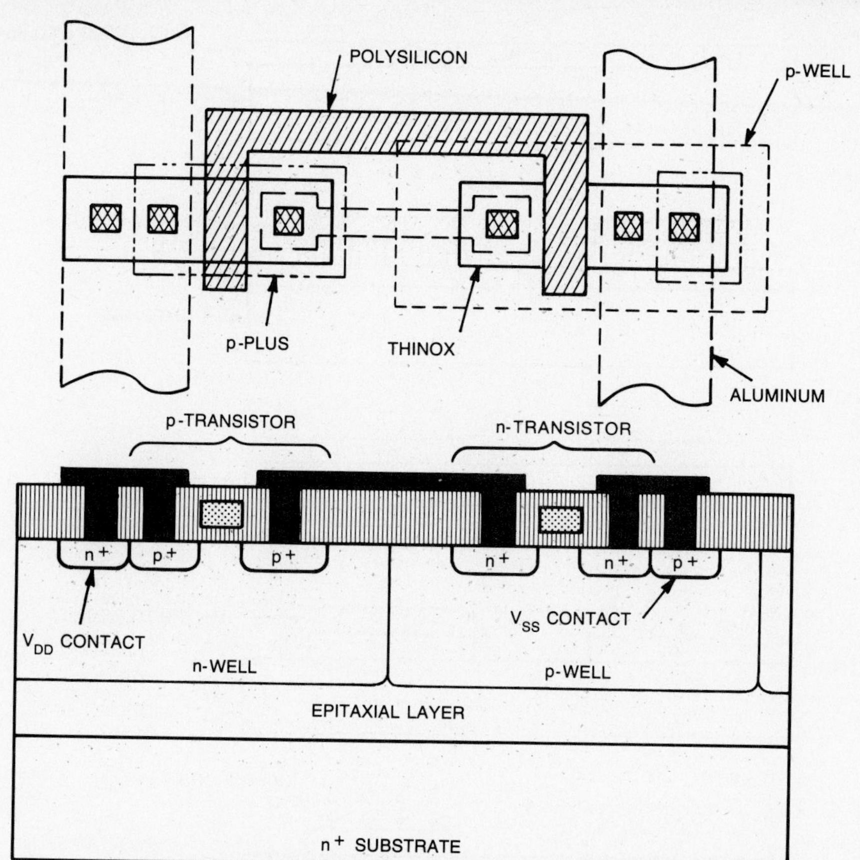

FIGURE 3.14. Twin-tub process cross-section and layout of an inverter

capacitances. In the SOI process a thin layer of single crystal silicon film is epitaxially grown on an insulator such as sapphire or magnesium aluminate spinel. Various masking and doping techniques (Fig. 3.15 on page 90) are then used to form p-channel and n-channel devices. Unlike the more conventional CMOS approaches, the extra steps in well formation do not exist in this technology.

The steps used in typical SOI CMOS processes are:

- A thin film (7–8 μm) of very lightly-doped n-type Si is grown over an insulator. Sapphire is a commonly used insulator (Fig. 3.15a).
- An anisotropic etch is used to etch away the Si except where a diffusion area (n or p) will be needed. The etch must be anisotropic since the thickness of the Si is much greater than the spacings desired between the Si "islands" (Fig. 3.15b, 3.15c).
- The p-islands are formed next by masking the n-islands with a photoresist. A p-type dopant, boron, for example, is then im-

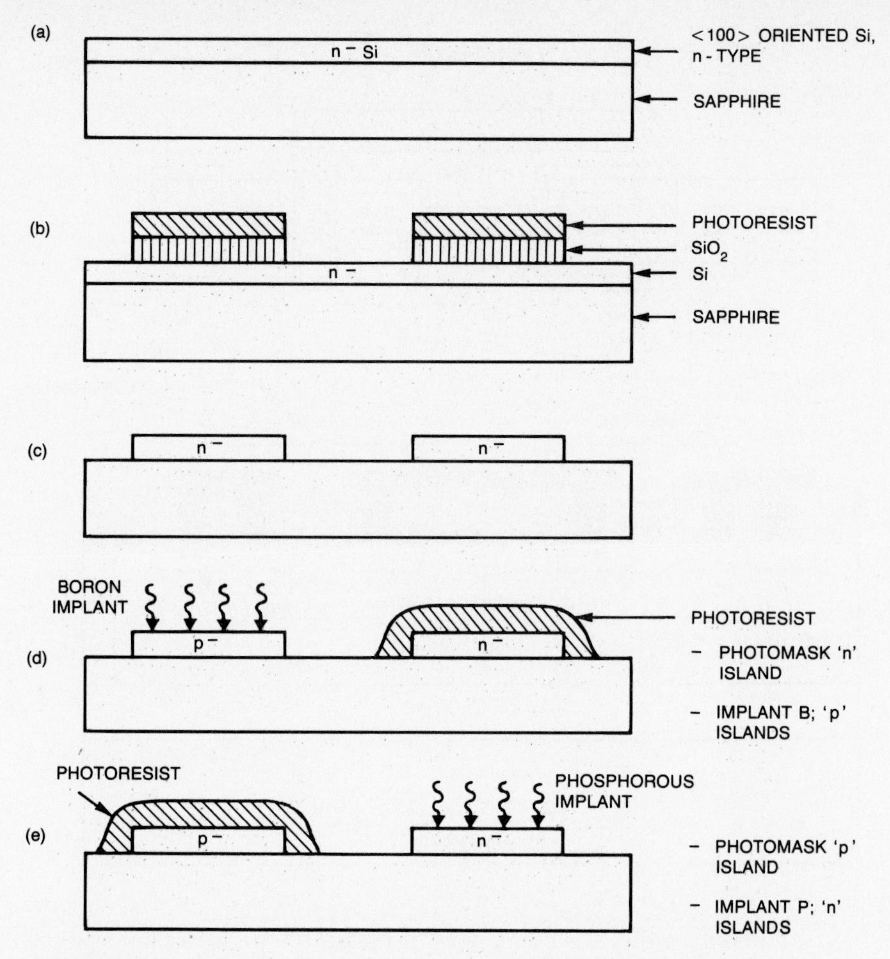

FIGURE 3.15. Typical silicon on insulator (SOI) process flow

planted. It is masked by the photoresist, but forms p-islands at the unmasked islands. The p-islands will become the n-channel devices (Fig. 3.15d).

- The p-islands are then covered with a photoresist and an n-type dopant, phosphorus, for example, is implanted to form the n-islands. The n-islands will become the p-channel devices (Fig. 3.15e).
- A thin gate oxide (around 500–600 Å) is grown over all of the Si structures. This is normally done by thermal oxidation.
- A polysilicon film is deposited over the oxide. Often the polysilicon is doped with phosphorus to reduce its resistivity (Fig. 3.15f).

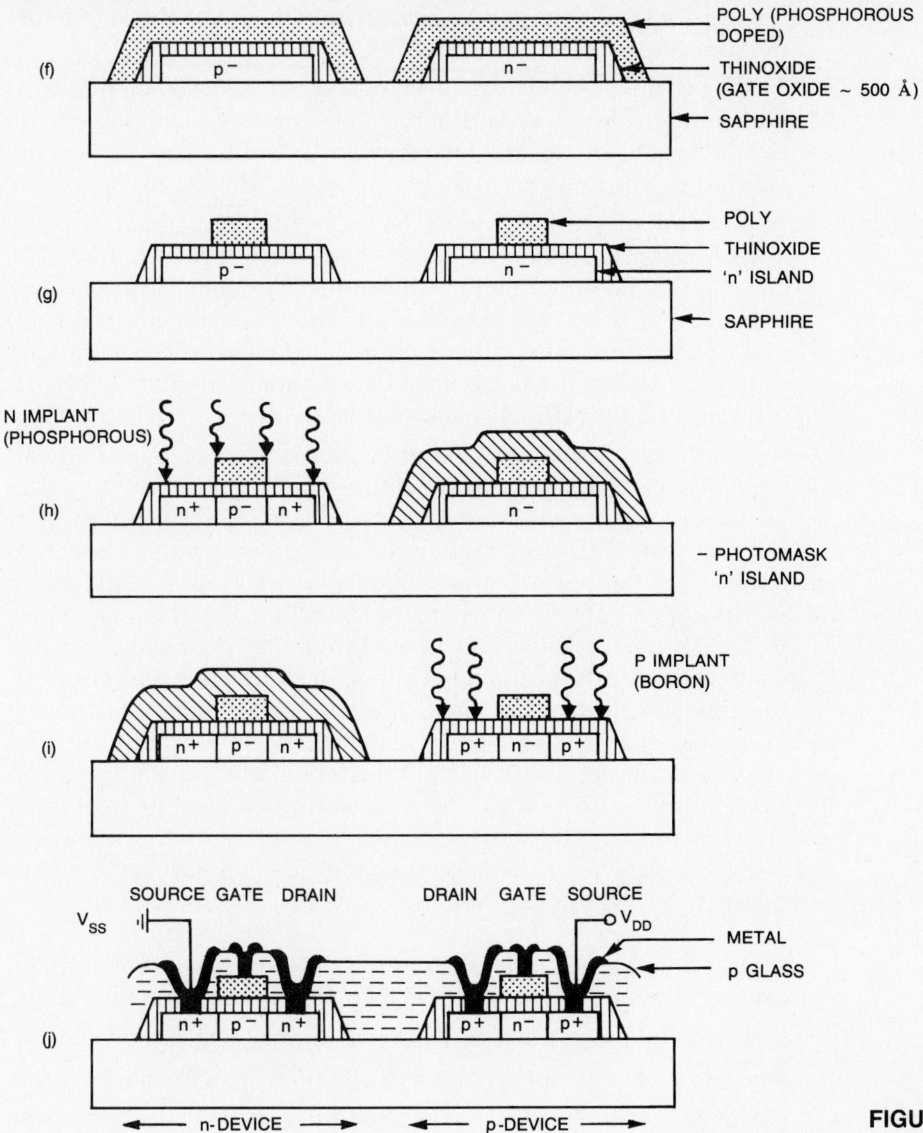

FIGURE 3.15. (Continued)

- The polysilicon is then patterned by photomasking and is etched. This defines the polysilicon layer in the structure (Fig. 3.15g).
- The next step is to form the n-doped source and drain of the n-channel devices in the p-islands. The n-islands are covered with a photoresist and an n-type dopant, normally phosphorus, is implanted. The dopant will be blocked at the n-islands by the photoresist and it will be blocked from the gate region of

the p-islands by the polysilicon. After this step the n-channel devices are complete (Fig. 3.15h).

- The p-channel devices are formed next by masking the p-islands and implanting a p-type dopant such as boron. The polysilicon over the gate of the n-islands will block the dopant from the gate, thus forming the p-channel devices (Fig. 3.15i).
- A layer of phosphorus glass or some other insulator such as silicon dioxide is then deposited over the entire structure. The glass is etched at contact cut locations. The metallization layer is formed next by evaporating aluminum over the entire surface and etching it to leave only the desired metal wires. The aluminum will flow through the contact cuts to make contact with the diffusion or polysilicon regions (Fig. 3.15j).
- A final passivation layer of a phosphorus glass is deposited and etched over bonding pad locations.

Because the diffusion regions extend down to the insulating substrate, only "sidewall" areas associated with source and drain diffusions contribute to the parasitic junction capacitance. Since sapphire is an extremely good insulator, leakage currents between transistors and substrate and adjacent devices are almost eliminated.

In order to improve the yield, some processes use "preferential etch" in which the island edges are tapered. Thus aluminum or poly runners can enter and leave the islands with a minimum step height. This is contrasted to "fully anisotropic etch" in which the undercut is brought to zero as shown in Fig. 3.16. An "isotropic etch" is also shown in the same diagram for comparison. The advantages of SOI technology are:

- Due to the absence of wells, denser structures than bulk silicon can be obtained. Also direct n to p connections may be made.
- Low capacitances provide the basis of very fast circuits.
- No field-inversion problems exist (insulating substrate).
- No latch-up due to isolation of n- and p-transistors by insulating substrate.
- As there is no conducting substrate, there are no body effect problems.
- Enhanced radiation tolerance.

However, on the negative side, due to absence of substrate diodes, the inputs are somewhat more difficult to protect. As device gains are lower, I/O structures have to be larger. Single crystal sapphire or spinel substrates are considerably more expensive than silicon and processing techniques tend to be less developed than bulk

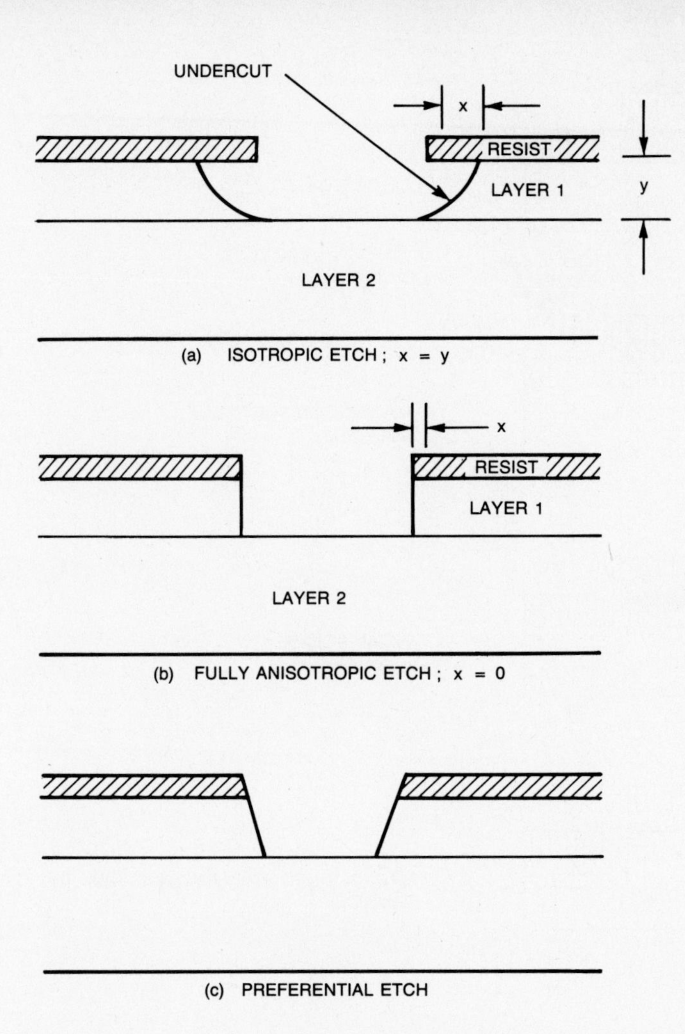

FIGURE 3.16. Classification of etching process

silicon techniques. Thus although SOI has the potential to be the fastest MOS technology, it is also the most expensive.

3.2.5 CMOS process enhancements

A number of enhancements may be added to the CMOS processes, primarily to increase routability of circuits, provide high quality capacitors for analog circuits and memories, or provide resistors of variable characteristics.

These enhancements include:

- double or triple level metal
- double or triple level poly
- combinations of the above.

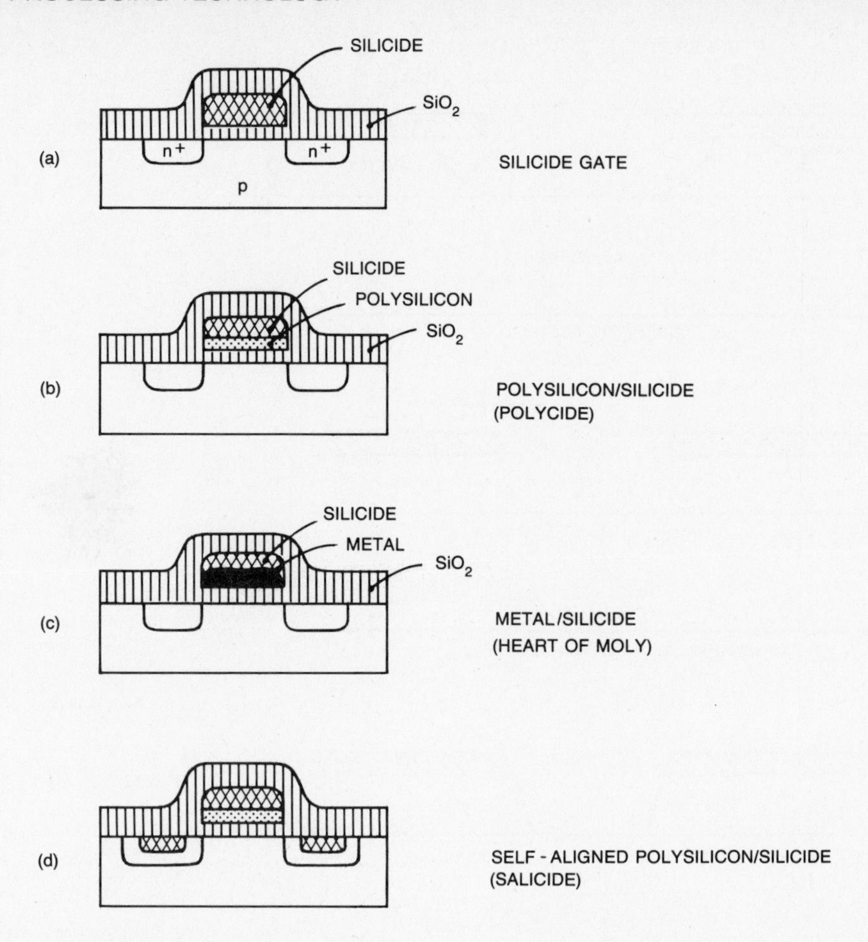

FIGURE 3.17. Refractory gates and interconnect © IEEE 1983 ([Chow83])

For example, a second level of good quality interconnect is almost mandatory in modern processes. One method that requires no extra mask levels is to reduce the polysilicon resistance by combining it with a refractory metal. Four such approaches are illustrated in Fig. 3.17 [Chow83]. In Fig. 3.17a, a *silicide* (e.g., silicon and tantalum) is used as the gate material. Sheet resistances of the order of 1–2 Ω/□ may be obtained. This is called the silicide gate approach. Silicides are mechanically strong and may be dry etched in plasma reactors. Tantalum silicide is stable throughout standard processing and has the advantage that it may be retrofitted into existing process lines. Fig. 3.17b uses a sandwich of silicide upon polysilicon, which is commonly called the *polycide* approach. A molybdenum gate, capped with silicide yields a metal/silicide sandwich or *heart of moly* structure (Fig. 3.17c). Finally, the silicide/polysilicon approach may be extended to include the formation of source and drain regions

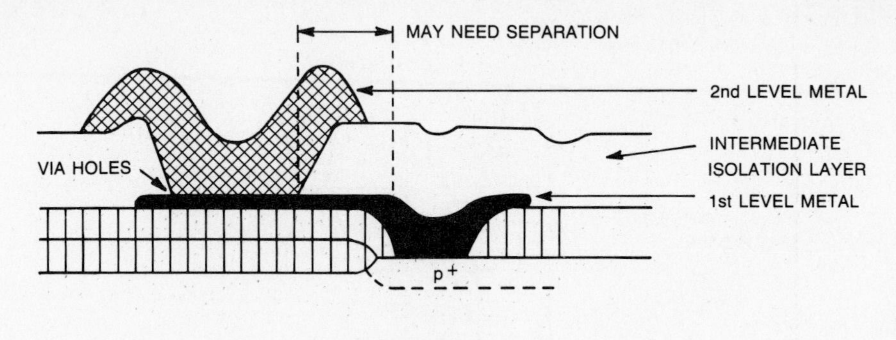

FIGURE 3.18. Cross-section of a second metal via

using the silicide. This is called the *salicide* process (Fig. 3.17d). The effect of all of these processes is to reduce the "second layer" interconnect resistance, allowing the gate material to be used as a reasonable long distance interconnect. This is achieved by minimum perturbation of an existing process.

A second approach is to just use a second layer of metal as discussed above. As a rule, second level metal layers have a coarser pitch as the topology of the silicon surface is more varied. Usually, contacting the second layer metal to first layer metal is achieved by a *via*, as shown in Fig. 3.18. If further contact to diffusion or polysilicon is required, a separation between the via and contact cut is usually required (although not in advanced processes). This requires a first level metal *tab* to bridge between metal 2 and the lower level conductor. It is important to realize that in contemporary processes first level metal must be involved in any contact to underlying areas. Probably as processes mature, this rule will be relaxed. A number of contact geometries are shown in Fig. 3.19 on page 96. Processes may require metal borders around the via on both levels of metal, on second level metal only, or no borders on either level, in which case the via normally overlaps the intersection of the two layers. Fig. 3.19b shows an example of the second instance. Aggressive processes allow one to stack vias on top of contacts, as shown in Fig. 3.19c. Consistent with the relative large thickness of the intermediate isolation layer, the vias may be larger than contact cuts and second layer metal needs to be thicker and requires larger via overlap. The process steps for a two metal process are briefly as follows:

- The oxide below the first metal layer is deposited by atmospheric chemical vapor deposition (CVD).
- The second oxide layer between the two metal layers is applied in a similar manner.

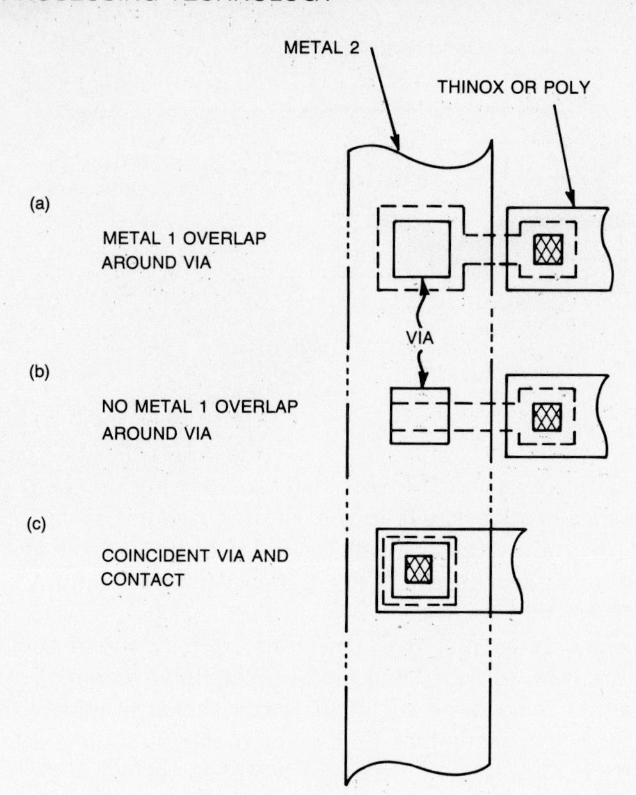

FIGURE 3.19. Second metal via geometries

- Depending on the process, removal of the oxide is accomplished using a plasma etcher designed to have a high rate of vertical ion bombardment. This allows fast and uniform etch rates. The structure of a via etched using such a method is shown in Fig. 3.18. Similarly, the bulk of the process steps for a double poly process are common to the processes described so far. With polysilicon, the oxide may be grown on top of the polysilicon to serve as isolation between polysilicon layers.

Recent innovations in research processes have included 3-D CMOS structures to reduce area and increase performance of circuits by using the vertical dimension in a silicon wafer [Gibb80] [ADYY83] [KSIM83]. Fig. 3.20a illustrates a cross-section of one particular 3-D SOI CMOS process [KSIM83], with an accompanying layout for an inverter shown in Fig. 3.20b. Trench isolation (Fig. 3.21a) improves latch-up and n to p spacing of transistors by including deep oxide

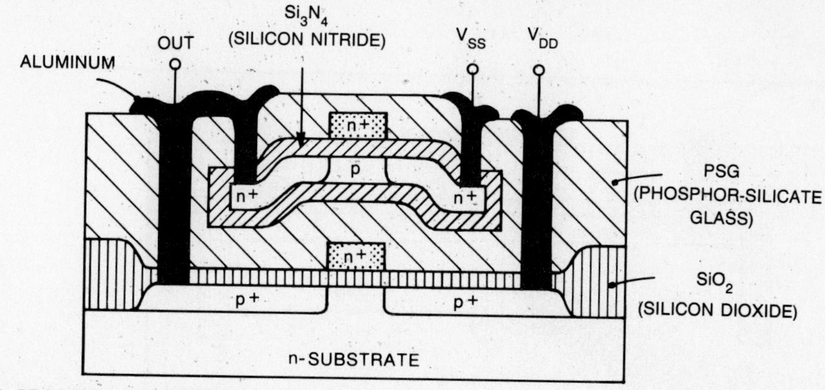

(a) PROCESS CROSS SECTION

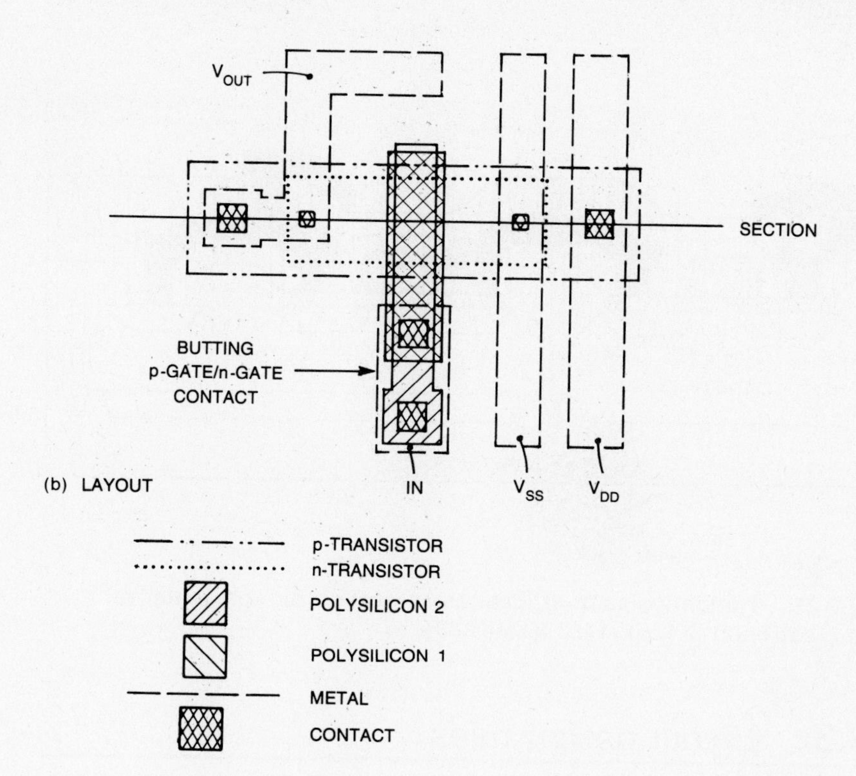

(b) LAYOUT

FIGURE 3.20. 3D CMOS process cross-section and layout of inverter © IEEE 1983 ([KSIM 83])

filled trenches between n- and p-transistors [IEDM83]. Fig. 3.21b shows a cross-section from an advanced twin-tub process developed by Tektronix that has silicide gates, trench isolation, and second level metal [YMKP83].

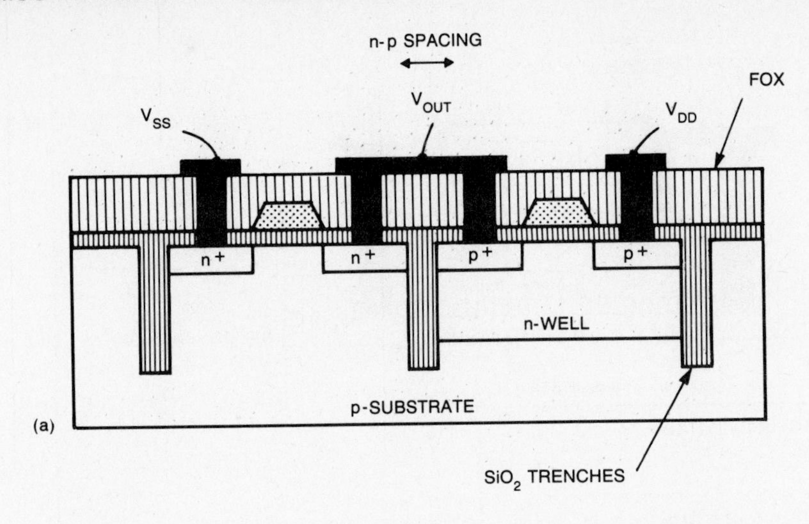

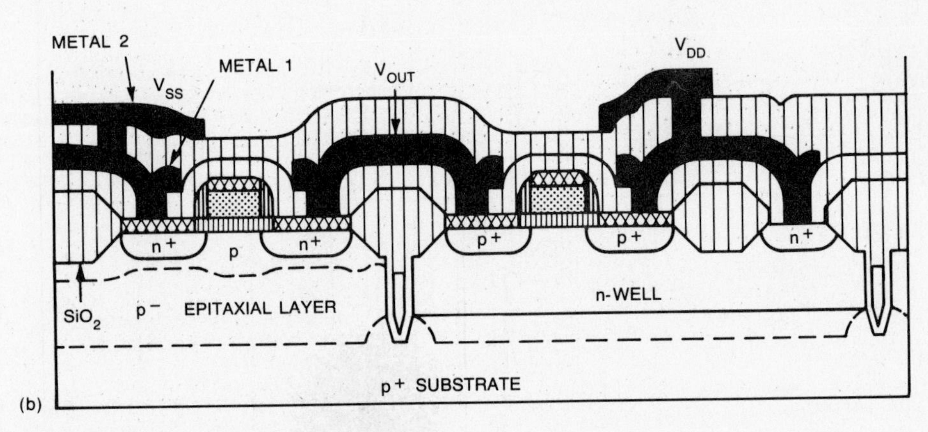

XXXXXXXXXX TITANIUM SILICIDE

FIGURE 3.21. Trench isolation a) ideal cross-section b) representative process (Tektronix) © IEEE 1983 ([YMKP83])

3.3 Layout design rules

Layout rules, also referred to as *design rules*, can be considered as a prescription for preparing the photomasks that are used in the fabrication of integrated circuits. The rules provide a necessary communication link between circuit designer and process engineer during the manufacturing phase. The main objective associated with the layout rules is to obtain the circuit with optimum yield in as small a geometry as possible without compromising reliability of the circuit.

In general, design rules represent the best possible compromise between performance and yield. The more conservative the rules are, the more likely it is that the circuit will function. However, the more aggressive the rules are, the greater the probability of improvements in circuit performance. This improvement may be at the expense of yield.

Design rules specify to the designer certain geometric constraints on the layout artwork so that the patterns on the processed wafer will preserve the topology and geometry of the designs. It is important to note that design rules do not represent some hard boundary between correct and incorrect fabrication. Rather, they represent a tolerance that ensures very high probability of correct fabrication and subsequent operation. For example, one may find that a layout that violates design rules may still function correctly and vice versa. Nevertheless, any significant or frequent departure from design rules will seriously prejudice the success of a design.

Two sets of design rule constraints in a process relate to line widths and interlayer registration. If the line widths are made too small it is possible for the line to become discontinuous. On the other hand, if the wires are placed too close to one another it is possible for them to merge together; that is, shorts can occur between two independent circuit nets. Furthermore, the spacing between two independent layers may be affected by the vertical topology of a process.

The design rules primarily address two issues: 1) the geometrical reproduction of features that can be reproduced by the mask masking and lithographical process, and 2) the interactions between different layers.

There are several approaches that can be taken in describing the design rules. These include 'micron' rules stated at some micron resolution, alpha (α) and beta (β) rules, and lambda (λ)-based rules.* Micron design rules are usually given as a list of minimum feature sizes and spacings for all the masks required in a given process. For example, the minimum thinox width might be specified as 4 μm. This is the normal style for industry. In α and β rules the basic feature size is defined in terms of β, while the minimum grid size that is needed is described by α. α and β may be related by a constant factor. The lambda-based design rules popularized by Mead and Conway [MeCo80] are based on a single parameter λ, which characterizes the linear feature — the resolution of the complete wafer implementation process — and permits first order scaling (which rarely applies). As a rule they can be expressed on a single page.

* The reader should not confuse this use of β and λ with that used in Chapter 2 for the transistor gain and channel length modulation parameters, respectively.

TABLE 3.1. Derivation of lambda-based rules from micron rules

MASK	FEATURE	DIMENSIONS: Micron rule	DIMENSIONS: λ rule
1: Thinox	Minimum thinox width	4 μm	2λ
	Minimum thinox spacing	4 μm	2λ
	Minimum p-thinox to n-thinox spacing	8 μm	4λ
3: Polysilicon	Minimum poly width	3.75 μm	2λ
	Minimum poly spacing	3.75 μm	2λ
	Minimum gate poly width (p)	4.5 μm	3λ
	Minimum gate poly width (n)	4.0 μm	2λ
	Minimum gate poly extension	3.5 μm	2λ
6: Aluminum	Minimum Al width	4.5 μm	3λ
	Minimum Al spacing	4.5 μm	3λ

The derivation of some λ rules based on a representative set of micron rules is illustrated in Table 3.1.

It should be noted that the degradation in circuit performance as well as the expected increase in silicon area could make the approach unsuitable for commercial circuits and even experimental circuits. In this text, we will use the λ rules to illustrate principles. By using symbolic techniques as described in Chapter 7, there is no reason why the actual micron rules cannot be used and it is the aim of this text to encourage this practice. The objective in this text is to illustrate approaches that completely hide the design rules from the designer.

3.3.1 Layer representations

The advances in the CMOS processes are generally complex and somewhat inhibit the visualization of all the mask levels that are used in the actual fabrication process. Nevertheless the design process can be abstracted to a manageable number of conceptual layout levels that represent the physical features one observes in the final silicon wafer. At a sufficiently high conceptual level all CMOS processes use the following features:

- two different substrates
- doped regions of both p- and n-transistor forming material
- transistor gate electrodes
- interconnection paths
- interlayer contacts.

TABLE 3.2. JPL/Mead Conway layer representation for *p*-well CMOS process

LAYER	JPL		
	Color	Symbolic	Comments
• p-well	Brown	—	Inside brown is p-well, outside is n-type substrate.
• Thin oxide	Green	*n*-transistor	Thinox may not cross a well boundary.
• Poly	Red	Polysilicon	Generally n^+.
• p^+	Yellow	*p*-transistor	Inside is p^+.
• Metal1	Light blue	Metal1	—
• Metal2	Dark blue	Metal2	—
• Contact cut	Black	Contact	—
• Overglass	—	—	—

The layers for typical CMOS processes are represented in various figures in terms of:

- a color scheme proposed by JPL*
- a modified color scheme to differentiate between nMOS and CMOS structures (as used on the cover of this text)
- stipple patterns
- line styles
- or a mixture of these.

Some of these representations are shown in Table 3.2 and Table 3.3. Where diagrams are presented, a legend will be used to indicate

TABLE 3.3. Alternate layer representations for *p*-well CMOS process

LAYER	ALTERNATE COLOR	CIF CODE
• p-well	Brown	CW
• Thin oxide*	Red	CD
• Poly	Green	CP
• p^+	Purple	CS
• Metal1	Tan	CM
• Metal2	Dark blue	CN
• Contact	Black	CC
• Overglass	White stipples	CG

* This layer is referred by Mead & Conway as diffusion. n-thinox is light blue. p-thinox is purple or magenta.

* Jet Propulsion Laboratory, California Institute of Technology, Pasadena.

TABLE 3.4. First character representation of CIF for process description

PROCESS	CIF CHARACTER
n-channel MOS	N
p-channel MOS	P
bulk CMOS processes	C
silicon on insulator processes	S

layer assignments. At the mask level, some layers may be omitted for clarity. At the symbolic level only n- and p-transistors will be shown. This should be viewed as translating to the appropriate set of masks for whatever process is being considered.

For convenience, the CIF (Caltech Intermediate Form) layer names as used by JPL for bulk CMOS are also presented in Table 3.3. It should be noted that CIF version 2.0 uses up to four alpha-numeric characters to describe a mask level. Generally, the first letter is used to characterize the process class, e.g., 'C' for bulk CMOS processes, followed by a second character to identify the layer type. Such a format for a group of closely related processes is shown in Table 3.4.

The n-well and twin-tub bulk CMOS processes as well as the SOI process can be represented in a similar manner. For example, in n-well bulk CMOS the only difference in the resulting wafer structure is the reversal of the role of the well and the original substrate. Different process lines may use different combinations of the n^+, p^+, n-well, or p-well masks to define the process. It is very important to intimately understand what set of masks a particular process line uses if you are responsible for generating interface formats. For instance, an n^+ mask, which is the reverse of a p^+ mask, may be used. Thus n^+ thinox denotes n-transistors and so on. Conceptually, the mask levels in a silicon on insulator process are probably the simplest. The levels and visible geometry in this process correspond directly to the features that a designer has to deal with conceptually (i.e., n-regions and p-regions). Perhaps the significant difference between SOI and bulk CMOS processes, from the designer's point of view, is the absence of wells.

Our overall thrust to bring the simplicity of SOI to bulk CMOS technologies is to encourage the adoption of a *symbolic* level of design, in which the designer directly manipulates n- and p-transistors and other *circuit* features of interest. In turn, this allows the design rules to be completely removed from layout design and also paves the way for the next generation of automatically performance optimized circuits.

TABLE 3.5. Lambda-based layout rules

NO.	MASK	FEATURE	DIMENSION
		A1.Minimum thinox width	2λ
1	Thinox	A2.Minimum thinox spacing (n^+ to n^+, p^+ to p^+)	2λ
		A3.Minimum p-thinox to n-thinox spacing	8λ
		B1.Minimum p-well width	4λ
		B2.Minimum p-well spacing (wells at same potential)	2λ
2	p-well	B3.Minimum p-well spacing (wells at different potential)	6λ
		B4.Minimum distance to internal thinox	3λ
		B5.Minimum distance to external thinox	5λ
		C1.Minimum poly width	2λ
		C2.Minimum poly spacing	2λ
3	Poly	C3.Minimum poly to thinox spacing	λ
		C4.Minimum poly gate extension	2λ
		C5.Minimum thinox source/drain extension	2λ
		D1.Minimum overlap of thinox	1.5–2λ
4	p-plus	D2.Minimum p-plus spacing	2λ
		D3.Minimum gate overlap or distance to gate edge	1.5–2λ
		D4.Minimum spacing to unrelated thinox	1.5–2λ
		E1.Minimum contact area	2λ × 2λ
		E2.Minimum contact to contact spacing	2λ
		E3.Minimum overlap of thinox or poly over contact	λ
		E4.Minimum spacing to gate poly	2λ
5	Contact	E5.n^+ source/drain contact	See Figure 3.22
		E6.p^+ source/drain contact	See Figure 3.22
		E7. V_{SS} contact	See Figure 3.22
		E8. V_{DD} contact	See Figure 3.22
		E9.Split contact V_{SS}	See Figure 3.22
		E10.Split contact V_{DD}	See Figure 3.22
		F1.Minimum metal width	2–3λ
6	Metal	F2.Minimum metal spacing	3λ
		F3.Minimum metal overlap of contact	λ

3.3.2 Lambda-based p-well rules

Table 3.5 and Fig. 3.22 are a version of p-well rules loosely based on the JPL rules [Gris80]. Plate 3 illustrates these rules in color. It should be noted that these are only representative and are the result of averaging a large number of processes. From Fig. 3.22 it can be seen that the rules are defined in terms of:

- feature sizes
- separations and overlaps.

There are a number of issues which require some discussion. These are:

1 Well spacing and separation rules. The p-well is usually a deep diffusion and therefore it is necessary for the outside dimension to provide sufficient clearance between the p-well edges and the adjacent p^+ diffusions. In current processes, 5λ ensures that edges of the p-well do not short to p^+ diffusions in the n-substrate. The inside clearance is determined by the transition

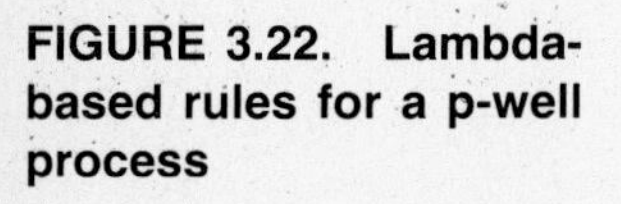

FIGURE 3.22. Lambda-based rules for a p-well process

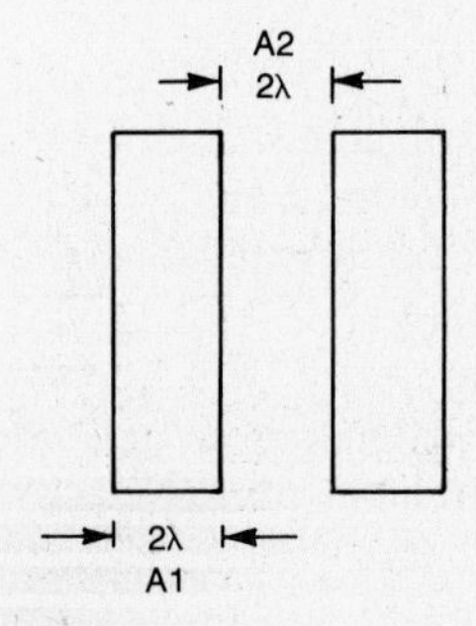

MASK 1: THINOX

A1. MINIMUM THINOX WIDTH 2λ

A2. THINOX SPACING (n^+ to n^+ or p^+ to p^+) 2λ

A3. p^+ to n^+ SPACING 8λ

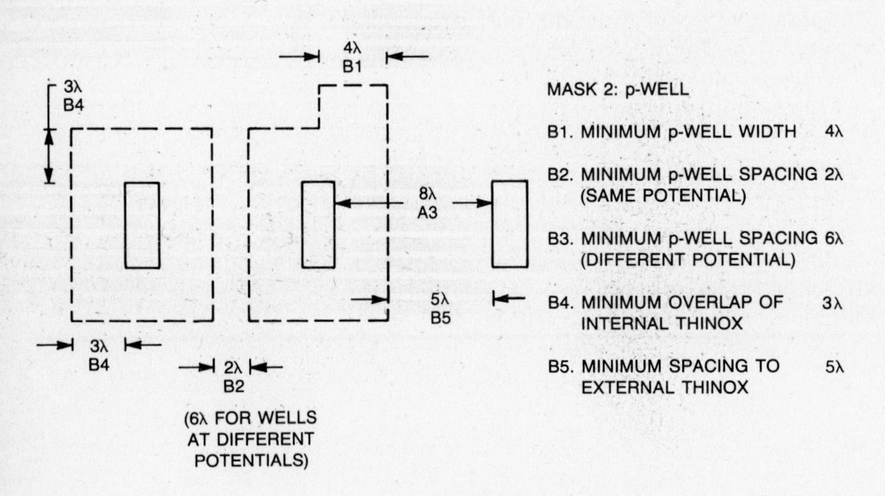

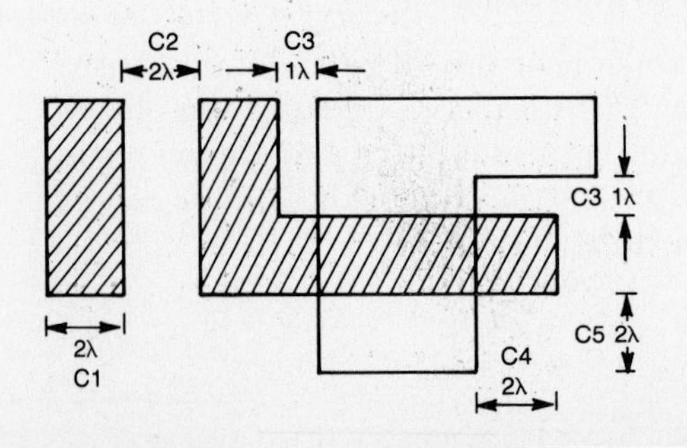

MASK 3: POLYSILICON

C1. MINIMUM POLY WIDTH 2λ

C2. MINIMUM POLY SPACING 2λ

C3. MINIMUM POLY-THINOX SPACING λ

C4. MINIMUM POLY GATE EXTENSION 2λ

C5. MINIMUM THINOX SOURCE/DRAIN EXTENSION 2λ

FIGURE 3.22. (Continued)

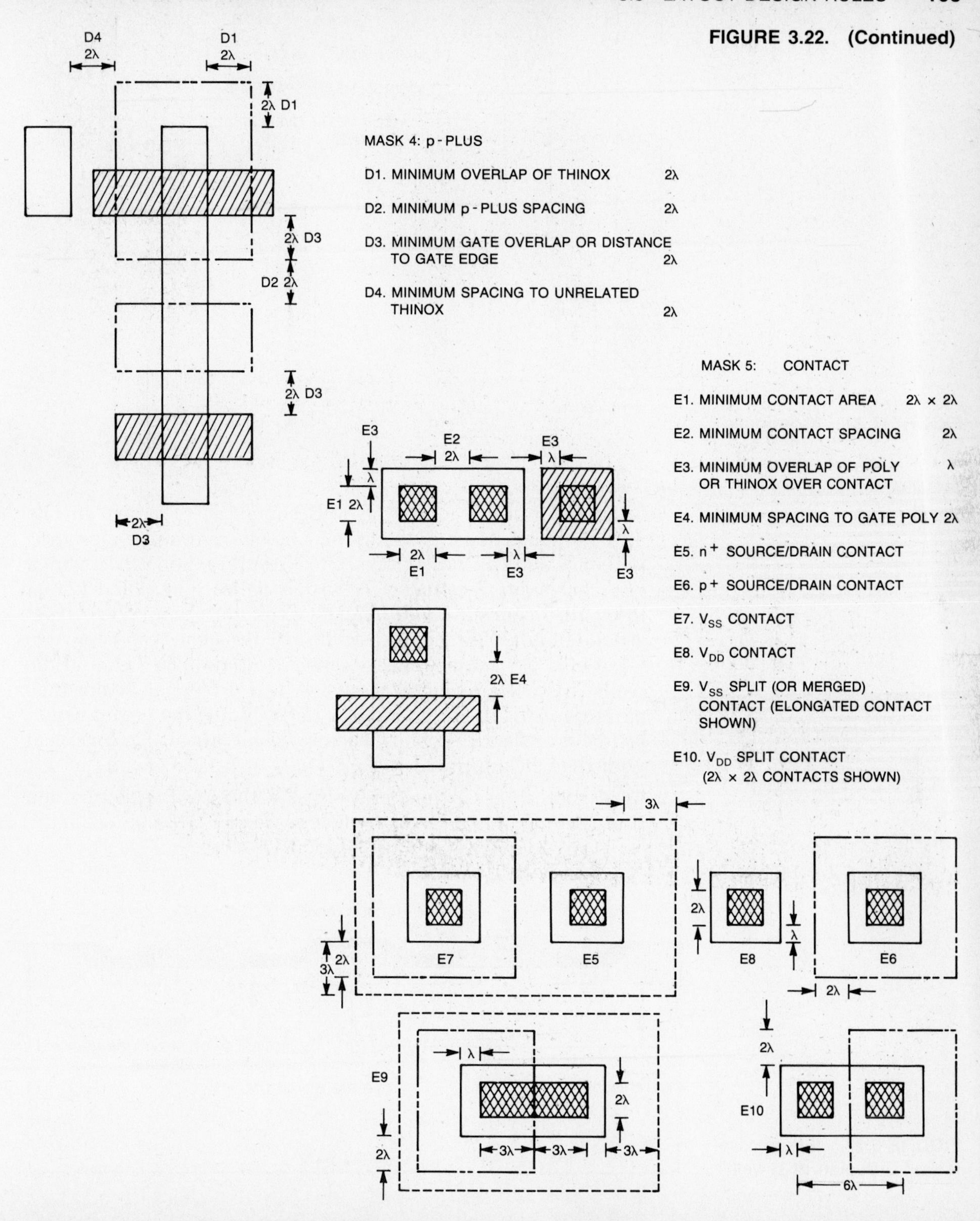
D4
2λ
D1
2λ
2λ D1
2λ D3
D2 2λ
2λ D3
2λ
D3
MASK 4: p-PLUS
D1. MINIMUM OVERLAP OF THINOX 2λ
D2. MINIMUM p-PLUS SPACING 2λ
D3. MINIMUM GATE OVERLAP OR DISTANCE TO GATE EDGE 2λ
D4. MINIMUM SPACING TO UNRELATED THINOX 2λ
MASK 5: CONTACT
E1. MINIMUM CONTACT AREA 2λ × 2λ
E2. MINIMUM CONTACT SPACING 2λ
E3. MINIMUM OVERLAP OF POLY OR THINOX OVER CONTACT λ
E4. MINIMUM SPACING TO GATE POLY 2λ
E5. n+ SOURCE/DRAIN CONTACT
E6. p+ SOURCE/DRAIN CONTACT
E7. VSS CONTACT
E8. VDD CONTACT
E9. VSS SPLIT (OR MERGED) CONTACT (ELONGATED CONTACT SHOWN)
E10. VDD SPLIT CONTACT (2λ × 2λ CONTACTS SHOWN)
E3
E2
2λ
E3
λ
λ
E1 2λ
2λ
E1
λ
E3
λ
E3
2λ E4
3λ
2λ
3λ
E7
E5
2λ
λ
E8
E6
2λ
E9
λ
2λ
3λ
3λ
3λ
2λ
2λ
E10
λ
6λ

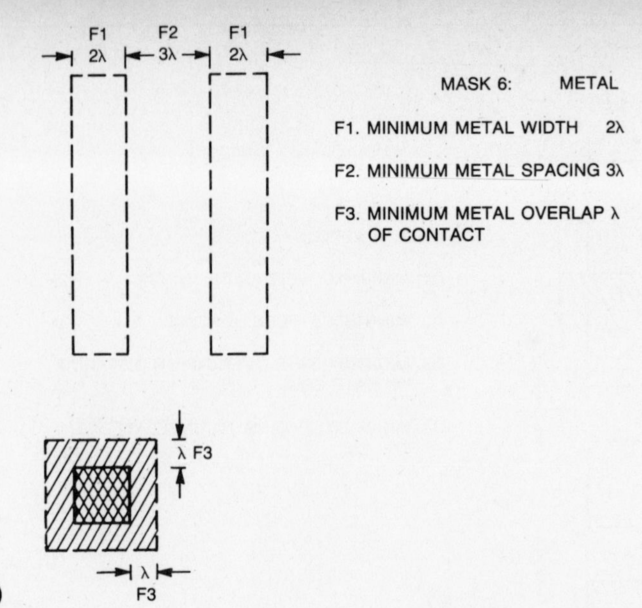

FIGURE 3.22. (Continued)

of the field oxide across the well boundary, as shown in Fig. 3.23. Although some processes may permit zero inside clearance, problems such as 'birds beaks' effect result in the 3λ rule, which is a conservative estimate. A further point to be noted is that to avoid a shorted condition, thinox is not permitted to cross a well boundary. Since the p-well sheet resistance can be several kilo-ohms per square, it is necessary to thoroughly ground the well. This will prevent excessive voltage drops due to substrate currents. Thus the rule to follow in grounding the p-well would be to put a substrate contact wherever space is available consistent with the rules outlined in Sec. 2.6.

2 Transistor rules. Where poly crosses thinox, the source and drain diffusion is masked by the poly region. The source, drain,

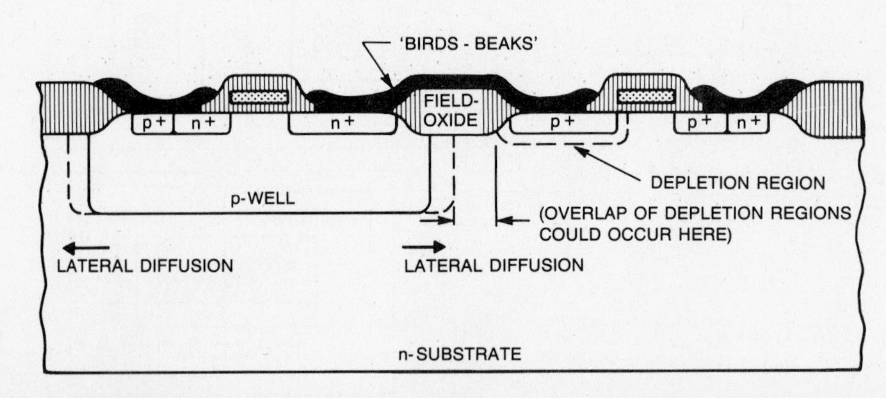

FIGURE 3.23. Influence of lateral diffusion of p-well

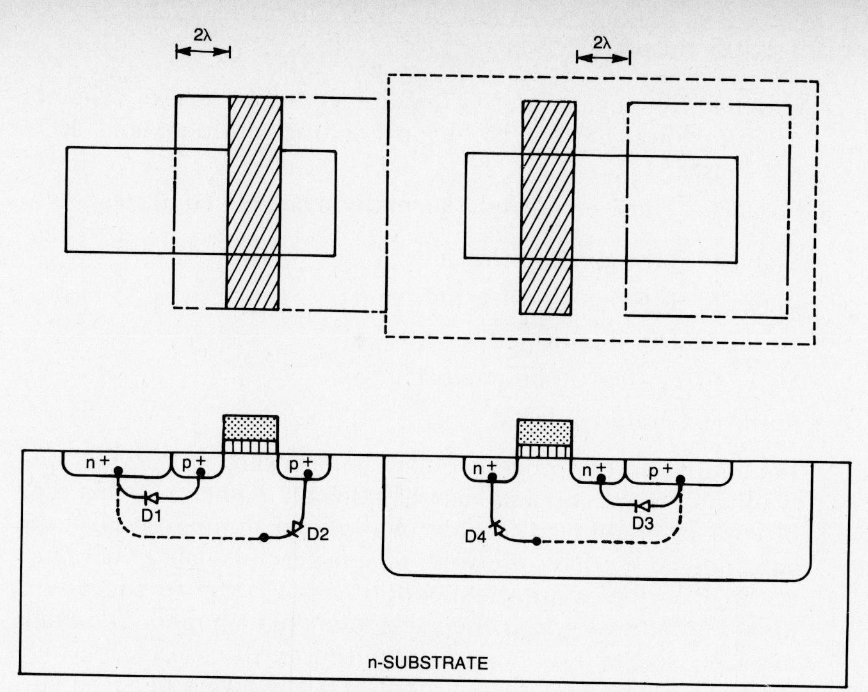

BV_{D1}	20 - 40 VOLTS	BV_{D3}	20 - 40 VOLTS
BV_{D2}	50 - 70 VOLTS	BV_{D4}	50 - 70 VOLTS

FIGURE 3.24. Placement of gate edges and p^+

and channel are thereby self-aligned to the gate. It is essential for the poly to completely cross thinox, otherwise the transistor that has been created will be shorted by a diffused path between source and drain. To ensure this condition is satisfied, 1.5–2λ of poly is necessary beyond the edges of the diffusion region. This is often termed the "gate extension." Thin oxide must extend beyond the poly gate so that diffused regions exist to carry charge into and out of the channel. 2λ of thinox extension is necessary to preserve the source and drain region. Poly and thinox regions that do not meet intentionally to form a transistor should be kept separated by .5–1λ. The rule for clearance and overlap of p^+ and thinox is 2λ. Fig. 3.24 demonstrates that both types of transistors have a thinox region (diffusion) and a polysilicon region. An n-device has a p-well region surrounding it, whereas a p-device has a p^+ region surrounding it. Thin oxide areas that are not covered by p^+ are n^+ and hence are n-devices or wires (within the p-well). Therefore a transistor is p-channel if it is inside a p^+ region; otherwise it is an n-channel device. From the above discussion it can be noted that there are two types of implant/diffusion used to form the p- and n-transistors. What is important to note is that p^+ diffusion is obtained by "logical anding" of thinox and p^+ masks, whereas n^+ diffusion is derived by "logical anding" of thinox and (NOT p^+) masks.

3 Contacts. There are several generally available contacts:

- metal to p-thinox (p-diffusion)
- metal to n-thinox (n-diffusion)
- metal to polysilicon
- V_{DD} and V_{SS} (substrate contacts)
- split (substrate contacts)

Depending on the process, other contacts such as "buried" polysilicon-thinox contacts may be allowed. Sometimes this type of contact is allowed to only one type of thinox. Because the substrate is divided into 'well' regions, each isolated well must be 'tied' to the appropriate supply voltage; that is, the p-well must be tied to V_{SS} and the substrate (what amounts to n-well) must be tied to V_{DD}. This is achieved by the use of substrate contacts. One needs to note that every n-device must be surrounded by a p-well and that the p-well must be connected to V_{SS} via a V_{SS} contact. Furthermore, every p-device must have access to a V_{DD} contact. The split or merged contact is equivalent to two separate metal-diffusion contacts that are strapped together with metal. This structure is used to tie transistor sources to either substrate or p-well. In order to ensure the p^+/n^+ doping boundary remains within 1λ of the center of the cut, 4–6λ rule is applied to the cut length if an elongated cut is used. This is shown for the V_{SS} merged contact in Fig. 3.22. Due to requirements as processes are scaled, the separated contact structure used for the V_{DD} merged contact in Fig. 3.22 is preferred. This results in the ability to make all contact cuts identical for the complete design, which can be beneficial in processing. A rectifying contact can be created as the result of missing the cut. This can be fatal although some circuits might appear to work.

4 Poly doping. In a number of current p-well CMOS processes, the poly layer is usually doped n^+. This means the p^+ doping step somewhat reduces the n^+ doping of polysilicon. Thus an increase in the sheet resistance of the poly is normally encountered within the regions. If this is a problem, then the rule to follow would be to place poly wires (as much as possible) outside the p^+ region.

5 p^+ and gate edges. The 2λ rule for the separation between the gate edges and p-plus provides the basis for a doping change and enables the creation of lateral diodes as illustrated by Fig. 3.24. As a general rule in the fabrication process, the transition from n^+ to p^+ doping is not controlled. Thus different junction gradients having different breakdown voltages can be expected for different processes.

6 Guard rings. Guard rings that are p^+ diffusions in the n-substrate and n^+ diffusions in the p-well are used to collect injected minority carriers. If they are implemented in a structure, then p^+ guard rings must be tied to V_{SS}, while n^+ guard rings must be tied to V_{DD}. An n^+ diffusion with p^+ guard ring is shown in Fig. 3.25a, while a p^+ diffusion with n^+ guard ring

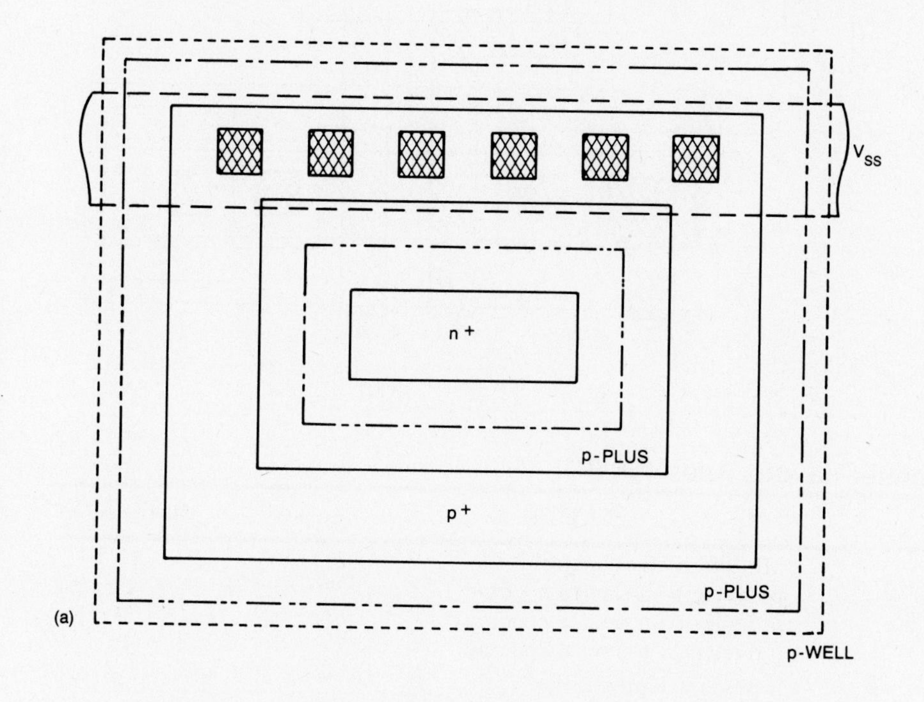

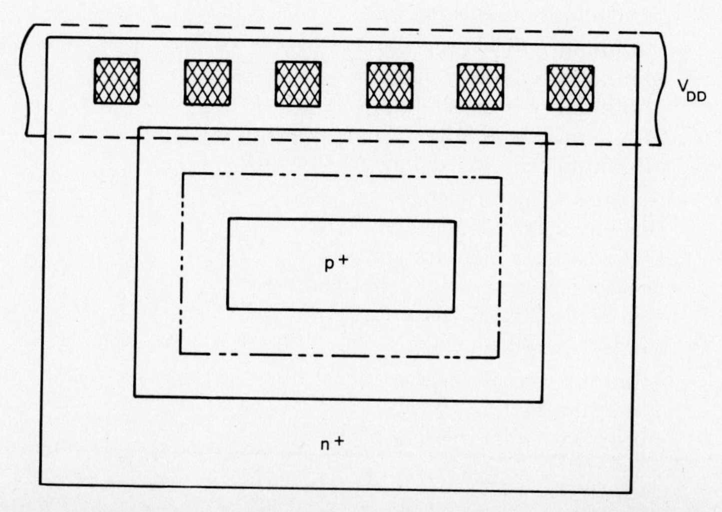

FIGURE 3.25. Realization of n^+ guard ring and p^+ guard ring

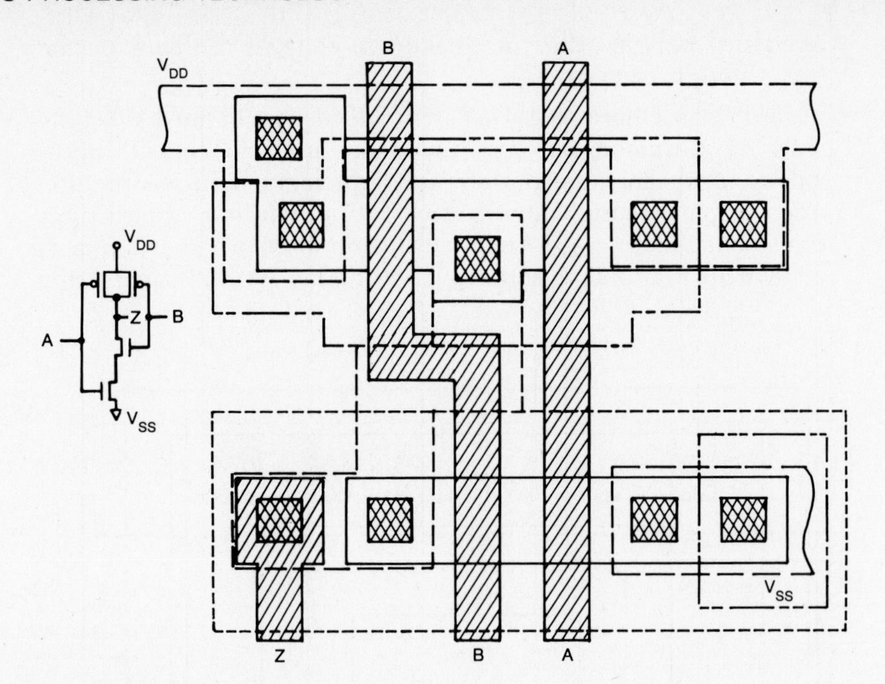

FIGURE 3.26. 2-input NAND gate layout using lambda rules for p-well CMOS

TABLE 3.6. Lambda-based layout rules for SOI

NO.	MASK	FEATURE	DIMENSION
1	Island	Minimum island width	2λ
		p-device to n-device spacing	2λ
		n-device to n-device spacing	3λ
		p-device to p-device spacing	3λ
2	Implant	Implant/island overlap	λ
		Implant/island spacing	λ
3	Poly	Minimum poly width	2λ
		Minimum poly-poly spacing	2λ
		Minimum poly to island	2λ
		Minimum island edge to poly spacing	2λ
		Minimum poly extension over island	2λ
4	Contact	Distance over poly edge	λ
		Distance over island edge	λ
		Distance from island edge	λ
		Distance from noncontacted feature	2λ
		Contact width on island	2λ
		Contact width on poly	2λ
5	Metal	Minimum metal width	3λ
		Minimum metal spacing	2λ
		Minimum metal overlap of contact	λ

is shown in Fig. 3.25b. A typical layout for a 2-input NAND gate using the lambda-based design rules is illustrated in Fig. 3.26 and Plate 4.

3.3.3 Lambda-based SOI rules

Table 3.6 and Fig. 3.27 (continued on page 112) show a set of lambda rules for CMOS silicon-on-insulator. The interesting feature about this set of rules is that apart from the n-device to n-device spacing rule, implant rules, metal-to-metal spacing rule, and contact rules, 2λ is the only value that needs to be remembered (but who needs to remember design rules?). The 2λ spacing rule between island edge and unrelated poly is used to ensure against shorts between the poly and island edges. This can be caused by thin or faulty oxide covering over the islands.

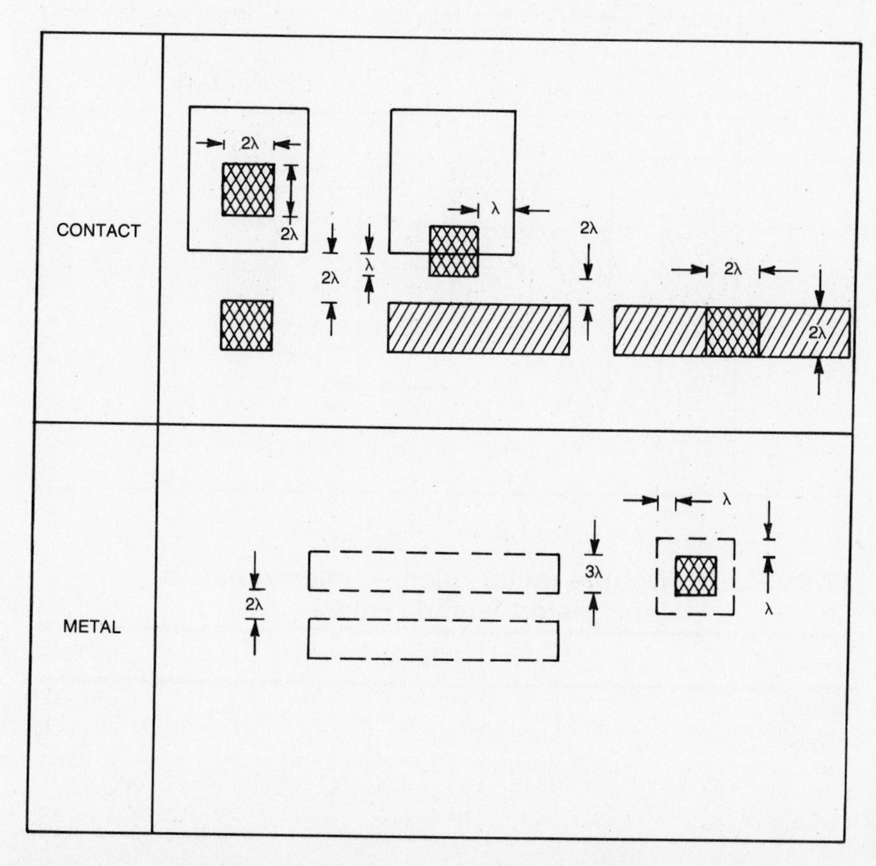

FIGURE 3.27. Lambda-based rules for SOI

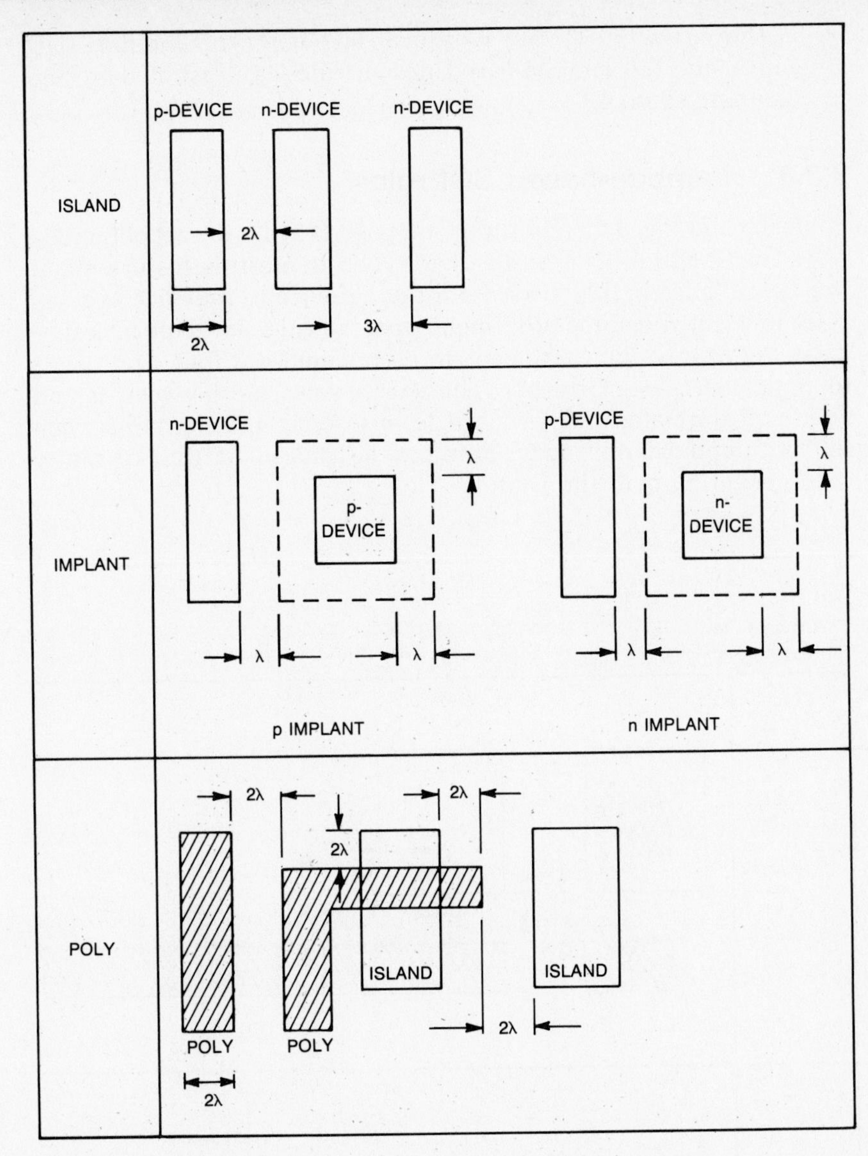

FIGURE 3.27. (Continued)

TABLE 3.7. Double metal rules — micron based (suggested lambda rules)

LAYER	WIDTH	SPACING
Metal1	3 μm (2λ)	4 μm (3λ)
Metal2	5 μm (4λ)	5 μm (4λ)
Via	3 μm × 3 μm (2λ × 2λ)	3 μ (2λ)
Cut	3 μm × 3 μm (2λ × 2λ)	3 μ (2λ)
Cut-via space	—	3 μm (2λ)

TABLE 3.8. Metal1, metal2, via constructs

LAYER	SIZE
Metal1	5 μm × 5 μm (4λ × 4λ)
Via	3 μm × 3 μm (2λ × 2λ)
Metal2	7 μm × 7 μm (5λ × 5λ)

3.3.4 Double metal design rules

Tables 3.7 and 3.8 show some typical rules relating second layer metal to first layer metal for a typical two layer metal process. The increase in width and separation of second level metal ensure against broken conductors or shorts between adjoining wires due to the vertical topology.

3.3.5 Design rules — summary

In commercial designs, lambda rules are rarely sufficient to describe high performance circuits. Some additional rules that might be present in some processes are as follows:

- Extension of polysilicon in the direction that metal wires exit a contact.
- Differing p- and n-transistor gate lengths.
- Differing gate poly extensions depending on the device length or the device construction.

While all of these rules can be worst-cased, very inefficient designs result. A better approach is to implement systems that synthesize the correct geometry from an intermediate form. Therefore, symbolic styles of design appear to provide a solution for creating generic CMOS circuits that can be implemented with a wide range of fabrication processes. This is the underlying theme in this text and will be expanded upon in Chapter 7. For the time being we will examine some ways of capturing the key rules of a process.

3.4 Process parameterization

As automated tools become more available, the necessity for an individual designer to know detailed design rules becomes less important. However, the tool designer must have a form in which the design rules for a process can be unambiguously represented.

If rules are to be communicated between tools, then a data format has to be designed to provide the appropriate interface. We will examine some of these ideas in this section. The main idea is to identify structures of interest and demonstrate algorithms that might be used to construct these. Spacing of these structures from other structures is achieved by applying the normal spacing rules.

3.4.1 Abstract layers

An important concept in the synthesis (and analysis) process is the definition of abstract layers. For instance, n-diffusion in a p-well process consists of the thin oxide mask "anded" with the p-well mask without p-plus present, while transistor p-diffusion consists of thin oxide "anded" with the p-plus mask in the absence of the p-well. We might state this for a p-well process in some pseudo-language represented by the following:

```
NDIFF = N_DIFFUSION = P_WELL AND THINOX AND NOT
   P_PLUS
PDIFF = P_DIFFUSION = P_PLUS AND THINOX AND NOT
   P_WELL
ACTIVE = ACTIVE_TRANSISTOR_AREA = THINOX AND
   POLYSILICON
VDDN = VDD_N_DIFFUSION = THINOX AND NOT P_WELL
   AND NOT P_PLUS
VSSP = VSS_P_DIFFUSION = THINOX AND P_PLUS AND
   P_WELL
```

3.4.2 Spacing rules

Using the abstract layers and regularly defined layers the spacing of layers may be specified. For instance, the following rules are according to Table 3.4:

```
ND_PD_SP = NDIFF TO PDIFF SPACING = 8*LAMBDA
ND_ND_SP = NDIFF TO NDIFF SPACING = 2*LAMBDA
PD_PD_SP = PDIFF TO PDIFF SPACING = 2*LAMBDA
CO_CO_SP = CONTACT TO CONTACT SPACING = 2*LAMBDA
CO_GP_SP = CONTACT TO GATE POLY SPACING =
   2*LAMBDA
```

However, the following rules are also needed:

```
ND_VP_SP = NDIFF TO VSSP SPACING = 2*LAMBDA
PD_VN_SP = PDIFF TO VDDN SPACING = 2*LAMBDA
ND_VN_SP = NDIFF TO VDDN SPACING = 8*LAMBDA
PD_VP_SP = PDIFF TO VSSP SPACING = 8*LAMBDA
```

3.4.3 Construction rules

Construction rules are used to build structures. The first level of these are the minimum width rules. For example:

```
TH_WID = MINIMUM THINOX WIDTH = 2*LAMBDA
CO_WID = MINIMUM CONTACT WIDTH = 2*LAMBDA
PO_WID = MINIMUM POLYSILICON WIDTH = 2*LAMBDA
```

In addition, extension rules may also be specified:

```
GP_A_EXT = EXTENSION GATE_POLY OVER ACTIVE =
   2*LAMBDA
PO_CO_EXT = EXTENSION POLYSILICON OVER CONTACT =
   LAMBDA
TH_GP_EXT = EXTENSION THINOX OVER GATE_POLY =
   2*LAMBDA
TUB_TH_EXT = EXTENSION PTUB OVER THINOX =
   3*LAMBDA
PP_TH_EXT = EXTENSION PPLUS OVER THINOX =
   2*LAMBDA
TH_CO_EXT = EXTENSION THINOX OVER CONTACT =
   LAMBDA
```

Using these parameters, the following is an example of the pseudo-code that may be used to build a minimum length transistor of variable width:

```
       type is transistor type
       x,y is transistor position
       w is transistor width

build_transistor(type,x,y,w)
{
     l = PO_WID + 2*TH_GP_EXT
     build_rectangle(THINOX, x-l/2: y-w/2,
     x+l/2, y+w/2)
     if(type == N_TRANSISTOR)
     {
        wp = w + 2*TUB_TH_EXT
        l = l + 2*TUB_TH_EXT
        build_rectangle(PTUB, x-l/2, y-wp/2,
        x+l/2, y+wp/2)
     }
     else
     {
        wp = w + 2*PP_TH_EXT
        l = l + 2*PP_TH_EXT
```

```
    build_rectangle(PPLUS, x-l/2, y-wp/2,
    x+l/2, y+wp/2)
}
wp = w + 2*GP_A_EXT
l = PO_WIDTH
build_rectangle(POLY, x-l/2, y-wp/2, x+l/2,
y+wp/2)
```

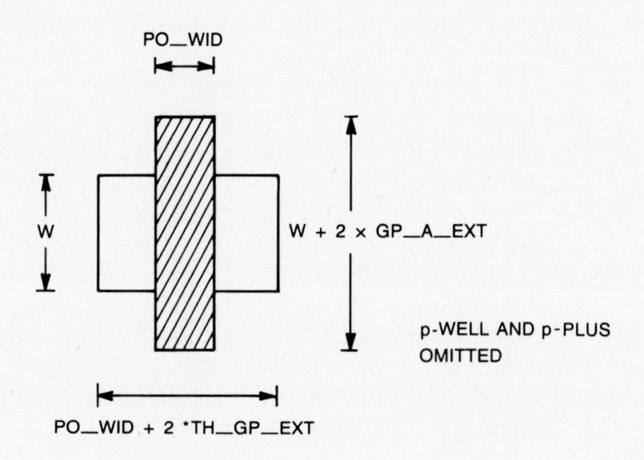

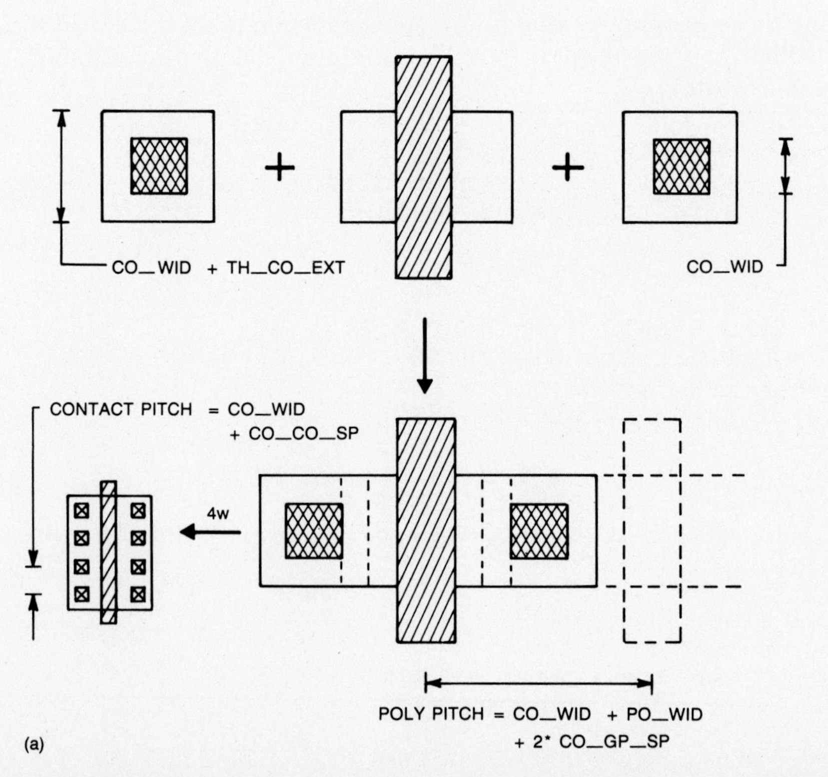

FIGURE 3.28. Algorithmically defined transistors (a)

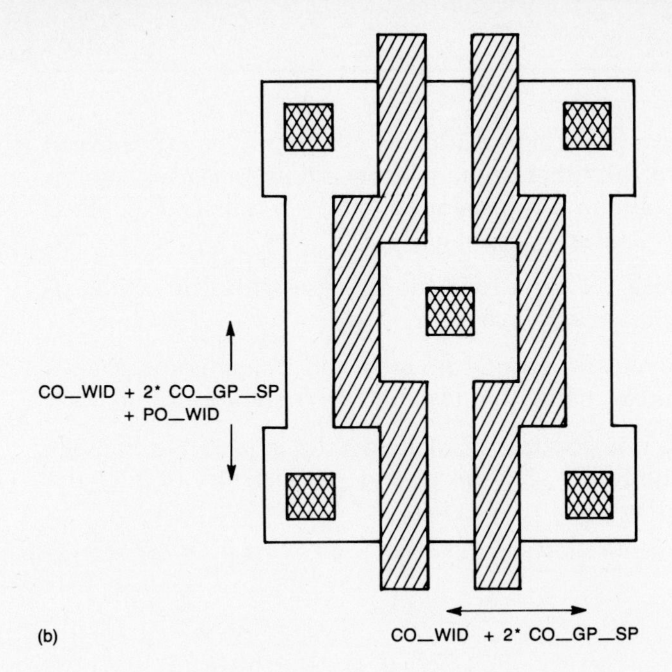

FIGURE 3.28. (Continued)

The resulting unconnected transistor is shown in Fig. 3.28a. The addition of some parameterized contacts completes the transistor. A larger transistor is shown with multiple source/drain contacts. The poly pitch and the source/drain contact pitch are illustrated. An alternative transistor structure is shown along with the associated dimensions in Fig. 3.28b. This has a reduced poly pitch at the expense of a reduced number of source/drain contacts.

3.5 Summary

This chapter has covered some of the more common CMOS technologies that are in current use. We have presented a set of lambda rules for a p-well process and an SOI process. Chapter 7 outlines symbolic layout techniques that allow the CMOS circuit designer to construct circuits without direct reference to such rules. Of course the design automation programmer still has to be cognizant of geometric design rules. The last section in this chapter dealt with some more obvious ways of parameterizing layout software.

3.6 Exercises

3.1 An n-well process has thin oxide, n-well and n-plus mask layers, in addition to the other regular layers. Draw the mask combinations to obtain an n-transistor contact, a p-transistor contact, a V_{DD} contact, and a V_{SS} contact.

3.2 Repeat Exercise 3.1 for a twin-tub process with thin oxide, p-well, and n-plus mask layers.

3.3 Design the layout for an SOI inverter and transmission gate 2-input multiplexer using the design rules included in Fig. 3.27.

3.4 Write a program to generate a CMOS inverter in p-well technology that can individually size the n- and p-transistors to alter the threshold of the gate.

CIRCUIT CHARACTERIZATION AND PERFORMANCE ESTIMATION

4.1 Introduction

In previous chapters we established that an MOS structure is created by superimposing a number of layers of conducting, insulating, and transistor-forming materials. It was further demonstrated that in a conventional silicon gate process an MOS device requires a gate forming region and a source/drain forming region, which consists of diffusion, polysilicon, and metal layers separated by insulating layers. Each layer has both a resistance and capacitance that are fundamental components in estimating the performance of a circuit or system. They also have inductance characteristics that we will assume to be negligible.

In this section we are primarily concerned with the development of simple models that will assist us in the understanding of system behavior and will provide the basis whereby systems performance, in terms of signal delays and power dissipation, can be estimated.

The issues to be considered in this section are:

- resistance and capacitance calculations
- delay estimations
- determination of conductor size for power and clock distribution
- power consumption
- charge storage mechanism
- effects of scaling.

4.2 Resistance estimation

The resistance of a uniform slab of conducting material may be expressed as

$$R = \left(\frac{\rho}{t}\right)\left(\frac{l}{w}\right) \qquad \text{(ohms)}, \tag{4.1}$$

where

$$\begin{aligned} \rho &= \text{resistivity} \\ t &= \text{thickness} \\ l &= \text{conductor length} \\ w &= \text{conductor width.} \end{aligned}$$

This expression may be rewritten as

$$R = R_s\left(\frac{l}{w}\right) \qquad \text{(ohms)}, \tag{4.2}$$

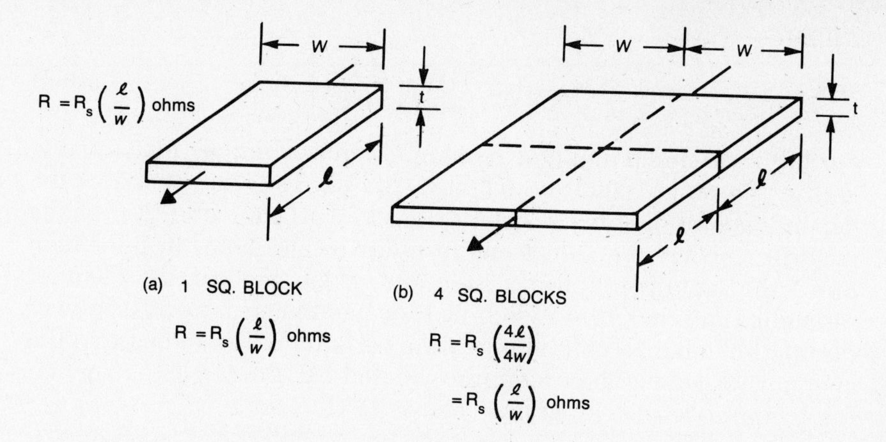

FIGURE 4.1. Determination of layer resistance

where R_s is the sheet resistance having units of ohm/square. Thus to obtain the resistance of a layer one would simply multiply the sheet resistance R_s, by the ratio of the length to width of the conductor. For example, the resistances of the two shapes shown in Fig. 4.1 are equivalent. Table 4.1 shows typical sheet resistances that can be expected in 3 μm to 5 μm MOS processes. Note that for metal having a given thickness t, the resistivity is known, while for poly and diffusion the resistivities are significantly influenced by the concentration density of the impurities that have been introduced into the conducting regions during implantation. This means that the process parameters have to be known to accurately estimate these quantities.

Although the voltage-current characteristic of an MOS transistor is generally nonlinear, it is sometimes useful to approximate its behavior in terms of a "channel" resistance to estimate performance. From Chapter 2, Eq. (2.9), one may express the channel resistance (in the linear region) R_c by

$$R_c = k\left(\frac{L}{W}\right), \qquad \textbf{(4.3)}$$

TABLE 4.1. Typical sheet resistances for conductors

	SHEET RESISTANCE OHM/SQ.		
Material	Min.	Typical	Max.
Metal (Al)	0.03	0.05	0.08
Silicides	2	3	6
Diffusion (n+ and p+)	10	25	50
Polysilicon	15	50	100

where

$$k = \left[\mu\left(\frac{\varepsilon_o \varepsilon_r}{t_{ox}}\right)(V_{gs} - V_t)\right]^{-1}. \quad \textbf{(4.4)}$$

For both the n-channel and p-channel devices, k may take a value within the range 5,000 to 30,000 $\Omega/sq.$. Eq. (4.4) demonstrates the dependence of channel resistance on the surface mobility μ of the majority carriers (i.e., electrons in n-device and holes in p-device). Since the mobility is also a function of temperature, the channel resistance and therefore switching time parameters, as well as power dissipation, change with temperature variations. The increase in the channel resistance may be approximated by +0.25 percent per °C for an increase in temperature above 25°C.

4.2.1 Resistance of nonrectangular regions

Many times during the course of a layout nonrectangular shapes are used (for instance, the corners of wires). The resistance of these shapes requires more elaborate calculation than that for simple rectangular regions. One method of calculating the resistance is to break the shape in question into simple regions, for which the resistance may be calculated [HoDu83]. Fig. 4.2a summarizes the resistance of a number of commonly encountered shapes. Fig. 4.2b shows some shapes that are commonly encountered in practice. Table 4.2 presents the results of a study [HoDu83] to calculate the resistances of these shapes for different dimension ratios. This shape information may also be used to estimate the effective w/l of odd shaped transistors [GiBo83]. A few precautions need to be taken, however, especially concerning which side of a shape is the source or drain. The values shown in Table 4.2 best approximate the linear region of operation of an MOS transistor. Note that contacts and vias also have a resistance associated with them. As contacts are reduced in size, the associated resistance increases. Typical values for processes currently in use range from .25Ω to 100Ω.

TABLE 4.2. Resistance of test shapes

SHAPE	RATIO	RESISTANCE
A	1	1
A	5	5
B	1	2.5
B	1.5	2.55
B	2	2.6
B	3	2.75
C	1.5	2.1
C	2	2.25
C	3	2.5
C	4	2.65
D	1	2.2
D	1.5	2.3
D	2	2.3
D	3	2.6
E	1.5	1.45
E	2	1.8
E	3	2.3
E	4	2.65

4.3 Capacitance estimation

The dynamic response (e.g., switching speed) of MOS systems are very much dependent on the parasitic capacitances associated with the MOS device and interconnection capacitances that are formed by metal, poly, and diffusion wires (often called "runners") in concert with transistor and conductor resistances. The total load capacitance on the output of an MOS gate is the sum of:

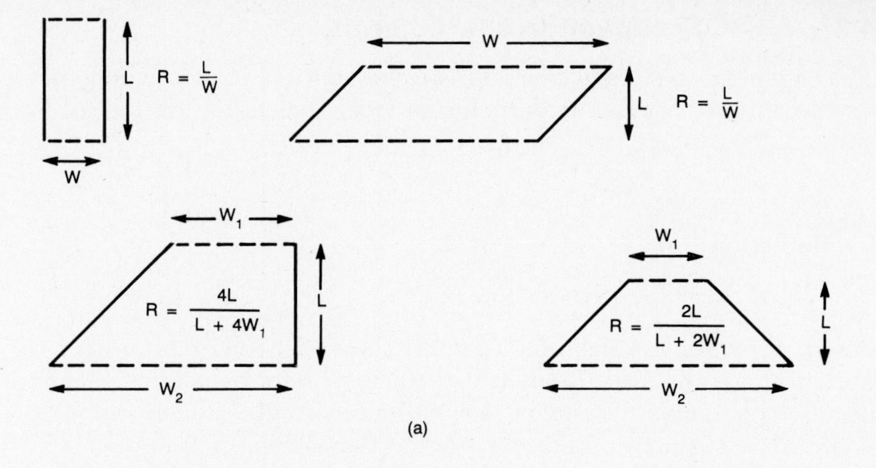

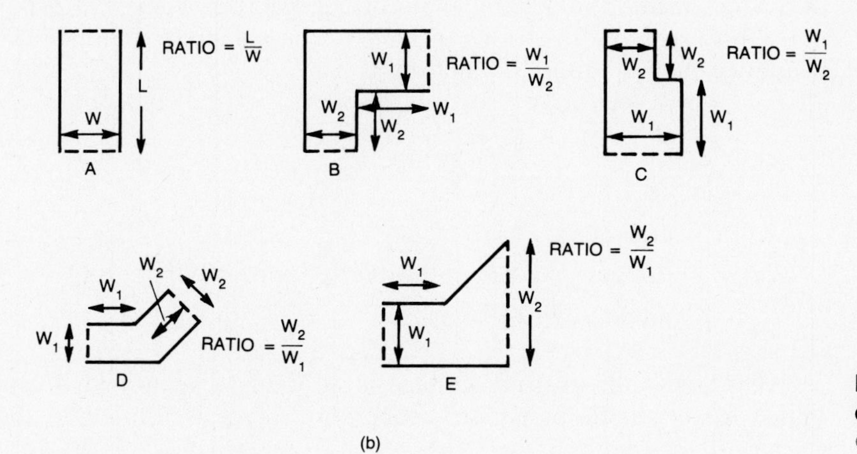

FIGURE 4.2. Resistance of nonrectangular shapes © IEEE 1983 ([HoDu83])

- gate capacitance (of other inputs connected to the output of the gate)
- diffusion capacitance (of the drain regions connected to the output)
- routing capacitance (of connections between the output and other inputs).

Understanding the source of parasitic loads and their variations is essential in the design process, where system performance in terms of the speed of the system form part of the design specification.

We will first examine the characteristics of an MOS capacitor. Following this, the MOS transistor gate capacitance, source/drain capacitance, and routing capacitance will be estimated.

4.3.1 MOS capacitor characteristics

The capacitance-voltage characteristics of an MOS structure depend on the state of the semiconductor surface. Depending on the gate voltage, the surface may be in:

- accumulation
- depletion
- inversion.

Referring to the p-substrate structure shown in Fig. 4.3, an accumulation layer is formed when $V_G < 0$ ($V_G > 0$ for n-substrate). The negative charge on the gate attracts holes toward the silicon surface. When an accumulation layer is present the MOS structure behaves like a parallel plate capacitor. The gate conductor forms one plate of the capacitor. The high concentration of *holes* in a p-substrate (n-device) forms the second plate of a capacitor. Since the *accumulation* layer is directly connected to the substrate, the gate capacitance may be approximated by

$$C_o = \left(\frac{\varepsilon_{SiO_2}\varepsilon_o}{t_{ox}}\right) \cdot A, \tag{4.5}$$

where

A = area of gate

ε_{SiO_2} = dielectric constant (or relative permittivity of SiO_2 taken as 3.9).

When a small positive voltage is applied to an n-device gate with respect to the substrate, a *depletion* layer is formed in the

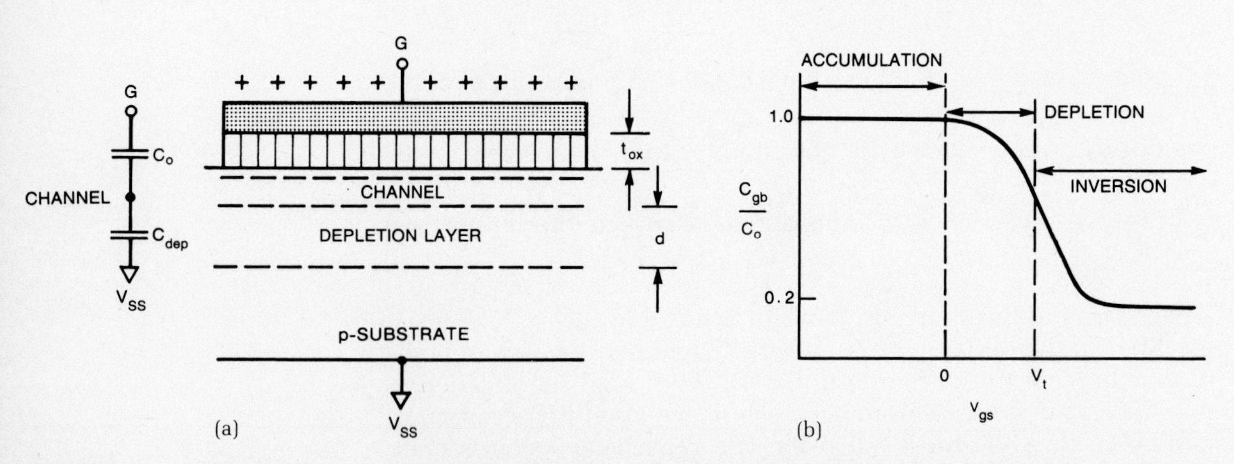

FIGURE 4.3. MOS capacitance (a) physical structure and (b) variation as a function of V_{gs}

p-substrate directly under the gate. The positive gate voltage repels holes leaving a negatively charged region depleted of carriers. A corresponding effect occurs in an n-substrate device for a small negative gate voltage.

Since the magnitude of the charge density per unit area in the surface depletion region is dependent on the doping concentration (N), electronic charge (q), and the depth of the surface depletion region (d), increasing the gate to substrate voltage also increases d. C_{dep}, the depletion capacitance, is given by

$$C_{dep} = \left(\frac{\varepsilon_o \varepsilon_{Si}}{d}\right) \cdot A, \qquad \textbf{(4.6)}$$

where

d = depletion layer depth

ε_{Si} = dielectric constant of silicon taken as 12.

Thus as the depth of the depletion region increases, the capacitance from gate to substrate will decrease. The total capacitance from gate to substrate under depletion conditions can be regarded as that being due to the gate oxide capacitance, C_o in series with C_{dep}; specifically

$$C_{gb} = \frac{C_o C_{dep}}{C_o + C_{dep}}. \qquad \textbf{(4.7)}$$

As the gate voltage is further increased, minority carriers (electrons for the p-substrate) are attracted toward the surface. This effectively inverts the silicon at the surface and creates an n-type channel. Surface inversion yields a relatively high conductivity layer under the gate, which restores the low frequency capacitance to C_o. Because of the limited supply of carriers (electrons) to the inversion layer, the surface charge is not able to track fast moving gate voltages. Hence the dynamic capacitance remains the same as for the maximum depletion situation.

$$C_{gb} = C_o; \qquad \textit{low frequency } (<100\text{Hz})$$

$$= \frac{C_o C_{dep}}{C_o + C_{dep}}; \ \textit{dynamic}$$

Fig. 4.3b summarizes dynamic gate capacitance as a function of gate voltage.

4.3.2 MOS device capacitances

So far we have considered the MOS gate in isolation. Fig. 4.4 on page 126 is a diagrammatic representation of the parasitic capacitances of an MOS transistor. In this model and in the subsequent analysis the overlap of the gate over the drain and source is assumed to be

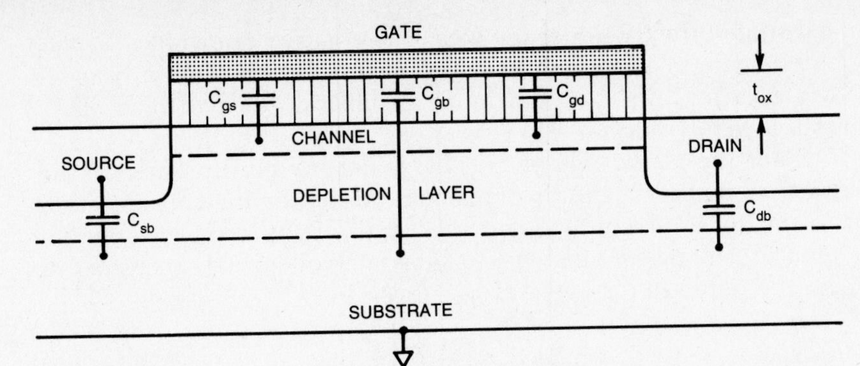

FIGURE 4.4. Representation of parasitic capacitance for an MOS transistor

zero, a simplification that is valid to a first order in self-aligned silicon gate processes.

In Fig. 4.4, the following capacitive components have been identified:

C_{gs}, C_{gd} = gate to channel capacitances, which are lumped at source and drain regions of channel, respectively.

C_{sb}, C_{db} = source and drain diffusion capacitances to bulk (or substrate).

C_{gb} = gate to bulk capacitance.

It is now possible to view the model in terms of circuit symbols. This is illustrated in Fig. 4.5. The total gate capacitance C_g of an MOS transistor is given by

$$C_g = C_{gb} + C_{gs} + C_{gd}. \tag{4.8}$$

The behavior of the gate capacitance of an MOS device can be explained in terms of the following simple models in the three regions of operation:

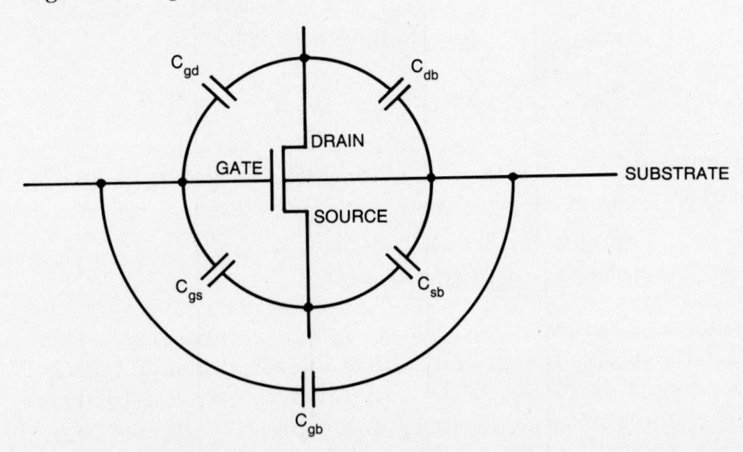

FIGURE 4.5. Circuit symbols for parasitic capacitance

TABLE 4.3. Approximation of intrinsic MOS gate capacitance

PARAMETER	CAPACITANCE		
	Off	Linear	Saturation
C_{gb}	$\frac{\varepsilon A}{t_{ox}}$	0	0
C_{gs}	0	$\frac{1}{2}\left(\frac{\varepsilon A}{t_{ox}}\right)$	$\frac{2}{3}\left(\frac{\varepsilon A}{t_{ox}}\right)$
C_{gd}	0	$\frac{1}{2}\left(\frac{\varepsilon A}{t_{ox}}\right)$	0
$C_g = C_{gb} + C_{gs} + C_{gd}$	$\frac{\varepsilon A}{t_{ox}}$	$\frac{\varepsilon A}{t_{ox}}$	$\frac{2}{3}\left(\frac{\varepsilon A}{t_{ox}}\right)$

Note: $\varepsilon = \varepsilon_0 \varepsilon_{SiO_2}$

1 Off region, where $V_{gs} \leq V_t$. When the MOS device is "OFF", there is no channel, and hence $C_{gs} = C_{gd} = 0$. C_{gb} can be modeled as the series combination of the two capacitors (C_o and C_{dep}), as shown in Fig. 4.3.

2 Linear region, where $V_{gs} - V_t \geq V_{ds}$. In this region, the depletion layer depth remains relatively constant. C_{gb}, therefore, remains constant. As a result of the formation of the channel, gate to channel capacitances C_{gs} and C_{gd} now become significant. These capacitances are dependent on gate voltage. Their values can be conservatively estimated as

$$C_{gd} = C_{gs} = \frac{1}{2}\left(\frac{\varepsilon_o \varepsilon_{SiO_2}}{t_{ox}}\right) \cdot A. \tag{4.9}$$

3 Saturation region, where $V_{gs} - V_t < V_{ds}$. In this mode the channel is heavily inverted. The drain region of the channel is pinched-off, causing C_{gd} to be zero. C_{gs} increases to approximately

$$\frac{2}{3}\left(\frac{\varepsilon_o \varepsilon_{SiO_2}}{t_{ox}}\right) \cdot A.$$

The behavior of the input capacitances in the three regions of operation can be approximated as shown in Table 4.3. If the gate does overlap the drain and source (as in a metal gate process), then a fixed parasitic contribution due to the overlap area and separation must be added to C_{gs} and C_{gd}. Note that although some of the components of the gate capacitance are highly voltage dependent, the overall gate capacitance (for an n-device), as shown in Fig. 4.6 on page 128, is approximately equal to the intrinsic "gate-oxide" capacitance for all values of gate voltage. The only region where

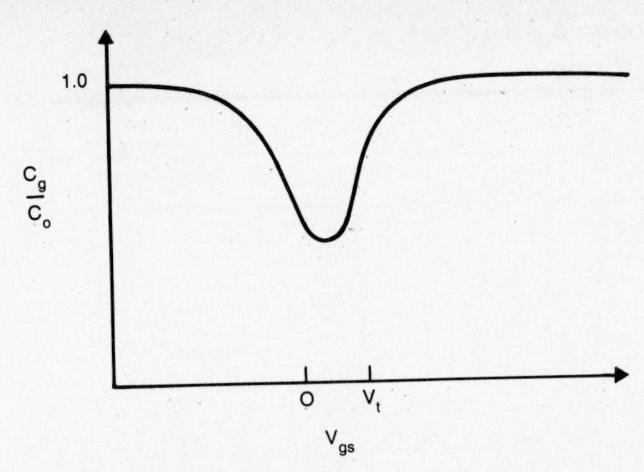

FIGURE 4.6. Total gate capacitance as a function of V_{gs}

this does not hold is around the threshold voltage of the transistor (V_t). Since transistors in digital circuits switch through this region rapidly, we can conservatively approximate $C_g = C_o$.

Another way of stating this approximation is

$$C_g = C_{ox} \cdot A, \tag{4.10}$$

where C_{ox} is in the "thin oxide" capacitance per unit area given by

$$C_{ox} = \frac{\varepsilon_o \varepsilon_{SiO_2}}{t_{ox}}. \tag{4.11}$$

With a thin-oxide thickness in the order of 500–1000 Å, and the relative permittivity for SiO_2 approximated as 4, the value of C_{ox} is

$$C_{ox} = \frac{4 * 8.854 * 10^{-4}}{(500 - 1000) * 10^{-8}}$$
$$\approx (8 - 4) * 10^{-4}\ pF/\mu m^2.$$

Approximation of the gate capacitance may now be undertaken by simply taking the above value and multiplying it by the gate area. For example, the input (or gate) capacitance for a typical MOS transistor shown in Fig. 4.7, with $\lambda = 2\ \mu m$ and $t_{ox} = 1000$ Å, is

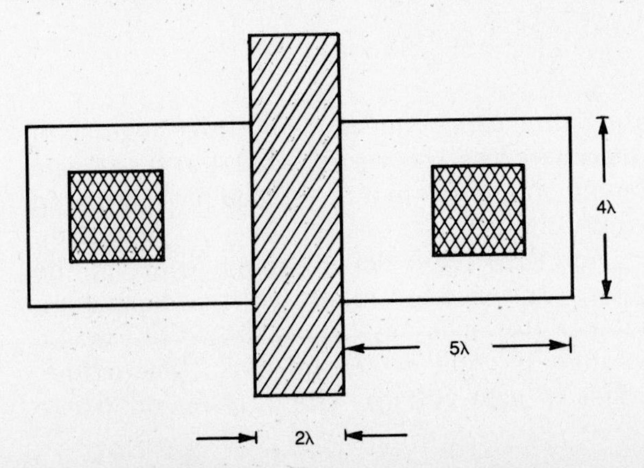

FIGURE 4.7. Physical layout of an MOS transistor for capacitance estimation

$$C_g = 3.5 * 10^{-4} * 8\lambda^2 pF$$
$$= 11.2\, fF.$$

We will refer to this transistor as a "unit transistor" — a transistor that can be conveniently connected to metal at source and drain. It is the same width as a metal-diffusion contact.

4.3.3 Diffusion capacitance

Shallow n^+ and p^+ diffusions form the source and drain terminals of n- and p-channel devices. Diffusion regions are also used as wires. All diffusion regions have a capacitance to substrate that depends on the voltage between the diffusion regions and substrate (or well), as well as on the effective area of the depletion region separating diffusion and substrate (or well). The diffusion capacitance C_d is proportional to the total diffusion to substrate junction area. As shown in Fig. 4.8, this is a function of "base" area and also of the

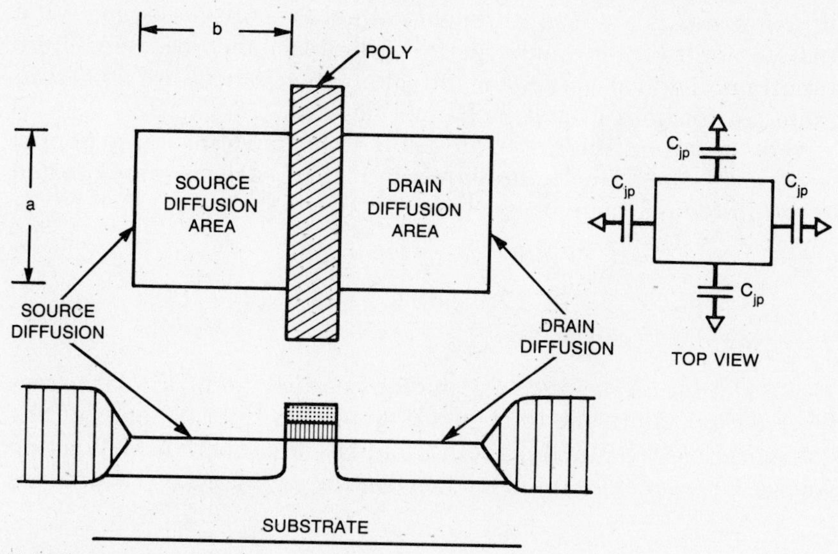

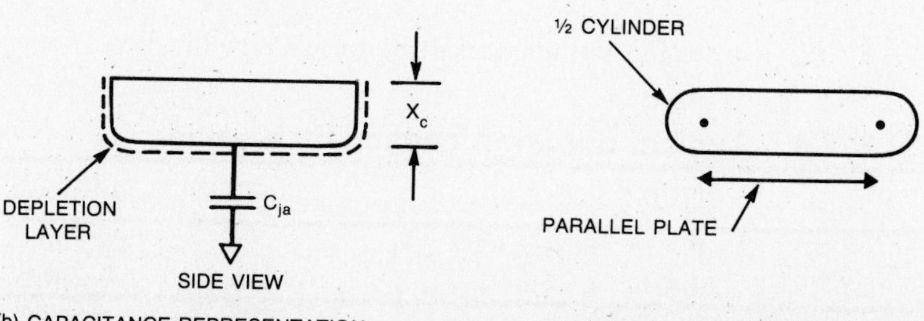

FIGURE 4.8. Area and peripheral components of diffusion capacitance

area of the "sidewall" periphery. The latter occurs because the diffusion region has a finite depth. Sidewall capacitance can be characterized (assuming constant depth diffusion) by a periphery capacitance per unit length. The model generally used is shown in Fig. 4.8c. Total C_d can be represented by

$$C_d = C_{ja} * (ab) + C_{jp} * (2a + 2b), \tag{4.12}$$

where

C_{ja} = junction capacitance per sq. μm

C_{jp} = periphery capacitance per μm

a = width of diffusion region

b = extent of diffusion region.

Note that the capacitance contributed by the sidewall facing the channel will be reduced somewhat by the presence of the channel depletion region.

An obvious factor that emerges from Eq. (4.12) is that as the diffusion area is reduced (through scaling, to be discussed later) the relative contribution of the peripheral capacitance becomes more important. Typical values for diffusion capacitances are shown in Table 4.4 for both n- and p-channel devices.

For example, the drain diffusion capacitance of an n-channel device with the dimensions shown in Fig. 4.7 may be approximated in the following manner:

$$\begin{aligned} C_d &= C_{ja} * (ab) + C_{jp} * (2a + 2b) \\ &= 1 * 10^{-4} (10 * 8) + 9 * 10^{-4} (20 + 16) \\ &\approx 40\, fF. \end{aligned}$$

These simple capacitance calculations assume zero DC bias across the junction. Since the thickness of depletion layer depends on the voltage across the junction, both C_{ja} and C_{jp} are functions of junction voltage V_j. A general expression that describes the junction capacitance is

$$C_j = C_{j0}\left(1 - \frac{V_j}{\phi_B}\right)^{-m}, \tag{4.13}$$

where

V_j = junction voltage (negative for reverse bias)

TABLE 4.4. Typical diffusion capacitance values

	n-DEVICE (OR WIRE)	p-DEVICE (OR WIRE)
C_{ja}	$1 * 10^{-4} pF/\mu m^2$	$1 * 10^{-4} pF/\mu m^2$
C_{jp}	$9 * 10^{-4} pF/\mu m$	$8 * 10^{-4} pF/\mu m$

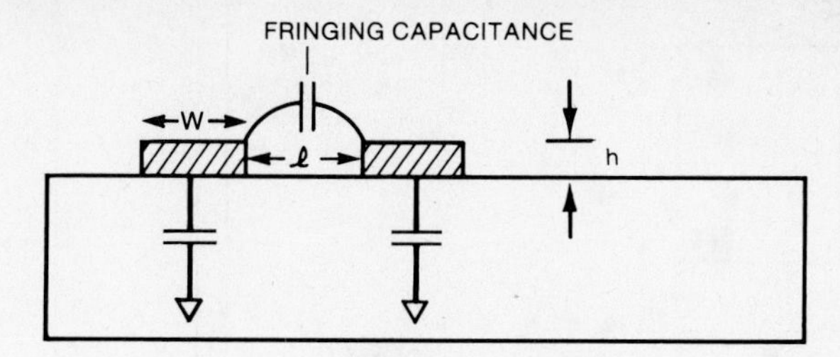

FIGURE 4.9. Effect of fringing fields on capacitance

C_{j0} = zero bias capacitance; (V_j = 0)

ϕ_B = built-in junction potential ≈ 0.6 volts

and m is a constant, which depends on the distribution of impurities near the junction, and has a value of the order of 0.3 to 0.5.

4.3.4 Routing capacitance

Routing capacitances between metal and poly layers and the substrate can be approximated using a parallel plate model ($C = \frac{\varepsilon}{t}A$), where A is area of the parallel plate capacitor, t is the insulator thickness, and ε is the dielectric constant of the insulating material between the plates. The parallel-plate approximation, however, ignores fringing fields. The effect of fringing fields is to increase the effective area of the plates. Consequently, poly and metal lines will actually have a higher capacitance (up to twice as large) than that predicted by the model. Interlayer capacitance such as metal-poly capacitance is also enhanced by fringing. Fig. 4.9 illustrates this effect. As line widths are scaled, the width (w) and heights of wires tend to reduce less than their separations (l). Accordingly, this fringing effect increases in importance. For current processes, a factor of 1.5–3 should be used. Methods for more accurately computing the fringing factor can be found in [RuBr73].

Another factor, which should be taken into account for small geometries when using the parallel plate model, is that a drawn shape (on mask) will not be the same as the actual physical shape produced on silicon. This effect, shown in Fig. 4.10 on page 132, is most pronounced for diffusion regions and the extent may be determined analytically or empirically through experimentation.

4.3.5 Distributed RC effects

The propagation of a signal along a wire depends on many factors, including the distributed resistance and capacitance of the wire, the impedance of the driving source, and the load impedance. For

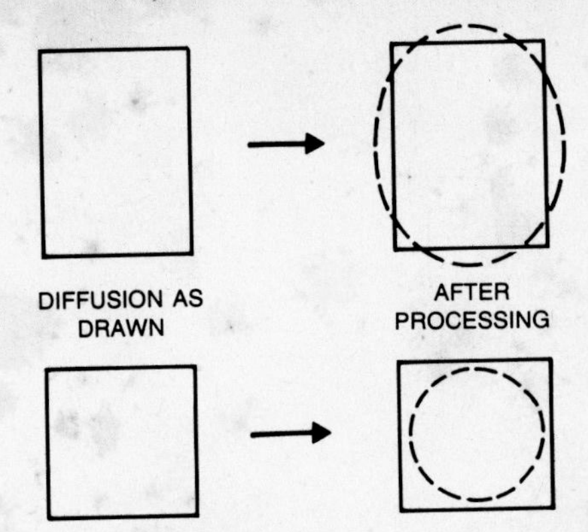

FIGURE 4.10. Effect of processing on drawn geometry

very long wires propagation delays caused by distributed resistance-capacitance (RC) in the wiring layer tend to dominate. This transmission line effect is particularly severe in poly wires because of the relatively high resistance of this layer. A long wire can be represented in terms of several RC sections, as shown in Fig. 4.11.

The response at node V_j with respect to time is then given by

$$C\frac{dV_j}{dt} = (I_{j-1} - I_j) \tag{4.14}$$

$$= \frac{(V_{j-1} - V_j)}{R} - \frac{(V_j - V_{j+1})}{R}.$$

As the number of sections in the network becomes large (and the sections become small), the above expression reduces to the differential form:

$$rc\frac{dV}{dt} = \frac{d^2V}{dx^2}, \tag{4.15}$$

where

x = distance from input

r = resistance per unit length

c = capacitance per unit length.

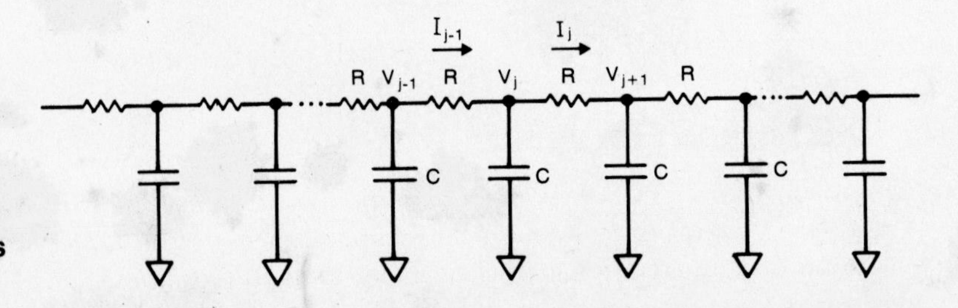

FIGURE 4.11. Representation of long wire in terms of distributed RC sections

The form of this relation is that of the well known diffusion equation. The solution for the propagation of a voltage step along the wire shows that the propagation time t_x over a wire of length x is

$$t_x = k.x^2, \tag{4.16}$$

where k is a constant. Alternatively, a discrete analysis of the circuit shown in Fig. 4.11 yields an approximate signal delay of

$$t_n = \frac{RC\, n(n + 1)}{2}, \tag{4.17}$$

where

n = number of sections.

As n becomes very large (i.e., the individual sections become very small), this reduces to

$$t_l = \frac{rc\, l^2}{2}, \tag{4.18}$$

where

r = resistance per unit length
c = capacitance per unit length
l = length of the wire.

The l^2 term in Eq. (4.18) shows that signal delay will be totally dominated by this RC effect for very long signal paths. In order to optimize speed in a long poly line, one possible strategy is to segment the line into several sections and insert buffers within these sections. Fig. 4.12 shows a poly bus of length 2 mm that has been divided into two 1 mm sections. For $r = 12\ \Omega/\mu m$ and $c = 4 \times 10^{-4}\ pf/\mu m$, Eq. (4.18) yields

$$t_l = 2.4 \times 10^{-15}\ l^2.$$

Assuming that the delay associated with the buffer is τ_{buf}, the total delay for this bus is

$$\begin{aligned} t_p &= 2.4 \times 10^{-15} \times (1000)^2 + \tau_{buf} \\ &= 2.4\ ns + \tau_{buf}. \end{aligned}$$

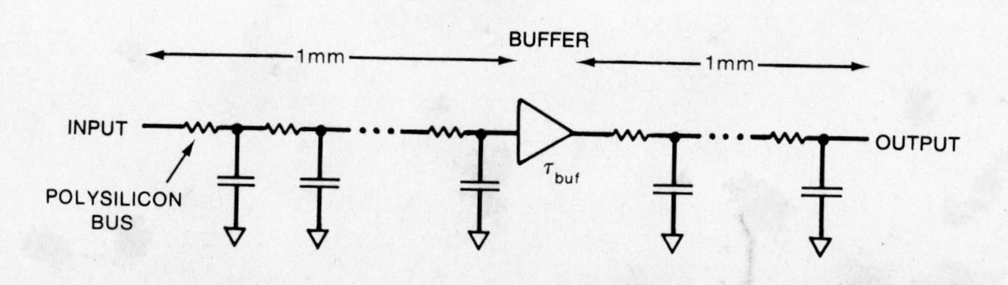

FIGURE 4.12. Segmentation of polysilicon line

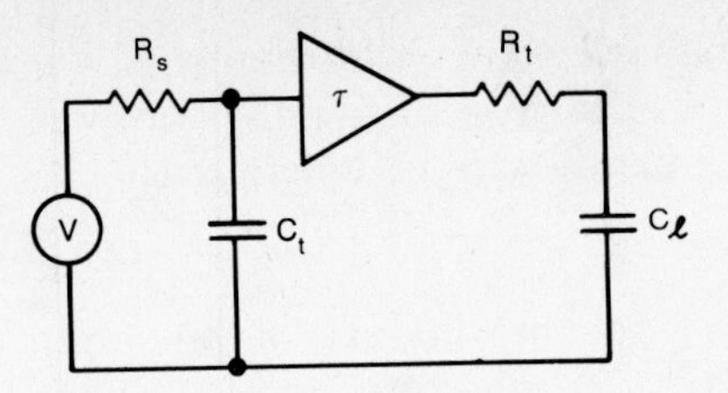

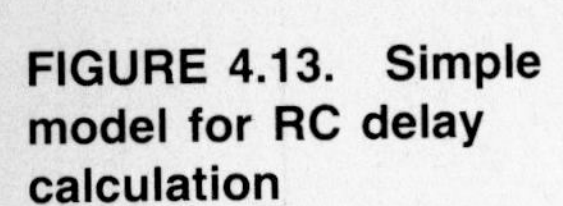
FIGURE 4.13. Simple model for RC delay calculation

This may be contrasted to the situation in which the buffer is missing, which yields

$$t_p = 9.6 \ ns.$$

Thus by keeping τ_{buf} small, significant gain can be obtained through appropriately segmenting the bus. τ_{buf} does, in fact, depend on the resistance of the first section of the bus and on the capacitance of the second section of the bus. The relative importance of these two terms depends on other circuit parameters such as final load capacitance. In some situations, it may be preferable to use a wide poly wire to reduce overall series resistance at the expense of capacitance.

When the RC product of a long line is high, it might not be possible to compensate by simply adding drivers. This is where technologies such as double metal layer prove useful. Here, the second metal layer may be arranged to run in place of the polysilicon line. Poly is still used for local gate connections. Another approach that is also effective in reducing the influence of RC effects is to use silicides (tantalum, molybdenum) with sheet resistances in the order of 2–4 ohm/sq. This approach provides considerable improvement in the RC delay associated with a line and is argued to be as effective as a second metal layer while maintaining compatibility with a single metal process.

A model for the distributed RC delay, which takes driver and receiver loading into account, is shown in Fig. 4.13. R_s is the output resistance of the driver. C_l is the receiver input capacitance. R_t and C_t are the total lumped resistance and capacitance of the line. τ is the RC delay calculated using Eq. (4.18). Such a model yields results that are very economical in terms of computation and, more importantly, are accurate enough for most purposes. The concept of using RC time constants for delay estimations is based upon the assumption that the time taken for a signal to reach 63 percent of its final value approximates the switching point of an inverter.

4.3.6 Capacitance design guide

As a guide to the design process and, in particular, to the choice of layers, Table 4.5 is provided. It shows a range of capacitance values (no fringing) for a typical 4 μm ($\lambda = 2 \ \mu m$) silicon gate

TABLE 4.5. Typical 4 μm silicon gate CMOS process capacitances

PARAMETER	MIN.	MAX.	COMMENTS
$C_g(pF/\mu m^2)$	$4.0 * 10^{-4}$	$5.0 * 10^{-4}$	Gate
$C_p(pF/\mu m^2)$	$0.4 * 10^{-4}$	$0.6 * 10^{-4}$	Polysilicon over field
$C_{mp}(pF/\mu m^2)$	$0.4 * 10^{-4}$	$0.6 * 10^{-4}$	Metal over poly
$C_{mf}(pF/\mu m^2)$	$0.15 * 10^{-4}$	$0.3 * 10^{-4}$	Metal over field
$C_{md}(pF/\mu m^2)$	$0.8 * 10^{-4}$	$1.0 * 10^{-4}$	Metal over diffusion; (p^+ *AND* n^+)
$C_{ja_n}(pF/\mu m^2)$	$0.8 * 10^{-4}$	$1.0 * 10^{-4}$	n-diffusion
$C_{ja_p}(pF/\mu m^2)$	$0.8 * 10^{-4}$	$1.0 * 10^{-4}$	p-diffusion
$C_{jp_n}(pF/\mu m)$	$7.0 * 10^{-4}$	$9.0 * 10^{-4}$	n-channel device
$C_{jp_p}(pF/\mu m)$	$6.0 * 10^{-4}$	$8.0 * 10^{-4}$	p-channel device

CMOS process. Table 4.6 shows typical capacitance values for the second level metal in a two metal process. These are included for comparison purposes. Since the second level metal capacitance is governed by the planarization layer thickness, it is strongly dependent upon the process. Furthermore, fringing fields are quite small and generally can be neglected for current technologies. An example showing the manner in which detailed parasitic capacitances may be calculated is shown in Fig. 4.14 on page 136. Substituting in some typical values ($\lambda = 2\ \mu m$) yields:

$$\text{metal; } C_{mf} = (3\lambda * 100\lambda)0.3 * 10^{-4} = 0.036pF$$

$$\text{poly; } C_p = [(4\lambda * 4\lambda) + (\lambda + 2\lambda) * 2\lambda]0.6 * 10^{-4} = 0.0053pF$$

$$\text{gate; } C_g = [2\lambda * 2\lambda]5.0 * 10^{-4} = 0.008pF.$$

Therefore the total capacitance is

$$C_T = C_{mf} + C_p + C_g$$
$$\approx 0.049\text{pF}.$$

It is important to be able to estimate capacitances before any detailed layout is completed. For each process, it is useful to have an approximate figure for the gate capacitance of a unit size n- and p-transistor, the capacitance of 100 μm of poly wire, etc. In this

TABLE 4.6. Typical 4 μm second level metal CMOS process capacitances

PARAMETER	MIN.	MAX.	COMMENTS
$C_{m2f}(pF/\mu m^2)$	$0.1 * 10^{-4}$	$0.15 * 10^{-4}$	Metal 2 to substrate
$C_{m2p}(pF/\mu m^2)$	$0.2 * 10^{-4}$	$0.3 * 10^{-4}$	Metal 2 to poly
$C_{m21}(pF/\mu m^2)$	$0.3 * 10^{-4}$	$0.5 * 10^{-4}$	Metal 2 to metal 1

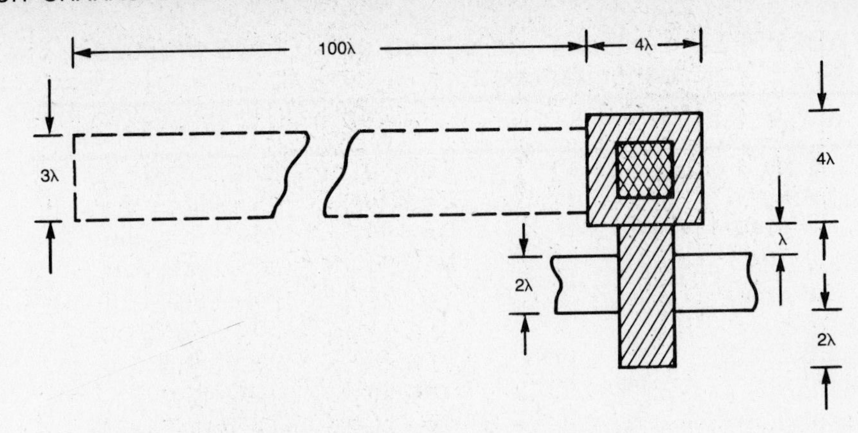

FIGURE 4.14. Example of parasitic capacitance calculation for λ = 2μm

way, bus loadings and other critical parasitics can be estimated to a first order without resorting to any detailed analysis.

4.3.7 Wire length design guide

For the purposes of timing analysis, an electrical node may be defined as that region of connected paths in which the delay associated with signal propagation is small in comparison with gate delays. For sufficiently small wire lengths, RC delays can be ignored. Wires can then be treated as one electrical node and modeled as simple capacitive loads. It is therefore useful to define simple electrical rules that can be used as a guide in determining the maximum length of communication paths for the various interconnect levels. To do this we require that wire delay τ_ω and gate delay τ_g satisfy the following condition:

$$\tau_\omega \ll \tau_g. \quad \textbf{(4.19)}$$

On substituting Eq. (4.18) into Eq. (4.19), we obtain the result

$$l \ll \sqrt{\frac{2\tau_g}{rc}}. \quad \textbf{(4.20)}$$

This establishes an upper bound on the allowable length of the interconnects where the above approximations are valid. The electrical rules governing interconnect paths for a typical process are illustrated in Table 4.7 in terms of λ (as used in specification of design rules). This table assumes gate delays of the order of 1.5 nS to 2.0 nS and uses the values in Table 4.1 and Table 4.5 in conjunction with Eq. (4.20). For example, for aluminum,

$$l = \sqrt{\frac{2 * 2.0 * 10^{-9}(\tau_g)\,\lambda^2}{.03(r) * .3 * 10^{-16}(c)}}$$
$$\approx 60000\lambda.$$

TABLE 4.7. Guidelines for ignoring RC wire delays

LAYER	MAXIMUM LENGTH
Metal	20,000 λ
Silicide	2,000 λ
Poly	200 λ
Diffusion	20 λ

So, conservatively,

$$l < 20000\lambda.$$

The significant factor that emerges from the table is the difference in tolerable communication distance between metal and polysilicon. Although the rules shown in Table 4.7 are very conservative, they are simple to remember and satisfy most of the CMOS processes that are currently available. They may be simply rederived, for future processes, using Eq. (4.20).

4.4 Switching characteristics

The switching speed of a CMOS gate is limited by the time taken to charge and discharge the load capacitance C_L. An input transition results in an output transition that either charges C_L towards V_{DD} or discharges C_L towards V_{SS}.

In this section, we develop simple models that describe the switching characteristics of a CMOS inverter. Before proceeding, however, we need to define some terms. Referring to Fig. 4.15, on page 138:

- Rise time; t_r = time for a waveform to rise from 10 percent to 90 percent of its steady-state value.
- Fall time; t_f = time for a waveform to fall from 90 percent to 10 percent of its steady-state value.
- Delay time; t_d = time difference between input transition (50 percent level) and the 50 percent output level. (This is the time taken for a logic transition to pass from input to output.)

Fig. 4.15a shows the familiar CMOS inverter with a capacitive load C_L that represents the load capacitance (input of next gates, output of this gate and routing). Of interest is the voltage waveform $V_o(t)$ when the input is driven by a step waveform $V_{in}(t)$, as shown in Fig. 4.15b.

4.4.1 Fall time determination

Fig. 4.16, on page 138, shows the trajectory of the n-transistor operating point as the input voltage, $V_{in}(t)$, changes from zero volts to V_{DD}. Initially, the n-device is cut-off and the load capacitor, C_L, is charged to V_{DD}. This is illustrated by X_1 on the characteristic curve. Application of a step voltage (i.e., $V_{gs} = V_{DD}$) at the input of the inverter changes the operating point to X_2. From there onwards, the trajectory moves

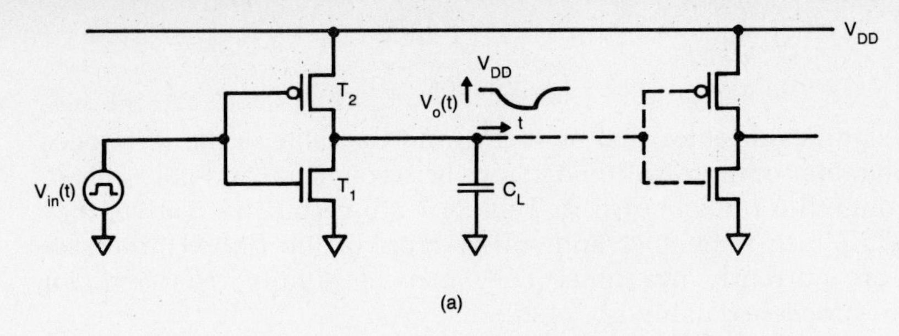

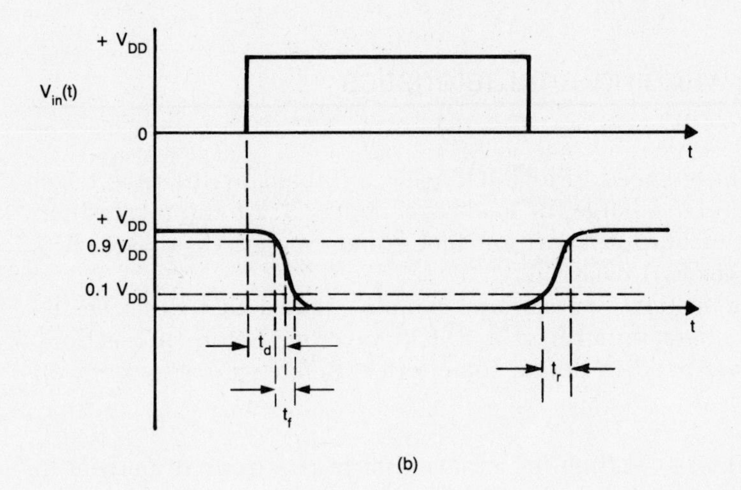

FIGURE 4.15. Switching characteristic for CMOS inverter

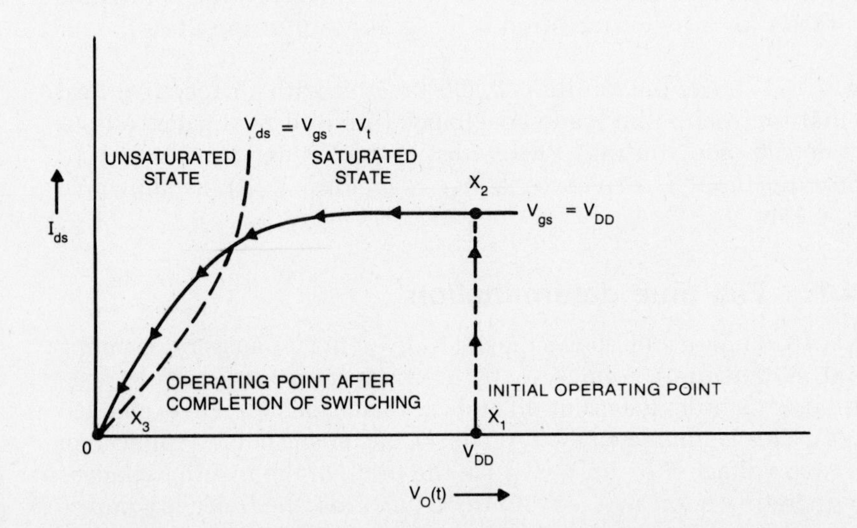

FIGURE 4.16. Trajectory of n-transistor operating point during switching

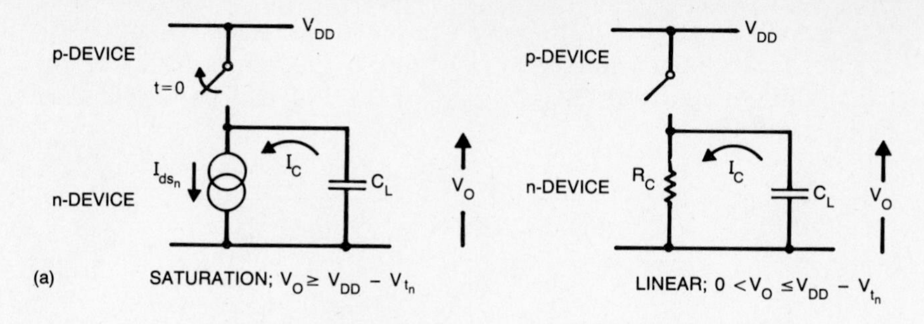

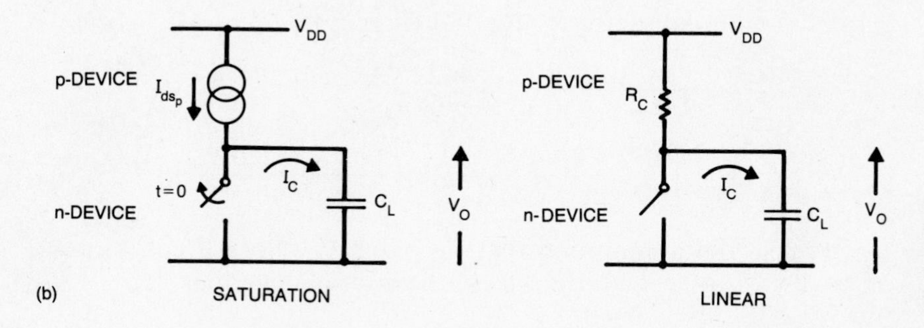

FIGURE 4.17. Equivalent circuits for fall and rise time determination

on the $V_{gs} = V_{DD}$ characteristic curve towards point X_3 at the origin. From the switching characteristics shown in Fig. 4.15, it is evident that the fall time, t_f, consists of two intervals:

1 t_{f1} = period during which the capacitor voltage, V_o, drops from $0.9\ V_{DD}$ to $(V_{DD} - V_{t_n})$.
2 t_{f2} = period during which the capacitor voltage, V_o, drops from $(V_{DD} - V_{t_n})$ to $0.1\ V_{DD}$.

The equivalent circuits that illustrate the above behavior are shown in Fig. 4.17. From Fig. 4.17a, while in saturation

$$C_L \frac{dV_o}{dt} + \frac{\beta_n}{2}(V_{DD} - V_{t_n})^2 = 0; \qquad V_o \geqslant V_{DD} - V_{t_n}. \tag{4.21}$$

Integrating from $t = t_1$, corresponding to $V_o = 0.9\ V_{DD}$, to $t = t_2$ corresponding to $V_o = (V_{DD} - V_{t_n})$ results in

$$\begin{aligned} t_{f1} &= 2\frac{C_L}{\beta_n(V_{DD} - V_{t_n})^2}\int_{V_{DD}-V_{tn}}^{0.9V_{DD}} dV_o \\ &= \frac{2C_L(V_{t_n} - 0.1\ V_{DD})}{\beta_n(V_{DD} - V_{t_n})^2}. \end{aligned} \tag{4.22}$$

When the n-device begins to operate in the linear region, the discharge current is no longer constant. The time, t_{f2}, taken to discharge the capacitor voltage from $(V_{DD} - V_{t_n})$ to 0.1 V_{DD} can be obtained as before, giving

$$t_{f2} = \frac{C_L}{\beta_n(V_{DD} - V_{t_n})} \int_{0.1V_{DD}}^{V_{DD}-V_{tn}} \frac{dV_o}{\frac{V_o^2}{2(V_{DD} - V_{t_n})} - V_o} = \frac{C_L}{\beta_n(V_{DD} - V_{t_n})} \ln\left(\frac{19\ V_{DD} - 20\ V_{t_n}}{V_{DD}}\right). \quad (4.23)$$

Thus the complete term for the fall time, t_f is

$$t_f = 2\frac{C_L}{\beta_n(V_{DD} - V_{t_n})} \times \left[\frac{V_{t_n} - 0.1\ V_{DD}}{V_{DD} - V_{t_n}} + \frac{1}{2}\ln\left(\frac{19\ V_{DD} - 20\ V_{t_n}}{V_{DD}}\right)\right]. \quad (4.24)$$

If we make the assumption that $V_{t_n} \approx 0.2\ V_{DD}$ (in a 5 volt process $V_{t_n} \approx 1$ v $V_{t_p} \approx -1$ v), then t_f can be approximated as

$$t_f \approx 4\frac{C_L}{\beta_n V_{DD}}. \quad (4.25)$$

4.4.2 Rise time

Due to the symmetry of the CMOS circuit, a similar approach may be used to obtain the rise time, t_r (Fig. 4.17b). Thus

$$t_r = 2\frac{C_L}{\beta_p(V_{DD} - |V_{t_p}|)} \times \left[\frac{|V_{t_p}| - 0.1\ V_{DD}}{V_{DD} - |V_{t_p}|} + \frac{1}{2}\ln\left(\frac{19\ V_{DD} - 20|V_{t_p}|}{V_{DD}}\right)\right]. \quad (4.26)$$

As before, with $|V_{t_p}| \approx 0.2\ V_{DD}$, Eq. (4.26) reduces to

$$t_r \approx 4\frac{C_L}{\beta_p V_{DD}}. \quad (4.27)$$

For equally sized n- and p-transistors, where $\beta_n = 2\beta_p$

$$t_f = \frac{t_r}{2}. \quad (4.28)$$

Thus the fall time is faster than the rise time, primarily due to different carrier mobilities associated with the p- and n-devices (i.e.,

$\mu_n \approx 2\mu_p$). Therefore, if we want to have approximately the same rise and fall time for an inverter we need to make

$$\frac{\beta_n}{\beta_p} = 1.$$

This implies that the channel width for the p-device must be increased to approximately two times that of the n-device, so

$$W_p = 2W_n.$$

Note that to accurately specify the width ratio required to achieve equal rise and fall times, an accurate ratio of μ_n and μ_p must be known. These, in turn, depend on process parameters.

4.4.3 Delay time τ_d

In an MOS circuit, the delay of a single gate is dominated by the output rise and fall time. The delay is approximately given by

$$t_{dr} = \frac{t_r}{2}$$
$$t_{df} = \frac{t_f}{2}. \qquad (4.29)$$

The average gate delay for rising and falling transitions is then

$$\tau_{av} = \frac{t_{df} + t_{dr}}{2}$$
$$= \frac{t_r + t_f}{4}. \qquad (4.30)$$

4.5 CMOS gate transistor sizing

4.5.1 Similar stage loads

The discussions so far have led us to believe that if we want to have approximately the same rise and fall times for an inverter, we must make

$$W_p \approx 2W_n,$$

where W_p is the channel width of the p-device and W_n is the channel width of the n-device. This, of course, increases layout area and, as we shall see later, dynamic power dissipation. In some cascaded

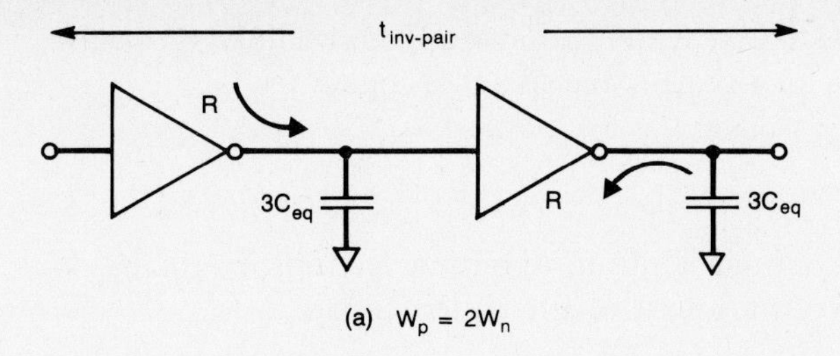

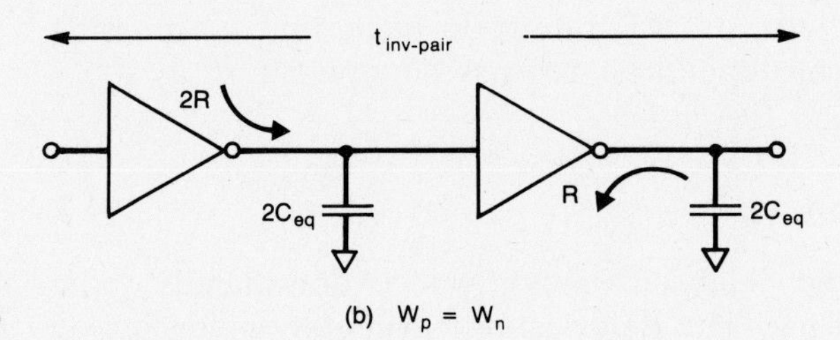

FIGURE 4.18. CMOS inverter pair timing response

structures it is possible to use minimum size devices without compromising the switching response. This is illustrated in the following analysis, in which the delay response for an inverter pair (Fig. 4.18a) with $W_p = 2W_n$ is given by

$$\begin{aligned} t_{inv-pair} &= t_{fall} + t_{rise} \\ &= R.3C_{eq} + 2\left(\frac{R}{2}\right).3C_{eq} \\ &= 3RC_{eq} + 3RC_{eq} \\ &= 6RC_{eq}, \end{aligned} \tag{4.31}$$

where R is the effective 'on' resistance of a unit-sized n-transistor and $C_{eq} = C_g + C_d$ is the capacitance of a unit-sized gate and drain region. The inverter pair delay with $W_p = W_n$ (Fig. 4.18b) is

$$\begin{aligned} t_{inv-pair} &= 4RC_{eq} + 2RC_{eq} \\ &= 6RC_{eq}. \end{aligned} \tag{4.32}$$

Thus we find similar responses are obtained for the two different conditions.

It is important to remember that changes in β ratio also affect inverter threshold voltage V_{inv}. From Eq. (2.15), the relation defining V_{inv} is given by

$$V_{inv} = \frac{V_{DD} + V_{t_p} + V_{t_n}(\beta_n/\beta_p)^{1/2}}{1 + (\beta_n/\beta_p)^{1/2}}. \tag{4.33}$$

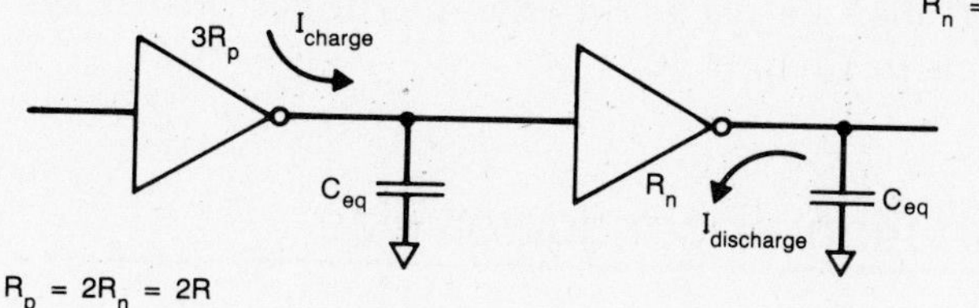

FIGURE 4.19. Pseudo nMOS inverter pair timing response

With $V_{DD} = 5$ volts, $V_{t_n} = 1$ volt, and $V_{t_p} = -1$ volt, we obtain

$$V_{inv} = \begin{cases} 2.5 \text{ volts; } W_p = 2W_n \\ 2.24 \text{ volts; } W_p = W_n, \end{cases} \tag{4.34}$$

which shows less than 10 percent variation in V_{inv} for these two β ratios. Based on all these results it is evident that it is usually preferable (density and power dissipation — see Sec. 4.7.2) to use $W_p = W_n$ when similar structures are cascaded.

4.5.2 Switching performance of the pseudo nMOS inverter

A simple timing model of the pseudo-nMOS inverter introduced in Chapter 2 is shown in Fig. 4.19. This uses the 3:1 transistor width ratios determined in that chapter. The approximate delay for a pair of inverters is

$$\begin{aligned} t_{inv-pair} &= 6R(C_g + 2C_d) + R(C_g + 2C_d) \\ &= 7RC_{eq}, \end{aligned} \tag{4.35}$$

where $C_{eq} = C_g + 2C_d$.

4.5.3 Cascaded stage loads

Often it is desired to drive large load capacitances such as long buses, I/O buffers, or, ultimately, pads and off-chip capacitive loads. A method of determining the transistor ratios needed is as follows:

- Calculate final stage size according to required t_r and t_f and C_L.
- Calculate size and number of intermediate stage sizes needed to drive output for a given optimization criteria (i.e., power, speed, density).

The *stage ratio* is the ratio used to multiply the size of transistor sizes in successive stages. For optimum speed, this ratio is 2.7, but values from 2 to 10 may be used. An example of cascaded stage optimization is given in Chapter 5.

4.6 Determination of conductor size

Electromigration is the transport of metal ions through a conductor resulting from the passage of direct current. It is caused by a modification of the normally random diffusion process to a directional one caused by charge carriers. This can result in the deformation of conductors and subsequent failure of circuitry. Factors that influence electromigration rate are:

- current density
- temperature
- crystal structure.

In determining the minimum size of conductors, particularly those for V_{DD} and V_{SS}, it is necessary to estimate the current density in the conductor. If the current density, J, of a current carrying conductor exceeds a threshold value, then one finds that the conductor atoms begin to dislocate and move in the direction of the current flow. If there is a constriction in the conductor, then the conductor atoms move at a faster rate in the region of the constriction. This results in a weakening of the constriction, which eventually blows like a fuse. For example, the limiting value for 1 μm thick aluminum is

$$J_{AL} \approx 1 \rightarrow 2mA/\mu m.$$

As a rule of thumb, one should use 0.5mA/μm to 1.0mA/μm of metal width for both V_{DD} and V_{SS} lines. Apart from electromigration, voltage drops can occur on power conductors due to IR drop during charging transients. While electromigration usually sets the minimum width of conductors, the need to supply correct V_{DD} and V_{SS} is often the driving consideration. Sometimes the supply conductors cannot be increased to the desired width. For such circumstances, other techniques such as adding extra supply pins to distribute the current flow could be considered. Another consideration is that the current density in a window (cut) periphery must be kept below about 0.1mA/μm. One finds that due to current crowding around the perimeter of a window, a chain of small windows, suitably spaced, generally provides just as much current carrying capacity as a single long, narrow cut.

4.7 Power consumption

There are two components that establish the amount of power dissipated in a CMOS circuit. These are:

1. Static dissipation — due to leakage current.
2. Dynamic dissipation — due to:
   a. switching transient current
   b. charging and discharging of load capacitances.

4.7.1 Static dissipation

Considering a complementary CMOS gate, as shown in Fig. 4.20, if the input = '0', the associated n-device is 'OFF' and the p-device is 'ON'. The output voltage is V_{DD} or logic '1'. When the input =

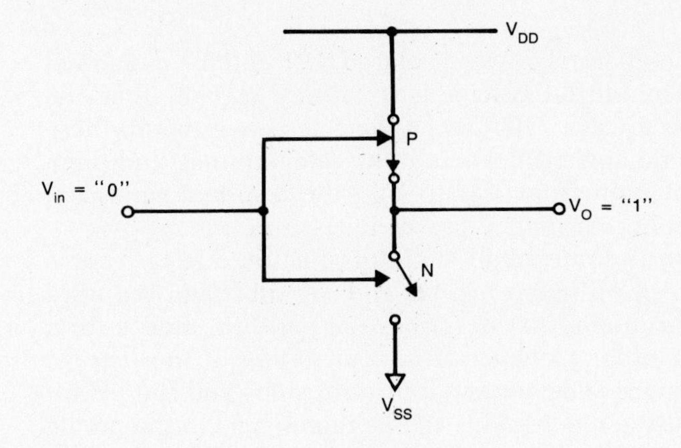

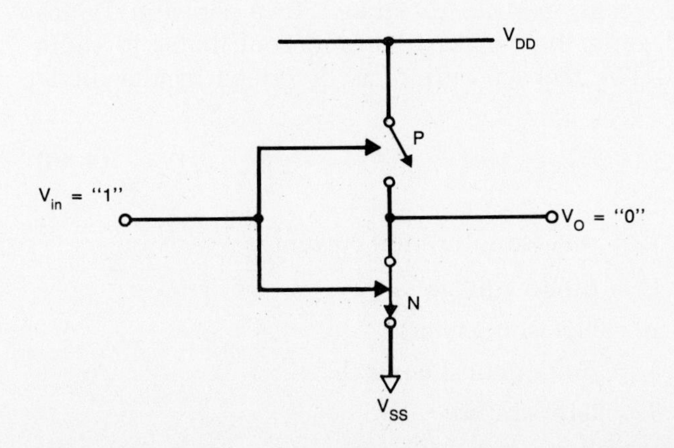

FIGURE 4.20. CMOS inverter states for static dissipation calculations

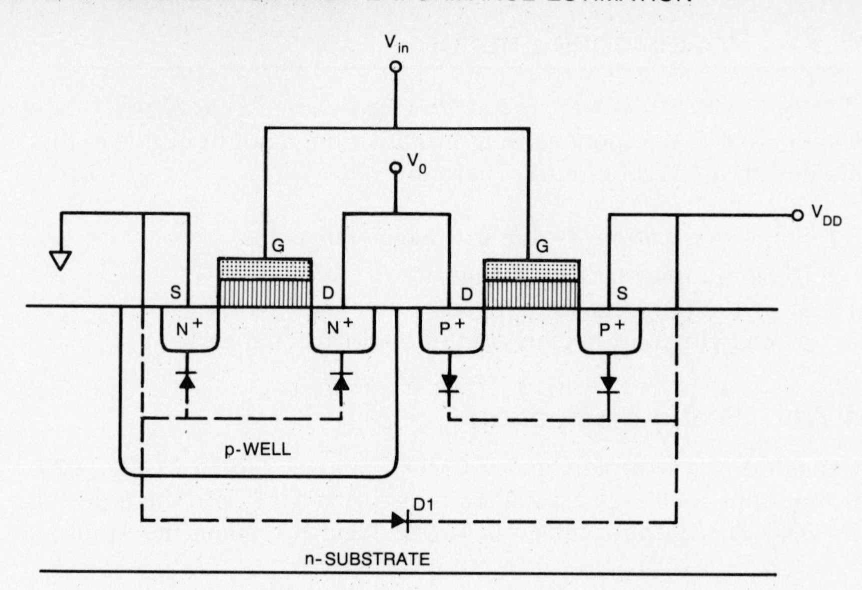

FIGURE 4.21. Model describing parasitic diodes

'1', the associated n-channel device is biased 'ON' and the p-channel device is 'OFF'. The output voltage is 0 volts (V_{SS}). Note that one of the transistors is always 'OFF' when the gate is in either of these logic states. Since no current flows into the gate terminal, and there is no D.C. current path from V_{DD} to V_{SS}, the resultant quiescent (steady state) current, and hence power P_s, is zero.

However, there is some small static dissipation due to reverse bias leakage between diffusion regions and the substrate. We need to look at a simple model that describes the parasitic diodes for a CMOS inverter in order to have an understanding of the leakage involved in the device. The source-drain diffusions and the p-well diffusion form parasitic diodes. This can be represented in the profile of an inverter shown in Fig. 4.21. In the model, diode D1 is a parasitic diode between p-well to substrate. Since parasitic diodes are reverse biased, only their leakage current contributes to static power dissipation. The leakage current is described by the diode equation

$$i_o = i_s(e^{qV/kT} - 1), \tag{4.36}$$

where

i_s = reverse saturation current
V = diode voltage
q = electronic charge
k = Boltzmann's constant
T = temperature.

The static power dissipation is the product of the device leakage current and the supply voltage. A useful estimate is to allow a leakage current of 0.1nA to 0.5nA per gate at room temperature. Then total static power dissipation P_S is obtained from

$$P_S = \sum_{1}^{n} \textit{leakage current} * \textit{supply voltage}, \tag{4.37}$$

where

$$n = \text{number of devices.}$$

For example, typical static power dissipation due to leakage for an inverter operating at 5 volts is between 1–2 nano-watts.

4.7.2 Dynamic dissipation

During transition from either '0' to '1' or, alternatively, from '1' to '0', both n- and p-transistors are on for a short period of time. This results in a short current pulse from V_{DD} to V_{SS}. Current is also required to charge and discharge the output capacitive load. This latter term is generally the dominant term. The current pulse from V_{DD} to V_{SS} results in a "short-circuit" dissipation which is dependent on the load capacitance and gate design. This is of relevance to I/O buffer design. Further discussion may be found in [Veen84].

The dynamic dissipation can be modeled by assuming the rise and fall time of the step input is much less than the repetition period. The average dynamic power, P_d, dissipated during switching for a square-wave input V_{in}, having a repetition frequency of $f_p = 1/t_p$, as shown by Fig. 4.22 (page 148), is given by

$$P_d = \frac{1}{t_p}\int_0^{t_p/2} i_n(t)V_o\,dt + \frac{1}{t_p}\int_{t_p/2}^{t_p} i_p(t)(V_{DD} - V_o)\,dt, \tag{4.38}$$

where

$$i_n = \text{n-device transient current}$$

$$i_p = \text{p-device transient current.}$$

For a step input and with $i_n(t) = C_L\,dV_o/dt$ (C_L = load capacitance)

$$P_d = \frac{C_L}{t_p}\int_0^{V_{DD}} V_o\,dV_o + \frac{C_L}{t_p}\int_{V_{DD}}^{0} (V_{DD} - V_o)\,d\,(V_{DD} - V_o)$$

$$= \frac{C_L V_{DD}^2}{t_p} \tag{4.39}$$

with $f_p = \dfrac{1}{t_p}$,

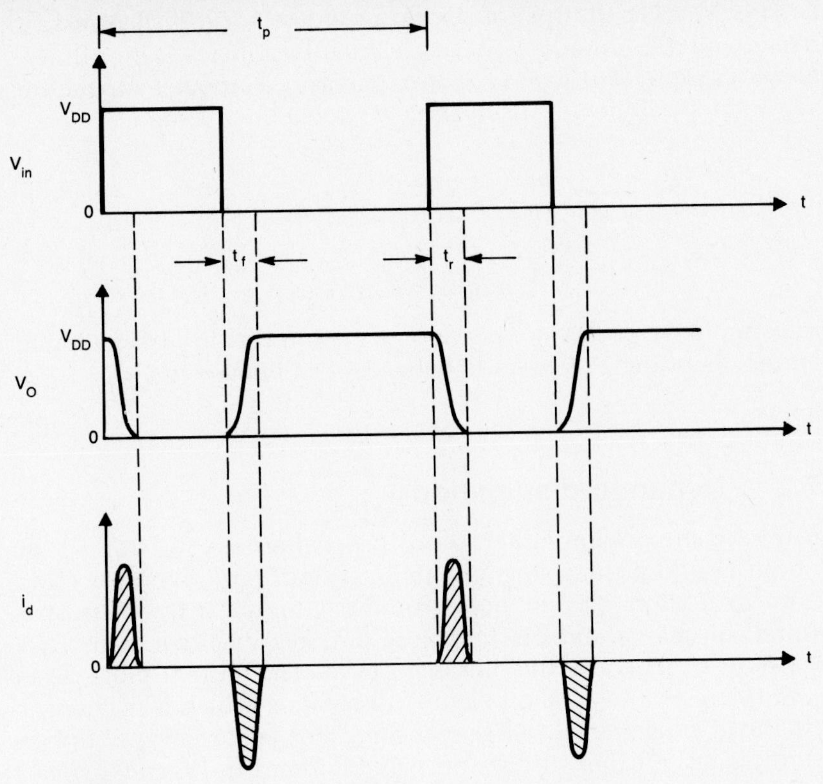

FIGURE 4.22. Waveforms for determination of dynamic power dissipation

resulting in

$$P_d = C_L V_{DD}^2 f_p. \tag{4.40}$$

Thus for a repetitive step input the average power that is dissipated is proportional to the energy required to charge and discharge the circuit capacitance. The important factor to be noted here is that Eq. (4.40) shows power to be proportional to switching frequency but independent of the device parameters.

Total power dissipation can be obtained from the sum of the two dissipation components, so

$$P_{total} = P_s + P_d. \tag{4.41}$$

When calculating the power dissipation, a rule of thumb is to add all capacitances operating at a particular frequency and calculate the power. Then the power from other groups operating at different frequencies may be summed. The dynamic power dissipation may be used to estimate total power consumption of a circuit and also the size of V_{DD} and V_{SS} conductors to minimize transient induced voltage drops. The latter design decision is increasingly important in large CMOS designs (See Section 9.7).

Example:

Determine the power dissipated for a typical system using N inverters when operated at a frequency of 10 MHz and $V_{DD} = +5$ volts.

$$\text{output capacitance} = 2C_d$$

$$\text{input capacitance} = 2C_g$$

$$P_s = N(0.1 \cdot 10^{-9} \cdot 5)watts$$
$$= N(0.5 \cdot 10^{-9})watts$$
$$P_d = \frac{N(2C_d + 2C_g)25}{100} \cdot 10^9$$
$$= N(25 \cdot 10^{-6})watts \quad \begin{matrix}(C_d = 40\text{fF} & \text{Fig. 4.7)} \\ (C_g = 11.2\text{fF} & \text{Fig. 4.7).}\end{matrix}$$

4.8 Charge sharing

In many structures a bus can be modeled as a capacitor C_b, as shown in Fig. 4.23. Sometimes the voltage on this bus is sampled (latched) to determine the state of a given signal. Frequently, this sampling can be modeled by the two capacitors, C_s and C_b, and a switch. In general, C_s is in some way related to the switching element. The charge associated with each of the capacitances prior to closing the switch can be described by

$$Q_b = C_b V_b \tag{4.42}$$

and

$$Q_s = C_s V_s.$$

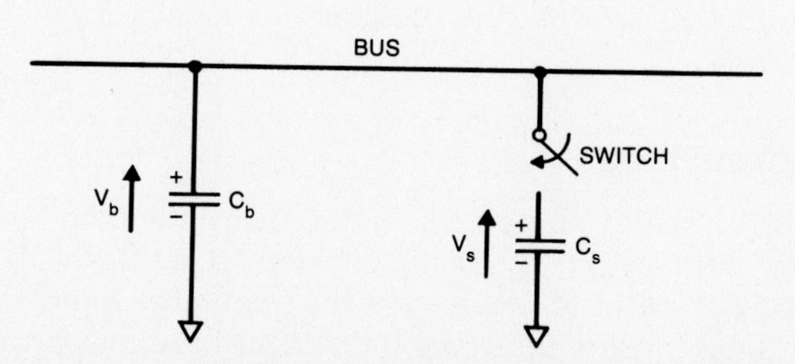

FIGURE 4.23. Charge sharing mechanism

The total charge Q_T is then given by

$$Q_T = C_b V_b + C_s V_s. \tag{4.43}$$

The total capacitance C_T is given by

$$C_T = C_b + C_s. \tag{4.44}$$

Therefore, when the switch is closed, the resultant voltage V_R (not shown in Fig. 4.23) is

$$V_R = \frac{Q_T}{C_T} = \frac{C_b V_b + C_s V_s}{C_b + C_s}. \tag{4.45}$$

For example, if

$$V_b = V_{DD}$$

and

$$V_b \gg V_s,$$

then

$$V_R = V_{DD}\left[\frac{C_b}{C_b + C_s}\right]. \tag{4.46}$$

To ensure reliable data transfer from C_b to C_s, it is necessary to ensure $C_s \ll C_b$. A useful rule to follow is $C_b > 10 * C_s$.

4.9 Scaling of MOS transistor dimensions

So far in this chapter we have examined some electrical design issues and formulated some electrical design rules that should be taken into account when building high performance circuits with current CMOS processes. As CMOS processes are improved and device dimensions are reduced, these rules will change. In this section, we take a look at the effect that these reduced dimensions will have on electrical circuit behavior. The model we use is a simple first-order "constant field" scaling. Although it is unlikely that CMOS processing will scale in such a simple fashion, the results presented give some idea as to what the designer can expect from future fine line processes.

4.9.1 Scaling principles

First-order MOS scaling theory, based upon the "constant field" model formulated by Dennard et al. [Denn73][Denn74], indicates that the characteristics of an MOS device can be maintained and the basic operational characteristics preserved if the critical parameters of a device are scaled in accordance to a given criterion. Such an approach has shown to be very effective in scaling from the range 5 μm to 10 μm minimum features to the range 1 μm to 3 μm minimum feature size.

Although first-order scaling does not give optimized device performance at small dimensions, the technique is very powerful in providing the necessary guidelines to identify the improvements (or otherwise) that can be expected as processes are scaled.

Basically the scaled device is obtained by applying a dimensionless factor α to

- all dimensions, including those vertical to the surface
- device voltages
- the concentration densities.

The resultant effect of the first-order scaling process is illustrated in Fig. 4.24 and Table 4.8. Table 4.8 shows that if device dimensions (which include channel length L, channel width W, oxide thickness t_{ox}, junction depth X_j, applied voltages, and substrate concentration density N) are scaled by the constant parameter α, then the depletion

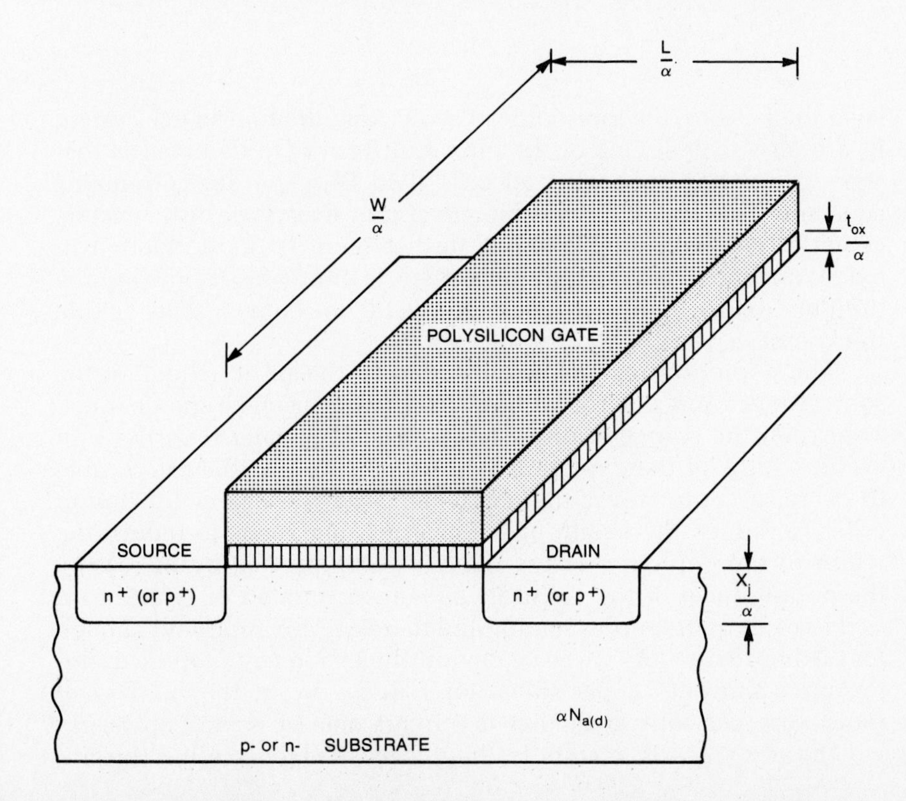

FIGURE 4.24. Basic scaled MOS device

TABLE 4.8. Influence of first-order scaling on MOS device characteristics

	PARAMETERS	SCALING FACTOR
DEVICE PARAMETERS	Length; L	$1/\alpha$
	Width; W	$1/\alpha$
	Gate oxide thickness; t_{ox}	$1/\alpha$
	Junction depth; X_j	$1/\alpha$
	Substrate doping; $N_{a\ (or\ d)}$	α
	Supply voltage; V_{DD}	$1/\alpha$
RESULTANT INFLUENCE	Electric field across gate oxide; E	1
	Depletion layer thickness; d	$1/\alpha$
	Parasitic capacitance; WL/t_{ox}	$1/\alpha$
	Gate delay; (VC/I)	$1/\alpha$
	DC power dissipation; P_s	$1/\alpha^2$
	Dynamic power dissipation; P_d	$1/\alpha^2$
	Power-speed product	$1/\alpha^3$
	Gate area	$1/\alpha^2$
	Power density; (VI/A)	1
	Current density; (I/A)	α
	Transconductance; g_m	1

layer thickness d, the threshold voltage V_t, and drain-to-source current I_{ds} are also scaled. One of the important factors to be noted is that since the voltage is scaled, electric field E in the device remains constant. This has the desirable effect that many nonlinear factors essentially remain unaffected. A further point is that reduction in oxide thickness would require the fabrication process to provide thinner oxides with comparable yield to conventional oxide thicknesses.

The depletion regions associated with the pn junctions of the source and drain determine how small we can make the channel. As a rule, the source-drain distance must be greater than the sum of the widths of the depletion layers to ensure that the gate is able to exercise control over the conductance of the channel. Thus in order to reduce the length of the channel one needs to reduce the width of the depletion layers. This is accomplished by increasing the doping level of the substrate silicon. A number of issues arise as the result of first-order scaling and therefore are important enough for further review. As we scale device dimensions by $1/\alpha$, the drain-to-source current I_{ds} per transistor reduces by α, the number of transistors per unit area; that is, circuit density scales up by α^2, which subsequently results in the current density scaling linearly with α. Thus wider metal conductors will be necessary for densely packed structures.

A second characteristic illustrated in Table 4.8 is power density. Both the static power dissipation P_s and frequency dependent dissipation P_d decrease by $1/\alpha^2$ as the result of scaling. However, since the number of devices per unit area increases by α^2, the resultant effect is that the power density remains constant.

An estimation of the limit in power density is derived from the thermodynamic relationship given by

$$T_j = T_{amb} + \theta_{jA}.P,$$

where

$$\begin{aligned} T_j &= \text{temperature of silicon chip} \\ T_{amb} &= \text{ambient temperature} \\ \theta_{jA} &= \text{thermal resistance of the package} \\ P &= \text{power dissipation.} \end{aligned}$$

Generally, the thermal resistance is expressed as Δ°C per watt, which means one watt of heat energy will raise the temperature by Δ°C. For a 40-pin ceramic package, this value is in the range of 30 to 40°C per watt. If we assume an ambient temperature of 75°C, and the maximum allowed silicon junction temperature is about 175°C, then the maximum power dissipation that does not require special cooling is

$$\begin{aligned} P_{\max} &= \frac{T_j - T_{amb}}{\theta_{jA}} \\ &= \frac{175 - 75}{40} \\ &= 2.5 \textit{ watts}. \end{aligned}$$

As the temperature increases, the carrier mobility falls, thus reducing the gain of devices. This, in turn, would reduce the speed of circuits. If high temperature, high speed circuits are required, then special consideration during design is necessary.

It is necessary to recognize that the variables shown in Table 4.8 are only first-order approximations. A more rigorous analysis would modify some of the values. For example, scaling of the substrate doping level by α, causes the mobility to decrease slightly. Therefore the propagation delay, as a rule, does not improve by as much as the predicted factor of $1/\alpha$. However, power dissipation will decrease by somewhat more than the expected value of $1/\alpha^2$. Thus the power-speed product remains at $1/\alpha^3$.

One of the limitations of first-order scaling is that it gives the wrong impression of being able to scale proportionally to zero di-

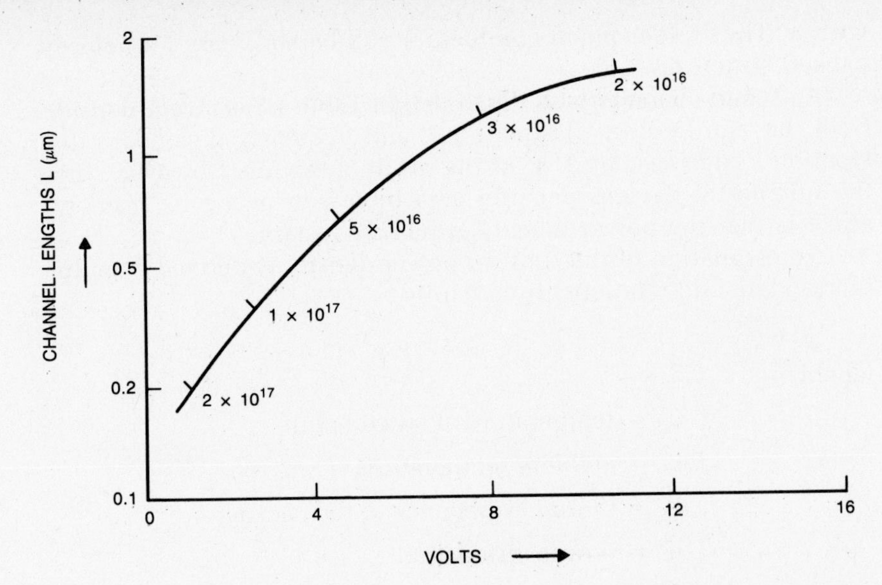

FIGURE 4.25. Relationship between channel length L, voltage and doping level (N)

mension, or to zero threshold voltages. In reality, both theoretical and practical considerations do not permit such behavior. This is highlighted when the surface concentrations become larger than $1 * 10^{19}$ cm^{-3}, above which the gate oxide breaks down before surface inversion can take place for the formation of the channel. Fig. 4.25 is a good example that shows some of the practical limitations. From the figure it is possible to estimate the maximum supply voltage and minimum channel length that can be used with each doping level. Both junction breakdown and oxide breakdown also limit the extent of scaling.

As memories scale, they involve special consideration. As diffusion areas are reduced in size they become susceptible to alpha radiation and special circuit techniques must be employed to detect the small stored charges.

4.9.2 Interconnect layer scaling

Although constant-field (first-order) scaling gives a number of improvements, there are a number of circuit parameters such as voltage drop, line propagation delay, current density, and contact resistance that exhibit significant degradation with scaling. For example, scaling the thickness and width of a conductor by α, reduces the cross-sectional area by α^2. The scaled line resistance R' is given by

$$R' = \frac{\rho}{t/\alpha}\left[\frac{L/\alpha}{W/\alpha}\right] \qquad (4.47)$$
$$= \alpha R,$$

where ρ is the conductivity term, which is related to sheet resistance by $R_s' = \rho/\alpha t$, and t is conductor thickness. The voltage drop along such a line can now be expressed as

$$\begin{aligned} V_d' &= (I/\alpha)(\alpha R) \\ &= IR, \end{aligned} \tag{4.48}$$

which is a constant. However, for constant chip size, the length of some of the signal paths that traverse across the chip, as a rule, do not scale down. This gives the principal result that voltage drops along communication paths are larger by a factor of α with respect to the scaled voltages. In a similar manner, we can derive the line response time as

$$\begin{aligned} \tau_s' &= (\alpha R)(C/\alpha) \\ &= RC, \end{aligned} \tag{4.49}$$

which is a constant. However, as before, for a constant chip size many of the communication paths do not scale. Thus the line response time normalized to scaled line response is larger by a factor of α. The significance of this result is that it is somewhat difficult to take the full advantage of the higher switching speeds inherent in scaled devices when signals are required to propagate over long paths. Thus the distribution and organization of clocking signals becomes a major problem as geometries are scaled.

The influence of scaling on interconnection paths is summarized in Table 4.9. As can be seen from Table 4.9, fine line metallization brings new problems. We find in the first instance that metal lines must carry a higher current with respect to cross-sectional area; thus electron migration becomes a major factor to consider. Thus new metallization schemes will be required to accommodate the higher current density. The second problem relates to an increase in the capacitance of wiring. As the level of integration increases, the average line length on a chip tends to increase also. However, the power dissipation per gate decreases, which diminishes the ability

TABLE 4.9. Influence of scaling on interconnect media

PARAMETERS	SCALING FACTOR
Line resistance; r	α
Line response; rc	1
Normalized line response	α
Line voltage drop; V_d	1
Normalized line voltage drop	α
Current density; J	α
Normalized contact voltage drop; V_c/V	α^2

of gates driving wiring capacitances. Under such conditions, average gate delay is determined by the interconnection rather than the gate itself.

Many of these limitations are being overcome by scaling lateral dimensions while keeping vertical dimensions approximately constant. This, however, imposes more constraints on the vertical topography of the process.

4.10 Yield

An important issue in the manufacture of VLSI structures is the yield [SaAr82]. Although yield is not a performance parameter, it is influenced by such factors as:

- technology
- chip area
- layout [Rung81].

Yield is defined as

$$Y = \frac{\textit{No. of Good Chips on Wafer}}{\textit{Total Number of Chips}}\,100\%$$

and may be described as a function of the chip area and defect density. Two common equations are used.

1 Seed's model [Seed67], which is given by

$$Y = e^{-\sqrt{AD}}, \qquad \textbf{(4.50)}$$

where

A = chip area

D = defect density (defined as lethal defects per cm^2).

This model is used for large chips and for yields less than about 30 percent.

2 Murphy's model [Murp64], which is described by

$$Y = \left(\frac{1 - e^{-AD}}{AD}\right)^2. \qquad \textbf{(4.51)}$$

This model is used for small chips and for yields greater than 30 percent.

From these relations it is obvious that yield decreases dramatically as the area of the chip is increased. One can easily encounter a situation in which all of the chips on a wafer are found defective. Modern fabrication lines using dry etching techniques generally yield a D value of around 4 defects/cm^2. In order to improve yield it is possible to incorporate redundancy into the structure. In random logic, yield improvement is minimal due to increase in area. However, in memory structures, it is possible to gain dramatic improvement in yield through incorporation of redundant cells.

4.11 Summary

In this chapter we have developed simple models to allow us to estimate circuit performance. We will use these models in subsequent chapters to evaluate different circuit approaches to different problems. The effects of scaling based on a first-order model yield some insight into where CMOS technology is heading.

4.12 Exercises

4.1 If the sheet resistance of aluminum is .05 Ω/sq, and the sheet resistance of polysilicon is 30 Ω/sq, calculate the resistance of a 5 mm run of each material, assuming a 4 μm wire width. If the sheet resistance remains the same, calculate the resistance for a 5 mm run based on a 0.5 μm wire width.

4.2 A transistor is built by overlaying a 2 μm by 6 μm polysilicon gate centrally over a 4 μm by 14 μm thinoxide region. $C_{ox} = 10^{-3}\ pf/\mu m^2$, $C_{ja} = 10^{-4}\ pf/\mu m^2$, $C_{jp} = 10^{-3}\ pf/\mu m$, and $C_p = 5.10^{-5}\ pf/\mu m^2$. Calculate the capacitance of the polysilicon region and the diffusion regions.

4.3 Calculate the approximate worst-case rise and fall times for a 3-input NAND gate using $\beta_n = 40\ \mu AV^{-2}$ ($w = 9\ \mu m$, $l = 3\ \mu m$) and $\beta_p = 20\ \mu AV^{-2}$ ($w = 9\ \mu m$, $l = 3\ \mu m$). What would you do to equalize the rise and fall time?

4.4 Design a four-stage minimum delay pad buffer to drive a 200 pf load with a 20 ns rise and fall time, using the parameters in Exercise 4.3. Calculate a) the dynamic power dissipation at 10 MHz; b) the aluminum power bus width if $J_{al} = 1.5\ mA/\mu m$. What worst-case supply rail noise spikes would you expect to see if the buffer is located 2 mm. from the supply point?

4.5 A dynamic memory node has a capacitance of 0.1 pf and is charged to 5 volts. The leakage current from the node is 1 nA. How often must the node be refreshed? What is the average static power dissipation for an array of 64K such nodes?

4.6 The node in Exercise 4.5 is switched onto a bit-line with a capacitance of 3 pf, initially charged to 2.5 volts. What is the change in the bit-line if the memory node is initially charged to 0 or 5 volts?

4.7 If D = 4 defects/cm^2, calculate the yield for a 1 mm by 1 mm chip (chip A), a 7 mm by 7 mm chip (chip B), and a 1.2 by 1.2 cm chip (chip C). How many good chips would you expect to get from a 4 inch wafer and a 6 inch wafer for each? Packaging costs are A — $2, B — $15, and C — $50. A 6 inch wafer costs $600 to fabricate. A system can use 50 As, 4 Bs, or 1 C. Which is the most economic solution based on these processing and packaging costs?

CMOS CIRCUIT AND LOGIC DESIGN

5.1 Introduction

In Chapter 1, CMOS logic was introduced with the assumption that MOS transistors act as simple switches. We have seen in subsequent chapters that certain limitations pertain to MOS transistors that detract from this idealized viewpoint. Furthermore, we have only considered fully complementary logic structures and the ratioed CMOS inverter.

In this chapter we first examine alternative CMOS logic configurations to the fully complementary CMOS logic gate. We then examine the effects of non-ideal switch behavior on circuits. Finally, we compare the alternate logic structures that are available. As we are interested in designing physical layouts and creating performance optimized designs, two areas have to be addressed in order to achieve a prescribed behavior:

1. circuit (structural) design
2. layout (physical) design.

As we will see, these two phases of design are intimately meshed. The behavior of many circuits may have a direct impact on any high level architectural decisions. For this reason it is important for the system designer to have some idea of low level circuit options.

5.2 CMOS logic structures

In some situations, the area taken by a fully complementary static CMOS gate may be greater than that required, the speed may be too slow, or the function may just not be implementable as a purely complementary structure (as in the case of a large PLA). In these cases, it is desirable to implement smaller and faster gates at a cost of increased design and operational complexity and, possibly, decreased circuit stability. There are a number of alternate CMOS logic structures that can be used. These structures will be summarized in this section.

5.2.1 CMOS complementary logic

For review, the complementary CMOS inverter, NAND, and NOR gates are shown in Fig. 5.1. All complementary gates may be designed as *ratioless* circuits. That is, if all transistors are the same size the

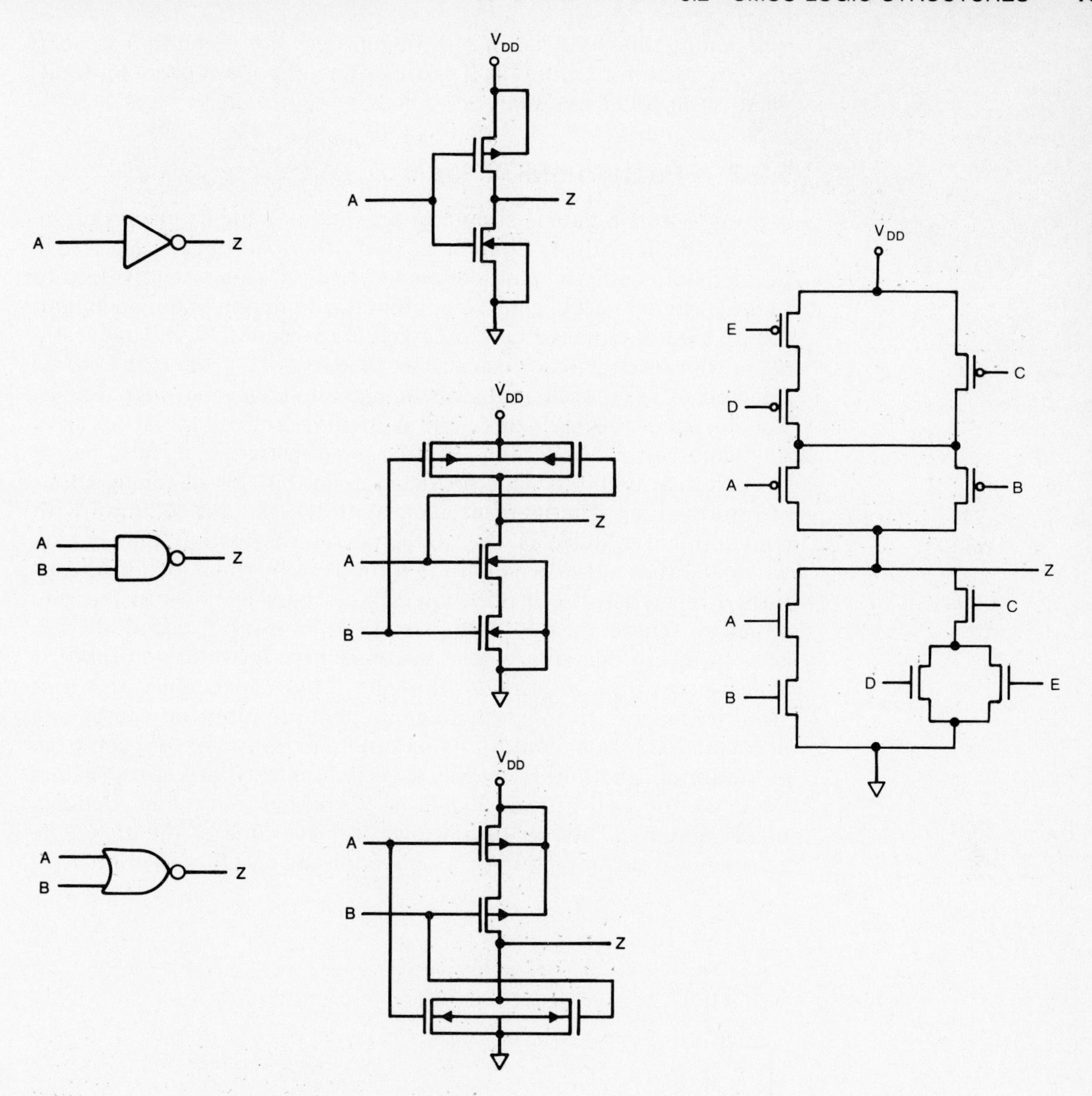

FIGURE 5.1. CMOS complementary logic

circuit will function correctly. Later in this chapter, methods of optimizing the speed of such circuits by using different sized transistors will be presented. In addition, a complex gate that will form the basis for comparison between logic families is shown. It implements the function $Z = \overline{(A.B) + C.(D + E)}$. In these schematics the substrate connection has been shown, although in subsequent schematics this connection will be omitted. It is important to keep this connection in mind, as it causes some important behavior modification in MOS circuits.

5.2.2 Pseudo-nMOS logic

A pseudo-nMOS gate is shown in Fig. 5.2. It is the extension of the inverter dealt with in Chapter 2. Here, the load device is a single p-transistor, with the gate connected to V_{SS}. This is equivalent to a conventional nMOS gate except that the depletion or enhancement nMOS load is replaced by a p-device. As with nMOS, the gain ratio of the p-transistor load to n-driver transistors $\beta_{load}/\beta_{driver}$ has to be selected to yield sufficient gain to generate consistent logic levels. The design of this style of gate thus involves *ratioed* transistor sizes to ensure correct switching. That is, the effective β_n/β_p ratio has to be consistent with the value predicted in Eq. 2.37 for all combinations of input values. The main problem with the gate (in common with conventional nMOS) is the static power dissipation that occurs whenever the pull-down chain is turned on. As the p load is always turned on, when the n pull-down is on, current flows in the gate structure. There are n + 1 transistors in an n-input pseudo-nMOS gate. In a complementary gate, the capacitive load on each input is at least two *unit gate loads* (the gate input capacitance of a unit sized transistor). In this type of gate, the minimum load can be one unit gate load, as a result of using only one transistor for each term of the input function. However, if minimum sized driver transistors are used, the pull-up gain has to be decreased to provide adequate noise margins. This, in turn, slows the rise time of the gate. The gate has no advantage over a conventional nMOS depletion load

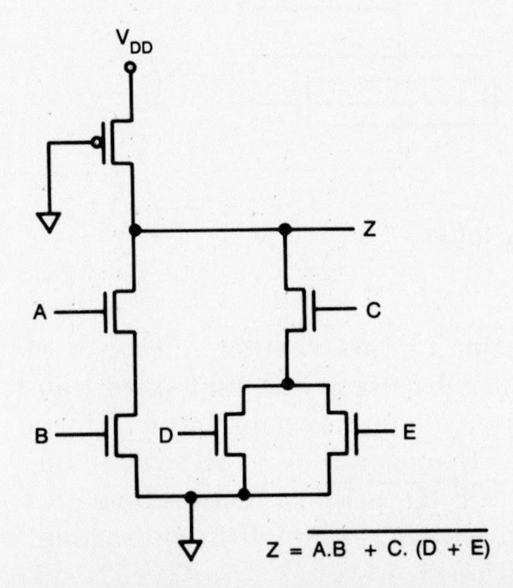

FIGURE 5.2. Pseudo-nMOS logic

gate, save that in a CMOS process it gives one a method of emulating nMOS circuits. One possible advantage of the pMOS load is that it does not suffer from body effect as the nMOS depletion load does. A gate so implemented may have a density advantage over a fully complementary gate.

5.2.3 Dynamic CMOS logic

A basic dynamic CMOS gate is shown in Fig. 5.3 [HeHo76]. It consists of an n-transistor logic structure whose output node is precharged to V_{DD} by a p-transistor (precharge) and conditionally discharged by an n-transistor (evaluate) connected to V_{SS}. (Alternatively, an n-transistor precharge to V_{SS} and p-transistor discharge to V_{DD} and p logic block may be used.) ϕ is a single phase clock. For the former case, the precharge phase occurs when $\phi = 0$. The path to the V_{SS} supply is closed via the n-transistor "ground switch" during $\phi = 1$. The input capacitance of this gate is the same as the pseudo-nMOS gate. The pull-up time is improved by virtue of the active switch but the pull-down time is increased due to the ground switch. Note that the ground switch may be omitted if the inputs are guaranteed to be zero during precharge.

A number of problems are manifest in this structure. Firstly, the inputs can only change during the precharge phase. If this condition is not met, charge redistribution effects can corrupt the output node voltage. This is expanded upon in Section 5.3.6. Simple single phase dynamic CMOS gates cannot be cascaded. For instance, consider the circuit in Fig. 5.4 on page 164. When the gates are precharged, the output nodes are charged to V_{DD}. During the evaluate phase, the output of the first gate will conditionally discharge. However, some delay will be incurred due to the finite pull-down time. Thus

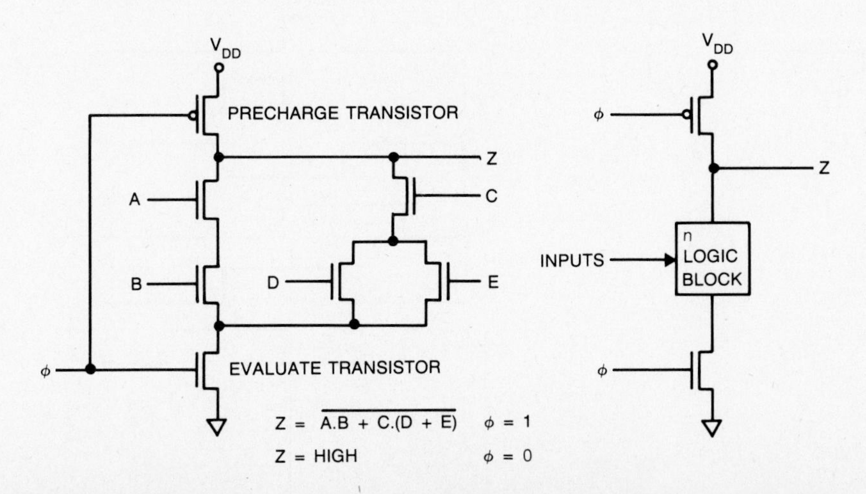

FIGURE 5.3. Dynamic CMOS logic

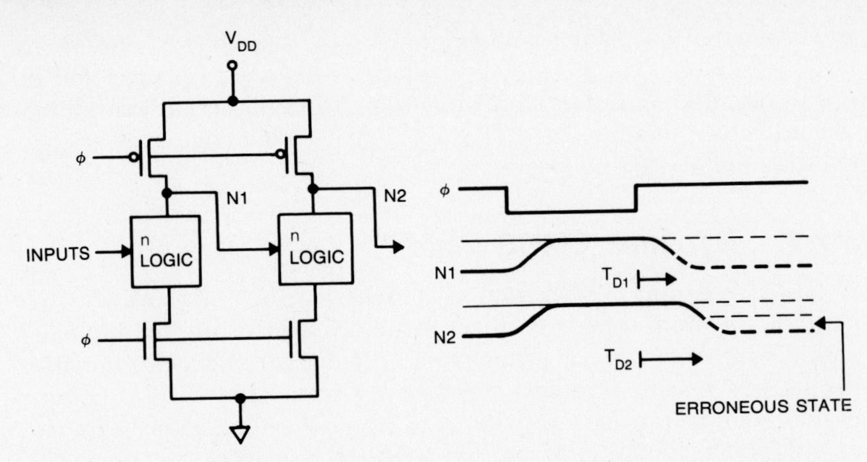

FIGURE 5.4. Cascaded dynamic gates

the precharged node can discharge the output node of the following gate before the first gate is correctly evaluated.

Improvements on this structure use the forms of two and four phase logic that have been developed for earlier types of MOS design [PeLa73]. These gates add a sample and hold clock phase to the precharge and evaluate cycles. Fig. 5.5a shows one version of a gate implemented using the clock relationships shown in Fig. 5.5b. The composite clocks $\phi 12$ and $\phi 23$ are used in this example. During $\phi 1$, node PZ is precharged, while node Z is held at its previous value. When $\phi 2$ is true, node PZ remains precharged and, in addition, the transmission gate turns on, thus precharging node Z. When $\phi 3$ is

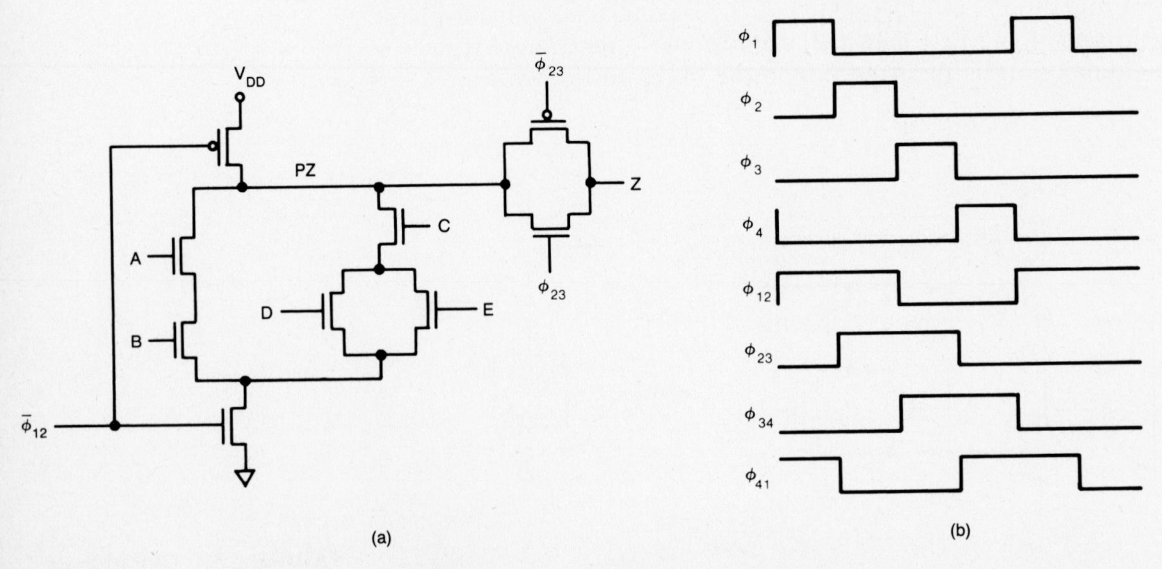

FIGURE 5.5. 4-phase logic — type A

asserted, the gate evaluates and node PZ conditionally discharges. Node Z follows node PZ as the transmission gate remains on. Finally, when $\phi 4$ is true, node Z will be held in the evaluated state. The state of node PZ is immaterial. There are four types of gates characterized by the phase in which evaluation occurs. When using such logic gates, they must be used in the appropriate sequence. The allowable connections between types are shown in Fig. 5.6.

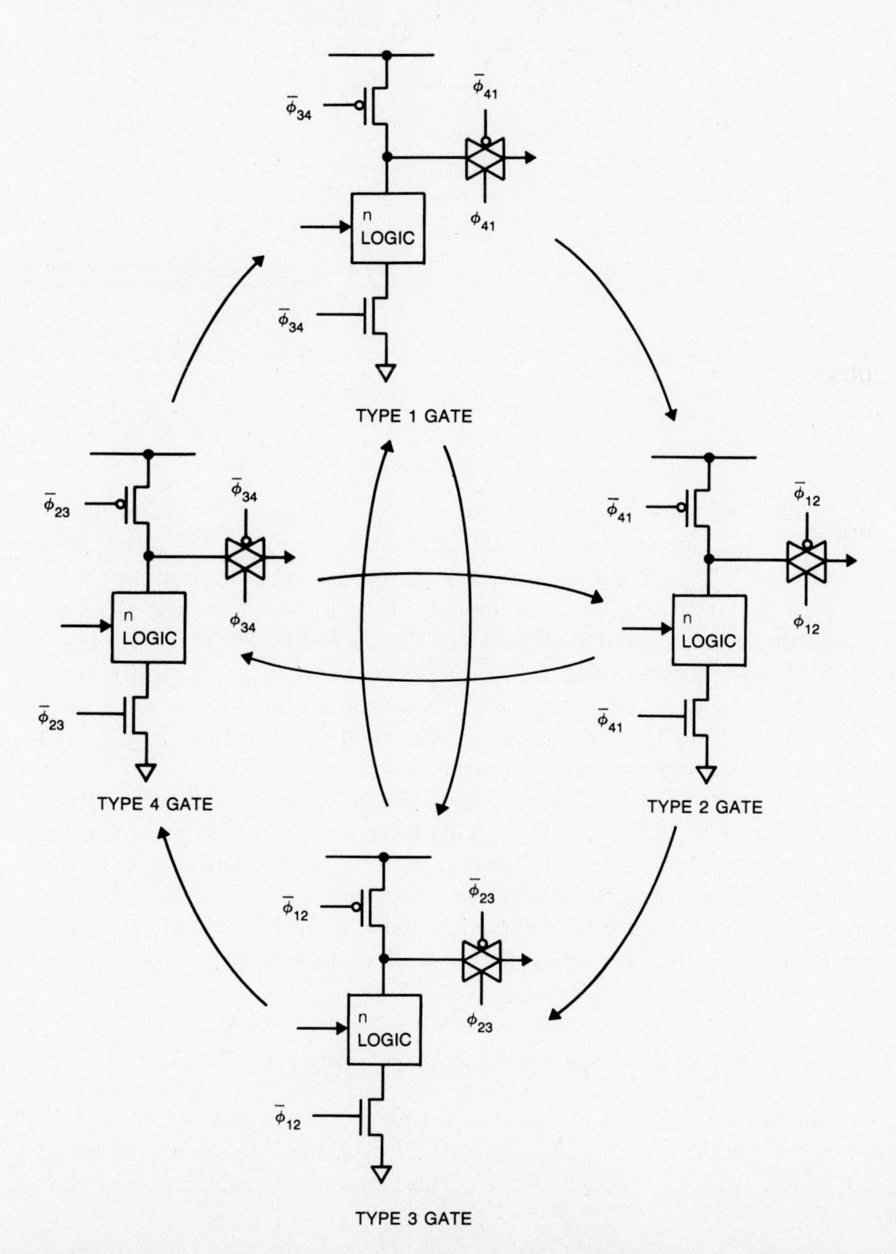

FIGURE 5.6. Allowable gate interconnections — type A

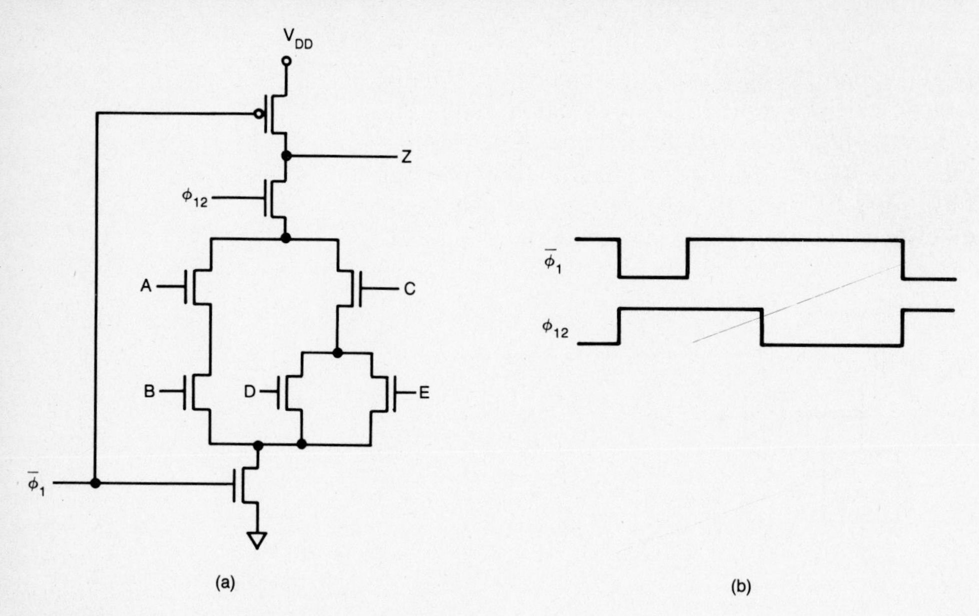

FIGURE 5.7. 4-phase logic — type B

Note that four levels of logic may be evaluated per bit time. Alternatively, a 2-phase logic scheme may be employed by using type 4 gates and type 2 gates or type 1 gates and type 3 gates.

An alternate 4-phase structure is shown in Fig. 5.7a. The clocking waveforms are shown in Fig. 5.7b. This structure has the intergate restrictions shown in Fig. 5.8. This gate type is more restrictive than the previous gate, but the circuit is simpler, the number of clocks is reduced, and the layout would be smaller. Similarly, a 2-phase system could employ gate types 2 and 4.

The number of transistors required for such logic gates is either n + 4 or n + 3 for an n-input gate. A problem that occurs with such gates is that the clock frequency must be long enough to allow for the slowest gate to evaluate. Thus fast gates tend to evaluate quickly and the remainder of the cycle is "dead time." Other system design problems arise when trying to distribute four or more clocks and synchronize them around a large chip.

5.2.4 Clocked CMOS logic (C^2MOS)

A clocked CMOS gate is shown in Fig. 5.9 on page 167. This form of logic was originally used to build low power dissipation CMOS logic [SuOA73]. The reasons for the reduced dynamic power dis-

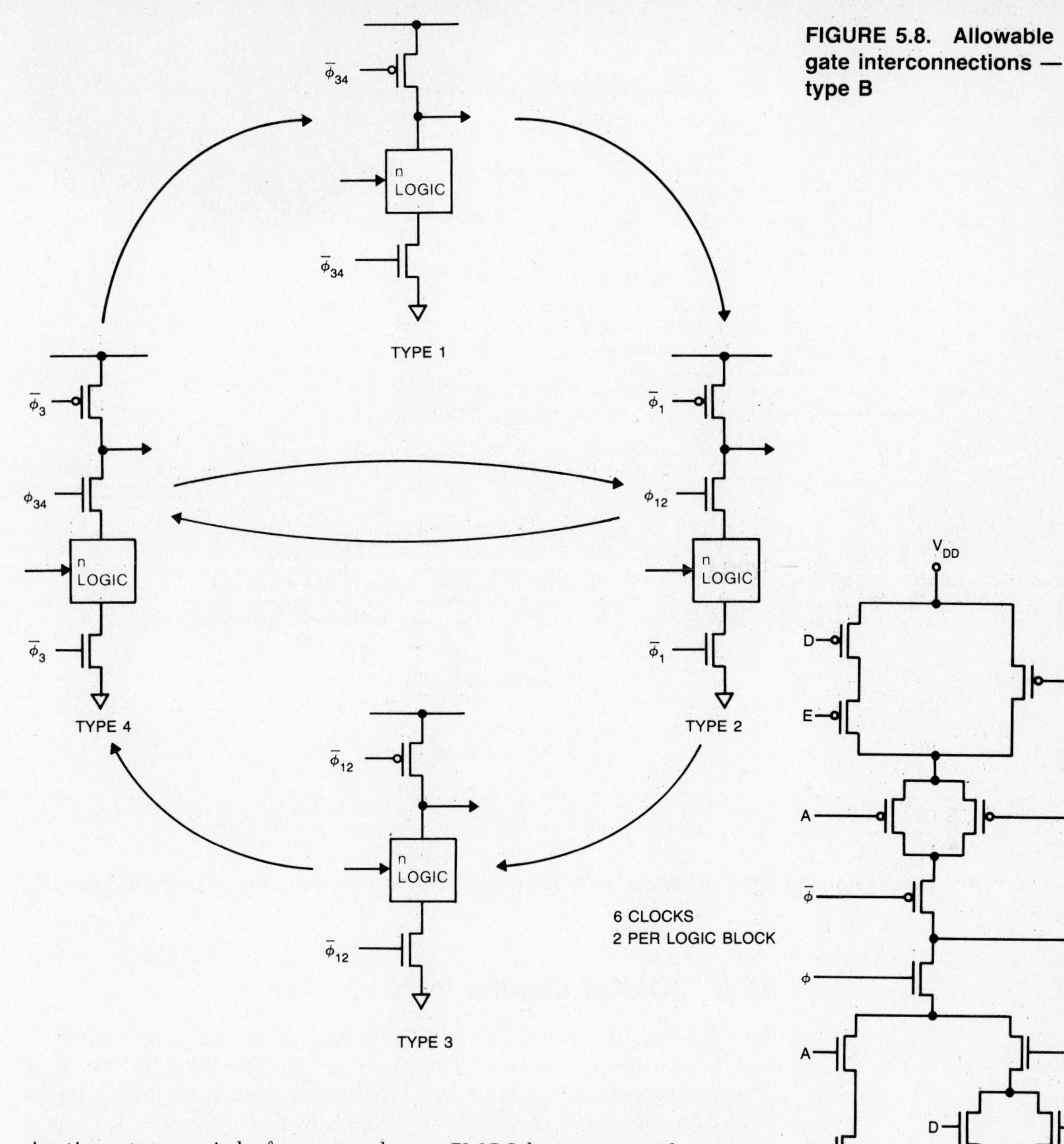

FIGURE 5.8. Allowable gate interconnections — type B

FIGURE 5.9. Clocked CMOS logic (C^2MOS)

sipation stem mainly from metal gate CMOS layout considerations. The main use of such logic structures at this time is to form clocked structures that incorporate latches or interface with other dynamic forms of logic (see Section 5.4.6). The gates have the same input capacitance as regular complementary gates but larger rise and fall times due to the series clocking transistors.

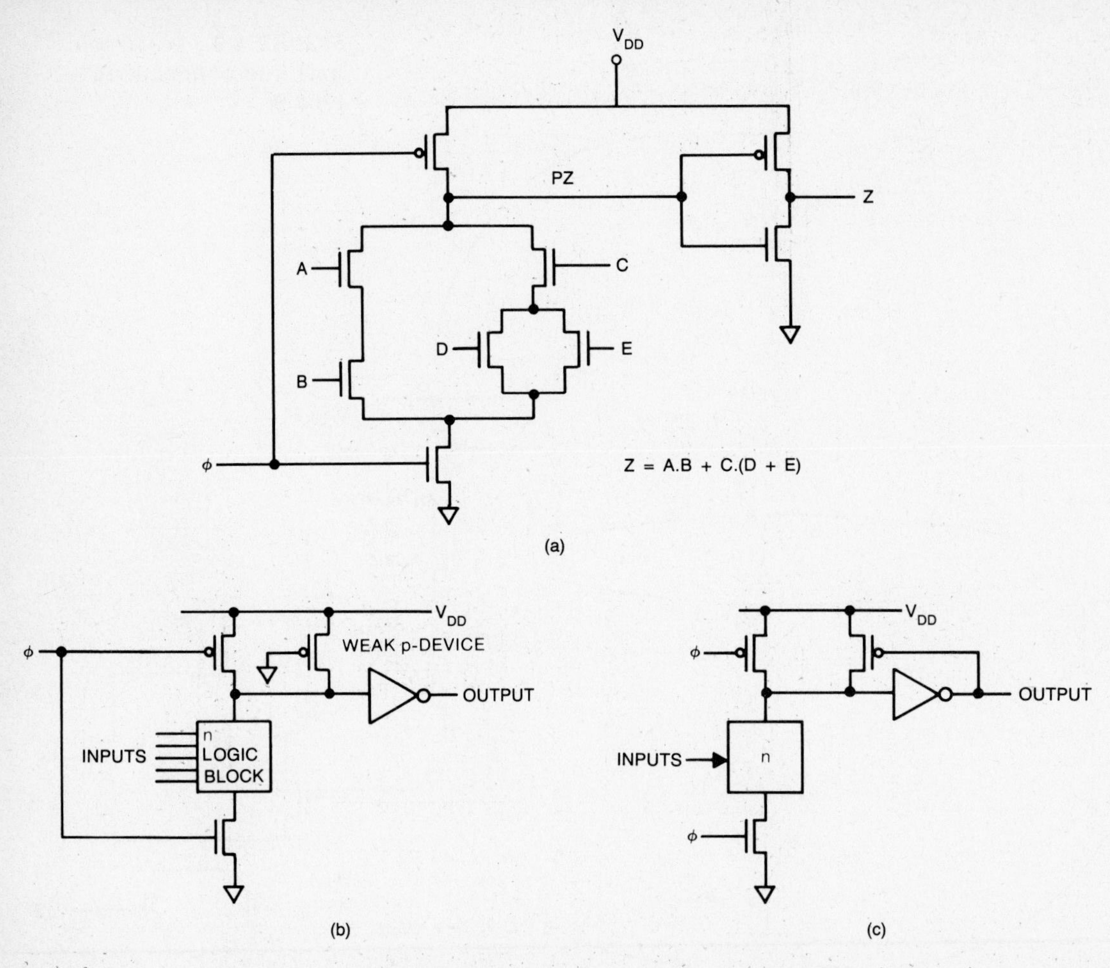

FIGURE 5.10. CMOS domino logic (a) basic gate, (b) static version (low frequency), and (c) latching version

5.2.5 CMOS domino logic

A modification to the clocked CMOS logic allows a single clock to precharge and evaluate a cascaded set of dynamic logic blocks. This involves incorporating a static CMOS buffer into each logic gate as shown in Fig. 5.10a [KrLL82]. During precharge ($\phi = 0$), the output node of the dynamic gate is precharged high and the output of the buffer is low. As subsequent logic stages are fed from this buffer, transistors in subsequent logic blocks will be turned off during the precharge phase. When the gate is evaluated, the output will conditionally discharge, causing the output of the buffer to conditionally go high. Thus each gate in sequence can make at most one transition ($1 \rightarrow 0$). Hence, the buffer can only make a transition from ($0 \rightarrow$

1). In a cascaded set of logic blocks, each state evaluates and causes the next stage to evaluate — in the same manner that a stack of dominos fall. Any number of logic stages may be cascaded, provided that the sequence can evaluate within the evaluate clock phase. A single clock can be used to precharge and evaluate all logic gates within a block.

Some limitations are evident with the structure. Firstly, only non-inverting structures are possible. Secondly, each gate must be buffered. Finally, in common with clocked-CMOS, charge redistribution can be a problem. Depending on the situation, the effect of these problems can be minimized. For example, in complex logic circuits, such as arithmetic logic units, the necessary XOR gates may be implemented conventionally (as complementary gates) and driven by the last domino circuit [KrLL82]. The buffer is often needed from circuit loading considerations and would be needed in any case.

The domino gate may be made static by including a *weak* p-transistor, as shown in Fig. 5.10b. A weak p-transistor is one that has low gain (small W/L ratio). It has to have a gain such that it does not *fight* the pull-down transistors, yet can balance the effects of leakage. This will allow low frequency or static operation when the clock is held high. In this case the pull-up time could be an order of magnitude slower than the pull-down speed. In addition, the current drawn by the gate during evaluation should be small enough so that the static power dissipation of a circuit is not impacted. A value of 10 μA is suggested [KrLL82]. Note that the precharge transistor may be eliminated if the time between evaluation phases is long enough to allow the weak pull-up to charge the output node. Authors have claimed the addition of the weak p pull-up transistor can moderate the charge redistribution problem and also improve the noise margin. These claims are not valid for high speed circuits, as the response time of the weak p-device is usually very slow. The addition of this transistor may impact the speed performance. Note that the gate may also be made latching by placing a weak p feedback transistor, as shown in Fig. 5.10c.

5.2.6 Cascade voltage switch logic (CVSL)

The basic form of this style of CMOS logic is depicted in Fig. 5.11a [HGDT84]. It is a differential style of logic requiring both true and complement signals to be routed to gates. Two complementary nMOS switch structures are constructed and then connected to a pair of cross-coupled p pull-up transistors. When the inputs switch, nodes Q and $\overline{Q}$ are either pulled high or low. Positive feedback applied to the p pull-ups causes the gate to switch. The logic trees may be further minimized from the full differential form using logic min-

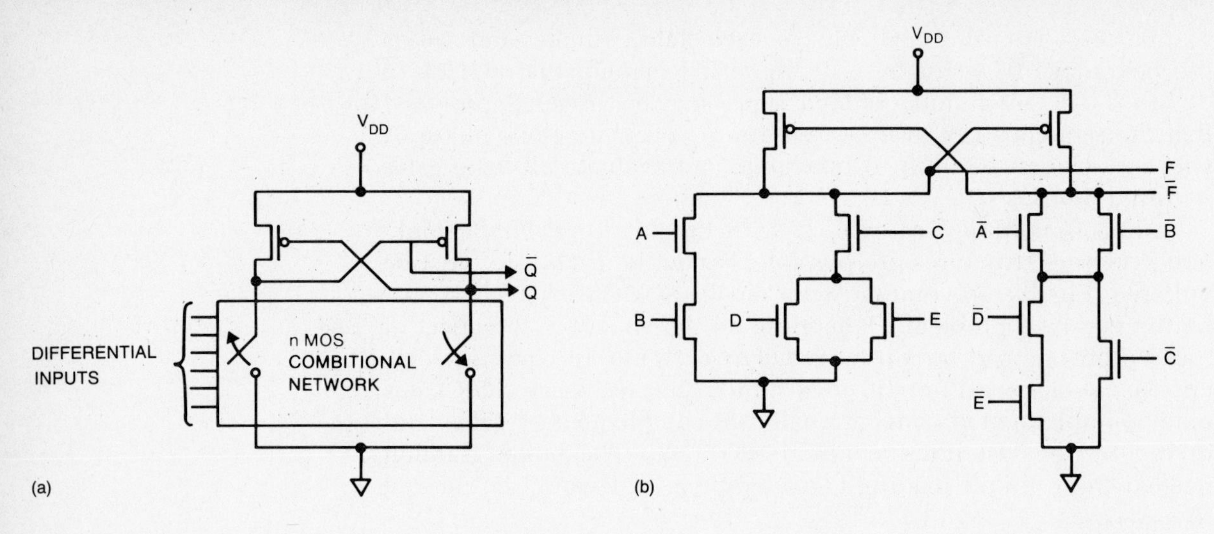

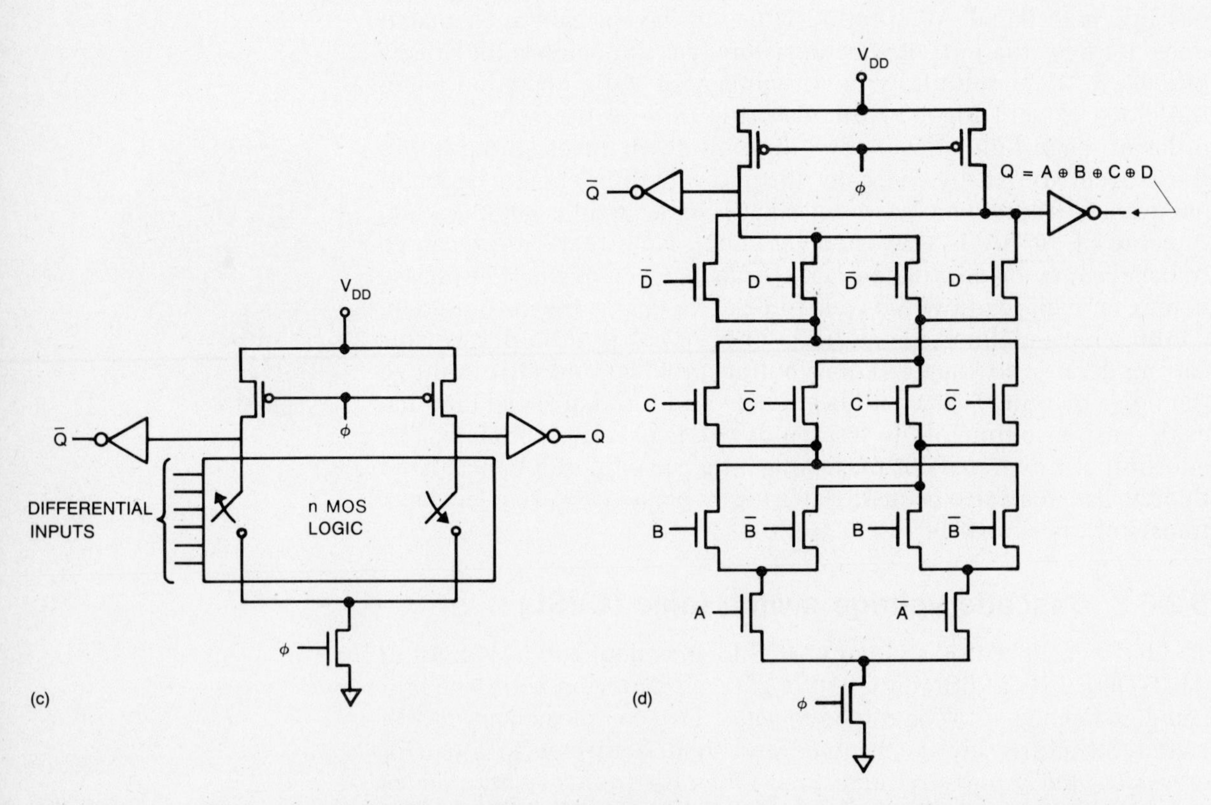

FIGURE 5.11. CVSL logic (a) static version, (b) example of static gate, (c) dynamic version, and (d) four input XOR CVSL gate

imization algorithms. This version, which might be termed a "static" CVSL gate, is slower than a conventional complementary gate employing a p-tree and n-tree. This is because during the switching action, the p pull-ups have to "fight" the n pull-down trees. Fig. 5.11b shows the implementation of the example gate. Note that this is not a very efficient implementation of this gate.

Further refinement leads to a clocked version of the CVSL gate (Fig. 5.11c). This is really just two "domino" gates operating on true and complement inputs with a minimized logic tree. The advantages of this style of logic over domino logic is simply the ability to generate any logic expression, making it a *complete* logic family. This is achieved at the expense of the extra routing, active area, and complexity associated with dealing with double rail logic. However, the ability to generate any logic function is of advantage where automated logic synthesis is required. A four-way XOR gate is shown in Fig. 5.11d [HGDT84].

5.2.7 Modified domino logic

A further refinement of the domino CMOS is shown in Fig. 5.12a on page 172. Basically, the domino buffer is removed, while cascaded logic blocks are alternately composed of p- and n-transistors [FrLi84] [GoDM83]. In the circuit in Fig. 5.12a, when $\phi = 0$, the first stage (with n-transistor logic) is precharged high. The second stage is precharged low and the third stage is precharged high. As the second logic stage is composed of p-transistors, these will all be turned off during precharge. Also, as the second stage is precharged low, the n-transistors in the third logic state will be off. Domino connections are possible as shown in Fig. 5.12b on page 172.

Problems that occur with this type of logic include poor speed response of the p-logic blocks, charge redistribution, and reduced noise margin. These characteristics will be enlarged upon in Section 5.3.6 and Section 5.4.6. The advantages are primarily due to the possibility of using only one clock and also the absence of buffers on the gate outputs.

Common advantages of the dynamic logic styles are as follows:

- Smaller area than fully static gates.
- Smaller parasitic capacitances, hence higher speed.
- Glitch free operation if designed carefully.

The last point is the catch. If you want to use dynamic circuits you must be prepared to invest the extra design effort to ensure correct operation under all circuit conditions.

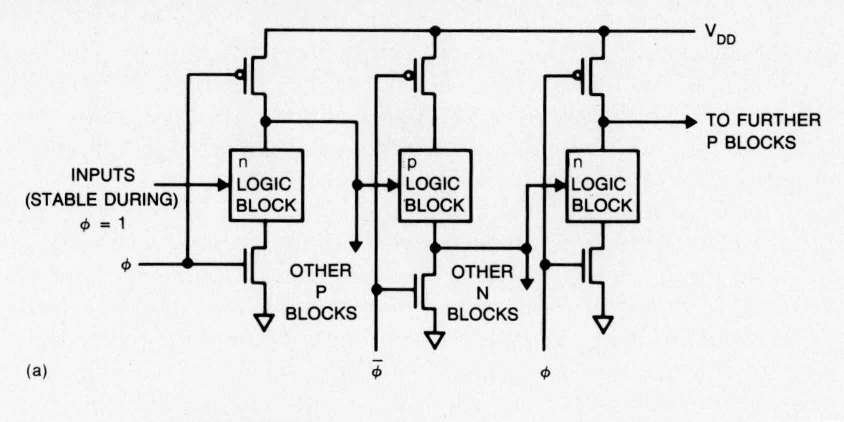

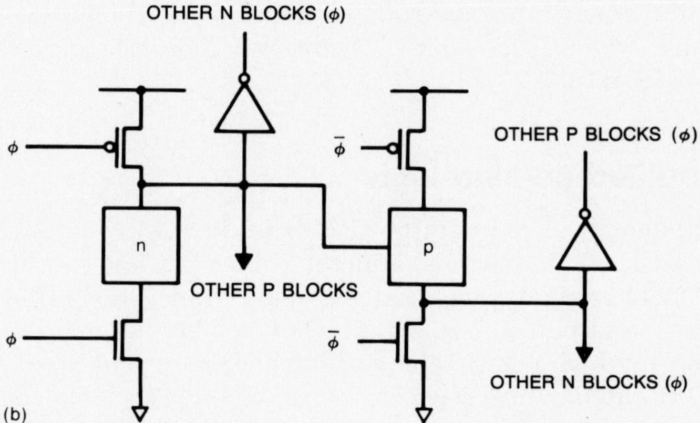

FIGURE 5.12. Alternating n and p domino logic blocks

5.2.8 Pass transistor logic

One form of logic that is popular in nMOS circuits is pass transistor logic, the simplest example probably being a 2-input multiplexer. A popular use of pass transistor logic is the "function unit" used in the ALU in the OM-1 computer [MeCo80]. The nMOS structure is shown in Fig. 5.13a. In CMOS, this structure can be replicated, as shown in Fig. 5.13b, by using a full transmission gate for each original n-transistor. A more realizable layout is possible by using the circuit shown in Fig. 5.13c. This alleviates many direct n- to p-transistor connections. A dynamic version is shown in Fig. 5.13d. In terms of speed, the nMOS version has the fastest fall time with the complementary version having the fastest rise time. Using larger p-transistors decreases the rise time but increases the fall time. The dynamic version is roughly the same speed as the nMOS version but requires a precharge period that may extend clock cycle times. An alternative to the dynamic approach is to include a buffer which

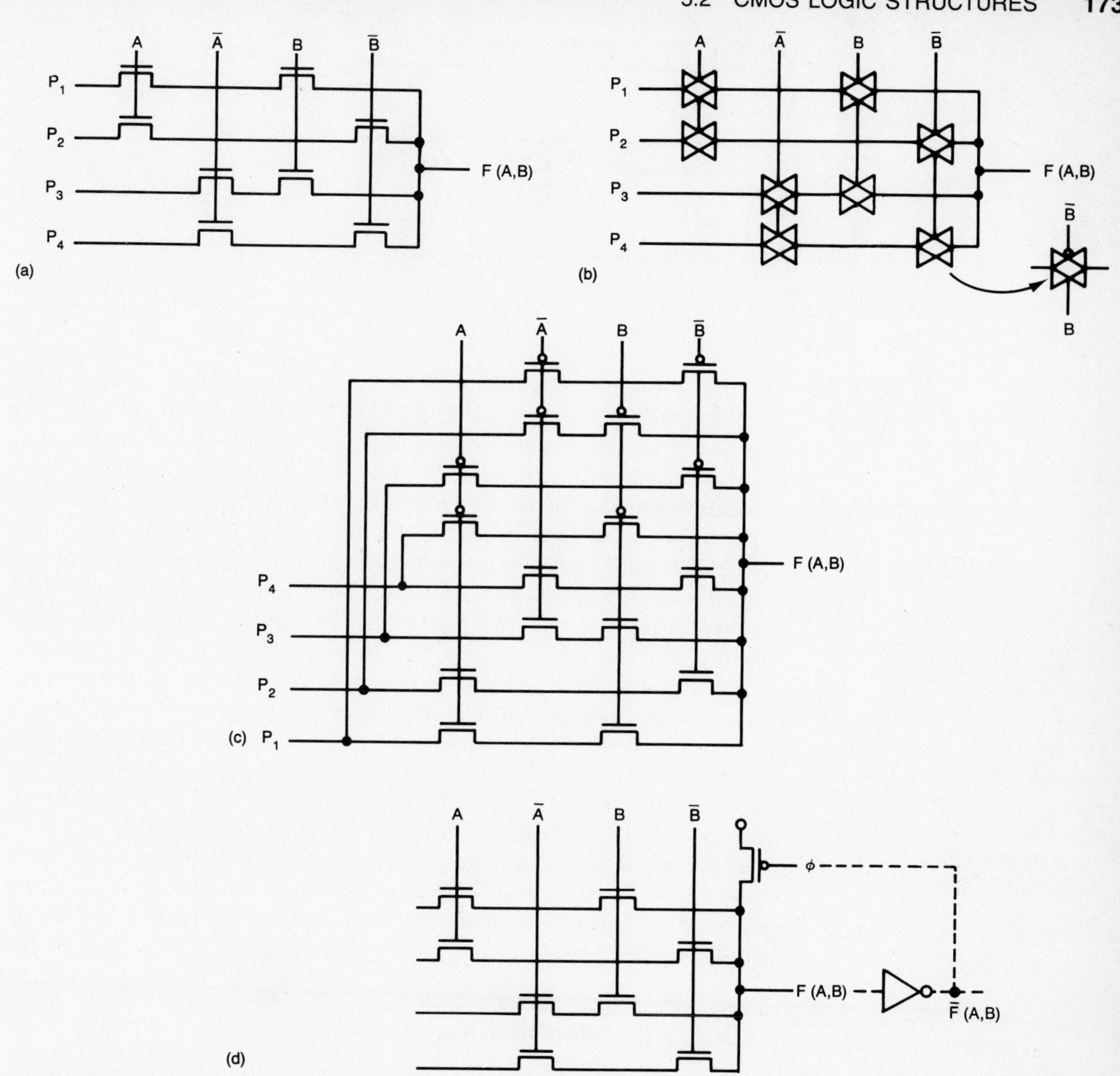

FIGURE 5.13. Pass logic function unit (a) nMOS, (b) full CMOS transmission gates, (c) modified CMOS for better layout, and (d) p-pullup version

is fed back to the p-transistor pull-up. This then yields a static gate with zero DC power dissipation. The p-transistor pull-up and n-transistor pull-downs must be ratioed in accordance with Eq. 2.37 to allow the output buffer to switch.

Formal methods for deriving pass-transistor logic have been presented for nMOS [Whit83]. They are based on the model shown

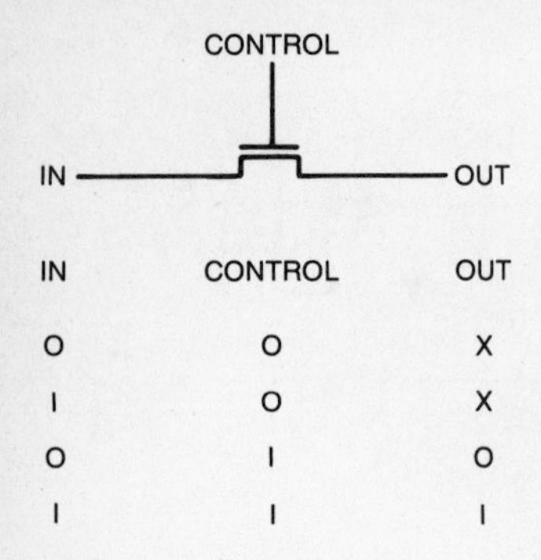

CONTROL
IN
OUT
IN CONTROL OUT
O O X
I O X
O I O
I I I

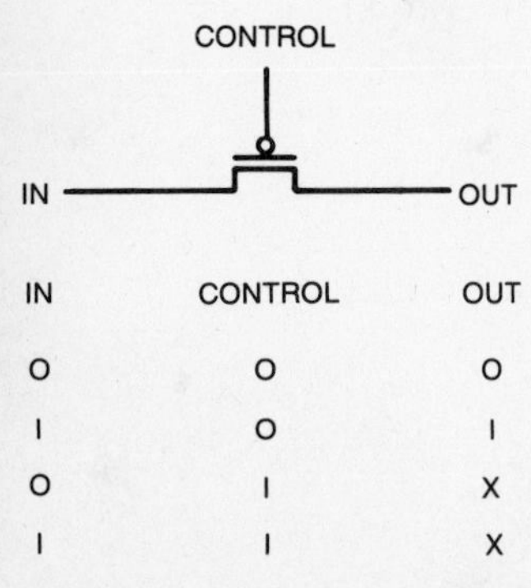

FIGURE 5.14. Pass transistor logic model. Reprinted from *ELECTRONICS*, September 22, 1983. Copyright © 1983, McGraw-Hill Inc., All rights reserved.

TABLE 5.1. XOR truth table

A	B	A ⊕ B	PASS FUNCTION
0	0	0	$A + B$
0	1	1	$\overline{A} + B$
1	0	1	$A + \overline{B}$
1	1	0	$\overline{A} + \overline{B}$

in Fig. 5.14, where a set of variables control a pass transistor network to which pass variables are applied. In the case of an exclusive-or gate, the truth table is shown in Table 5.1. The pass function column refers to the input variables, which *could* be passed to the output to achieve the function. For instance, in the first row of Table 5.1, *A* or *B* may be passed to the output to yield a '0' and hence satisfy the XOR function. A modified Karnaugh map may be drawn for the pass functions, as shown in Table 5.2. The input variables are grouped to appropriately steer the pass variables to the output under the influence of the control variables. In this case, $\overline{B}$ is a pass variable under the control of *A*, and *B* is a pass variable under the control of $\overline{A}$. The resulting structure is shown in Fig. 5.15. Note that groupings that pass both true and false input variables to the output must be made to avoid undefined states. In addition, if a complementary version is required the p pass function must also be constructed. This is just the dual of the n-structure.

The apparent advantages of pass transistor networks in CMOS should be studied carefully and judiciously utilized. A few points detract from the use of pass networks. To achieve good logic levels complementary pass networks are desirable but incur extra delay

TABLE 5.2. Modified Karnaugh map

		A = 0		*A* = 1	
B	0	*A*	*B*	*A*	$\overline{B}$
	1	$\overline{A}$	*B*	$\overline{A}$	$\overline{B}$

in pull-down. In comparison to regular gates, the merging of source and drain regions is difficult, leading to higher internal node capacitances. Finally, true and complement control variables are required. The best use that can be made of such networks is when the output nodes can be precharged, or the static implementation as illustrated in the function block implementation shown in Fig. 5.13d. The effectiveness of any pass transistor network must be assessed for any given situation by simulation and layout. Note that the pass networks derived here may be used with the CVSL logic mentioned previously in Section 5.2.6.

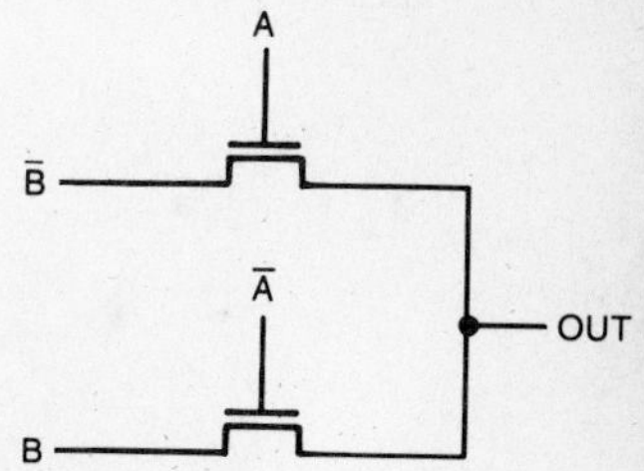

FIGURE 5.15. Pass transistor structure for XOR function. Reprinted from *ELECTRONICS*, September 22, 1983. Copyright © 1983, McGraw-Hill Inc. All rights reserved.

5.3 Electrical and physical design of logic gates

We have summarized some alternate styles of CMOS logic. In this section we will examine the physical layout of CMOS gates in a general sense to examine the impact of the physical structure on the behavior of the circuit. In addition, more detailed analyses of some of the detrimental effects mentioned in the first section will be completed. This section begins with an outline of different inverter layout forms. (To simplify layouts, "unit" sized transistors will generally be shown. In actual layouts, the correct dimension transistors would be arrived at via detailed circuit design. p-Transistors will often be shown double "unit" size.)

5.3.1 The inverter

In Chapter 1, we defined a symbol and connection strategy that may be used to indicate a transistor. By examining the circuit diagram for the inverter (Fig. 5.16a), we should be able to effect a physical

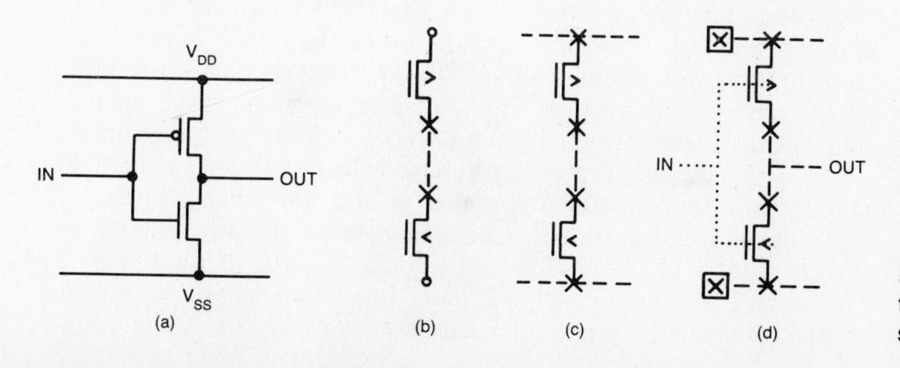

FIGURE 5.16. Schematic to symbolic layout conversion for inverter

layout by substituting this symbol for the schematic symbol. In a schematic, lines drawn between device terminals represent connections. Any nonplanar situation is dealt with by simply crossing two lines (i.e., the connection between the drain of the n-transistor and the drain of the p-transistor). However, in a physical layout, we have to concern ourselves with the interaction of physically different interconnection layers. We know from our consideration of the fabrication process, that the source and drain of the n-transistor are n-diffusion regions, while the p-transistor uses p-diffusion regions for these connections. Additionally, in a bulk CMOS process, we cannot make a direct connection from n-diffusion to p-diffusion. Thus we have to implement the simple interdrain connection in the structural domain, as at least one wire and two contacts in the physical domain. Assuming that the process does not have buried contacts, this connection has to be in metal. Substituting our layout symbols, the partial inverter shown in Fig. 5.16b results. By similar reasoning, the simple connections to power (V_{DD}) and ground V_{SS} could be made using metal wires and contacts (Fig. 5.16c). Power and ground are usually run in metal (for low resistance from circuit to power supply). The common gate connection may be a simple polysilicon wire. Finally, we must add substrate contacts that are not implied in the schematic. The resulting symbolic schematic is shown in Fig. 5.16d. Converting this to a symbolic layout yields the arrangement shown in Fig. 5.17a. An alternative layout is shown in Fig. 5.17b, where the transistors are aligned horizontally.

Note that there are some topology variations that may be used to enable nonplanar connection schemes to be implemented. For instance, if a metal line has to be passed through the middle of the cell from the left end of the cell to the right end, the layout shown in Fig. 5.17c could be used. Here, horizontal metal straps connect to a vertical polysilicon line, which in turn connects the drains of the transistors. Alternatively, if a metal line is to be passed from left to right at the top or bottom of the cell, the power and ground connections to the transistors may be made in the appropriate diffusion layer (Fig. 5.17d). This, in effect, makes the inverter transparent to horizontal metal connections that may have to be routed through the cell. From the considerations that affect performance, the previous deviations from the original layout have little effect. In the case of the vertical polysilicon drain connection, an extra connection resistance is incurred. This would be approximately $2R_{contact} + R_{poly}$, where $R_{contact}$ is the resistance of a metal-polysilicon contact and R_{poly} is the resistance of the polysilicon runner. In addition, a slight extra capacitance may be incurred. Usually, the result of both of these effects would be inconsequential. For the power and ground diffusion connections, the penalty is a series connection resistance and increased capacitance. As a rule of thumb,

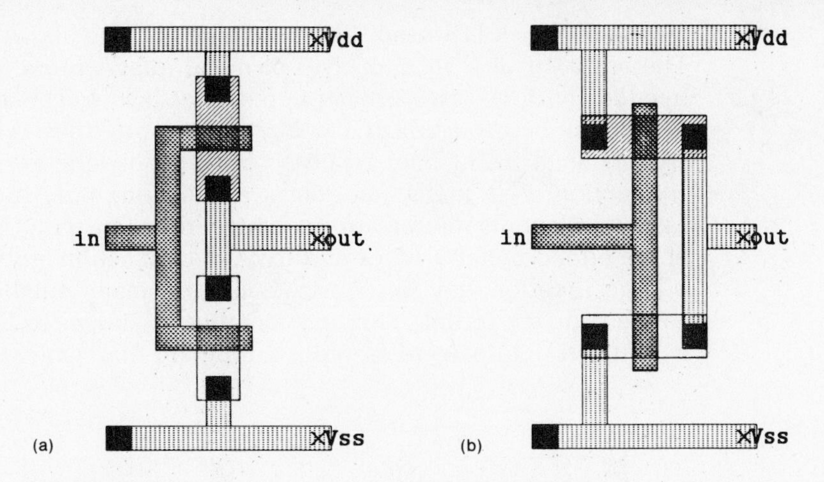

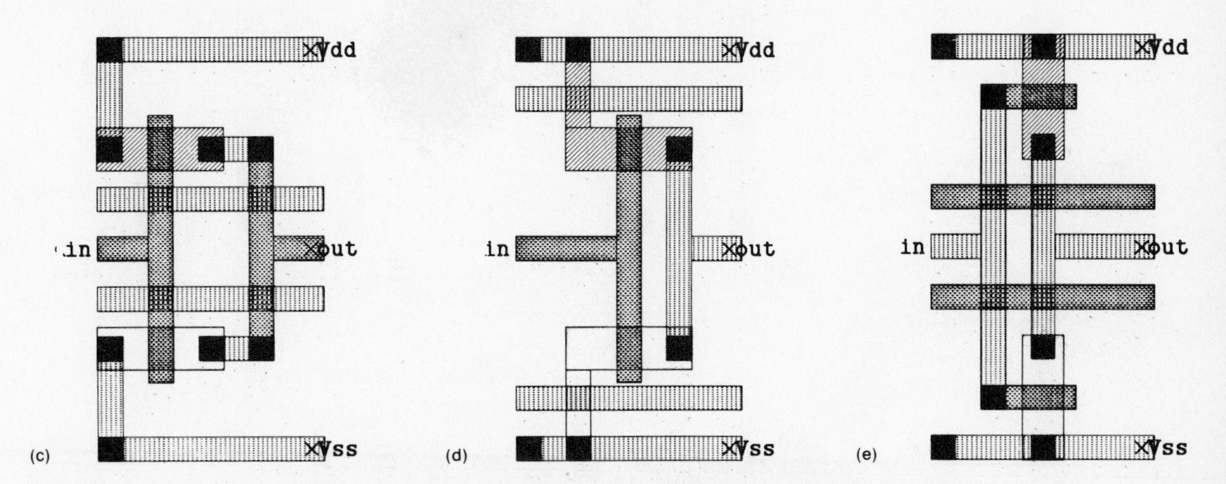

FIGURE 5.17. Various forms of an inverter layout

the resistance should be kept an order of magnitude below the transistor "on" resistance. The capacitance on supply connections does not normally affect performance.

Running a polysilicon connection from left to right must be completed below or above the transistors, with the transistors using metal connections to power and ground. Polysilicon passing from left to right through the middle of the cell requires a metal strap. Alternatively, the inverter layout may be reconstructed to use vertical oriented transistors, as originally proposed in Fig. 5.17a, and the layout in Fig. 5.17e used. These layouts are also shown in Plate 5. The addition of a second layer of metal allows more interconnect freedom with the two other interconnect layers. The second level metal may be used to run V_{DD} and V_{SS} supply lines. Alternatively, second level metal may be used to strap polysilicon in a parallel connection style to reduce delays due to long poly runs. In these cases, the layouts remain approximately the same with the exception of the added metal-2 wires and metal-1 connection stubs. Note that a large inverter may be constructed from many smaller inverters connected in parallel. This is symbolically shown in Fig. 5.18a. It is quite straightforward to write a program that generates a certain

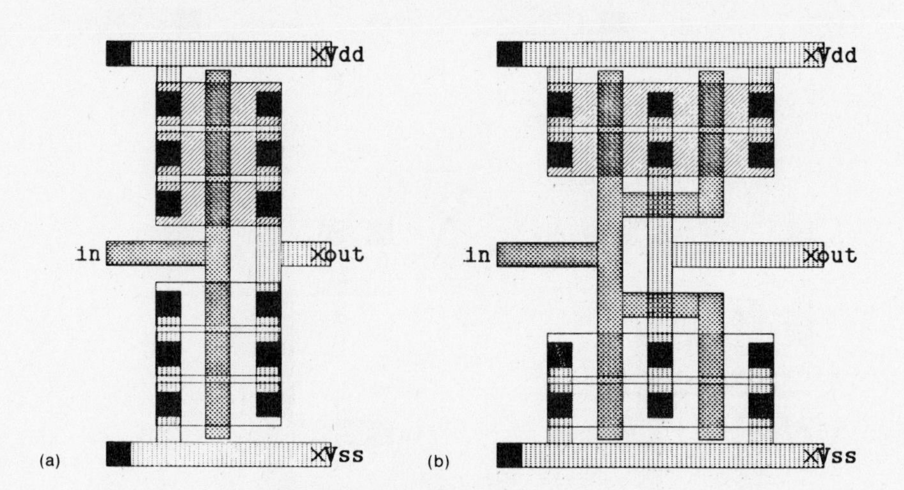

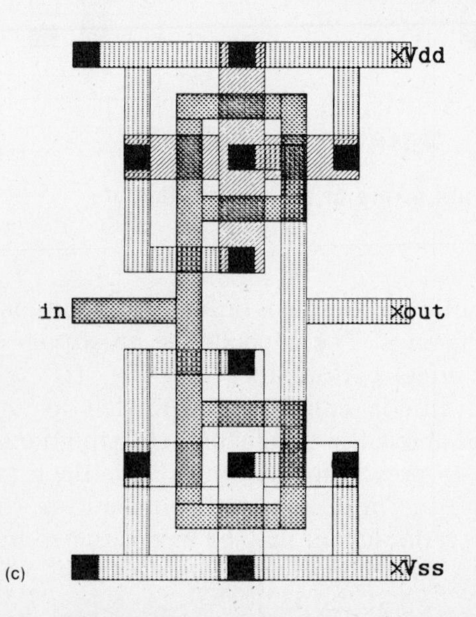

FIGURE 5.18. Paralleled inverter layouts

sized buffer in a given aspect ratio using this technique. The source and drain regions should be "stitched" with the contacts and metal to reduce source-drain resistance in large transistors. In addition adjacent diffusions are merged so that peripheral capacitance is reduced. Placing transistors back to back (Fig. 5.18b) yields a more optimum drain capacitance due to the merged diffusion regions. This results from the fact that the drain area does not increase in size much but the gain of transistors (β) is doubled. A further reduction in drain capacitance is achieved by using the star connection shown in Fig. 5.18c. Note that Fig. 5.18c represents the configuration symbolically. In mask the source and drain would be one continuous area with no corner gaps to increase gain and reduce peripheral capacitance. Here the β of the transistors is quadrupled, while the drain area is substantially the same as for a single inverter. Plate 6 shows these layouts in color.

In essence, these variations represent some "forms" for an inverter (and to an extent, other gates) that will be used in various situations in this text.

5.3.2 NAND and NOR gates

Similar reasoning can be applied to converting the 2-input NAND schematic to a layout. Fig. 5.19a shows a direct translation of the schematic. By orienting the transistors horizontally, the layout in Fig. 5.19b is possible.

Note that in the case of the NAND gate, the latter layout is much cleaner (and smaller). This is, in general, true for multiple input static gates, and we will adopt a style where transistors are oriented

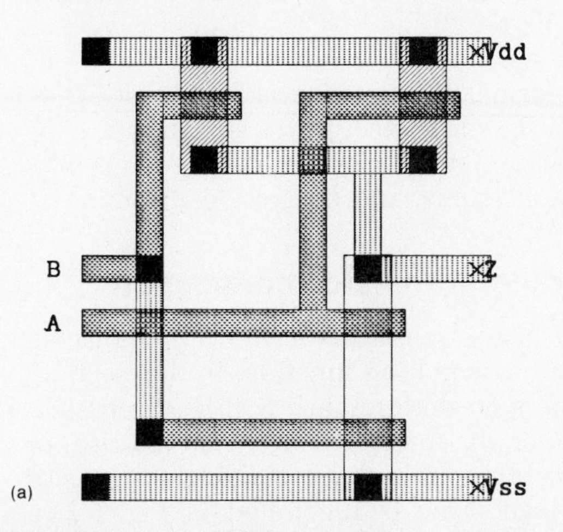

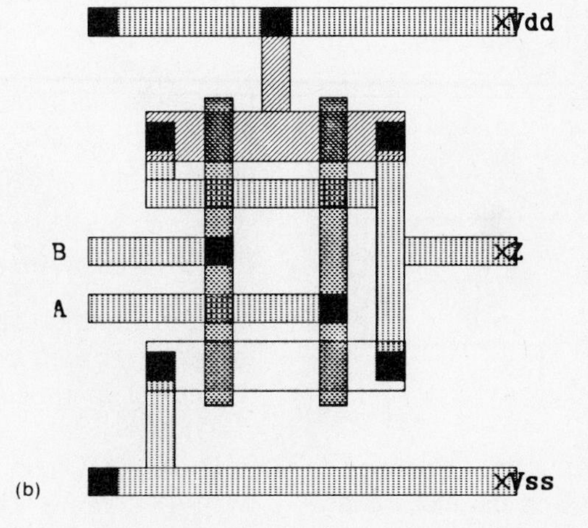

FIGURE 5.19. NAND layouts

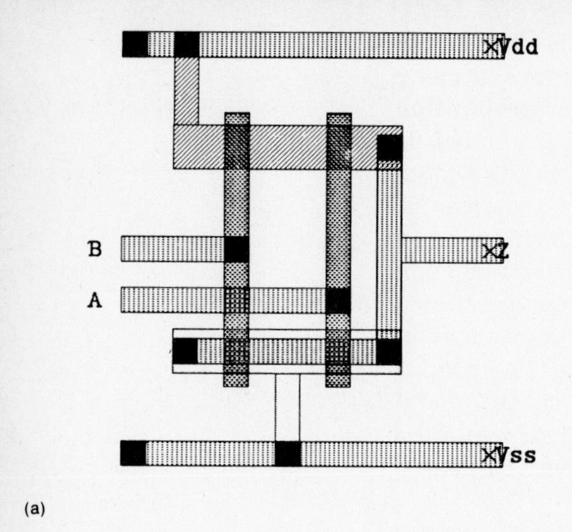

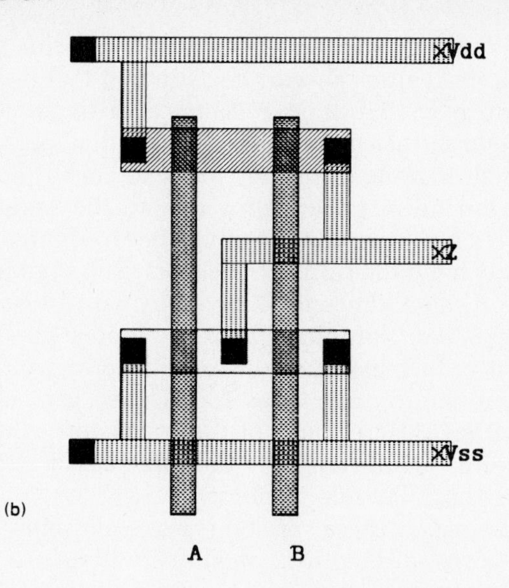

FIGURE 5.20. NOR layouts

horizontally and polysilicon gate signals run vertically. Where departures are made from this style, the reasons for doing so will be given. Note, of course, that the gate could be rotated 90° to obtain vertical metal and horizontal polysilicon connections.

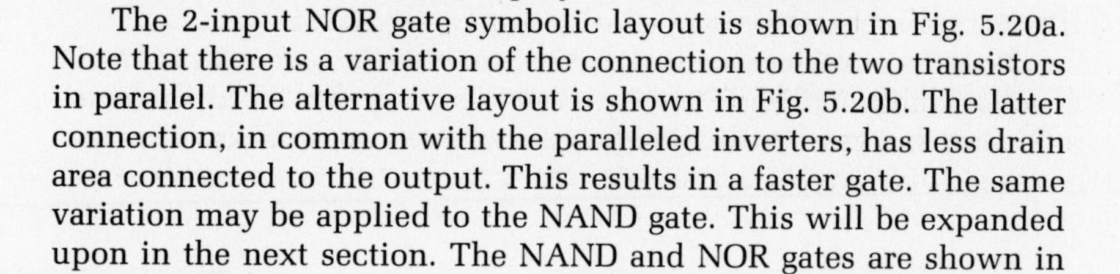

The 2-input NOR gate symbolic layout is shown in Fig. 5.20a. Note that there is a variation of the connection to the two transistors in parallel. The alternative layout is shown in Fig. 5.20b. The latter connection, in common with the paralleled inverters, has less drain area connected to the output. This results in a faster gate. The same variation may be applied to the NAND gate. This will be expanded upon in the next section. The NAND and NOR gates are shown in color in Plate 7.

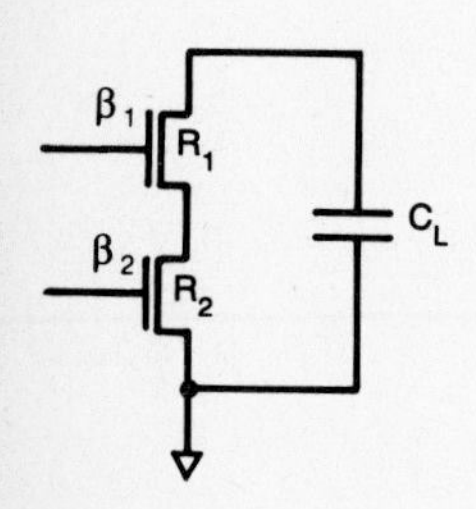

Complex gates are an extension of the gates so far treated. However, four factors affect the electrical, and hence, physical design of such gates. Specifically, these factors are, series transistor connection, body effect, source-drain capacitance, and charge redistribution.

5.3.3 Series and parallel transistor connection

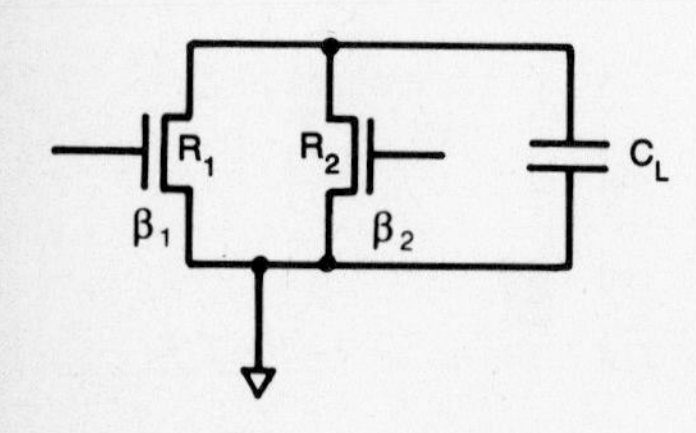

FIGURE 5.21. Effect of series and parallel transistor connections

In a number of examples to date, transistors have been connected in series and parallel to form a switching function. In this section, the effect of these connections on performance will be discussed.

If two identical transistors are connected in series the rise (or fall) time will be approximately double that for a single transistor with the same capacitive load. This is illustrated in Fig. 5.21a. Consider the two n-transistors in series. When the transistors are in their linear region, the fall time T_f, for capacitance load C_L, in an inverter is proportional to $C_L R_n$, where R_n is the *resistance* of the combined set of transistors. The resistance of a single transistor in its linear region of operation is approximately equal to $\frac{1}{\beta_n(V_{DD} - V_{t_n})}$. Thus T_f is proportional to $\frac{C_L}{\beta_n(V_{DD} - V_{t_n})}$. Combining two transistor resistances (R_1 and R_2) is achieved by evaluating $R_{total} = R_1 + R_2 = \frac{\beta_{n1} + \beta_{n2}}{\beta_{n1}\beta_{n2}}$. For equal β_n values this yields a fall time, T_f', twice that of an equivalently sized inverter. This approximation assumes that the transistors are in their linear region of operation for most of the switching time. In general, the fall time T_f' is mT_f for m n-transistors in series. Similarly the rise time T_r' for k p-transistors in series is kT_r.

In comparison, the fall time T_f'' for a parallel connection of transistors is T_f/m for m transistors in parallel, if all the transistors are turned on simultaneously. T_r'' for k p-transistors in parallel is T_r/k for k devices in parallel. These times are important if the fastest delay through a gate has to be evaluated.

When gates with large numbers of inputs have to be implemented, the best speed performance may be obtained by using gates where the number of series inputs ranges from about 2–5. To illustrate this point, a very simple analysis will be presented. We will consider T_R, the worst case rise time for an m input NAND (one p-device on) gate, to be (parasitic capacitances in transistors ignored)

$$T_R = R_p(mC_D + C_L), \qquad \textbf{(5.1)}$$

where

$$R_p = \frac{4}{\beta_p V_{DD}}$$ – *resistance of p-devices in gate*

C_D = *capacitance of a unit drain area*

C_L = *other load capacitance on gate (routing and fan-out).*

The fall time T_F is approximated by

$$T_F = mR_n(mC_D + C_L), \qquad \textbf{(5.2)}$$

where

$$R_n = \frac{4}{\beta_n V_{DD}}$$ – resistance of n-devices in gate

C_D and C_L as above. The equations for an m input NOR gate are similar in nature

$$T_R = mR_p(mC_D + C_L) \qquad \textbf{(5.3)}$$

$$T_F = R_n(mC_D + C_L). \quad \text{(one n-device on)} \qquad \textbf{(5.4)}$$

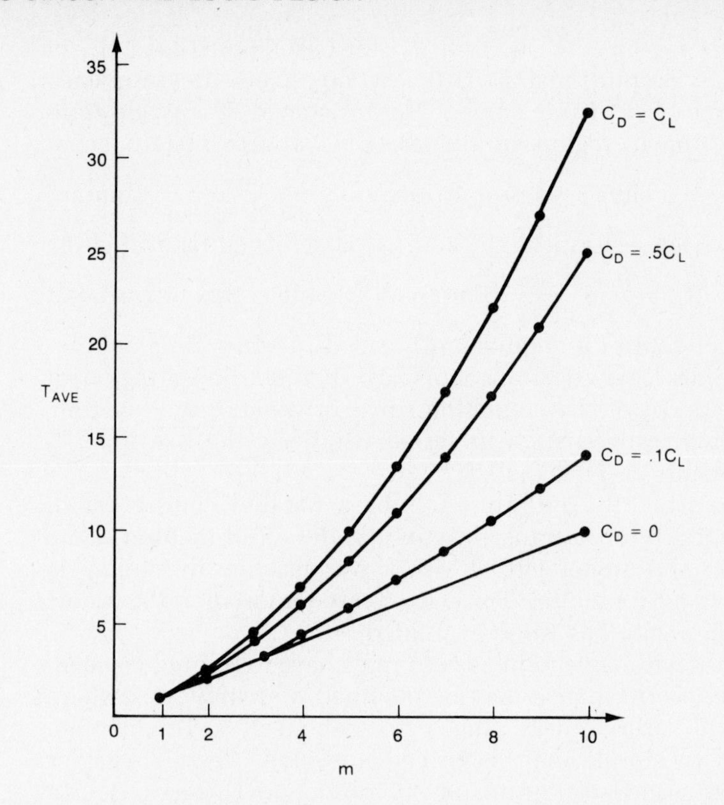

FIGURE 5.22. Delay characteristics of cascaded gates

If we normalize these times against the inverter time (m = 1), we get

$$\begin{aligned} T_{AVE(NOR)} &= T_{AVE(NAND)} \\ &= \frac{1}{2}\left(\frac{mC_D + C_L}{C_D + C_L} + \frac{m(mC_D + C_L)}{C_D + C_L}\right) \qquad \textbf{(5.5)} \\ &= \frac{m^2C_D + 2(m + 1)C_L + mC_D}{2(C_D + C_L)}. \end{aligned}$$

This relationship is graphed in Fig. 5.22 for a variety of ratios of C_D to C_L. This shows that as C_L increases, the relative contribution to delay caused by internal gate capacitance is reduced. For gates with a small routing load, an approximate ratio of $C_D = .5C_L$ will be used. If we consider the implementation of an 8-input AND gate, we may use the following (Fig. 5.23):

- an 8-input NAND and an inverter — approach 1
- two 4-input NANDs and a 2-input NOR — approach 2
- four 2-input NANDs, two 2-input NORs, and an inverter — approach 3.

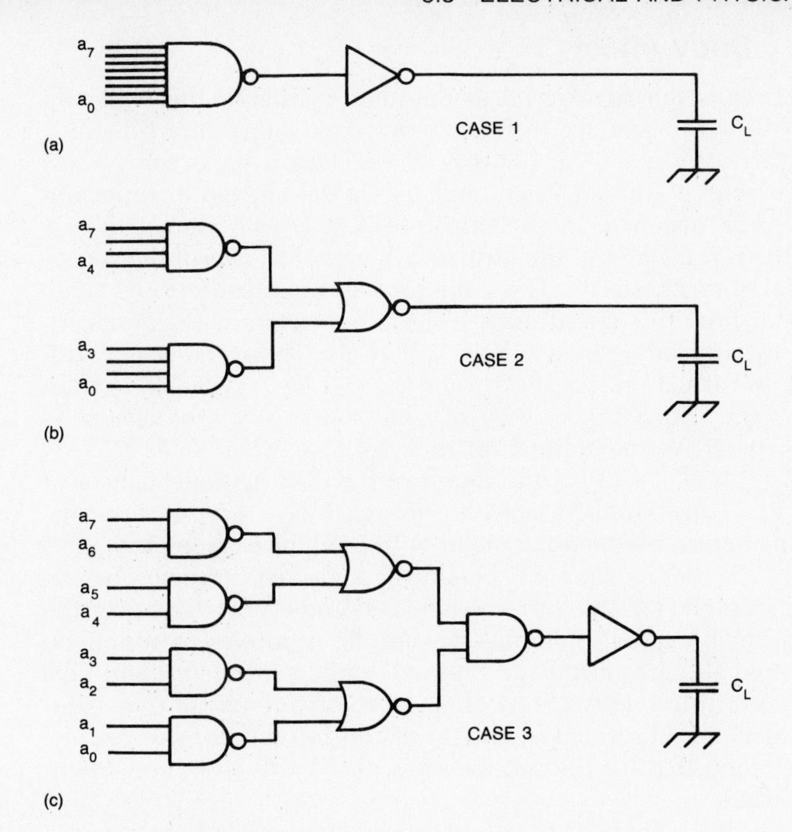

FIGURE 5.23. 8-input gate comparison

Using the graph we determine that approach (3) is the fastest, approach (2) the next fastest, and the 8-input gate is the slowest. For other values of C_D, where C_L dominates, approach (2) is the fastest, illustrating a final selection would have to be made on the basis of simulation. Simulation results carried out for the circuits shown with nominal loading predicted that approach (2) would be the fastest. This analysis is far from rigorous and is used for illustrative purposes only. A detailed simulation should be used to verify any design choices.

Note that any series resistance inserted in series with the charging or discharging path of a gate will affect switching speed. For instance, if an n-transistor is connected to the V_{SS} supply by a long resistive wire, the gate will be slower than necessary. Therefore, one should watch long resistive connections in gates. This also includes trying to connect power supplies to gates via resistive polysilicon and contacts.

5.3.4 Body effect

Body effect is the term given to the modification of the threshold voltage V_t with a voltage difference between source and substrate. Specifically, $\Delta V_t \propto \gamma\sqrt{V_{sb}}$ where γ is a constant, V_{sb} is the voltage between source and substrate, and ΔV_t is the change in threshold voltage. For instance, in the multiple NAND gate shown in Fig. 5.24a, the n-transistor at the output will switch slower if the source potential of this transistor is not the same as the substrate. Fig. 5.24b illustrates how this could occur. The n-transistors with inputs A–C are initially off ($V_{gsA} = V_{gsB} = V_{gsC} = 0$). The n-transistor with input D is turned on ($V_{gsD} = V_{DD}$ and then off ($V_{gsD} = 0$). This action charges the capacitance (C_1) at the source of n-transistor D ($V_{sbD} \neq 0$). If all the inputs are then set to a HIGH level ($V_{gsA} = V_{gsB} = V_{gsC} = V_{gsD} = V_{DD}$), the source of D will instantaneously be at $V_{DD} - V_{t_n}$. Thus n-transistors with gate signals A–C have to discharge this node to turn on the n-transistor with D on the gate. In particular, the fall time of this gate will be slower than that predicted by the approximations for the series connected transistors that was the basis for the previously completed example. To minimize this effect, gate design should minimize "internal" node capacitance and take into account the relative body effect of the two types of transistor. If, for instance, the relative impact of the n-transistor body effect is worse than that for the p-transistors, then NOR structures might be preferred.

Given that a number of series transistors may be required in a gate, a further optimization may be made. As the body effect is

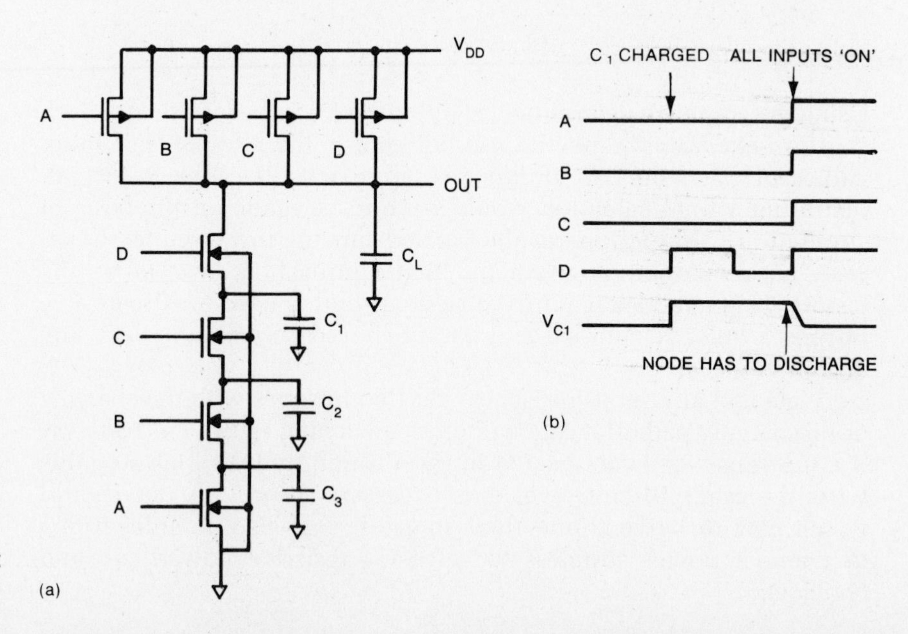

FIGURE 5.24. Body effect in a multiple input gate

essentially a dynamic problem involving the charging of parasitic capacitances, we can use the natural time sequencing of signals to offset the body effect. The first strategy is to place the transistors with the latest arriving signals nearest the output of a gate. The early signals, in effect, "discharge" internal nodes and the late arriving signals have to switch transistors with minimum body effect. The other strategy mentioned previously is to minimize the capacitance of internal nodes. Thus if a diffusion wire had to be used to minimize the geometric topology of a gate, then one would try to use it at the output of a gate rather than on some internal node. In the same vein, connections on internal nodes should be completed in metal or, if buried contacts are allowed, polysilicon. The diffusion attached to transistors should be made the minimum that design rules allow.

5.3.5 Source-drain capacitance

The third effect that leads to less than ideal gates occurs in the case of parallel connected transistors. This effect was encountered in the construction of parallel inverters and was also demonstrated in the 2-input NOR gate constructed previously. In the NOR schematic, the output is connected to one p-transistor drain and two n-transistor drains. However, in one of the NOR layouts (Fig. 5.20b), the drain connection between the two n-transistors is merged. This effectively means that only two drain connections are connected to the output, thus reducing the capacitance at the output. The parallel connection of two sources to the ground rail adds capacitance to the ground rail but this does not affect the output switching speed. Another example is seen in the gate that implements the function $F = \overline{(A + B + C).D}$ (Fig. 5.25a, page 186). The n-transistor connection for this gate is shown in Fig. 5.25b. The ground connection may be made at point 1 or 2. Point 1 would be preferred, as this connects three of the source regions to ground (V_{SS}). Actually, by merging the source-drain connections only two V_{SS} connections are made. In general, as a result of this effect and body effect, we try to assemble the most capacitive nodes closest to the supply and ground rails. Symbolic layouts for the function in Fig. 5.25a are shown in Fig. 5.25c and Fig. 5.25d, illustrating two approaches to implementing the gate. The gate in Fig. 5.25c has one "unit" output p-drain capacitance, one output n-drain capacitance, and four "internal" drain capacitances (two n and two p). Fig. 5.25d has four output drain capacitances and four internal drain capacitances. Thus the layout in Fig. 5.25c improves diffusion capacitance by at least one n and one p. Where this capacitance dominates, the optimized layout would result in a faster circuit.

Note that these strategies may coincide. For instance, in the complex gate shown in Fig. 5.25, the signal D may be delayed with

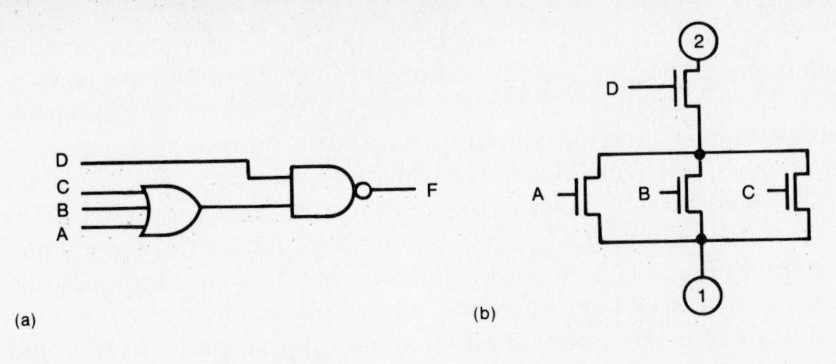

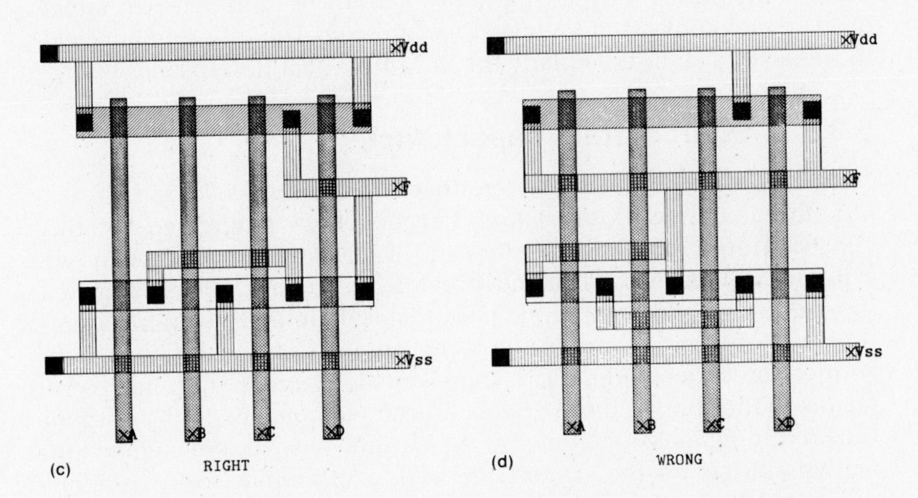

FIGURE 5.25. Optimization of multiple source-drain connections

respect to signals A, B, and C. This also indicates that connection point 1 should be grounded. If there is some doubt regarding the organization of a gate (i.e., signal A arrived first), a simulation should be done on the gate with appropriately timed inputs.

5.3.6 Charge redistribution (charge sharing)

We have frequently mentioned the fourth effect as a problem when considering dynamic gates. Consider the circuit shown in Fig. 5.26a, which shows a clocked inverter. Parasitic capacitances representing the source-drain capacitance and the external gate capacitance are indicated. A typical signal sequence is shown in Fig. 5.26b. If the clock ϕ pulses high and the input is low, then node 3 is discharged while the output is driven high. When the clock turns off the output remains charged high. If the input is now asserted, capacitor C_3 will charge, reducing the output voltage V_o by

$$V_o' = \frac{C_o}{C_3 + C_o} V_o.$$

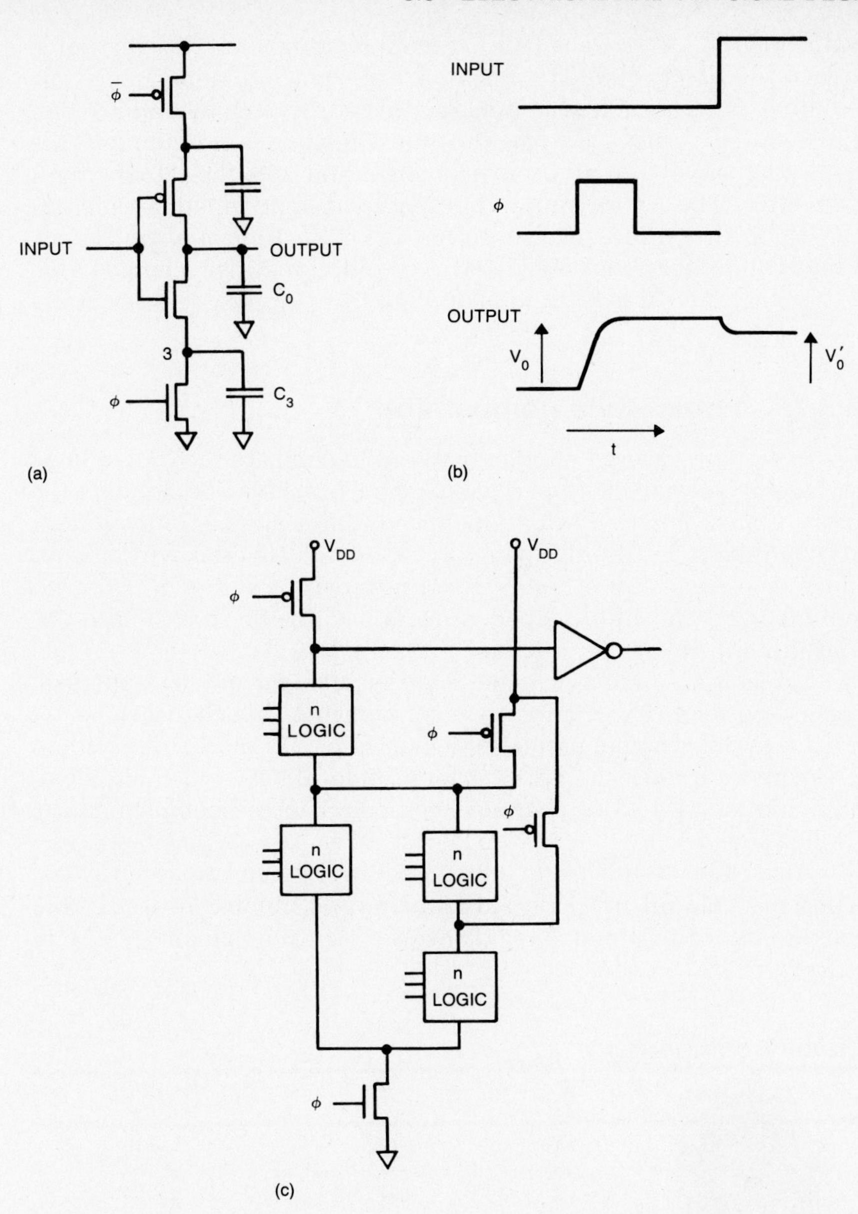

FIGURE 5.26. Charge redistribution in a clocked gate

Depending on the ratio of capacitances, this effect could switch the next gate. This means that a change in data has erroneously affected the output of the gate.

Solutions to charge redistribution involve carefully arranging the logic trees so that the effects are minimized. For example, in accordance with the formula derived above, the combined internal capacitance of a tree structure should be of such a value that the output voltage cannot cause a subsequent gate to falsely switch. A value of $C_{output}/C_{internal}$ of 10:1 is a good starting point. If a very large tree is necessary, added capacitance to the output node may be required. In forms of four-phase logic this was a common optimization process. To verify a design, the safest method is to simulate the gate exhaustively for all input sequences and with the actual circuit parasitics. The use of simple noncascaded logic functions such as the NOR structure can also aid in the reduction of such effects. Finally, internal nodes may have to be individually precharged. Fig. 5.26c illustrates this technique applied to a complex gate structure.

5.3.7 Logic style comparison

We have now covered enough material to compare the CMOS logic styles that we have summarized. Comparisons may be done on the basis of area, power dissipation, speed, noise margin, or even fault susceptibility. In the following, a "standard" inverter will be considered to consist of a "unit"-sized n-transistor and a "unit"-sized p-transistor. The input capacitance is 2 □ C_g, the rise time is 2τ, and the fall time is $\tau(\tau = 2R_n(C_g + C_d))$.

Table 5.3 summarizes the relationships for an m input gate (assuming load capacitance swamps parasitic capacitances).

All families except pseudo-nMOS have zero DC power dissipation. The zero level noise margin (V_{IL}) for pseudo-nMOS, dynamic CMOS, and domino CMOS is degraded compared to the complementary CMOS case.

Area can be estimated to a first order by the number of transistors. Thus the column in Table 5.3 showing the number of transistors may be used to estimate area. However, the complementary CMOS

TABLE 5.3. CMOS logic family comparison

LOGIC FAMILY	INPUT C	T_r	T_f	NO. TRANSISTORS
CMOS complementary (NAND)	2mCg	2τ	$m\tau$	2m
(NOR)	2mCg	$2m\tau$	τ	
Pseudo-nMOS (NAND)	mCg	6τ	$m\tau$	m + 1
(NOR)	mCg	6τ	τ	
Dynamic CMOS (NAND)	mCg	NA	$m\tau$	m + 2
(NOR)	mCg	NA	τ	
Clocked CMOS (NAND)	2mCg	4τ	$(m + 1)\tau$	2m + 2
(NOR)	2mCg	$2(m + 1)\tau$	2τ	
Domino CMOS (NAND) (N)	mCg	NA	$(m + 1)\tau$	m + 4
(including (NOR) (N)	mCg	NA	2τ	(m + 2
n-p CMOS) (NAND) (P)	mCg	$2(m + 1)\tau$	NA	for
(NOR) (P)	mCg	4τ	NA	n-p CMOS)

TABLE 5.4. CMOS logic style area comparison

LOGIC TYPE	AREA
CMOS	
complementary	1
Pseudo-nMOS	.95
Dynamic CMOS	.98
Clocked CMOS	1.5
Domino CMOS	1.3

gate has a very efficient layout style and layout density may be often determined by routing considerations. As an example the gates used in Figs. 5.1 through 5.10 were designed using a symbolic layout system with compaction. The area comparisons are shown in Table 5.4. Although this may not be representative of larger logic structures, it illustrates that the relationship between area and number of transistors is only an *estimate*. Usually routing dominates a design and it is often possible to sneak active devices under routing. Table 5.4 is included to illustrate that many simple-minded comparisons of area based on the number of devices may be erroneous in the system context. As a point of reference, CMOS designs are commonly quoted as being 20–30 percent larger than the nMOS equivalent designs.

Note that dynamic gates require a precharge interval. This may or may not impact performance. In general, small logic blocks should be implemented statically while large logic cascaded structures might be implemented with, say, domino dynamic logic. Many designers prefer to use static logic wherever possible. Of course, prior to commencing design a global clocking strategy should be formulated that defines what kind of dynamic logic can be used.

5.3.8 Physical layout of logic gates

All complementary gates may be designed using a single row of n-transistors above or below a single row of p-transistors, aligned at common gate connections. Most "simple" gates may be designed using an unbroken row of transistors in which abutting source-drain connections are made. This is sometimes called the "line of diffusion" rule, referring to the fact that the transistors form a line of diffusion intersected by polysilicon gate connections.

If we adopt this layout style, it has been shown that there are techniques for automatically designing such gates [UeVC79]. Those automated techniques that are applicable to static complementary gates are reviewed here. The CMOS circuit is converted to a graph where 1) the vertices in the graph are the source/drain connections,

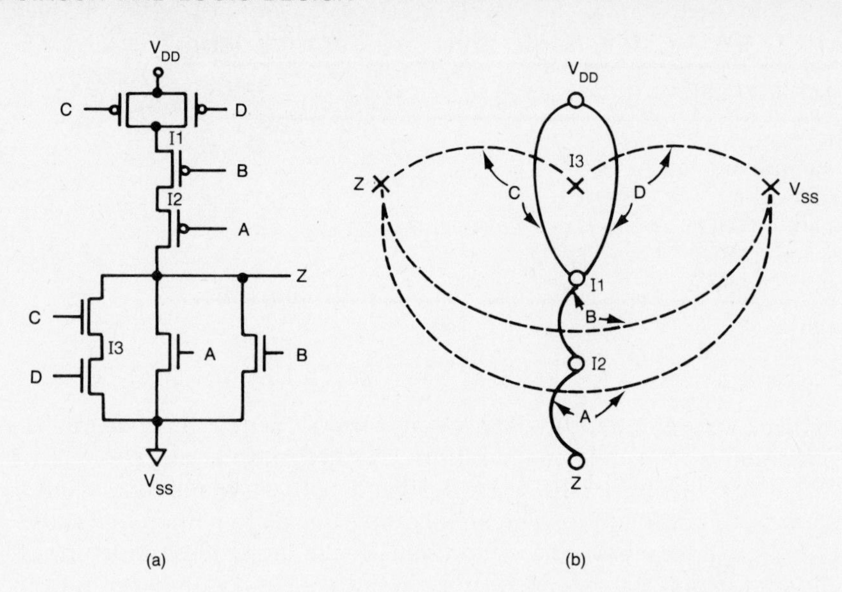

FIGURE 5.27. CMOS logic gate graph representation

and 2) the edges in the graph are transistors that connect particular source-drain vertices. Two graphs, one for the n-logic tree and one for the p-logic tree result. Fig. 5.27 shows an example of the graph transformation. The connection of edges in the graphs mirror the series-parallel connection of the transistors in the circuits. Each edge is named with the gate signal name for that particular transistor. Thus, for instance, the p-graph has four vertices: Z, I1, I2, and V_{DD}. It has four edges, representing the four transistors in the p-logic structure. Transistor A (A connected to gate) is an edge from vertex Z to I2. The other transistors are similarly arranged in Fig. 5.27b. Note that the graphs are the dual of each other as the p- and n-trees are the dual of each other. The n-graph overlays the p-graph in Fig. 5.27b to illustrate this point. If two edges are adjacent in the p- or n-graph, then they may share a common source-drain connection and may be connected by abutment. Furthermore, if there exists a sequence of edges (containing all edges) in the p-graph and n-graph that have identical labeling, then the gate may be designed with no breaks. This path is known as an Euler path. The main points of the algorithm [UeVC79] are as follows:

1 Find all Euler paths that cover the graph.

2 Find a p- and n-Euler path that have identical labeling (a labeling is an ordering of the gate labels on each vertex).

3 If **2**) is not found, then break the gate in the minimum number of places to achieve **2**) by separate Euler paths.

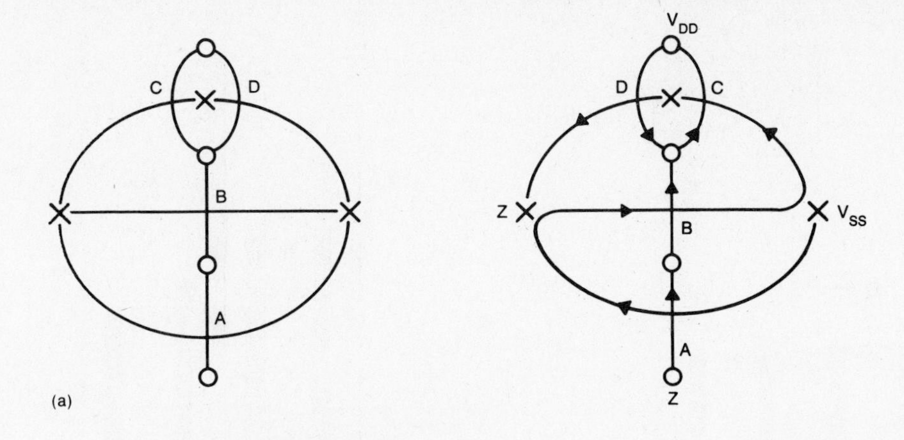

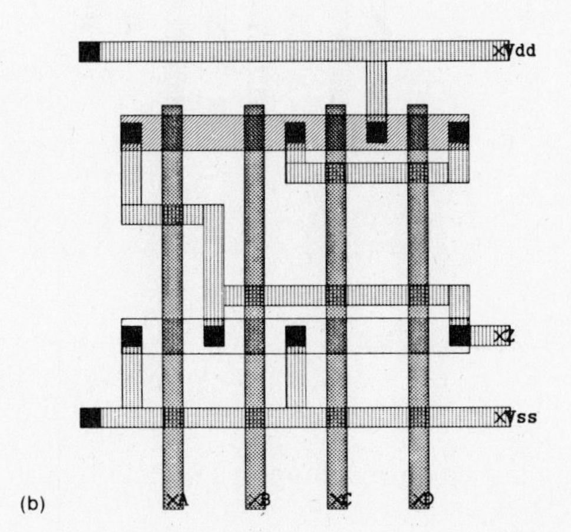

FIGURE 5.28. Euler paths in CMOS gate and the corresponding layout

In the example shown in Fig. 5.27, the original graph with a possible Euler path is shown in Fig. 5.28a. The sequence of gate signal labels in the Euler path is ⟨A,B,C,D⟩. Note that the graph for the n- and p-graph allow this labeling. To complete a layout the transistors are arranged in the ordering of the labeling, n- and p-transistors in parallel rows, as shown in Fig. 5.28b. Vertical polysilicon lines complete the gate connections. Metal routing wires complete the layout. This procedure may be followed when manually designing a gate.

A variation of the single line of n- and p-transistors occurs in logic gates where a signal is applied to the gates of multiple transistors. In this case, transistors may be stacked on the appropriate gate signal. This also occurs in cascaded gates that cannot be constructed from a single row of transistors. A good example of this is the complementary XNOR gate. The schematic for this gate is shown in Fig. 5.29a on page 192. According to the style of layout that we

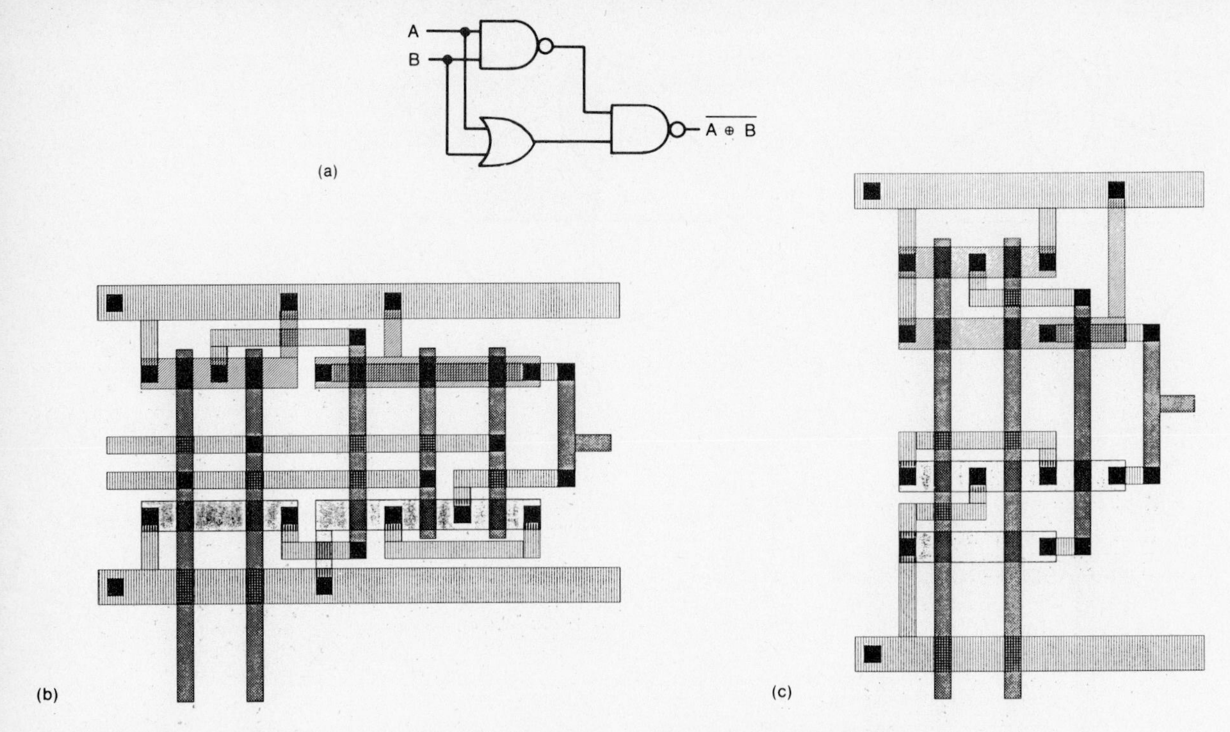

FIGURE 5.29. Complementary CMOS XNOR gate — alternative layout styles

have used to date, two possible layouts are shown in Fig. 5.29b and Fig. 5.29c.

The layout shown in Fig. 5.29b uses the single row of n- and p-transistors, with a break, while that in Fig. 5.29c uses a stacked layout. The selection of the styles would depend on the overall layout — whether a short, fat, or long thin cell was needed. Note that the gate segments that are maximally connected to the supply and ground rails are placed adjacent to these signals.

An automatic approach to achieve this style of layout that uses a graph-theoretic approach has been proposed [Wing83] [Wing82]. The approach is based on the use of interval graphs to optimally place transistors on vertical polysilicon lines in a gate matrix style (see Chapter 7). The layout style is similar to that used so far, with vertical polysilicon lines and horizontally arranged transistors. Power and ground run at the top and bottom of the cell. The approach is summarized in Fig. 5.30.

- Transistors are grouped in strips to allow maximum source/drain connection by abutment. To achieve better grouping, polysilicon columns are allowed to interchange to increase abutment.

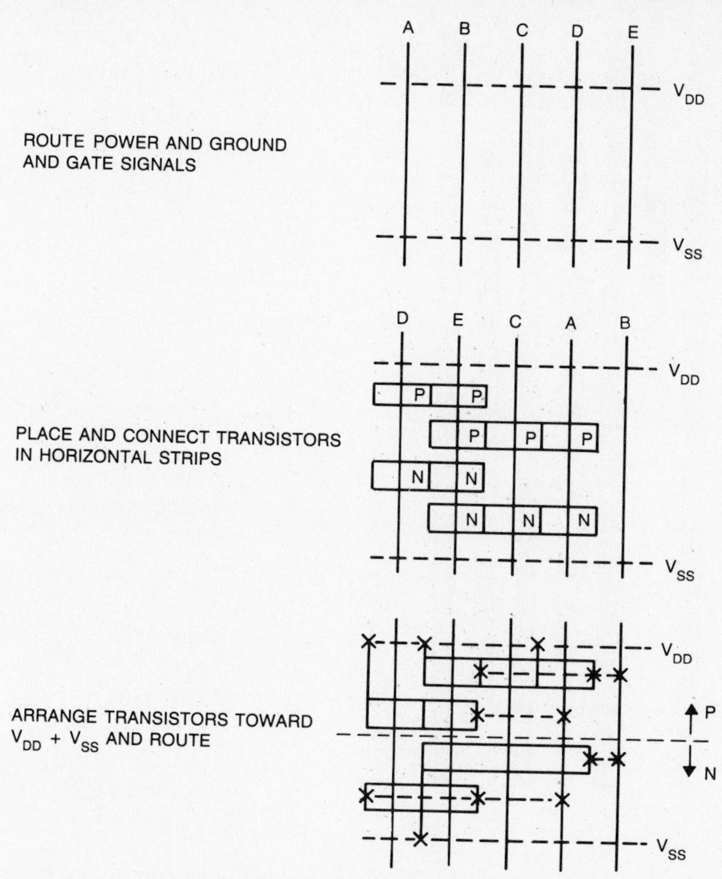

FIGURE 5.30. Outline of automated approach to CMOS gate layout

- The resultant groups are then placed in rows with groups maximally connected to the V_{SS} and V_{DD} rails placed towards these signals. Row placement is then based on the density of other connections.
- Routing is achieved by vertical diffusion or manhattan (horizontal and vertical) metal routing. This normally would require a maze router.

Other approaches to automating this task include the use of expert systems. This will be covered in Chapter 7.

5.3.9 CMOS standard cell design

When designing standard cells or polycells, geometric regularity is often required while maintaining some common electrical characteristics between cells in the library. A common physical limitation

FIGURE 5.31. Typical CMOS polycell layout ($F = \overline{(A.B.C) + D}$)

is to fix the physical height of the cell and vary the width according to the function. A typical standard cell is shown in Fig. 5.31. It is composed of a row of n-transistors of maximum height W_n, and a row of p-transistors of maximum height W_p, separated by a distance D_{np}, the design rule separation between n- and p-thinox. Power (V_{DD}) and ground (V_{SS}) busses traverse the cell at the top and bottom. The internal area of the cell is used for routing the transistors of specific gates.

A design objective in which W_p and W_n are selected may take into account such parameters as power dissipation, propagation delay, noise immunity, and area. Kang [Kang81] provides a good summary of the approach in the selection of W_p and W_n. The basic steps are as follows:

1 Identify a sample selection of gates (i.e., inverter, NAND, NOR) and compute an "average" delay time.
2 Calculate an objective function that relates worst-case propagation time to the ratio of W_p/W_n.

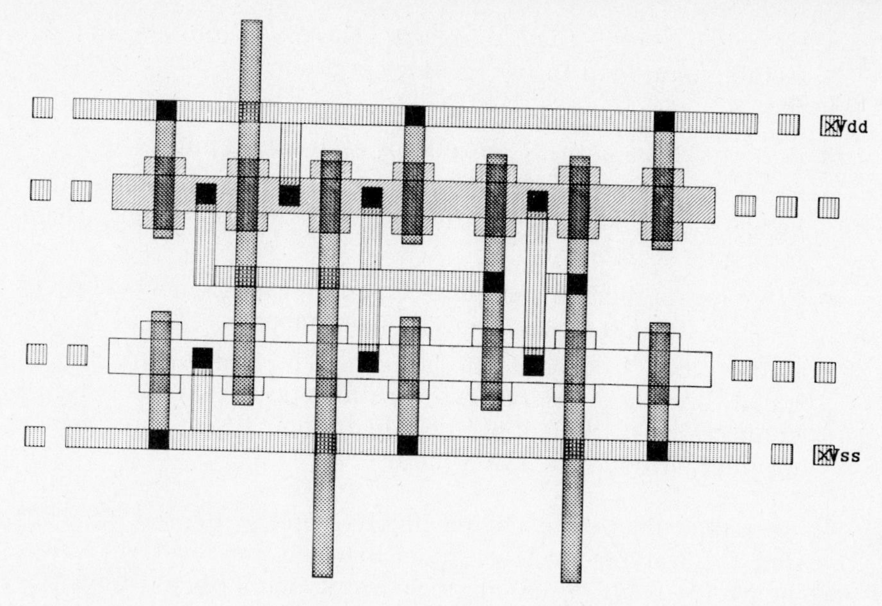

FIGURE 5.32. CMOS cell array (CCA) layout

3 Calculate an objective function relating the noise immunity to W_p/W_n.

4 Select an appropriate ratio that balances the required objective functions.

Techniques may then be employed to automatically generate these gates in a process independent manner from fairly straightforward intermediate forms [LeCJ81]. Note that in the above gate structure, all transistors of similar type were assumed to be the same size. One may further optimize a parameter such as noise immunity by adjusting individual transistor sizes to that below the maximum width allowed.

The general layout style used for the standard cell may be used to build an interesting gate-array type of circuit called a CMOS cell array [BGEK83]. In this array, gates are implemented in a continuous strip of n- and p-thinox. A gate is "isolated" from a neighboring gate by tying the gate of the end transistor to V_{SS} (n) or V_{DD} (p). This is shown in Fig. 5.32. Other gate array styles are discussed in Chapter 6.

5.3.10 General logic gate layout guidelines

From the considerations given to the layout of complementary gates, the following general layout guidelines may be stated:

1 Complete the electrical design of gate taking into account the factors mentioned in the previous sections.
2 Run V_{DD} and V_{SS} in metal at the top and bottom of the cell.
3 Run a vertical polysilicon line for each gate input.
4 Order the polysilicon gate signals to allow the maximal connection between transistors via abutting source-drain connections. These form gate segments.
5 Place n-gate segments close to V_{SS} and p-gate segments close to V_{DD}, as dictated by connectivity requirements.
6 Connections to complete the gate should be made in polysilicon, metal, or, where appropriate, in diffusion. (As in the case of connections to the supply rails or outputs.) Keep capacitance on internal nodes to a minimum.

Note that the style of layout involves optimizing the interconnection at the transistor level rather than the gate level. As a rule, better layouts result by taking logic blocks with 10–100 transistor complexities and following the rules above, rather than designing individual gates and trying to piece them together.

This improvement in density is due to a number of factors, which include the following:

1 More "merged" source-drain connections.
2 More usage of "white" space in sparse gates (blank area with no devices or connections).
3 Better use of all possible routing layers.
4 Use of optimum device sizes.

Improvements gained by optimizing at this level over a standard cell approach can be up to 60 percent in area. Furthermore, cells can be designed in such a way to provide "transparent routing" for cell to cell communication. This greatly reduces the global wiring problem.

5.3.11 Gate optimization

In this section a potpourri of optimization techniques will be presented. One technique that has been demonstrated to increase the speed of gates consisting of series combinations of transistors is to vary the size of the transistor according to the position in the series structure [Shoj82]. This is shown in Fig. 5.33 for a 4-input AND gate. The transistor closest to the output is the smallest, with transistors increasing in size the nearer they are to V_{SS}. The decreased switching times are attributed to the dominance of the capacitance term in

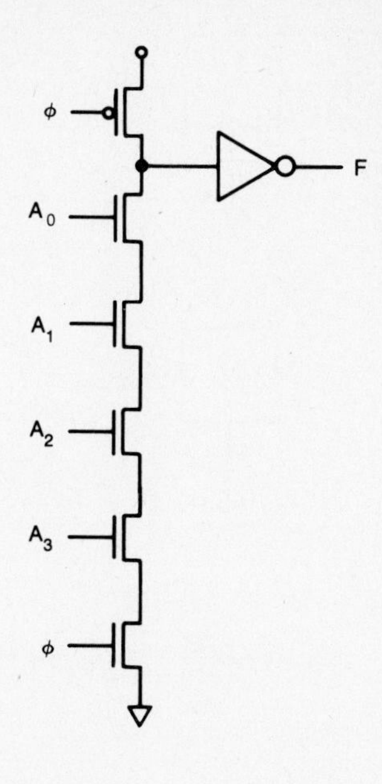

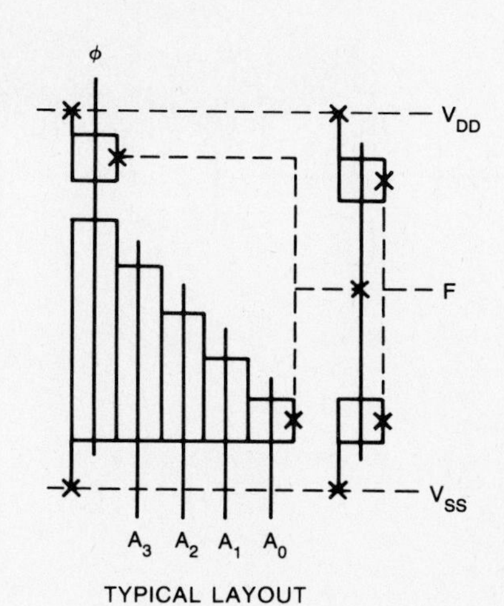

FIGURE 5.33. Sizing series transistors

the RC time constant of the gate. Increases in performance of 15–30 percent can be expected.

In this chapter, we have considered the design of gates that drive similar loads. Often in circuits the last gate in a succession of gates might have to drive a large load, such as a bus, an I/O driver, or, in the case of an I/O buffer, the external chip load. As noted in Chapter 4, if we consider a simple set of cascaded inverters we can define the *stage ratio* as the ratio by which successive transistor widths are multiplied. Fig. 5.34a shows a cascaded set of inverters. The optimum stage ratio for minimum delay is 2.7 [MeCo80]. However, ratios of 2–10 can be used to optimize other attributes such as size or power dissipation. To further illustrate this concept, Fig. 5.34b depicts an implementation of an XOR gate using a 2-input NAND gate, an OR-AND gate, and an inverter. The rise and fall times are approximately:

NAND gate

$$T_F = 2R_nC_L$$
$$T_R = R_pC_L$$

OR-AND gate

$$T_F = 2R'_nC'_L$$
$$T_R = 2R'_pC'_L$$

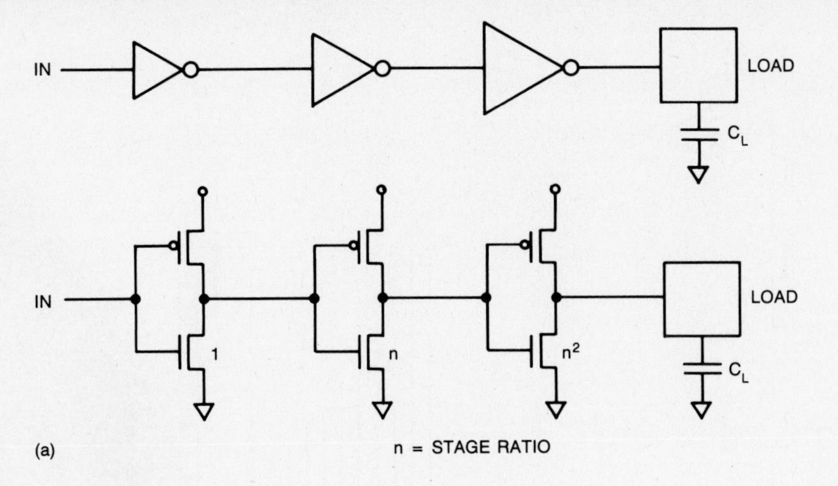

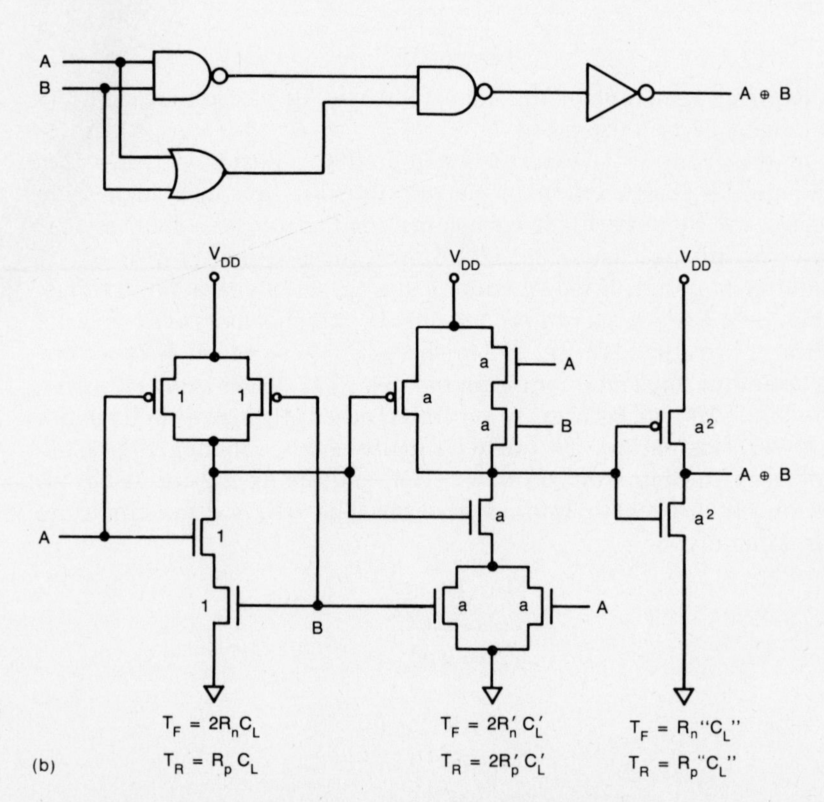

FIGURE 5.34. Stage ratios and the effect on the speed of cascaded gates

inverter

$$T_F = R_n''C_L''$$
$$T_R = R_p''C_L'',$$

where

C_L and primed versions are the individual stage load capacitances.

R_n and R_p and primed versions are the individual stage n- and p-transistor resistances.

Thus if the stage ratio is a, the average delay through the series of gates is

$$T_{AVERAGE} = \frac{1}{2}\left(2aR_nC_L + aR_pC_L + \frac{2a^2R_nC_L}{a} + \frac{2a^2R_pC_L}{a} + \frac{R_nC_{LOAD}}{a^2} + \frac{R_pC_{LOAD}}{a^2}\right), \quad \textbf{(5.6)}$$

where

$$C_{LOAD}$$

is the ultimate load capacitance. If

$$C_{LOAD} = 100C_L,$$

then with

$$a = 1$$
$$T_{AVERAGE} = 52R_nC_L + 51.5R_pC_L. \quad \textbf{(5.7)}$$

On the other hand, if

$$a = 5,$$

then

$$T_{AVERAGE} = 24R_nC_L + 19R_pC_L, \quad \textbf{(5.8)}$$

clearly illustrating the speed advantage that may be achieved.

Noise coupling can have a detrimental effect on the state of storage nodes that are driven by transmission gates. One form of this is illustrated in Fig. 5.35a on page 200. Here, a transfer gate is used to charge and discharge a storage node. From the figure it is evident that a short noise voltage spike comparable in magnitude to the threshold voltage V_t, may alter the logic state of a succeeding logic gate due to the charge leakage caused by the noise pulse. One should never place storage nodes on transistors whose sources are separated by series devices from the V_{DD} and V_{SS} lines. The result can also be unwanted discharge of storage nodes, as shown in Fig. 5.35b, due to capacitor divider action. A similar effect may occur in storage nodes where the parasitic capacitance is enhanced by

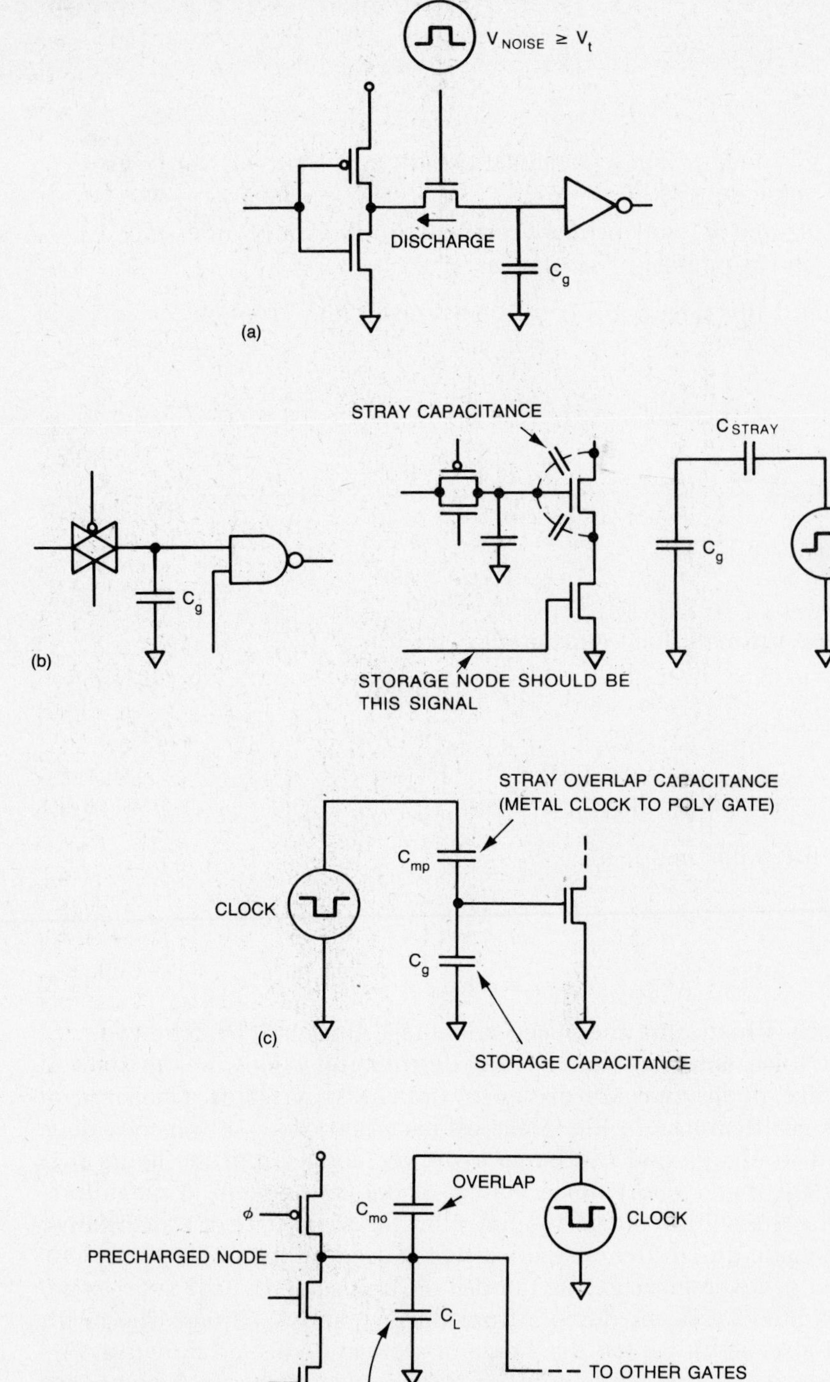

FIGURE 5.35. Storage node considerations and the effect of parasitic overlap capacitance

overlap of clock and storage nodes. This is illustrated in Fig. 5.35c. When using dynamic logic, the precharged nodes in dynamic gates must also be protected from unintentional discharge. An example of this situation is shown in Fig. 5.35d.

When designing circuits, one must balance the time to optimize such gates versus the overall effect on the system. Quite often, the 15 percent gained can be in a part of a circuit where the gain does not reflect in an overall gain in the system as a whole. For this reason a good starting point is to use unit-sized devices throughout and then optimize paths from a critical path analysis. Automatic programs linked with symbolic layout methods should provide such tools in the near future.

5.3.12 Transmission gate layout considerations

In the case of complementary gates, there is one point of contact between the n-transistors and the p-transistors. As we have seen, this can be completed with a metal strap or a combination of metal and polysilicon straps where metal routing transparency is required. When considering a transmission gate, the source and drain terminals of the p- and n-transistors are paralleled. According to the layout strategy presented, the two layouts shown in Fig. 5.36 would be suitable. Note that in Fig. 5.36a, no metal lines can pass from left to right. The layout shown in Fig. 5.36b is longer but does have horizontal metal transparency. The decision on which layout is more suitable would depend on the circuit being designed. For instance, in a shift register delay line, Fig. 5.36a might be preferred due to its small size. In a data path, where bus lines may have to pass horizontally, Fig. 5.36b would be preferred.

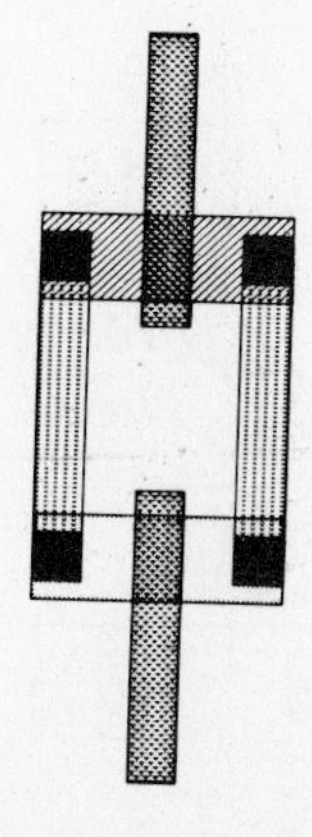

(a)

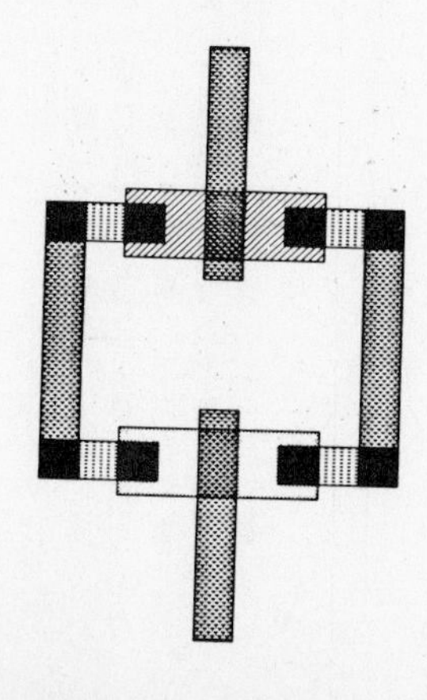

(b)

FIGURE 5.36. Transmission gate layout

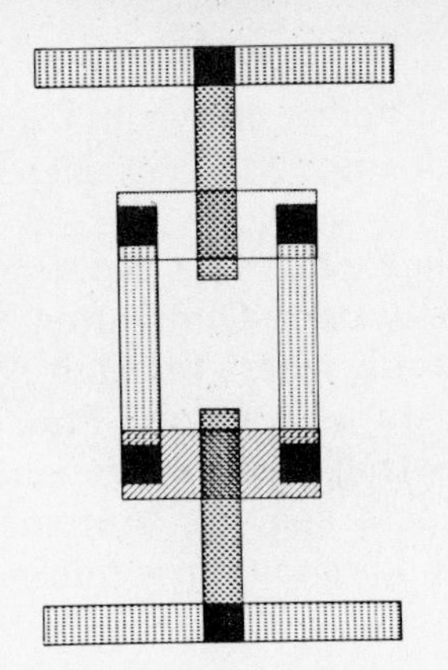

a)

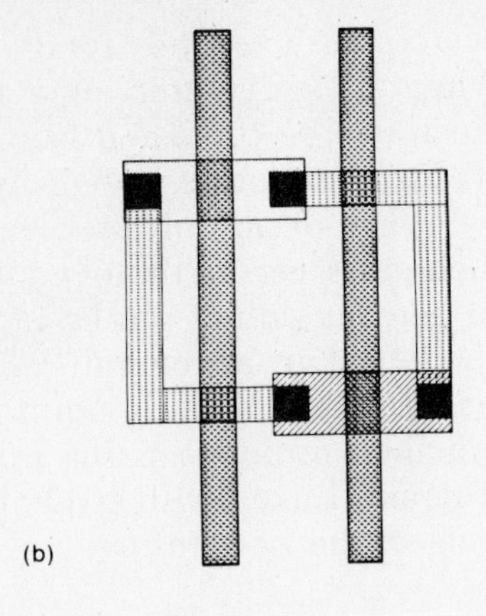

(b)

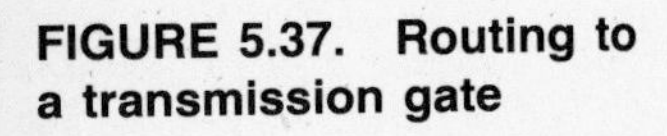

FIGURE 5.37. Routing to a transmission gate

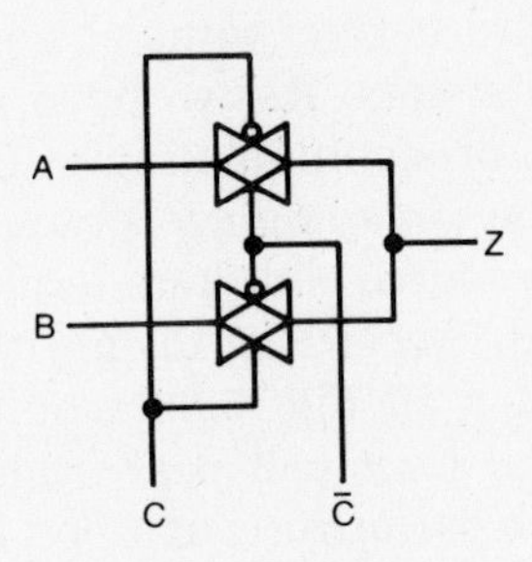

(a)

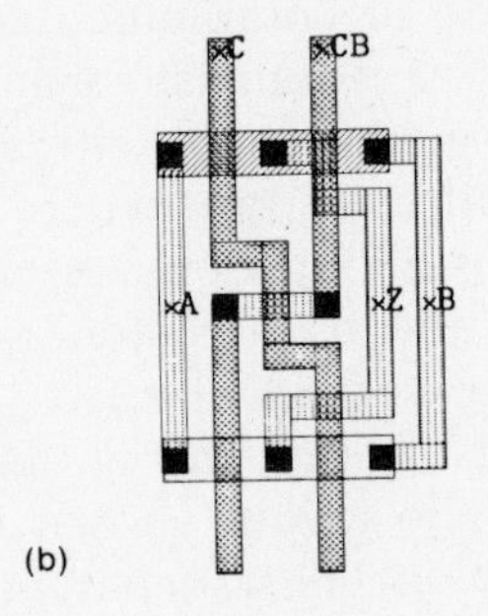

(b)

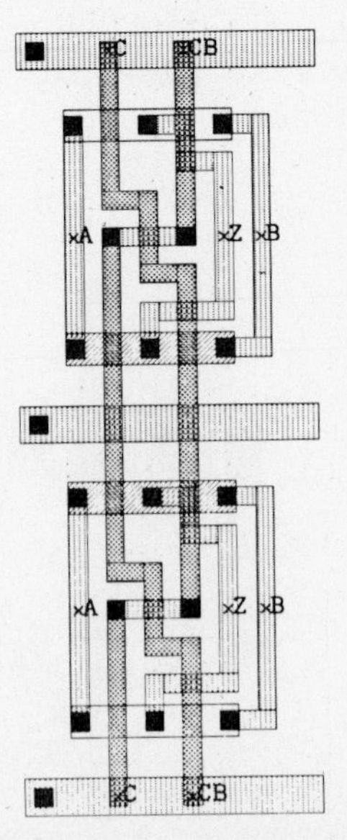

(c)

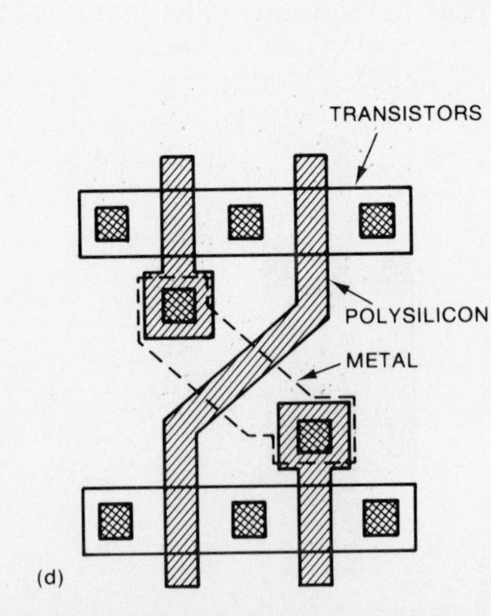

(d)

FIGURE 5.38. 2-input multiplexer layouts

The transmission gate has to be supplied with a switching signal and its complement. These signals may be generated at some distance and may have to be routed to the transmission gate (i.e., in an array of flip-flops controlled by a common signal). In these cases, it is necessary to consider the routing of the gate signals to the transmission gates. Two possibilities are shown in Fig. 5.37. In Fig. 5.37a, the control inputs are run horizontally in metal, outside the transistors. Note that in this case, polysilicon can be passed horizontally between the n- and p-transistors. In Fig. 5.37b, the control signals are routed vertically in polysilicon. In this case, the transistors are offset to allow the passage of the vertical control lines.

5.3.13 2-input multiplexer

The 2-input multiplexer in Fig. 5.38a is used frequently in flip-flops. A possible layout is shown in Fig. 5.38b. Note that the control lines are crossed in the middle of the cell. If this was to be a stacked structure, the number of contacts in each control line should be equalized, as shown in Fig. 5.38c. This equalizes any delay that might arise due to contact resistance. An alternative to the manhattan layouts shown is illustrated in the partial mask level design shown in Fig. 5.38d, which uses 45° shapes. As this is a common structure, manhattan-based symbolic systems can treat this as a special *cross-over* symbol.

5.4 Clocking strategies

In this chapter we have discussed various alternative forms of CMOS logic and the basic physical and electrical design of CMOS circuitry. Although we have studied logic gates in isolation, no global clocking strategy has been suggested. One of the most important decisions that may be made at the commencement of a design is that of the selection of the clocking strategy. In the Mead and Conway text [MeCo80], a simple 2-phase clocking scheme was selected. In this section we will look at this clocking scheme as applied to CMOS and also investigate some other clocking approaches. Suitable memory and logic elements for each clocking strategy will be summarized. Some layout guidelines are also given.

5.4.1 Pseudo 2-phase clocking

Pseudo 2-phase clocking takes the 2-phase nonoverlapping nMOS clocking scheme (as used in Mead and Conway) and adds the complementary clocks. Thus we can have $\phi 1$, $\phi 2$, $\overline{\phi 1}$ and $\overline{\phi 2}$, or up to

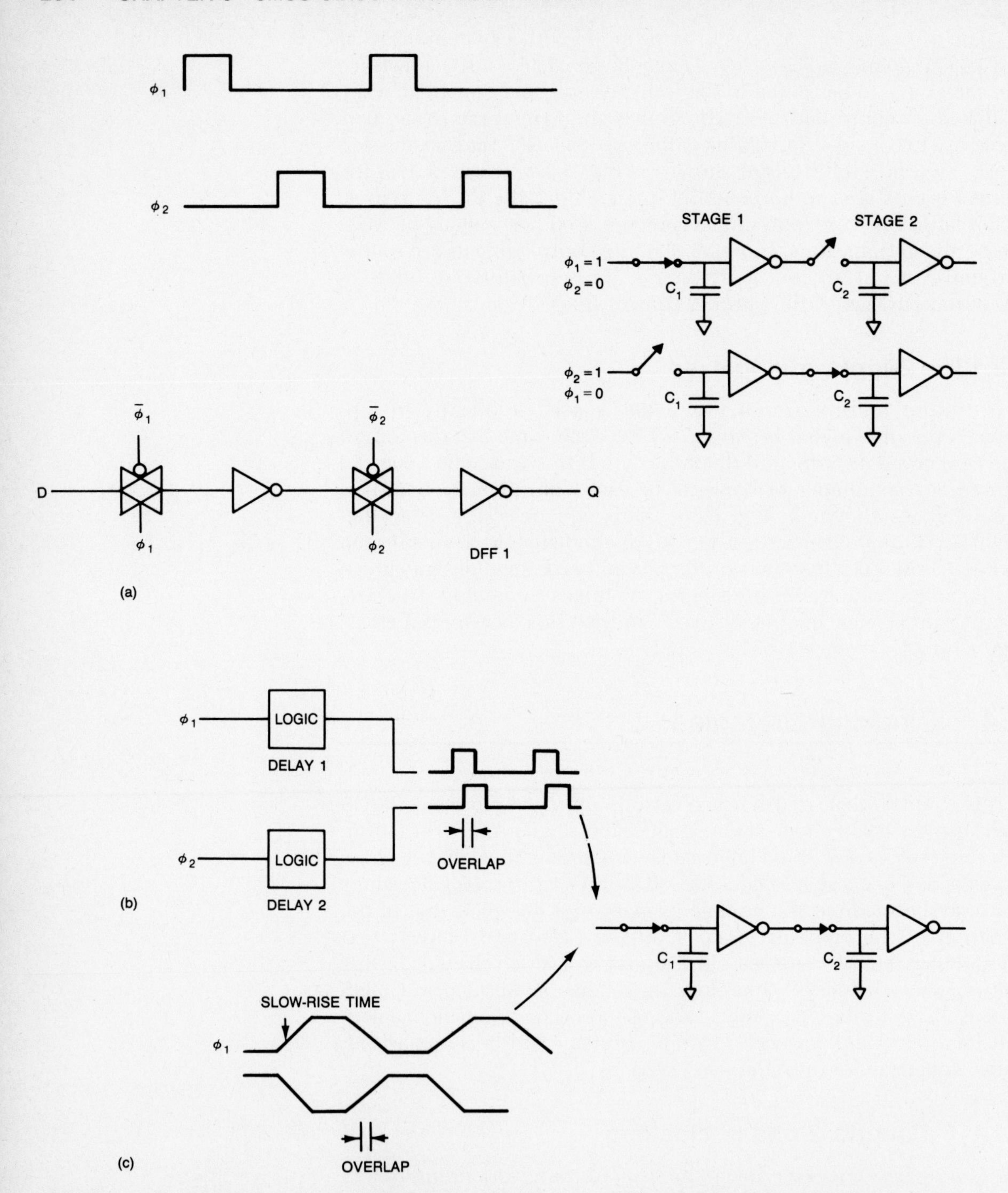

FIGURE 5.39. Pseudo 2-phase clocking (a) waveforms and simple latch, (b) clock skew, and (c) slow clock edges

four clock phases to route around a chip. Usually, two master clocks would be distributed with local buffers to generate local clocks. A typical set of clock waveforms and a simple latch (DFF1) are shown in Fig. 5.39a. Note that $\phi1\ (t).\phi2\ (t) = 0$ for all t. The operation of the latch is illustrated in Fig. 5.39a. During ϕ_1, the stage 1 transmission gate is closed, thereby storing the input level on the gate capacitance of the inverter and the output capacitance of the transmission gate (C_1). The state of stage 2 is stored on a similar capacitance C_2. During ϕ_2, the stage 1 transmission gate opens and the inverse of the stored value on C_1 is placed on C_2.

The selection of the actual clock relationships depends on the circuit. Some guidelines would be as follows. If $\phi1$ is used as a precharge clock, then it has to be of a duration to allow precharge of the worst case node in the circuit. Typically, this might be on a RAM bit line. The delay between clocks has to be chosen to ensure that for the worst case skew, the two clocks do not overlap. Clock skew can occur in two forms. The first is shown in Fig. 5.39b, where the clocks applied to a latch have travelled through different delay paths to arrive at the latch. The skew occurs while both clocks are simultaneously HIGH, causing the two transmission gates in the latch to be transparent. Another type of skew can occur even if the clocks are perfectly overlapping. This is shown in Fig. 5.39c. Here, the rise and fall times are so slow that the period of the transition region causes the latch transmission gates to couple. Both of these conditions can lead to incorrect values being stored on the C_1 and C_2 capacitances. Thus the period of the clocks must allow for the worst case logic propagation time in combinational blocks that are to be latched.

5.4.2 Pseudo 2-phase memory structures

An alternative implementation of the flip-flop shown in Fig. 5.39a is a flip-flop that makes use of a "clocked inverter." This is shown in Fig. 5.40a (DFF2). This is equivalent to DFF1, except that the connection between the n- and p-transistors at the "input" of the transmission gate has been deleted and the TG's have been moved to the output assuming storage on the next stage. This can result in a smaller layout as two contacts and at least one wire are omitted. Note that the operation is similar to the first latch. When $\phi1$ is high, node $n1$ will be conditionally pulled high or low by the p- or n-data transistors, respectively. When $\phi1$ is low, this value is stored, the clocked transistors being turned off. The second stage operates similarly. A historical variation of DFF2 (DFF3) is shown in Fig. 5.40b, in which the clocked transistors are placed between the inverter and supply rails. The charge redistribution problem is quite severe in this flip-flop. It is interesting to note that this flip-flop is mainly

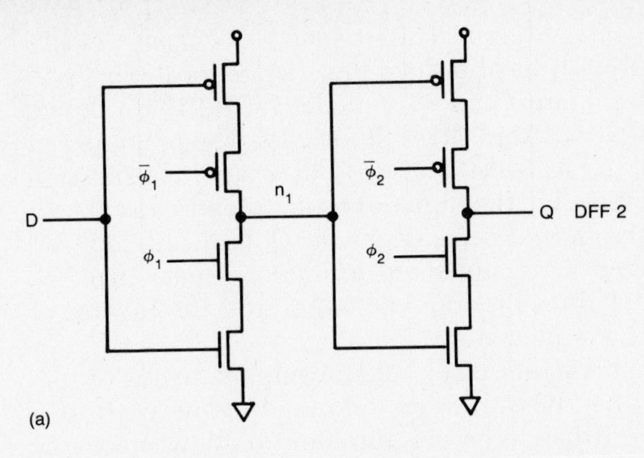

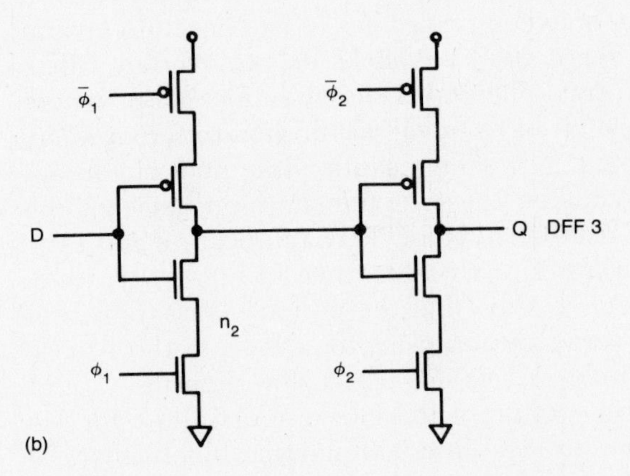

FIGURE 5.40. Pseudo 2-phase latches

of historical significance. In metal gate CMOS this flip-flop leads to a smaller layout compared to Fig. 5.40a. In silicon gate technologies this circuit/layout tradeoff is alleviated.

Considering DFF1, two representative layouts are shown in Fig. 5.41a and Fig. 5.41b. The layouts may be stacked vertically, with one layout allowing the vertical distribution of the clocks in polysilicon, while the other routes the clocks horizontally in metal. Layout variations representing the DFF2 configuration are shown in Fig. 5.41c and Fig. 5.41d, respectively. Note that routing clocks in polysilicon may lead to clock delay problems if long unbuffered clock lines are used. Fig. 5.41a–c are shown in color in Plate 8.

The abutment of DFF1 to form a four-by-four shift register is shown in Fig. 5.42 on page 208. Note that in this layout only 2.5 clock lines are needed per horizontal bit of layout, rather than four clock lines. This illustrates that the apparent disadvantage of routing four clock lines in a module may be less than expected. Plate 9 shows this layout in color.

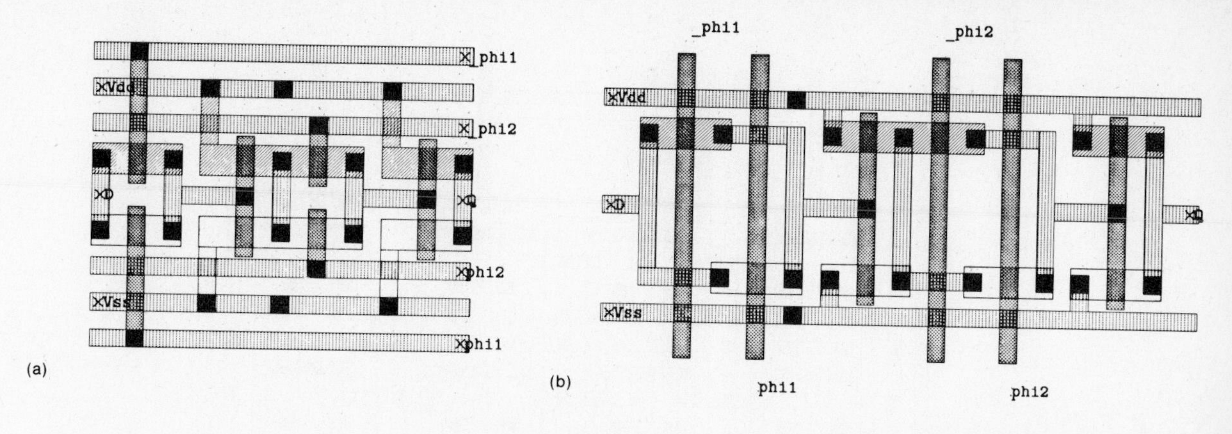

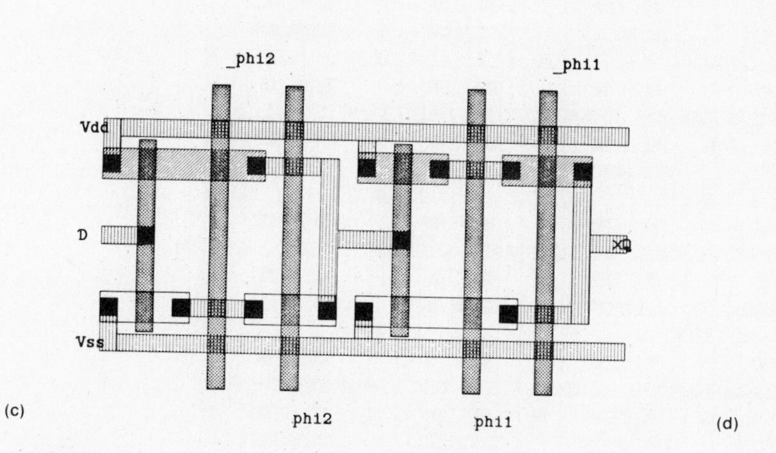

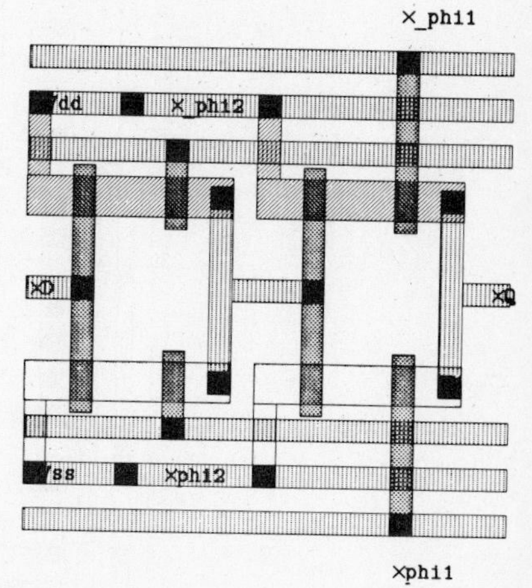

FIGURE 5.41. Pseudo 2-phase latch layouts

A reduction in the number of clock lines can be accommodated if only n-transistors are used in the transmission gate, as would be the case in an n-MOS design (Fig. 5.43, page 209). Two effects occur in this configuration. Firstly, the '1' level transferred to the input of the inverter is degraded to approximately $V_{DD} - V_{t_n}$. This has the effect of slowing down the low transition of the inverter. Furthermore, the high noise margin (NM_H) of the inverter is degraded. It also has the possible effect of causing static power dissipation. For instance, if $V_{t_p} < V_{DD} - V_{t_n}$, then the p-transistor in the inverter will be turned on when the inverter output is in the low state, thus causing current to flow through the inverter. This is consistent with the reduced NM_H. Although this is not catastrophic, it must be taken into account when calculating total power dissipation. The rising transition at the output of the inverter is faster because the capacitance

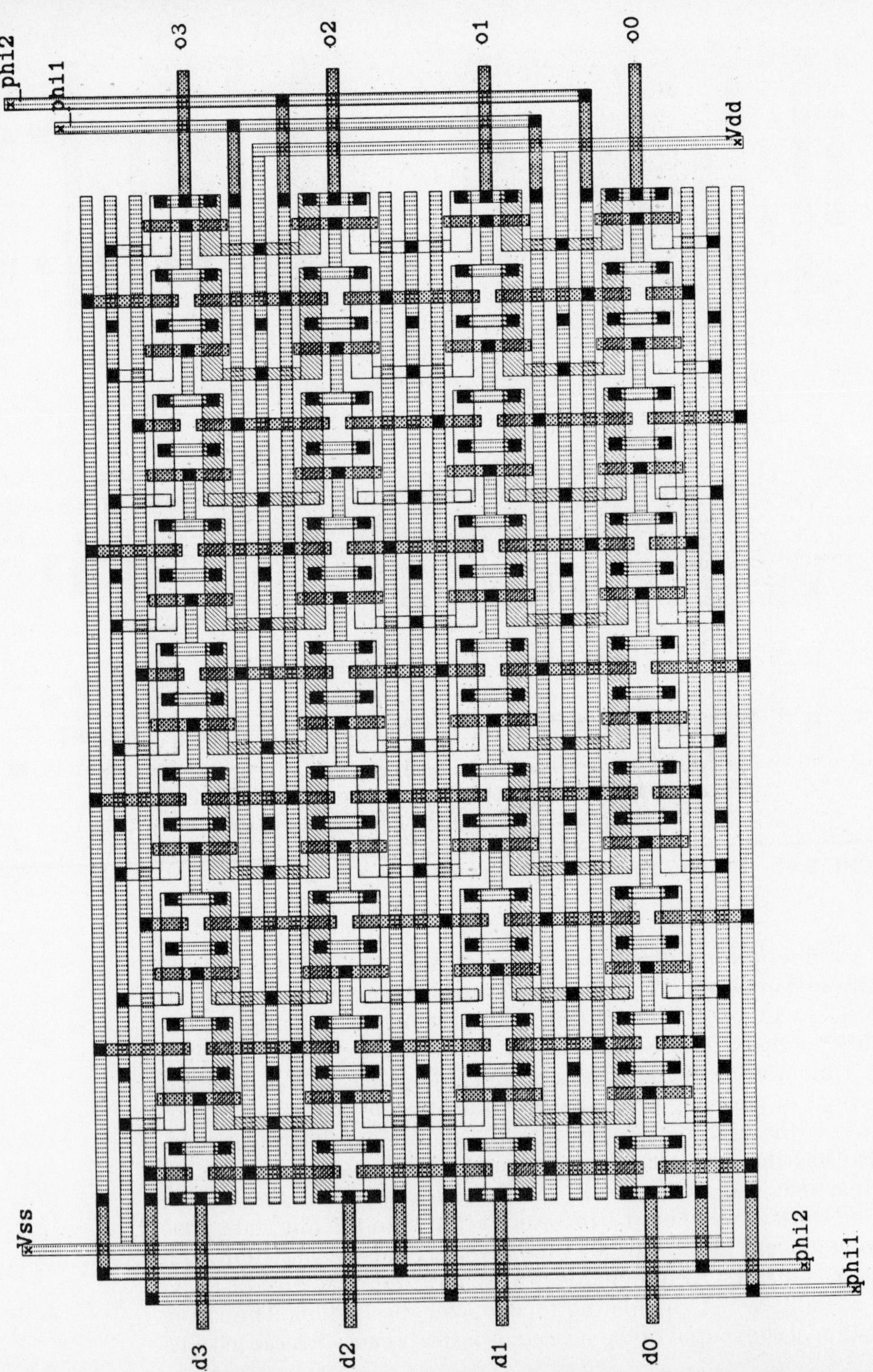

FIGURE 5.42. Shift register array layout

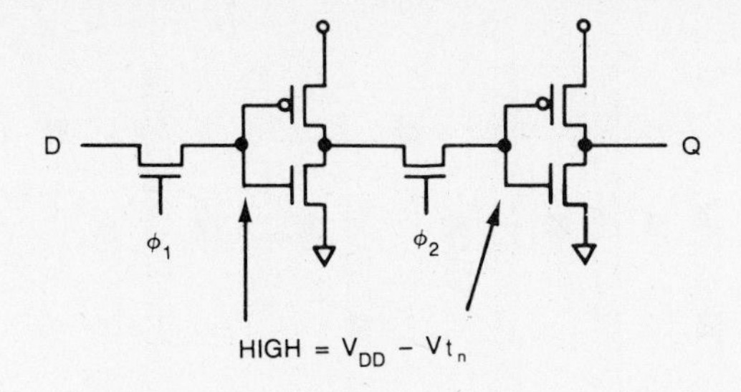

FIGURE 5.43. Reduced transistor count latch

at the output is reduced due to the absence of the p-transistor. There is no hard and fast rule for when the various flip-flop configurations should be used. Of course, if density is crucial the last mentioned flip-flop could be used, providing the speed was suitable and that static power dissipation was not a problem. This can only be reconciled by worst-case simulation and power-dissipation calculations.

A dynamic *D* latch is shown in Fig. 5.44a. If we denote the input by *D* and the output by *Q*, the characteristic equation may be written as

$$\begin{aligned} Q(t) &= D(t) \qquad LD = 1 \\ &= Q(t-1) \qquad LD = 0, \end{aligned}$$

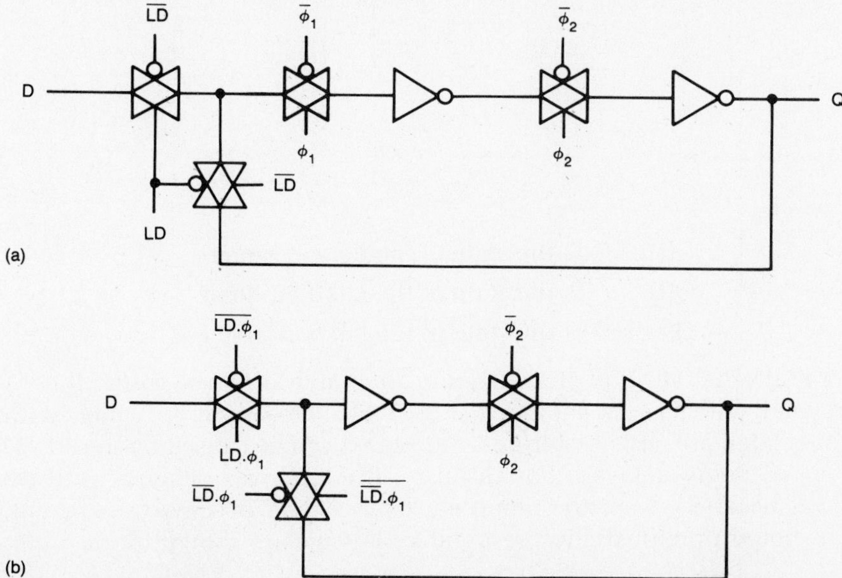

FIGURE 5.44. Dynamic D latches

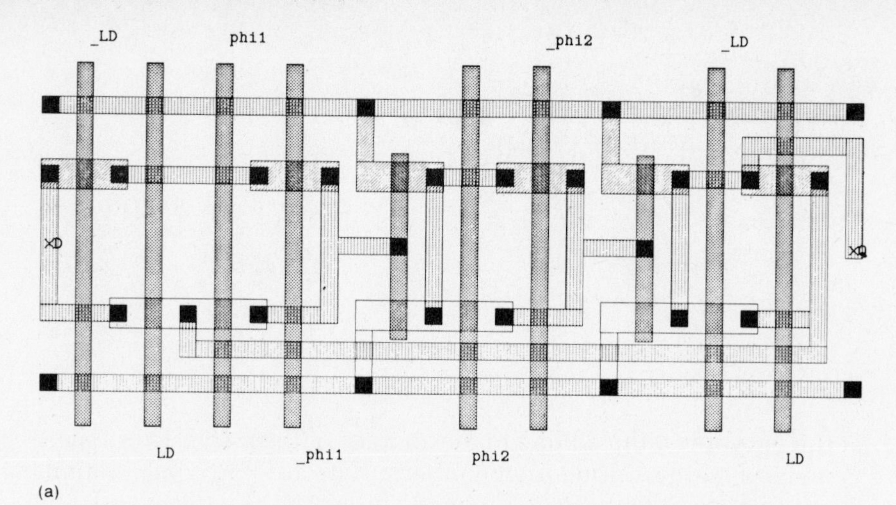

(a)

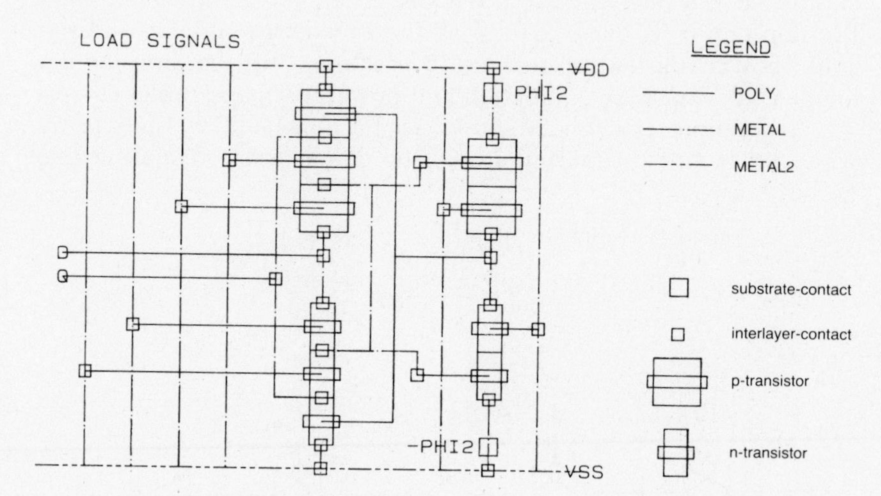

FIGURE 5.45. Dynamic D latch layouts

a) single metal version
b) two metal version

where

$D(t)$	is the state of the data at time t.
$Q(t)$	is the state of the latch at time t.
$Q(t - 1)$	is the state of the latch at time $t - 1$.

An alternative implementation of this latch involves gating the $\phi 1$ signal with LD, as shown in Fig. 5.44b. One thing to watch with this latch are any deleterious effects that can be caused by the $\phi 1.LD$ signal being delayed. For instance, if $\phi 1.LD$ was delayed until $\phi 2$ was asserted, then the complete latch would become transparent, as noted previously. Representative layouts are shown in Fig. 5.45.

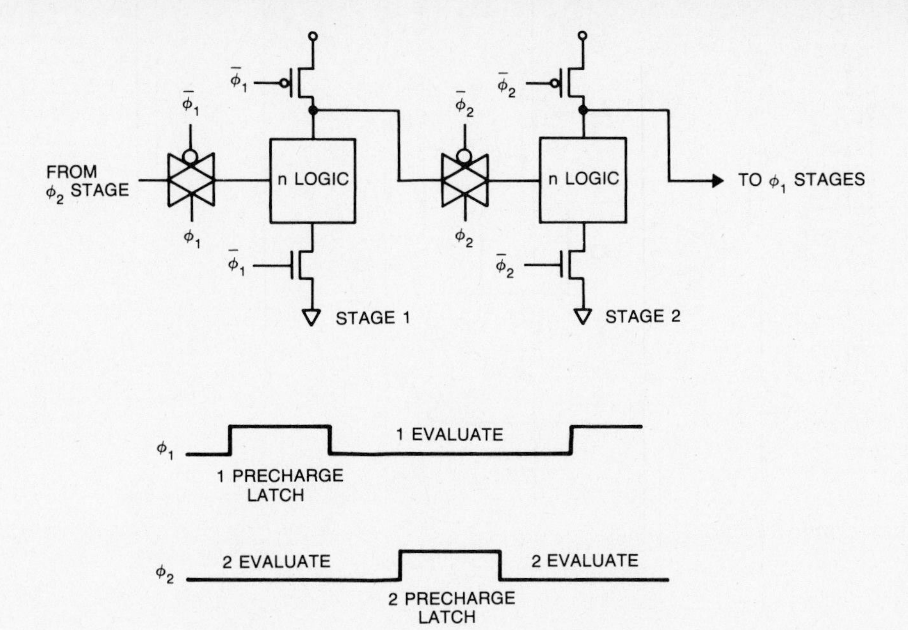

FIGURE 5.46. Pseudo 2-phase dynamic logic

5.4.3 Pseudo 2-phase logic structures

For pseudo 2-phase systems, conventional static logic may be used in conjunction with the memory elements that have been described in the last section. If dynamic logic is needed the 2-phase logic scheme outlined in Fig. 5.46 may be used. In this scheme, the first stage is precharged during $\phi 1$ and evaluated during $\phi 2$. While the first stage is evaluated, the second stage is precharged and the first stage outputs are stored on the second stage inputs. During $\phi 1$, the second stage is evaluated and latched in a succeeding $\phi 1$ stage.

Domino n-MOS gates may also be employed. A typical gate is shown in Fig. 5.47. Here, a single clock ($\phi 1$ *or* $\phi 2$) is used to precharge and evaluate the logic block. The succeeding stage is operated on the opposite clock phase, as illustrated in Fig. 5.47 on page 212. The difference between this logic structure and that previously shown in Fig. 5.46 is that in the domino logic, a number of logic stages may be cascaded before latching the result.

5.4.4 2-phase clocking

In some situations it is desirable to reduce the number of clock lines that are routed around the chip. One approach is to use a 2-phase clock that uses a ϕ, $\overline{\phi}$ type arrangement.

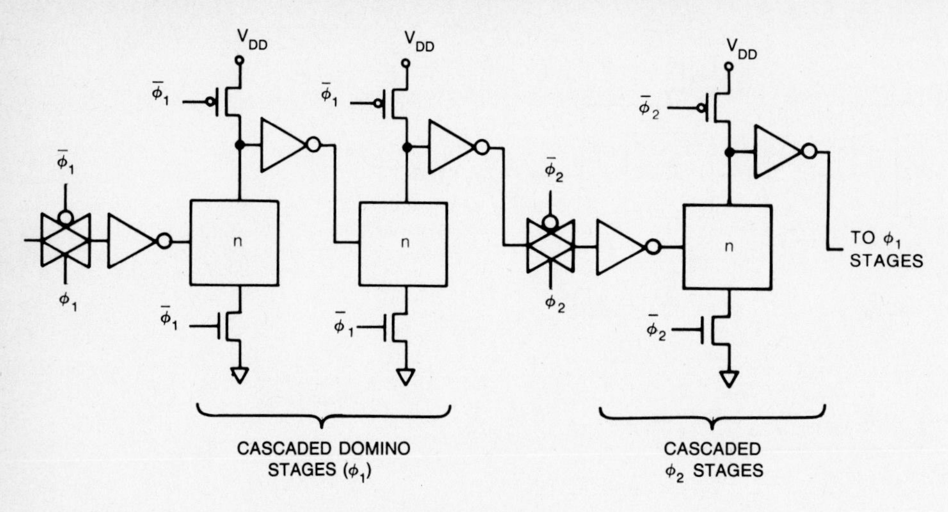

FIGURE 5.47. Pseudo 2-phase domino logic

5.4.5 2-phase memory structures

Applying this clocking strategy to the flip-flops used in the pseudo 2-phase clocking, the structure in Fig. 5.48a is constructed. A "clock race" condition similar to that encountered in the pseudo 2-phase latch can arise in this structure. This is, of course, an accentuated case of pseudo 2-phase clock skew mentioned previously. This effect is shown in Fig. 5.48b. Considering $\overline{\phi}$ delayed from ϕ, we see that the first transmission gate n-transistor can be turned on at the same time as the second transmission gate n-transistor. Hence the value on the input can ripple through the two transmission gates leading to invalid data storage. This problem means that close attention must be paid to the clock distribution to minimize the clock skew. One method for achieving this is shown in Fig. 5.48c [OhKD79]. A conventional clock buffer consisting of two inverters is shown. The buffered ϕ clock will always be delayed with respect to the buffered $\overline{\phi}$ clock. To reduce this undesirable delay, the ϕ signal may be passed through a transmission gate to equalize delay with respect to $\overline{\phi}$. The transmission gate should use similar sized transistors to those used in the inverters. Subsequent buffers may also be used to keep the initial buffers minimum size.

The chain latch [OhKD79] uses a 2-phase clock and single pass transistors to provide a static latch capability. The schematic for this latch is shown in Fig. 5.49a on page 214. The effective circuit configurations are shown for ϕ = high and ϕ = low, showing the "chain link" effect. The input threshold effects mentioned in Section 5.4.2 also apply here.

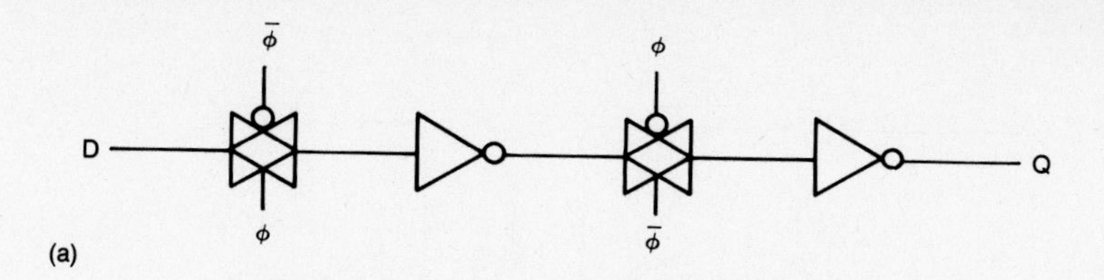

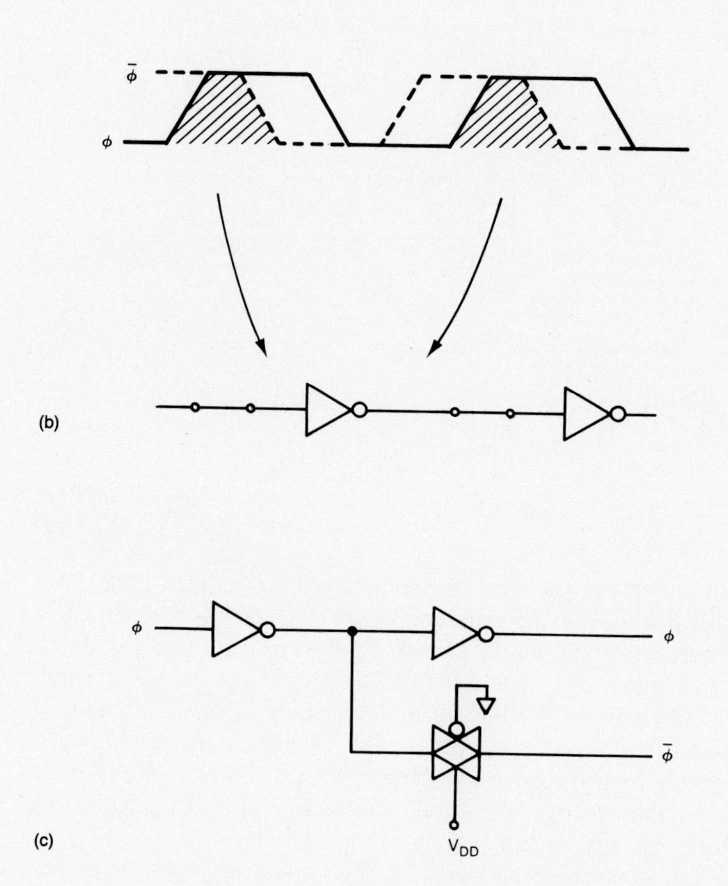

FIGURE 5.48. 2-phase flip-flop and skew reduction

An alternate 2-phase static latch is depicted in Fig. 5.50a on page 215. The race problem still manifests itself. When ϕ is high and $\overline{\phi}$ overlaps it due to skew, the D input and feedback signal will "fight" to determine the new value on the input of the latch. One method of reducing this effect is to make the feedback inverter a weak "trickle" inverter so that the input signal will override the effect of this signal. A "trickle" inverter is constructed by using transistors with a lower β than a regular "minimum" sized inverter.

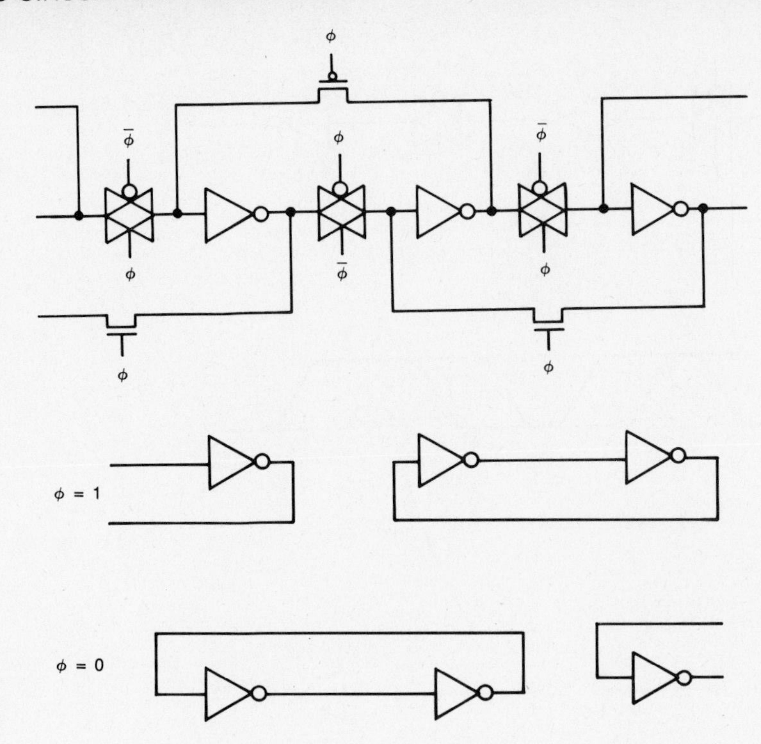

FIGURE 5.49. Chain latch

This is achieved by employing transistors with a length (L) greater than the minimum value. Alternatively, the W/L of the transmission gate transistors may be made larger than those in the feedback inverter. A gate-array optimized for this latch is shown in Fig. 6.2d. Appropriate ratios may be determined by simulation. Note that if the clock period is long compared to the overlap, the signals will eventually settle to their correct values. A master-slave flip-flop is constructed by cascading two such stages and operating them on alternate clock phases, as indicated in Fig. 5.50b.

An alternative static flip-flop is in Fig. 5.51a on page 216. This has made use of two clocked inverters and an additional inverter. It has the advantage that it may be easily constructed in such a way as to allow horizontal metal paths to cross the cell. Two final designs are shown in Fig. 5.51b and Fig. 5.51c. The first design uses only one clock phase and a gated RS flip-flop; it is clock race immune. It uses 14 transistors, compared to 8–10 transistors in the preceding flip-flops. The second design shows this circuit extended to build a master-slave design with set and reset. Only a single clock is required. Representative layouts are shown in Fig. 5.52 on page 218. (Fig. 5.52a represents Fig. 5.50a, Fig. 5.52b is Fig. 5.51a, Fig. 5.52c

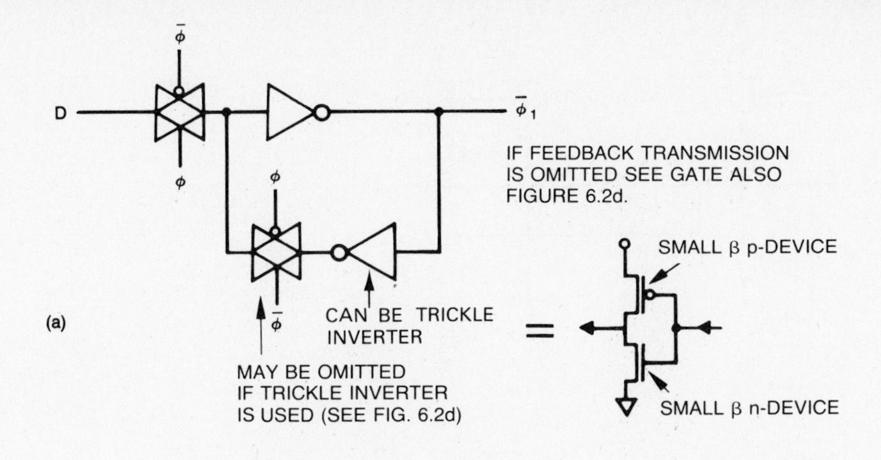

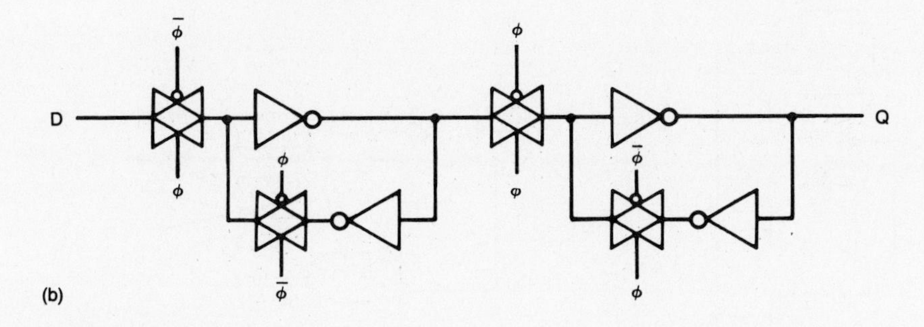

FIGURE 5.50. 2-phase static flip-flops

is Fig. 5.51b.) The static latch in Fig. 5.50 may be constructed in such a way (by adding vertical polysilicon connections) as to allow horizontal metal paths, but it then results in a longer cell. Plate 10 shows these layouts in color.

A simpler version of a static *D* flip-flop, with asynchronous reset and set, is shown in Fig. 5.53 on page 219. This flip-flop could have clock skew problems. The set and reset signals operate regardless of the clock state. The set/reset truth table is shown in Table 5.5 on page 219.

5.4.6 2-phase logic structures

Conventional static logic may be used with 2-phase clocking. In addition, domino nMOS logic may be used. However, it is difficult to pipeline such logic stages while using a single clock and complement. A logic family termed N-P CMOS dynamic logic (Fig. 5.54a, 5.54b on page 220), may be used to optimize speed and density at the expense of more detailed circuit and system design [GoDM83].

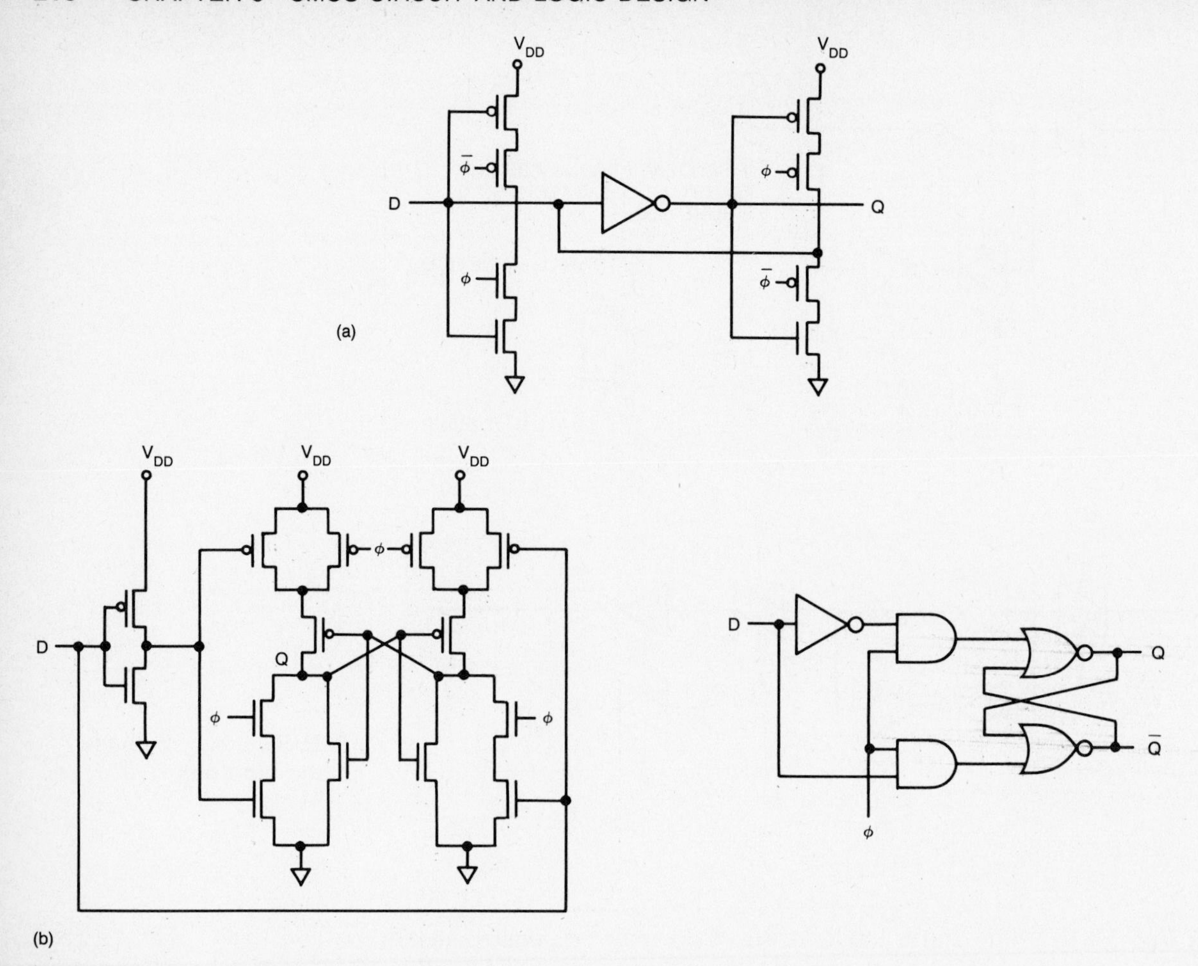

FIGURE 5.51. 2-phase static D flip-flops

This combines N-P sections of domino logic with a C^2MOS latch as the output stage. We can build ϕ blocks, which resolve (or evaluate) during ϕ, and $\bar{\phi}$ blocks, which resolve during $\bar{\phi}$. Cascading these N-P blocks is achieved using the structure in Fig. 5.54c. This yields a pipelined structure in which ϕ sections are precharged and $\bar{\phi}$ sections are evaluated when $\phi = 0, \bar{\phi} = 1$. Information to $\bar{\phi}$ sections is held constant by the clocked CMOS latch in the output of ϕ sections. When $\phi = 1, \bar{\phi} = 0$, ϕ sections are evaluated and $\bar{\phi}$ sections are precharged. Often it is desired to mix N-P dynamic sections with static logic or connect N-P sections with domino sections. If this is done, two problems must be avoided. Firstly, self-contained sections must be internally race free. Secondly, when different sections are cascaded to form pipelined systems, clock skew should result

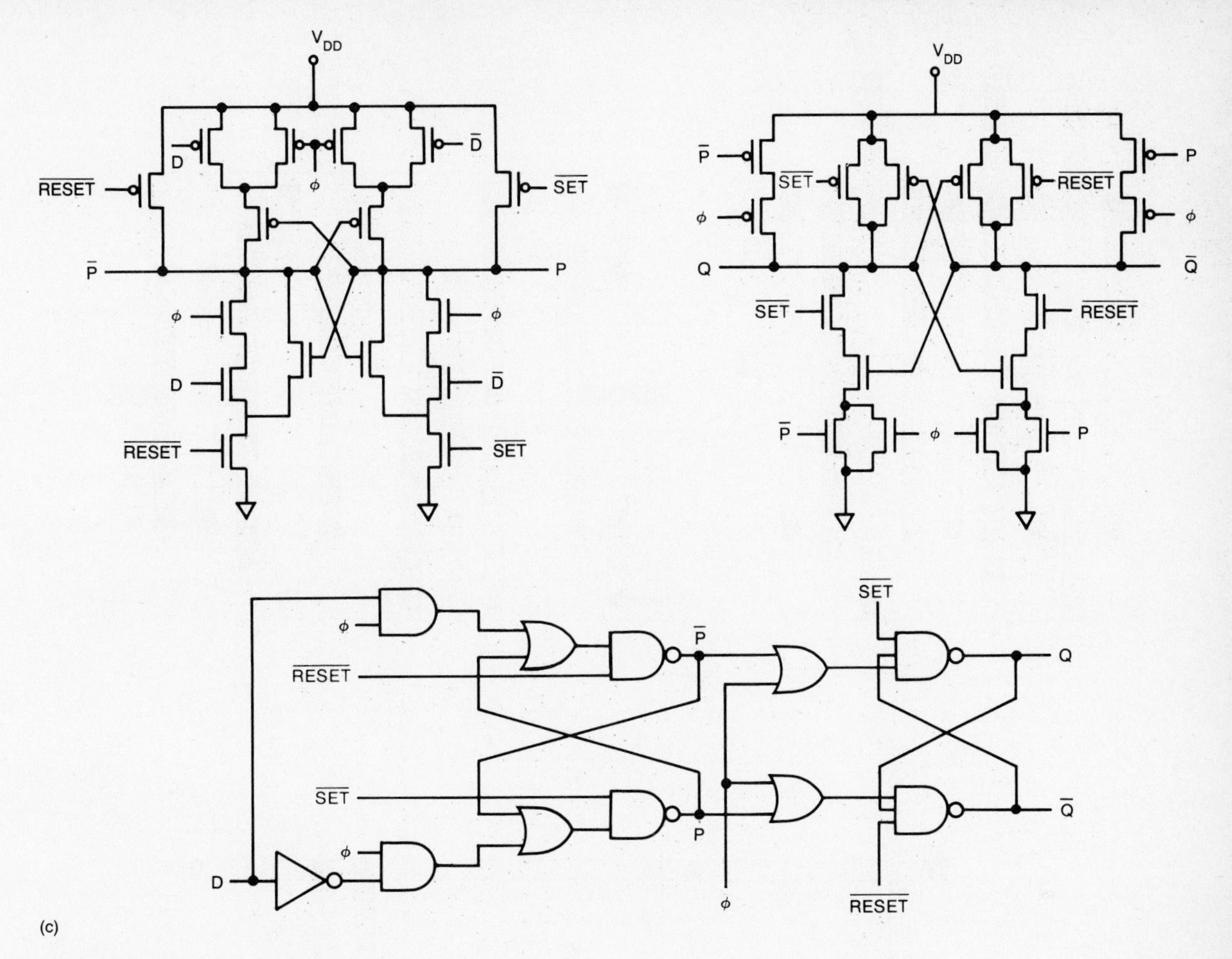

FIGURE 5.51. (continued)

in no deleterious effects. We will examine some rules that have been proposed to deal with both problems [GoDM83].

In the case of internal races, the basic rules for dynamic domino must be obeyed:

1. During precharge, logic blocks must be switched off.
2. During evaluation, the internal inputs can make only one transition.

This is achieved by using the N-P structures introduced earlier in this chapter. For complete dynamic blocks, either alternate n-p logic gates may be used or n-n or p-p blocks may be cascaded with buffer inverters between sections following the domino rules. Static logic

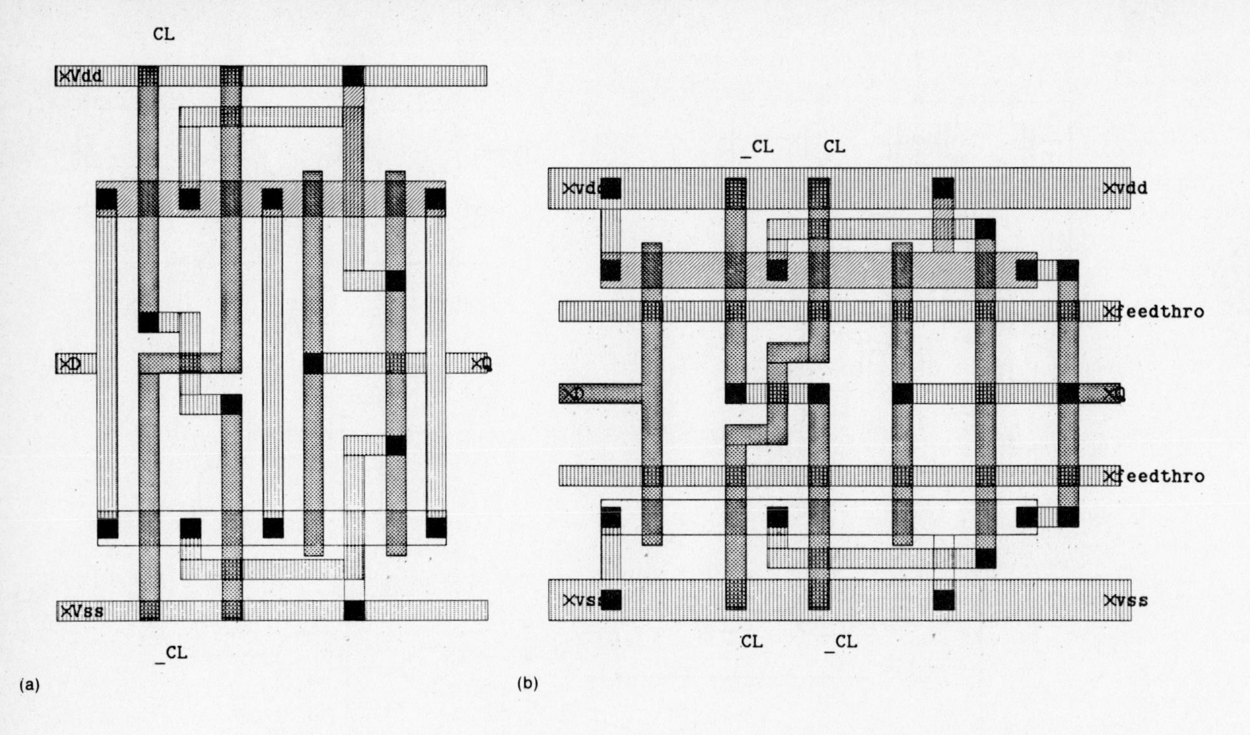

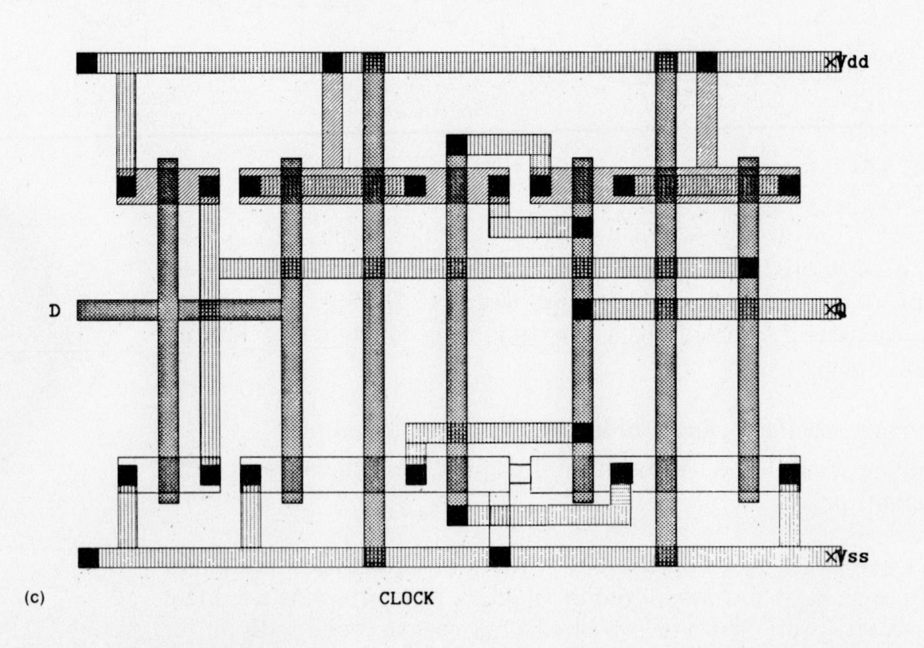

FIGURE 5.52. 2-phase D flip-flop layouts

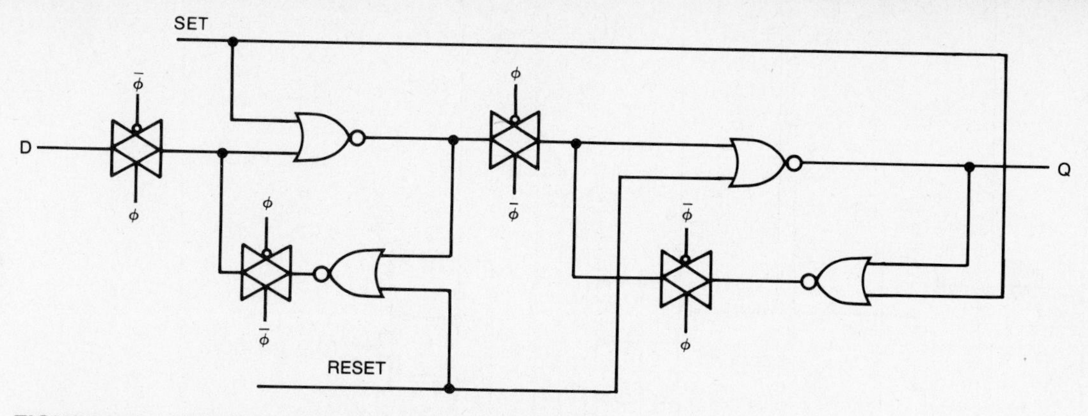

FIGURE 5.53. Static D flip-flop with set and reset

structures may be used. Where this is done, it is best to keep the logic static up to the C^2MOS latch, as the static structures generally can create glitches that violate the second condition mentioned above. When using the C^2MOS latches in conjunction with N-P logic sections, an additional rule guarantees race-free operation, even in the presence of clock skew. This requires that there be an even number of static inversions between the final dynamic gate and the C^2MOS output latch.

The ability to pipeline sections, as shown in Fig. 5.54, assumes that the output of a logic block (ϕ or $\overline{\phi}$) does not glitch (due to precharging or input variations) the input of the next stage in the pipeline while it is resolving its output. Under perfect clocking conditions, this assumption is met merely by following the ϕ–$\overline{\phi}$ logic block sequence. However, in the presence of clock skew, an early output transition on one block may glitch the next stage while it is still actively resolving its inputs. An additional rule ensures that glitches caused by clock skew will not be propagated from the output of one logic block through to the C^2MOS latch of the succeeding

TABLE 5.5. Static D flip-flop set/reset truth table

INPUTS				OUTPUT
CL	D	R	S	Q
X	X	1	0	0
X	X	0	1	1
X	X	1	1	NA

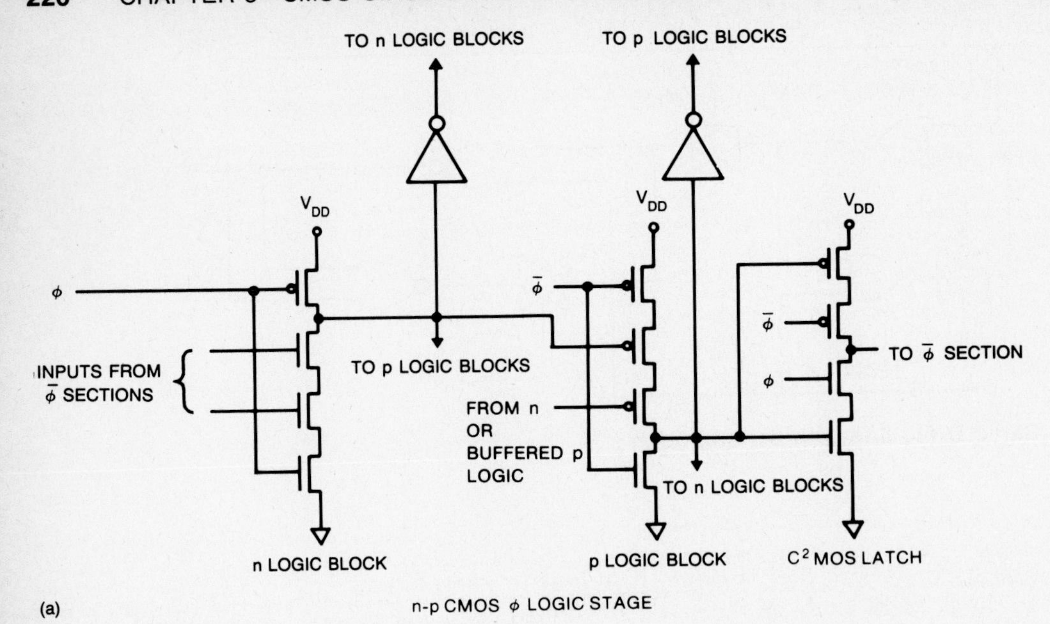

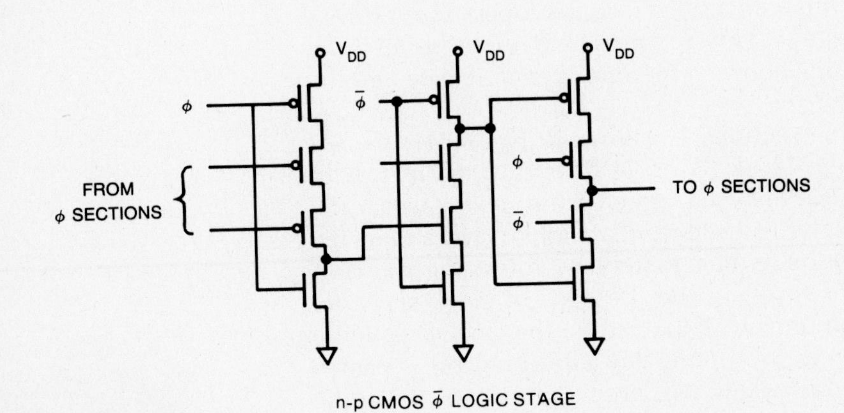

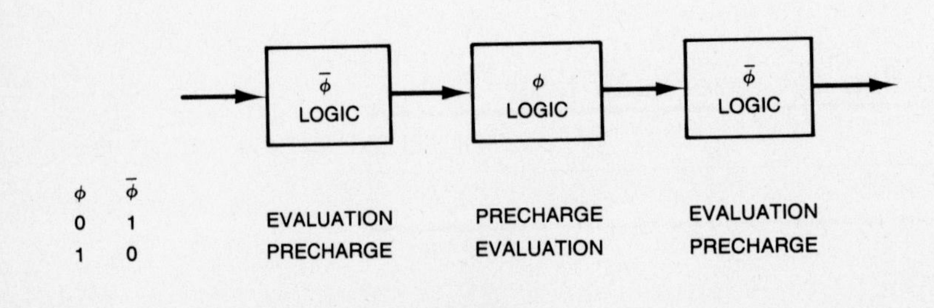

FIGURE 5.54. Pipelining 2-phase N-P logic

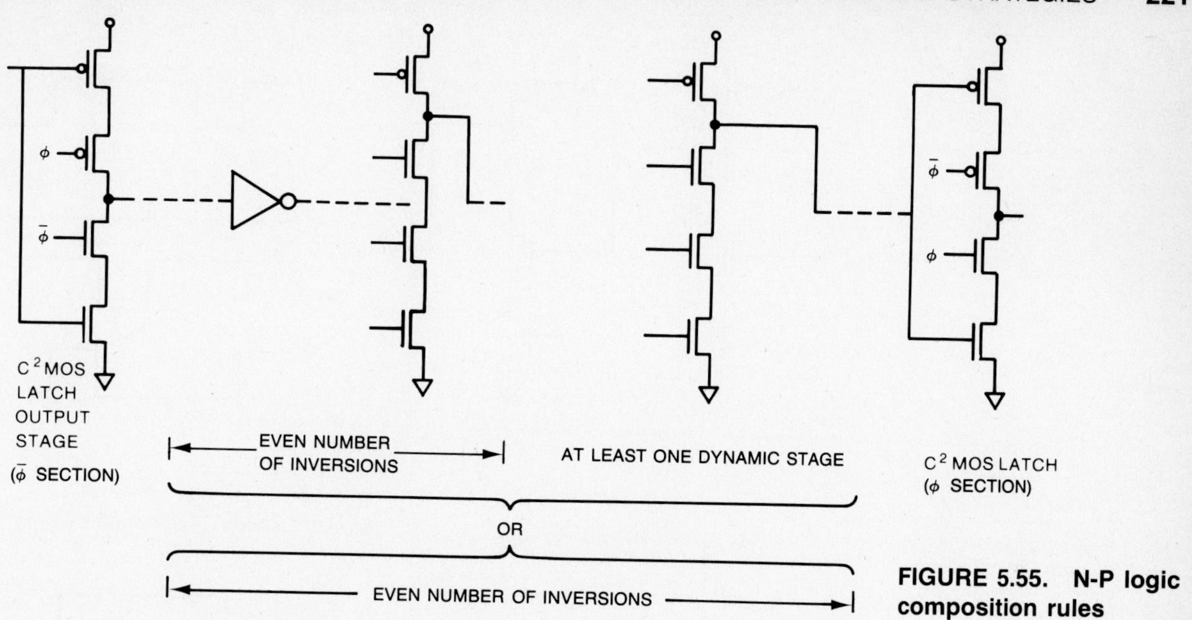

FIGURE 5.55. N-P logic composition rules

logic block. This rule states that either:

> There exists in each logic block at least one dynamic gate which is separated from the previous C^2MOS output stage by an even number of inversions.

or

> The total number of inversions between the C^2MOS stage and the previous C^2MOS stage is even.

Fig. 5.55 illustrates these rules.

5.4.7 4-phase clocking

The dynamic logic that has been described has a precharge phase and an evaluate phase. As we have learned, the addition of a "hold" phase simplifies dynamic circuit logic design. This primarily results from the elimination of charge sharing in the evaluation cycle. A disadvantage of 4-phase logic can be the number of clocks that may have to be generated.

5.4.8 4-phase memory structures

A 4-phase flip-flop is shown in Fig. 5.56a with its corresponding clock waveforms. During $\phi 1 = 0$, node $n1$ precharges. When $\phi 2 = 1$, node $n1$ conditionally discharges. When $\phi 2$ falls to 0, this value

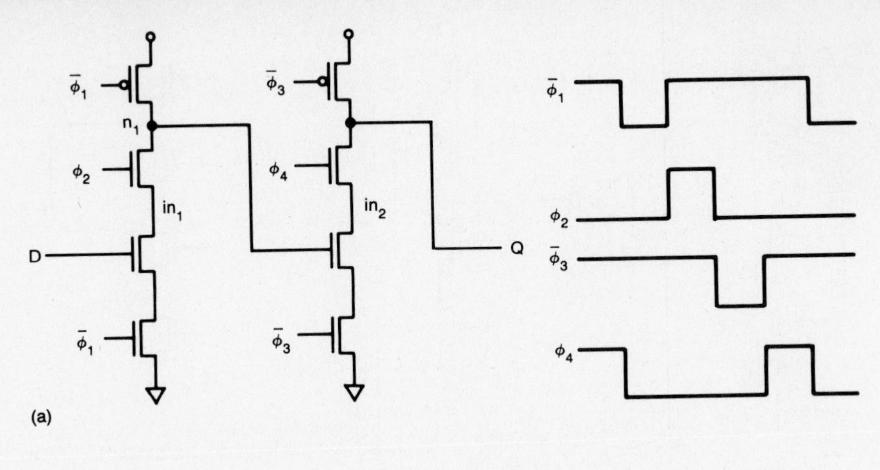

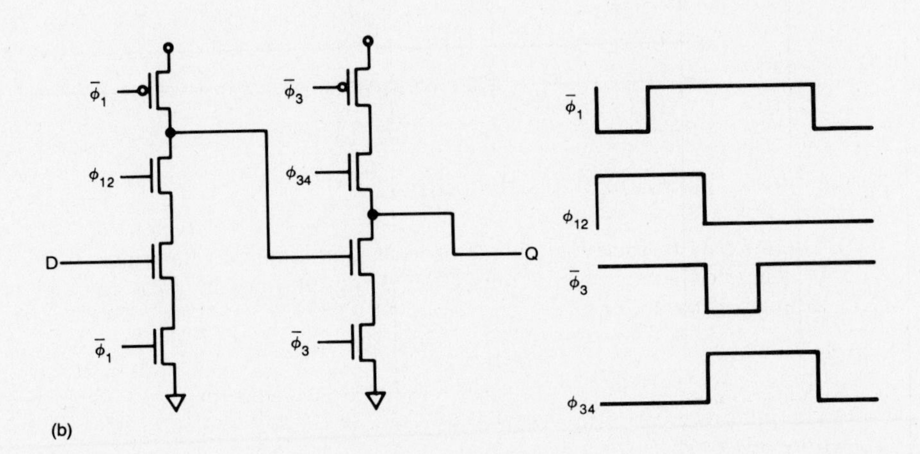

FIGURE 5.56. 4-phase flip-flops

is held on node $n1$ regardless of the state of the input D. During $\phi 3 = 0$, Q is precharged; during $\phi 4 = 1$, this node is conditionally discharged according to the state of node $n1$. This configuration can still have charge sharing problems as the intermediate nodes in the inverters ($in1$ and $in2$) may be corrupted due to charge sharing with outputs $n1$ and Q, respectively. This is solved by altering the clock waveforms so that $\phi 2$ is actually $\phi 12$ and $\phi 4$ is $\phi 34$, as shown in Fig. 5.56b. With these clocking waveforms the intermediate nodes are precharged uniformly.

5.4.9 4-phase logic structures

The main purpose in adopting a 4-phase clocking strategy is to enable the 4-phase logic gates to be built (although static gates may be also used). Illustrative gates are shown in Figs. 5.5, 5.6, 5.7, and 5.8. As discussed in the first part of this chapter, the sequence of gates has to be carefully controlled.

Arguments for using such a clocking strategy include the fact that no more clock lines are needed than for pseudo 2-phase clocking if certain 4-phase structures are used. In addition, a strict ratioless circuit technique may be applied, which can lead to very regular layouts.

5.4.10 Pseudo 4-phase clocking

It is also possible to use a 4-phase clock as a general clocking technique for domino circuits. This clocking scheme is covered in more detail in Chapter 9.

By use of the appropriate logic gate any combination of phases may be generated locally for circuits requiring different clocking strategies. $\phi 1$ may be used as a slave latch clock. $\phi 2$ is used as first-level logic evaluation. $\phi 3$ is used as the master latch clock, and $\phi 4$ is used as the second-level logic evaluation. This is shown in Fig. 5.57.

5.4.11 Recommended approaches

For first-time designs, where mostly static logic is to be used, the pseudo 2-phase clocking scheme is probably preferable. This also has the most commonality with the Mead and Conway text. The clock routing problem is minimal, especially in data path designs. Alternatively, a single phase clock with the latches shown in Fig. 5.51b and Fig. 5.51c could be used where density was not an issue. Dynamic gates may be employed using 2-phase dynamic logic outlined in this chapter. Methods for dealing with PLAs are treated in Chapters 8 and 9.

For bit-serial circuitry, where clock routing and flip-flop complexity is important, a 2-phase or 4-phase clocking scheme may be most suitable.

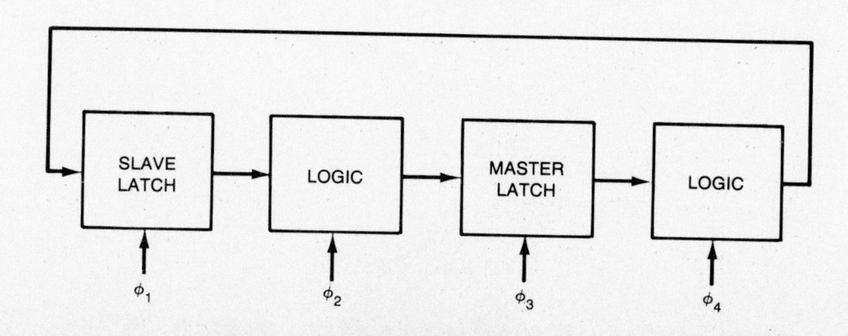

FIGURE 5.57. Pseudo 4-phase clocking strategy

Probably the most universal scheme is the pseudo 4-phase clocking scheme, which allows construction of any of the logic forms by suitable on-chip clock generation.

5.5 Input-output (I/O) structures

Of all the CMOS circuit structures that will be covered in this text, I/O structures require the most amount of circuit design expertise in association with detailed process knowledge. Thus it is probably inappropriate for a system designer to contemplate I/O pad design. Rather, well characterized library functions should be used for whatever process is being used. We will give basic outlines of how to design pads but will refrain from giving specific layouts.

5.5.1 Overall organization

It is often convenient to build I/O pads with a constant height and width, with connection points at prespecified locations. Pad size is defined usually by the minimum size to which a bond wire can be attached. This is usually of the order of 150 μm by 150 μm. Additionally, a constant position for V_{DD}, V_{SS}, and any other global control wires is an advantage. Fig. 5.58 illustrates some of these concepts. A variety of placement of components is shown. Power and ground bus widths may be calculated from a worst-case estimate of the power dissipation of a die and from a consideration of providing good supply voltages. Multiple power and ground pads may be used to reduce noise. Some designers advocate placing the lowest circuit voltage (V_{SS}) as the outermost track. With these points in mind, a *frame* generation program may be easily constructed. This takes a simple description of the pad ordering and produces a finished pad frame. A typical description might be as follows:

```
LEFT;
     INPUT  A
     INPUT  B
TOP ;
     VDD    VDD
     INPUT  C
RIGHT;
     OUTPUT Z
     OUTPUT Y
BOTTOM;
     OUTPUT W
     VSS    VSS
```

The resulting I/O frame is shown in Fig. 5.59a on page 226.

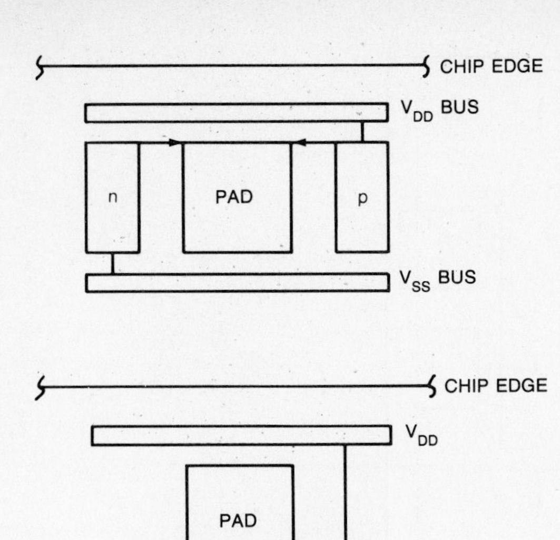

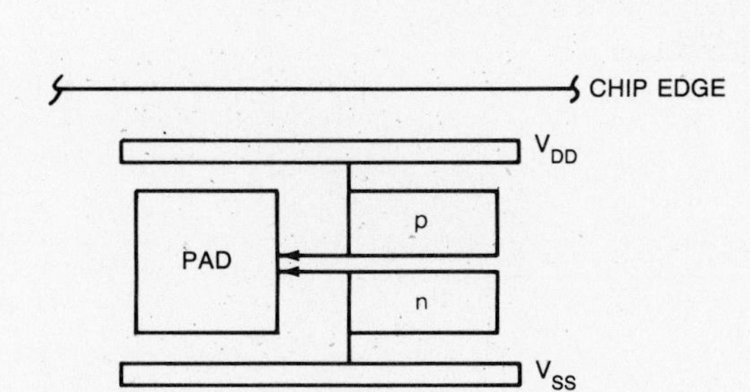

FIGURE 5.58. General pad layouts

5.5.2 V_{DD} and V_{SS} pads

These pads are easily designed and consist of a metal pad connected to the appropriate bus. A nonplanarity arises at one of the pads. The broken path may be completed in polysilicon, as shown in Fig. 5.59b. Alternatively, a two-level metal process affords good crossovers, providing that a large number of vias are used in the connection. There is no reason to reduce the size of the pad to power rail connection.

5.5.3 Output pads

First and foremost, an output pad must have sufficient drive capability to achieve adequate rise and fall times into a given capacitive load. If the pad drives non-CMOS loads, then any required DC charac-

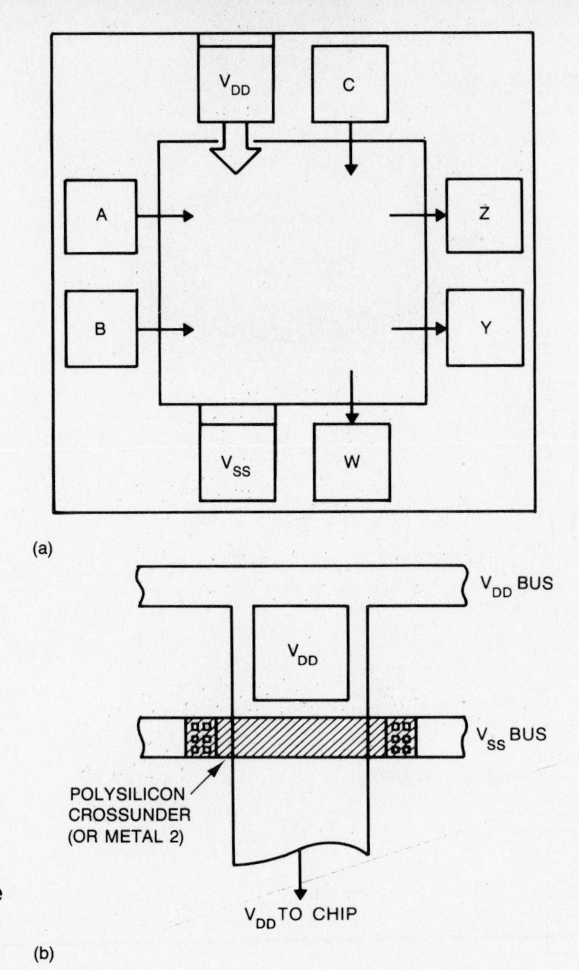

FIGURE 5.59. a) I/O frame generation, and b) V_{DD} pad design

teristics must also be met. In this discussion we will concentrate on pads to drive CMOS loads. Given a load capacitance and target rise and fall time, the output transistor sizes may be calculated from the equations derived in Chapter 4. One then generally needs buffering to present a lower load to internal circuitry. As previously discussed, a ratio of 2.7 is optimal for speed. However, a stage ratio of 2–10 will work adequately. Generally, in a pad, a two-stage inverter circuit is used to result in a non-inverting output stage.

Having estimated the transistor sizes, the layout may be commenced. As large transistors are typically used and I/O currents are high, the susceptibility to latch-up is highest in I/O structures. Hence, the layout guidelines given in Chapter 3 should be used. This means separating n- and p-transistors and using the appropriate guard rings tied to the supply rails. Latch-up will also occur when transients rise above V_{DD} or below V_{SS}. These conditions are more likely to occur at I/O pads due to the interface to external circuitry.

When driving TTL loads with CMOS gates, the different switching thresholds have to be considered. V_{IL} of a TTL gate is 0.4 volts. V_{OL} of a CMOS gate is 0 volts. Thus we have no problem in this respect. V_{IH} for a TTL gate is 2.4 volts. The V_{OH} for a CMOS gate is 5 volts (for a 5 volt supply) and hence there is no problem here. In the low state, the CMOS buffer must be capable of "sinking" 1.6 mA for a standard TTL load with a V_{OL} of $<.4$ v. For typical driver transistors, this is usually no problem.

5.5.4 Input pads

The design of input pads can parallel that of output pads with respect to transistor sizing. Often the transistors used in the output pad may be just "turned around." One extra precaution has to be taken. The gate connection of an MOS transistor has a very high input resistance (10^{12} to 10^{13} Ω). The voltage at which the oxide punctures and breaks down is about 40–100 volts. The voltage that can build up on a gate may be determined from

$$V = \frac{I\Delta t}{C_g}, \tag{5.9}$$

where

V is the gate voltage
I is the charging current
Δt is the time taken to charge the gate
C_g is the gate capacitance.

Thus if $I = 10\ \mu$A, $C_g = .03\ pF$, and $\Delta t = 1\ \mu$Sec, the voltage that appears on the gate is approximately 330 volts. Usually a combination of a resistance and diode clamps (electrostatic protection) are used to limit this potentially destructive voltage. A typical circuit is shown in Fig. 5.60 (on page 228), with approximate layouts. Clamp diodes D1 and D2 turn on if the voltage at node X rises above V_{DD} or below V_{SS}. Resistor R is used to limit the peak current that flows in the diodes in the event of an unusual voltage excursion. Values anywhere from 200Ω–3KΩ are used. This resistance, in conjunction with any input capacitance, C, will lead to an RC time constant, which must be watched in high speed circuits. A polysilicon resistor is preferable to a diffusion resistor in a p-well process, as it reduces the possibility of creating extra charge injection into the substrate which can contribute to latch-up. In an n-well process, all n-device I/O circuitry can be designed. In this case n^+ diffused protection resistors, as well as n "punch-through" devices, may be used. A

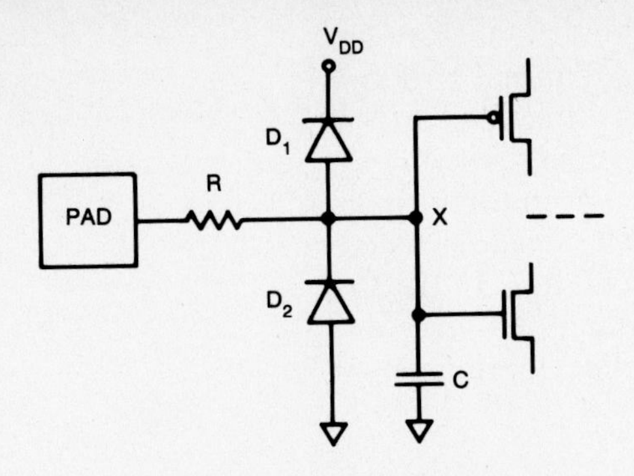

(a) TYPICAL INPUT PROTECTION CIRCUIT

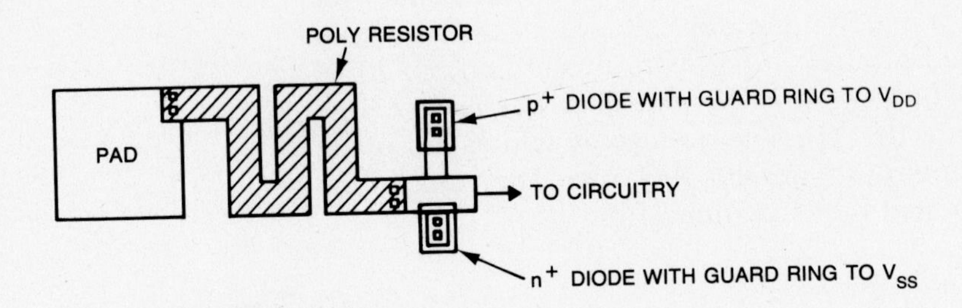

(b)

V_{DD}

p^+ DIODE

D1

n^+ GUARD RING

TO RESISTOR

TO CIRCUITRY

p^+ GUARD RING

D2

n^+ DIODE

V_{SS}

(c)

FIGURE 5.60. Input Pad Electrostatic Discharge (ESD) protection

"punch-through" device has closely spaced source and drain diffusions but no gate. The device affords protection by "avalanching" at around 50V. No wells need be included in this type of I/O [FGLM84].

When interfacing TTL logic to CMOS, it is advantageous to place the switching point of the input inverter in the middle of the TTL switching range. For TTL $V_{OL} = 0.4$ volts and $V_{OH} = 2.4$ volts. Thus the switching point should be set near 1.4 volts. This is achieved by ratioing the inverter transistors or using a reference voltage. Alternatively, the TTL output can use an additional resistor connected to the 5V supply to improve the TTL V_{OH}. This resistor may be included inside the pad in the form of a p-transistor.

5.5.5 Tri-state pads

A tri-state pad may be built modeled on the tri-state inverter structure shown in Chapter 2. Another circuit is shown in Fig. 5.61a. This is faster due to the reduced number of devices in series. Care must

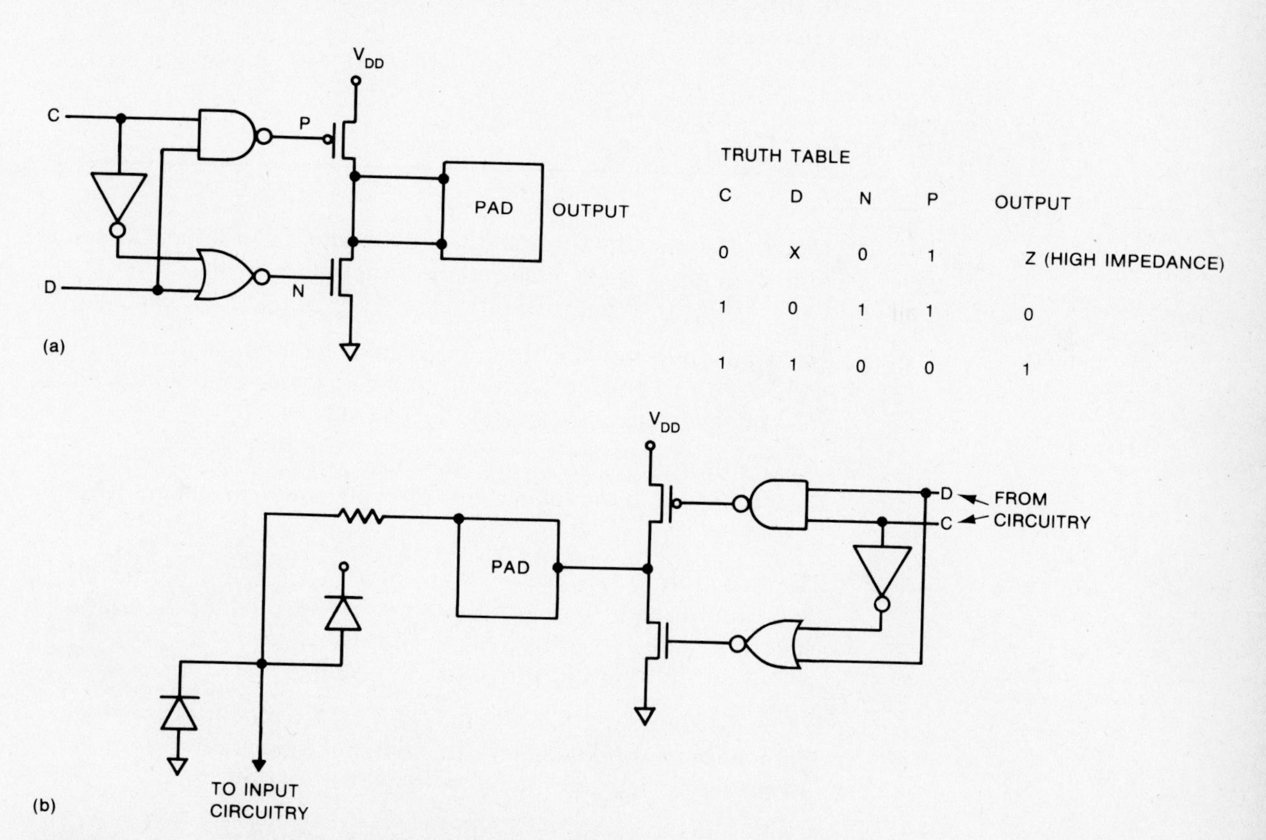

FIGURE 5.61. A Tri-state pad and a bi-directional pad

be taken to switch the buffer in such a way as to prevent large DC currents flowing during switching.

5.5.6 Bi-directional pads

By merging an input pad and a tri-state pad a bi-directional pad may be constructed. This is illustrated in Fig. 5.61b.

5.6 Summary

In this chapter some alternative CMOS logic circuits have been discussed. The layout and circuit design of CMOS gates was then treated. Clocking strategies were outlined indicating suitable memory elements and logic structures. Finally, the basics of I/O design were covered. Using this material as a base, Chapter 8 will examine some useful subsystems that use a variety of the techniques discussed in this chapter.

5.7 Exercises

5.1 Design the circuits for a static CMOS gate, pseudo-nMOS gate, and dynamic CMOS gate for the function $F = \overline{((C.(A + B + C)) + (D.E.F))}$. What precautions need to be taken for each gate?

5.2 Design layouts for the three circuits in Exercise 5.1, including a strategy for routing clocks to other modules if required. Can the transistor sizes be gracefully resized?

5.3 What circuit design techniques would you employ to minimize the effects of body effect in a p-well process? In an n-well process?

5.4 The function $F = (A.(B + C + D + E + F))$ is implemented as a domino gate with minimum sized transistors ($C_D = C_G$). Does this gate have any charge-sharing problems? If not, why not? If so, how is the problem alleviated?

5.5 Fig. 5.50a shows a 2-phase static D flip-flop with a trickle inverter. The feedback transmission gate switched on $\overline{\phi}$ may be omitted. Propose a design modification that would allow this.

5.6 A zero detect circuit is required on the output of a 32 bit ALU. Design a static and dynamic circuit to implement this function.

A 2-phase clock as shown in Fig. 5.48 is available. Design any clocking circuitry needed.

5.7 Design a layout for the chain latch shown in Fig. 5.49.

5.8 Design an input pad to convert TTL logic levels (V_{OL} = 0.4V, V_{OH} = 2.4V) into CMOS levels. The circuit is to be non-inverting and should include protection devices. Design a layout for the pad. Identify possible latch-up paths in the circuit.

PLATE CAPTIONS

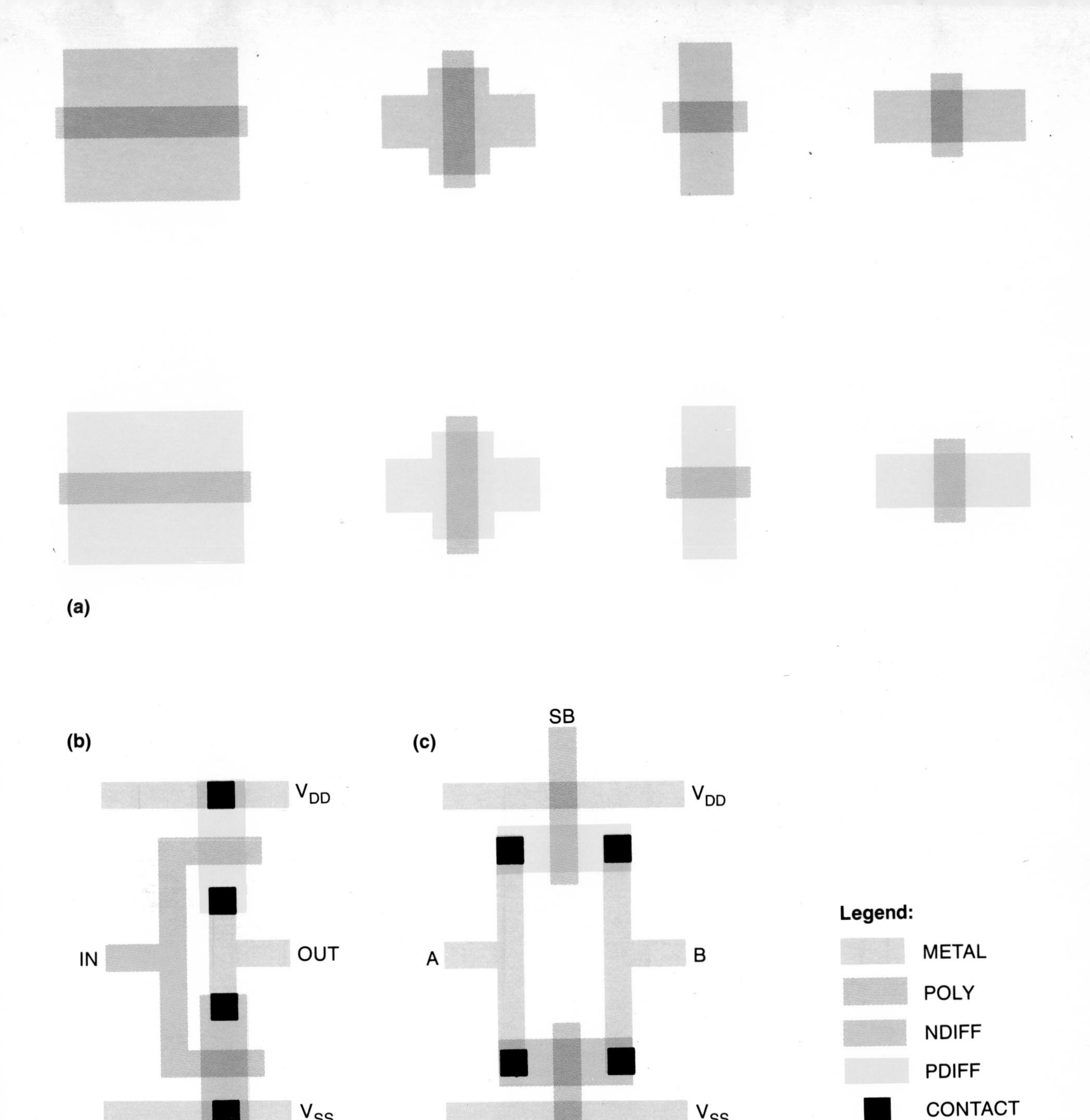

(a)
(b)
(c)
SB
VDD
VDD
IN
OUT
A
B
VSS
VSS
S
Legend:
METAL
POLY
NDIFF
PDIFF
CONTACT

Plate 1

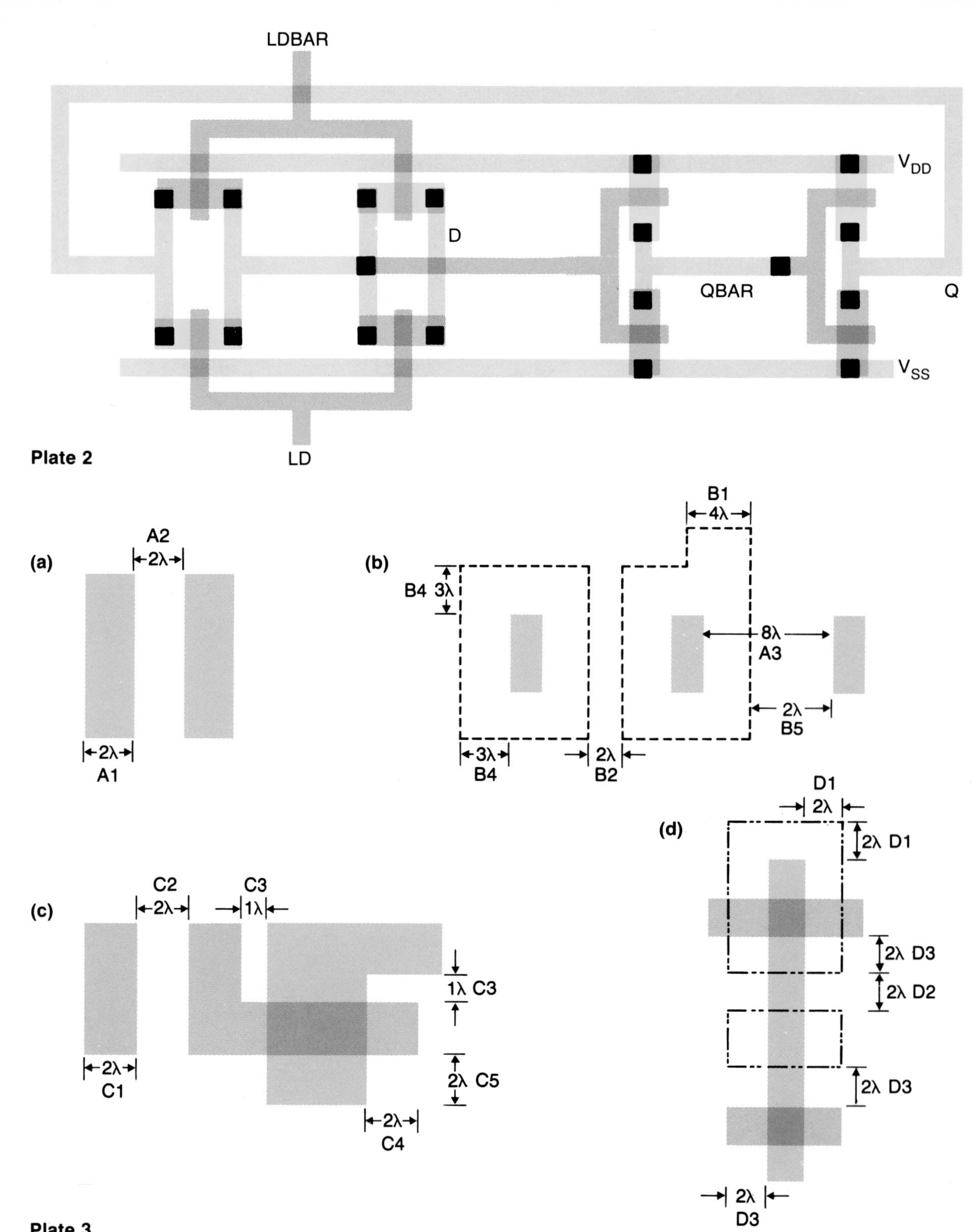

Plate 2

Plate 3

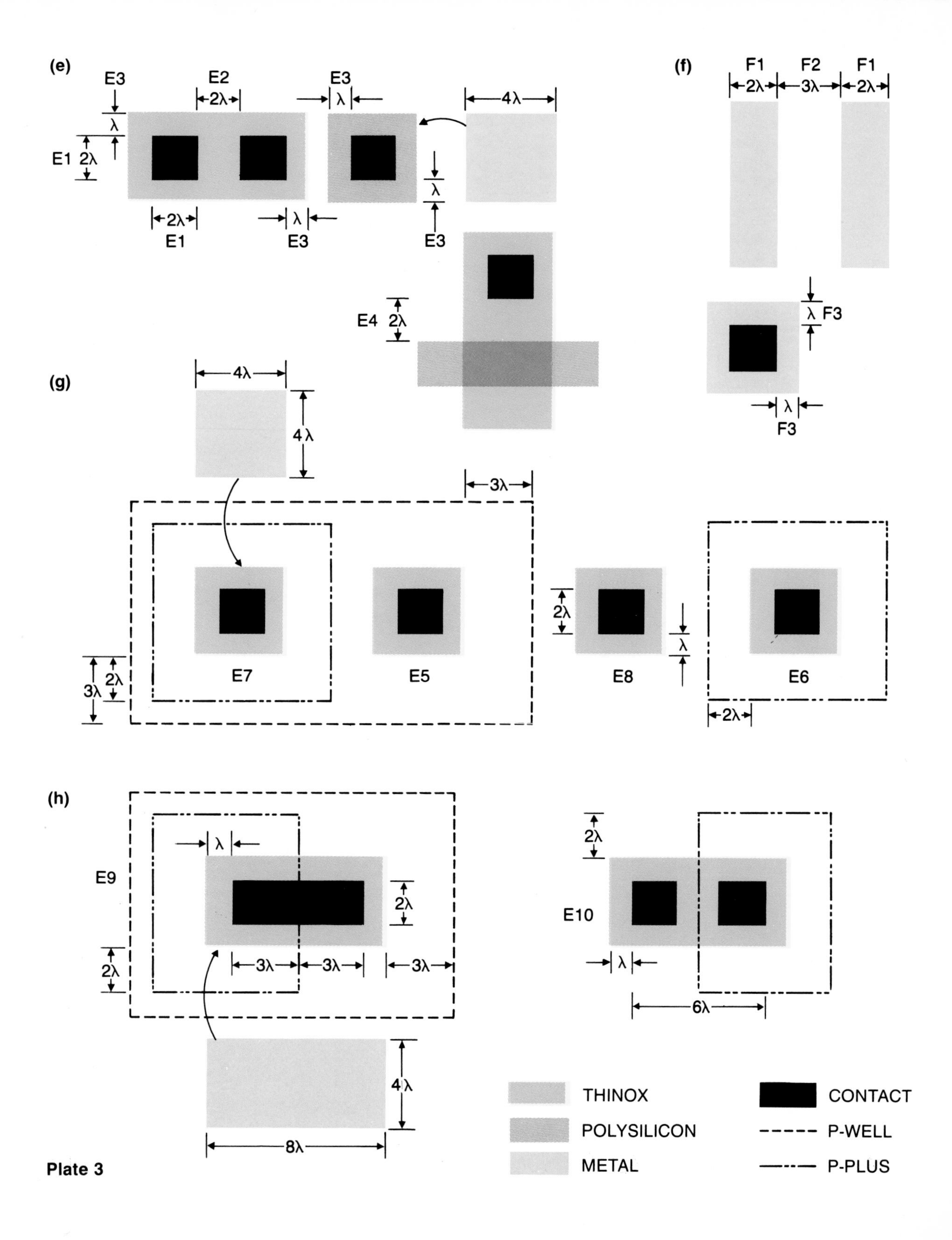

(e)
E1
E2
E3
E4
2λ
λ
4λ
(f)
F1
F2
F3
2λ
3λ
(g)
E5
E6
E7
E8
(h)
E9
E10
6λ
8λ
THINOX
POLYSILICON
METAL
CONTACT
P-WELL
P-PLUS

Plate 3

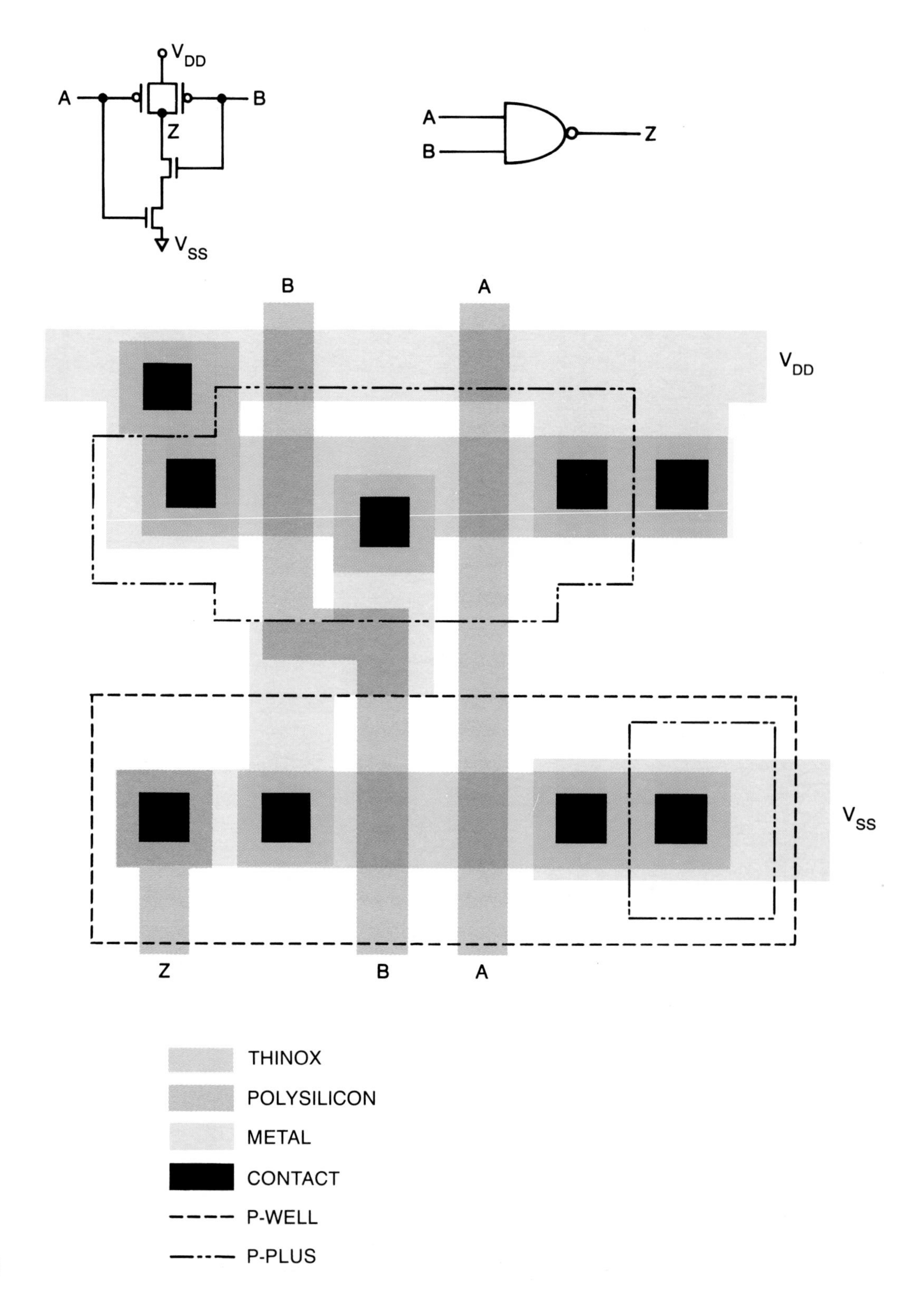

V_DD
A
B
Z
V_SS
A
B
Z
B
A
V_DD
V_SS
Z
B
A
THINOX
POLYSILICON
METAL
CONTACT
P-WELL
P-PLUS

Plate 4

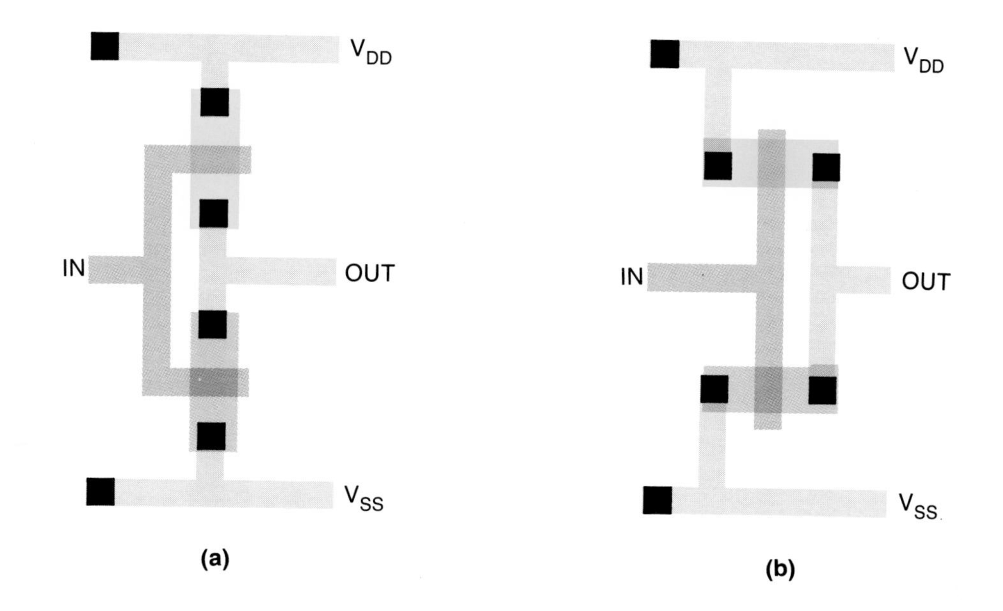
V_DD
IN
OUT
V_SS
(a)
V_DD
IN
OUT
V_SS
(b)

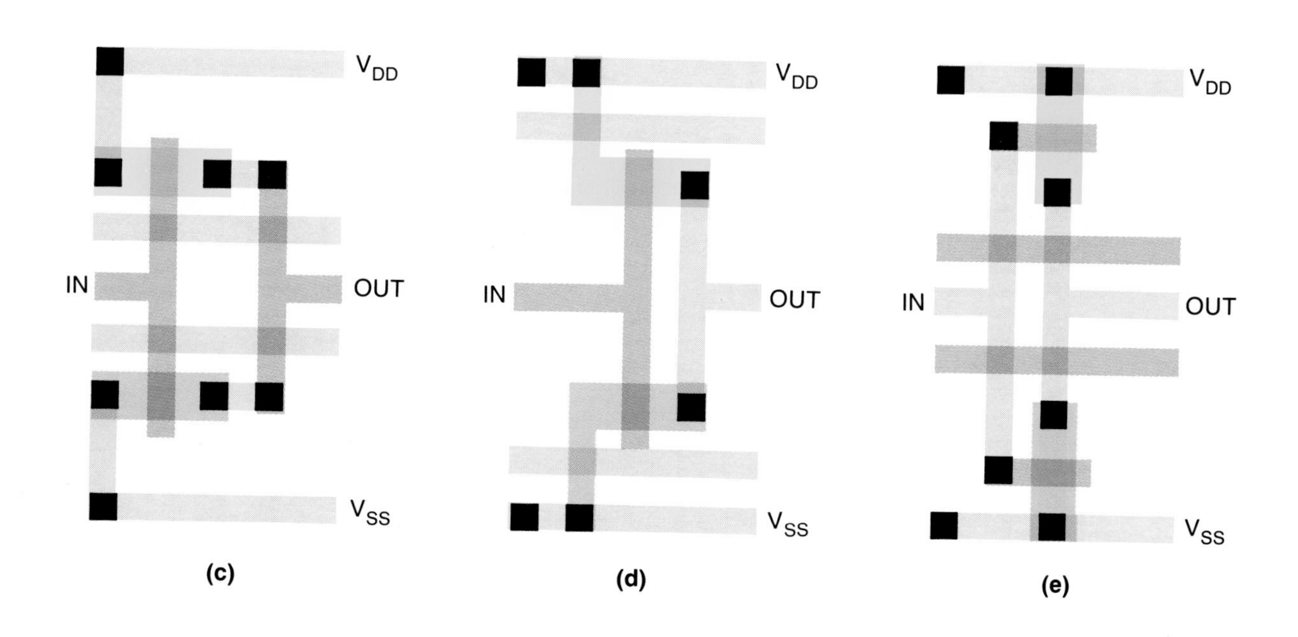
V_DD
IN
OUT
V_SS
(c)
V_DD
IN
OUT
V_SS
(d)
V_DD
IN
OUT
V_SS
(e)

Plate 5

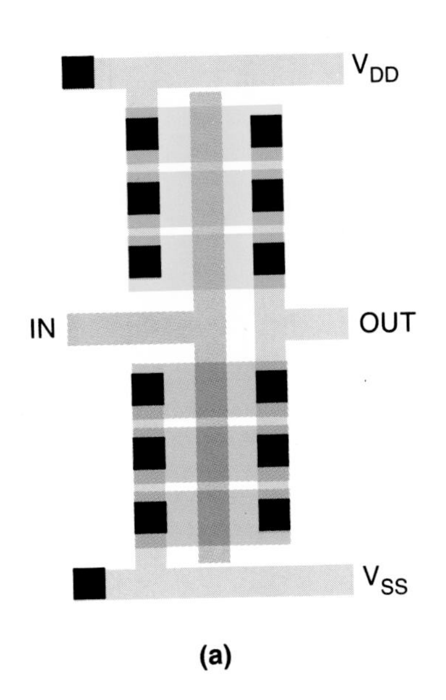

(a)

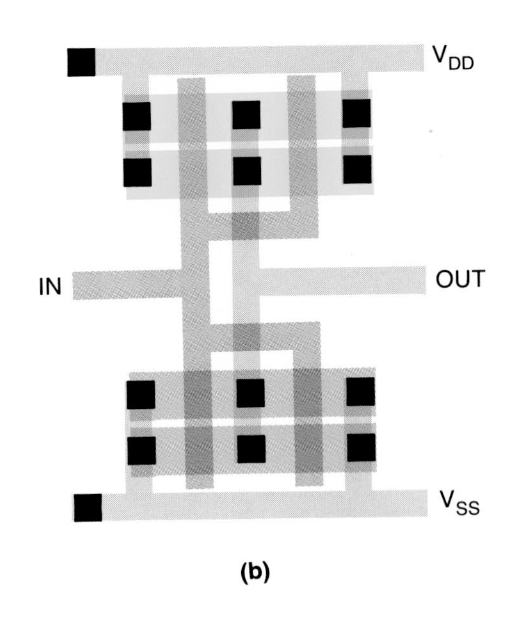

(b)

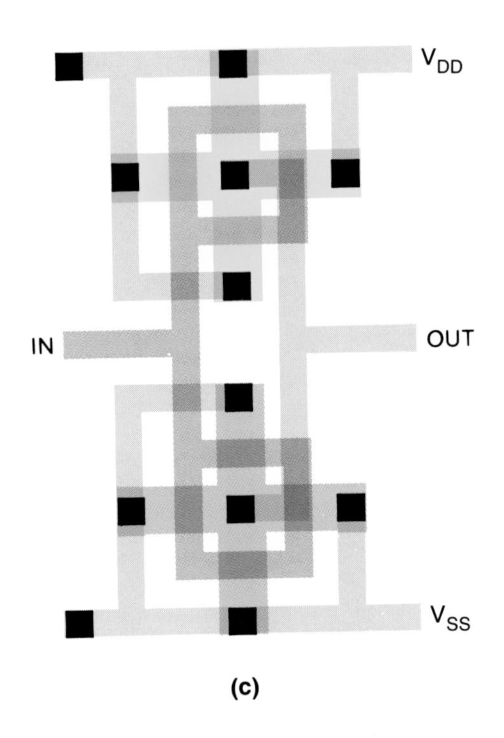

(c)

Plate 6

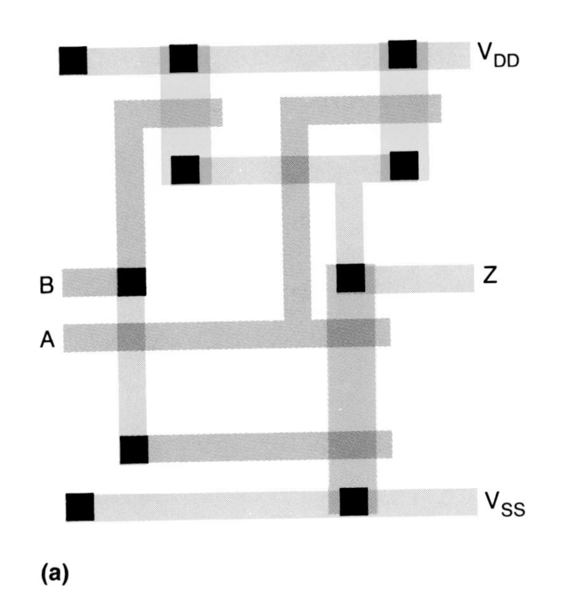

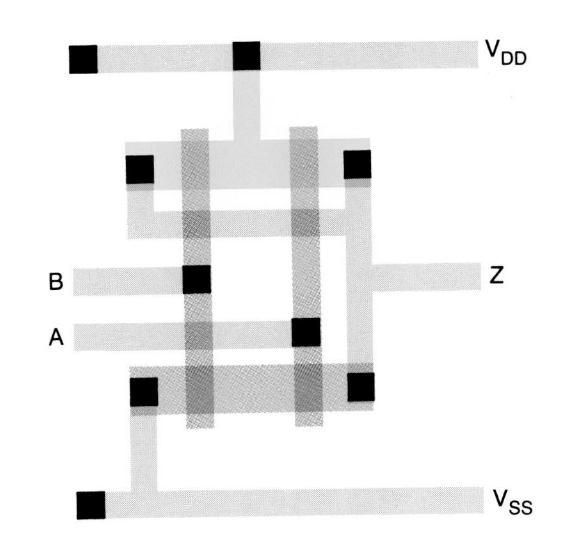

(a)

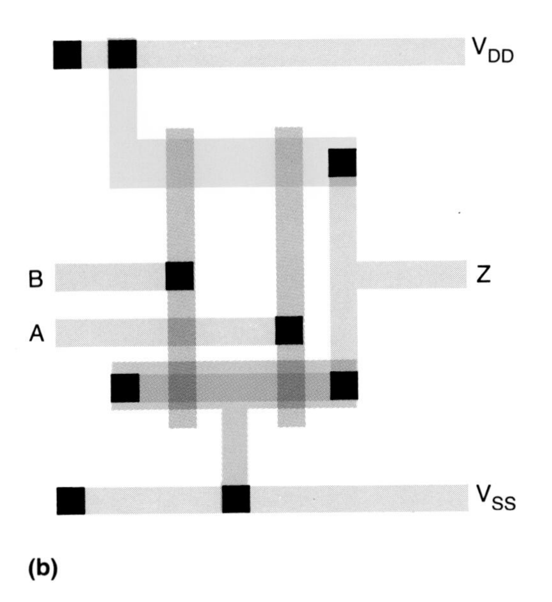

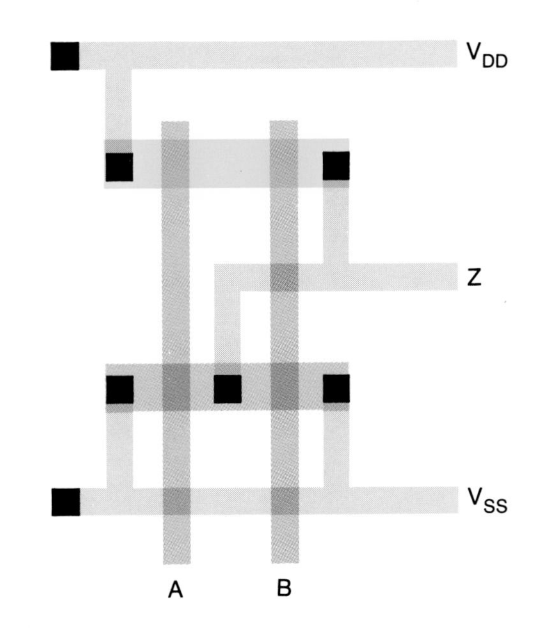

(b)

Plate 7

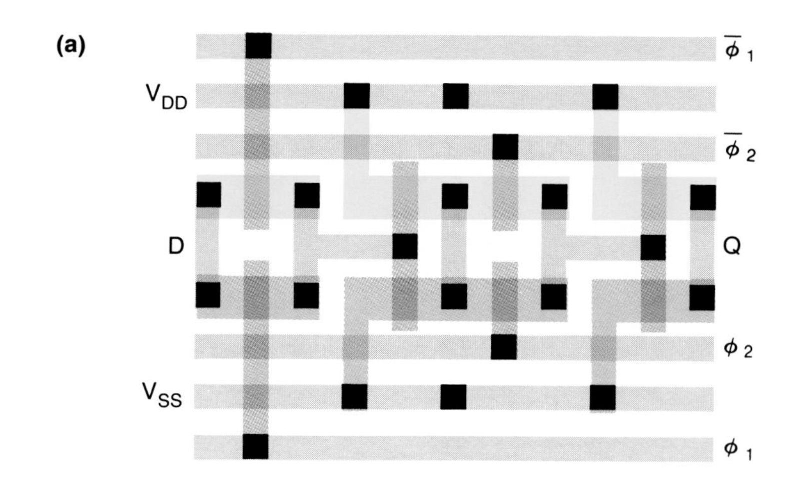
(a)
$\overline{\phi}_1$
V_{DD}
$\overline{\phi}_2$
D
Q
ϕ_2
V_{SS}
ϕ_1

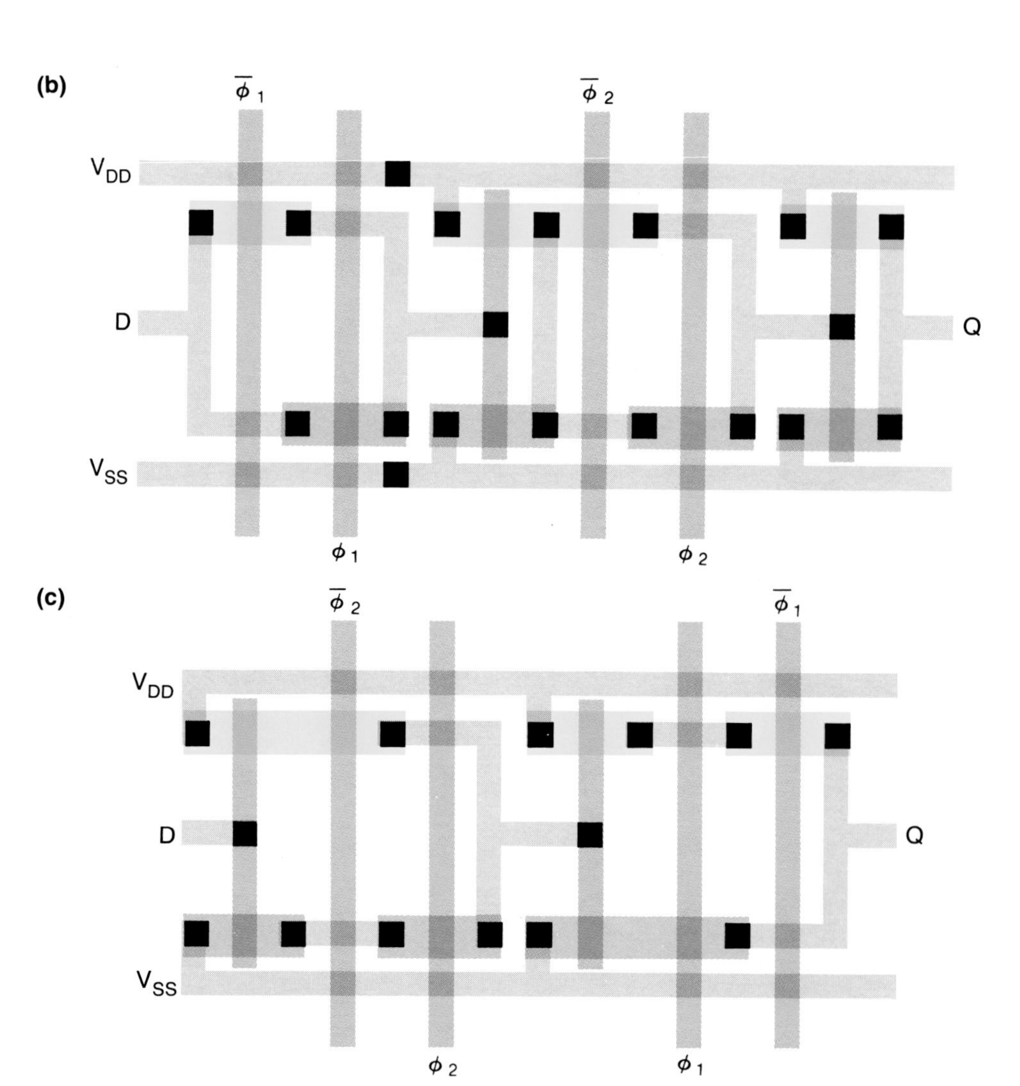
(b)
$\overline{\phi}_1$
$\overline{\phi}_2$
V_{DD}
D
Q
V_{SS}
ϕ_1
ϕ_2
(c)
$\overline{\phi}_2$
$\overline{\phi}_1$
V_{DD}
D
Q
V_{SS}
ϕ_2
ϕ_1

Plate 8

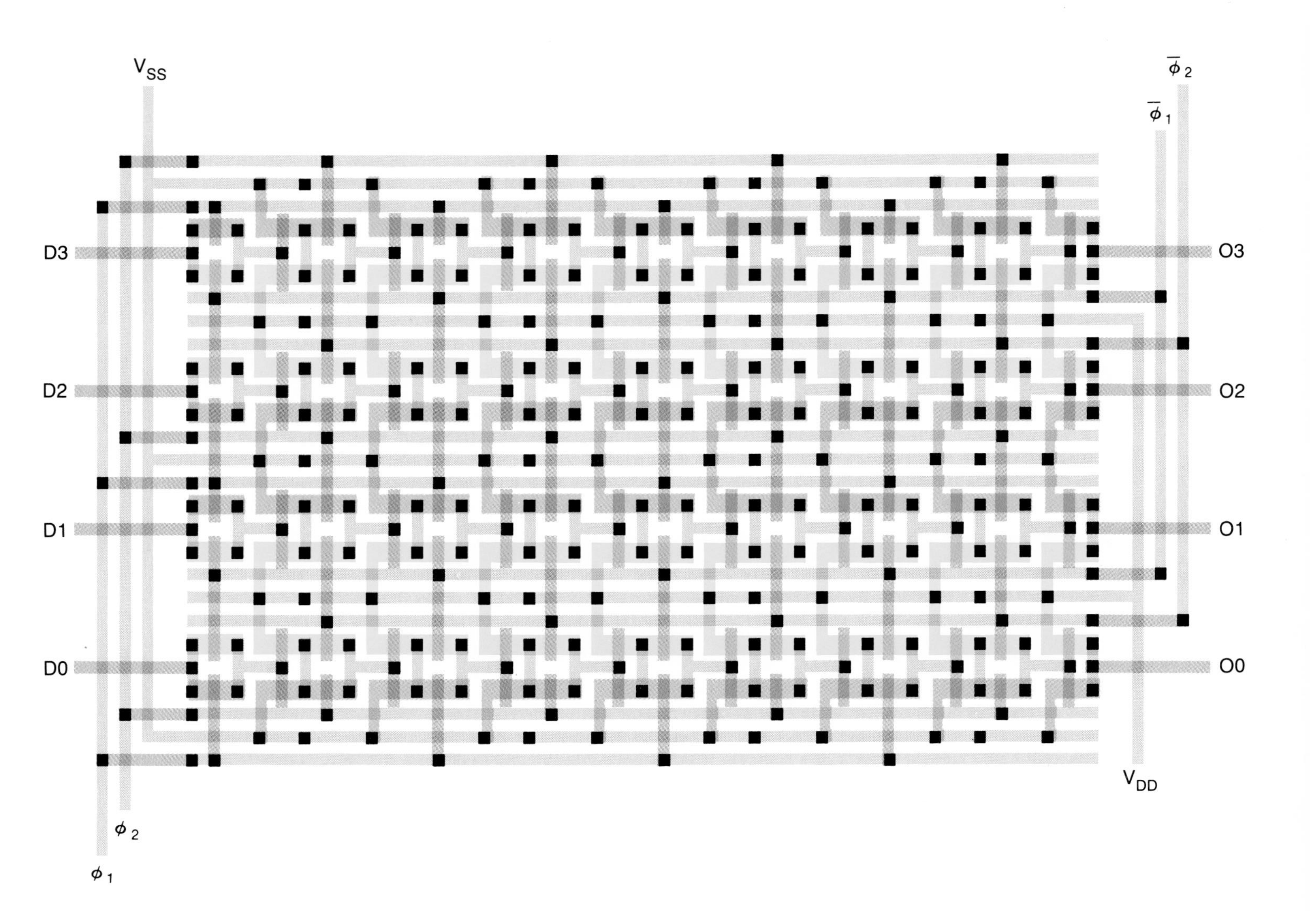

Plate 9

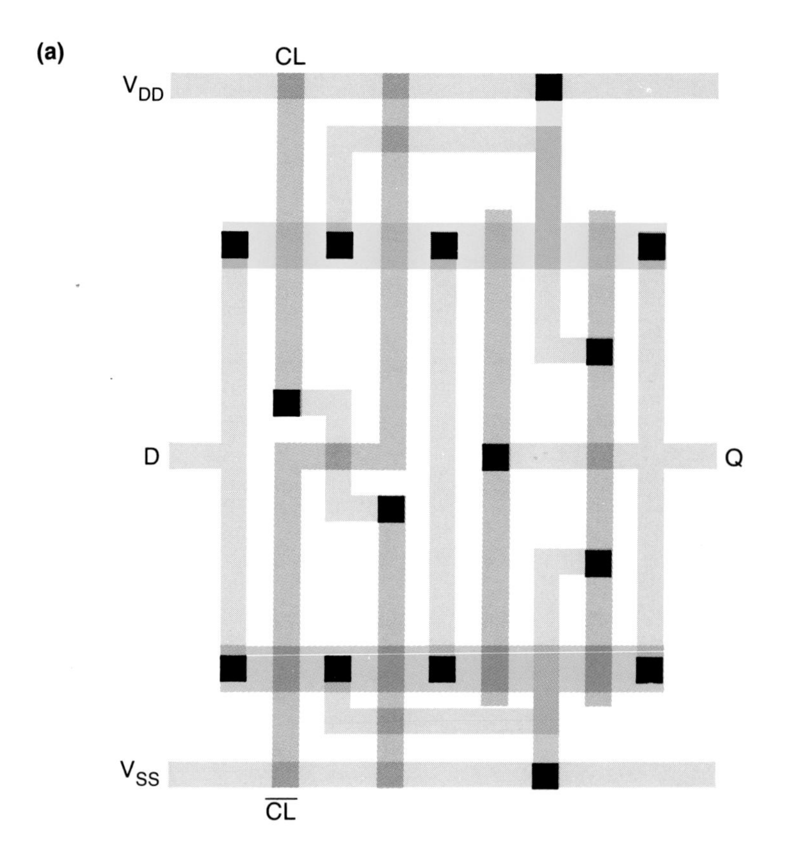

(a)
CL
V_{DD}
D
Q
V_{SS}
$\overline{CL}$

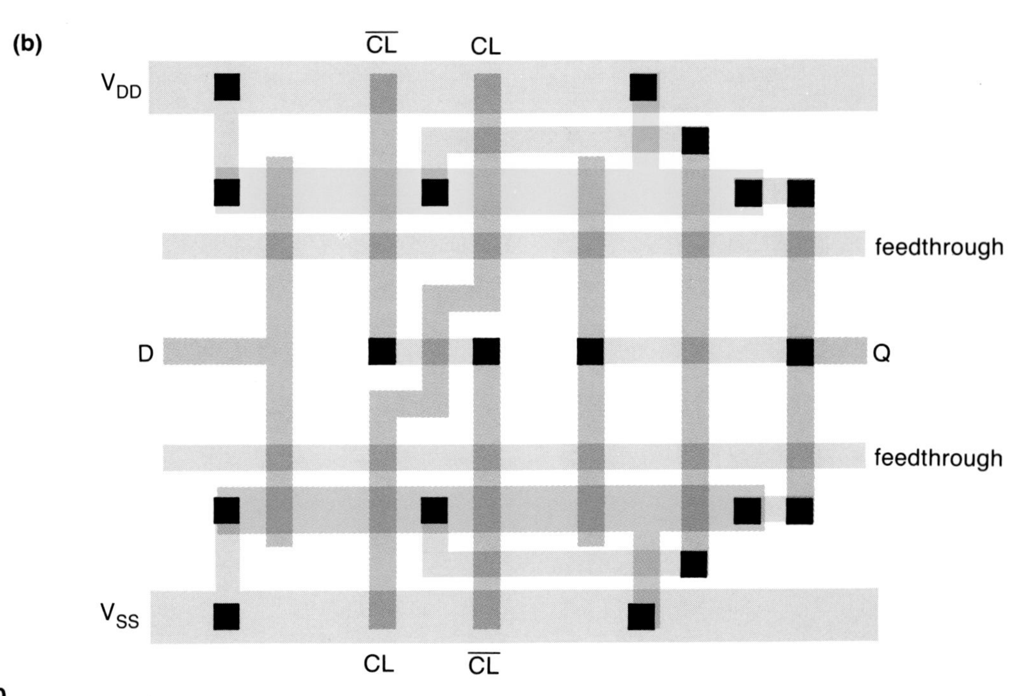

(b)
$\overline{CL}$
CL
V_{DD}
feedthrough
D
Q
feedthrough
V_{SS}
CL
$\overline{CL}$

Plate 10

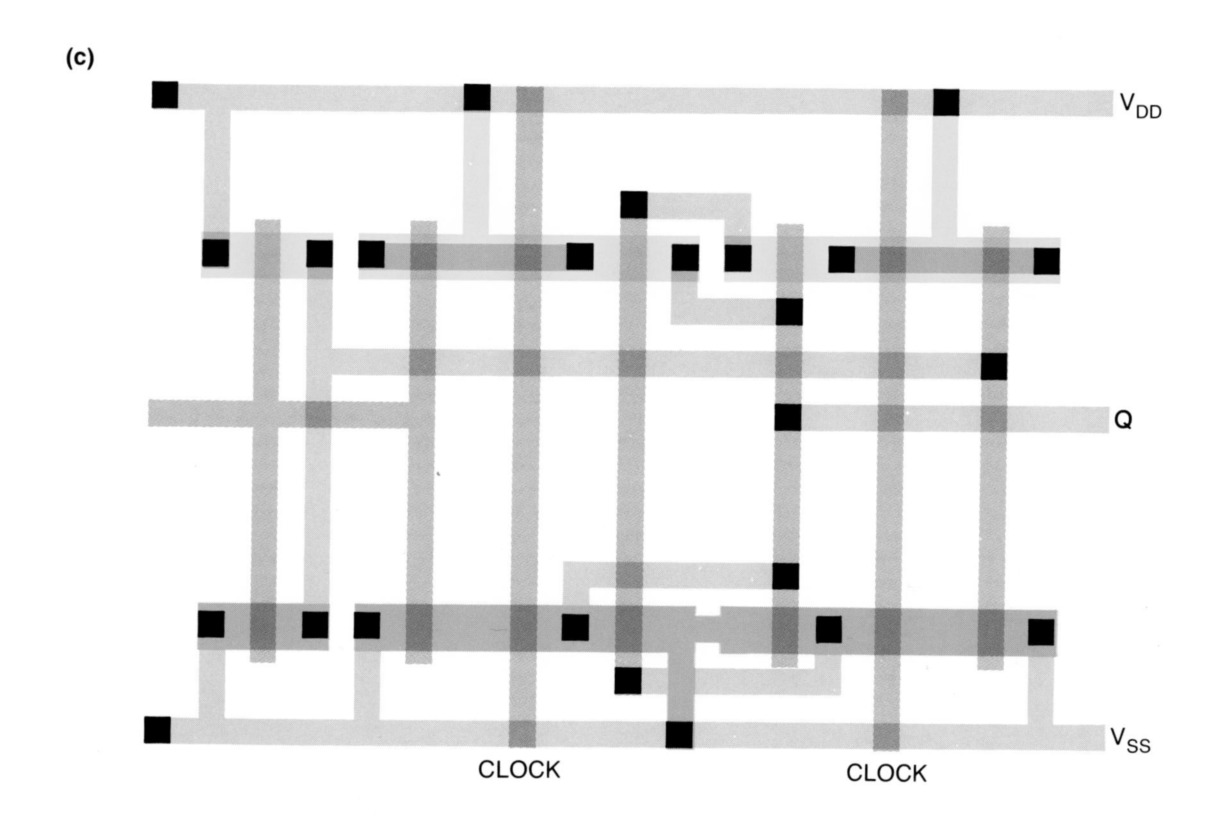
(c)
V_{DD}
Q
V_{SS}
CLOCK
CLOCK

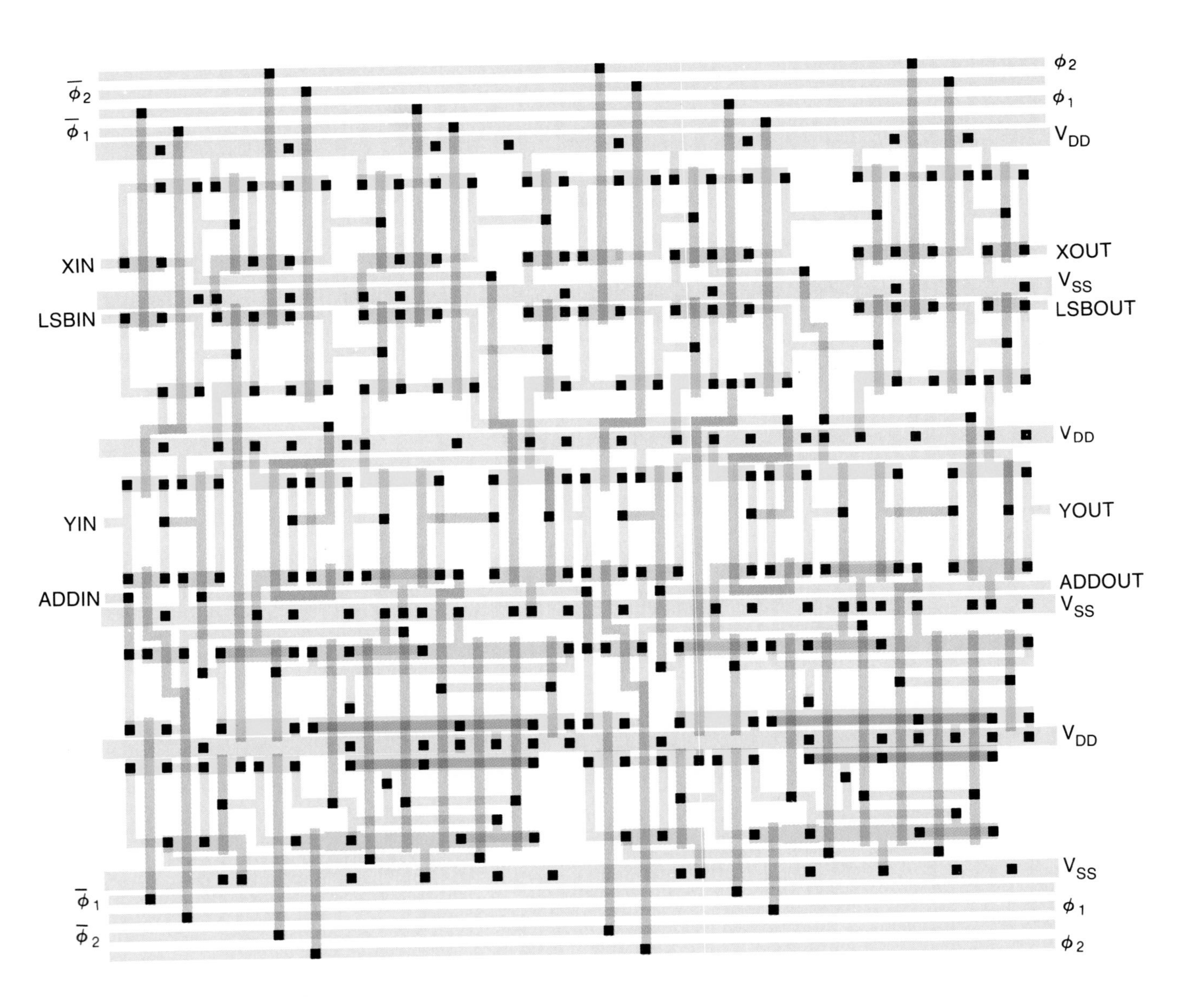
$\bar{\phi}_2$
$\bar{\phi}_1$
XIN
LSBIN
YIN
ADDIN
$\bar{\phi}_1$
$\bar{\phi}_2$
ϕ_2
ϕ_1
V_{DD}
XOUT
V_{SS}
LSBOUT
V_{DD}
YOUT
ADDOUT
V_{SS}
V_{DD}
V_{SS}
ϕ_1
ϕ_2

Plate 11

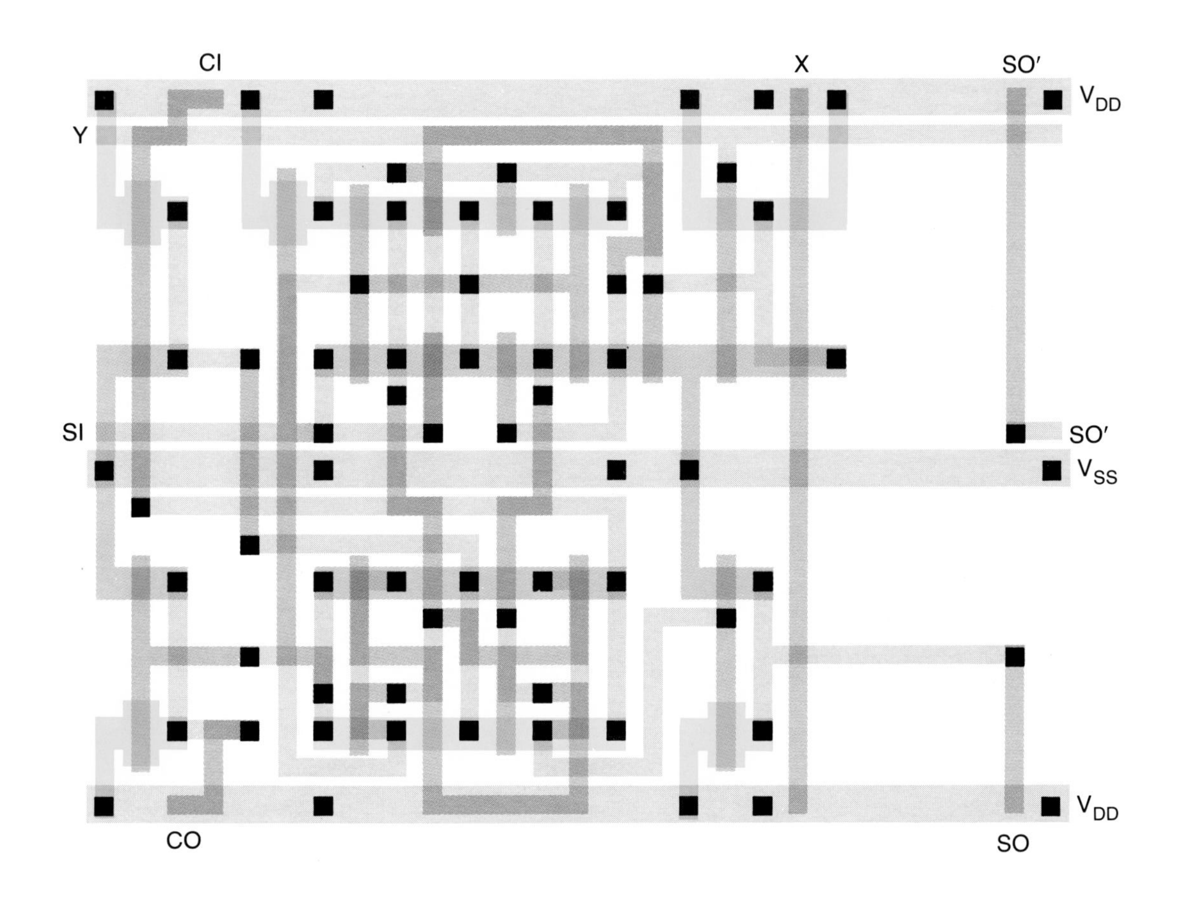

Plate 12

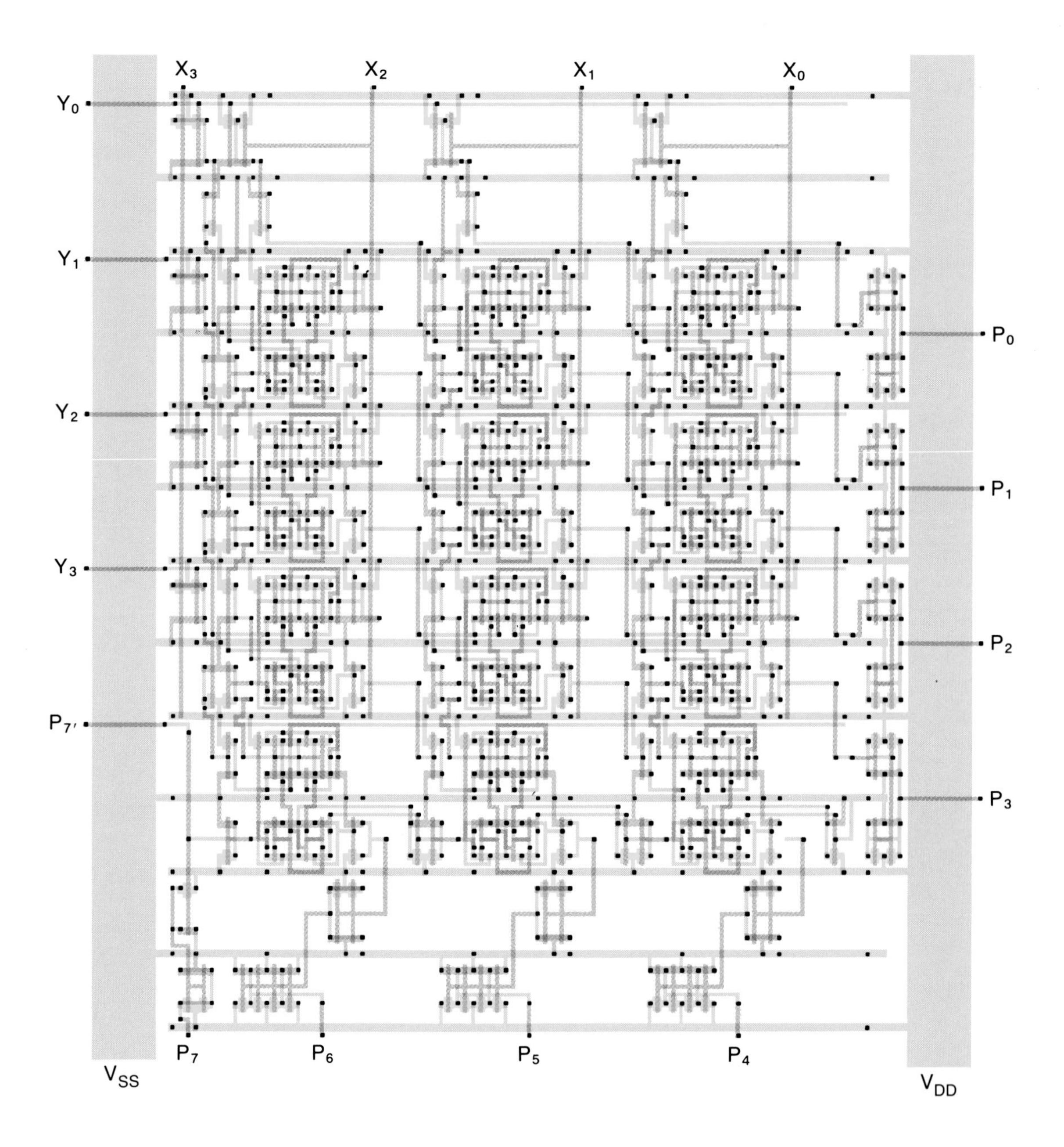

Plate 13

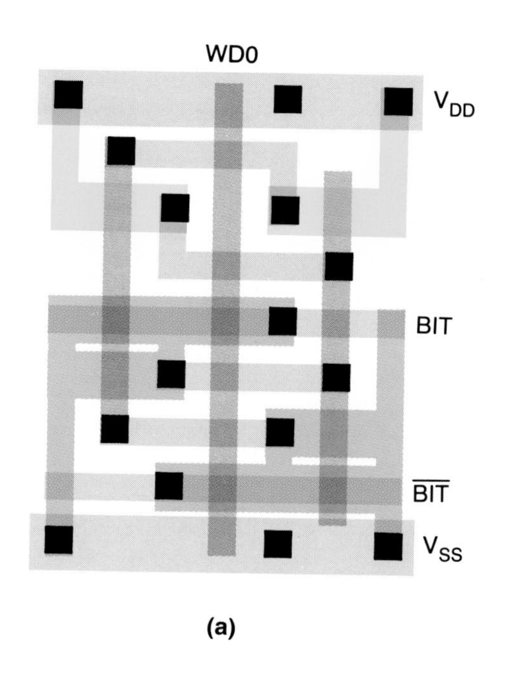

(a)

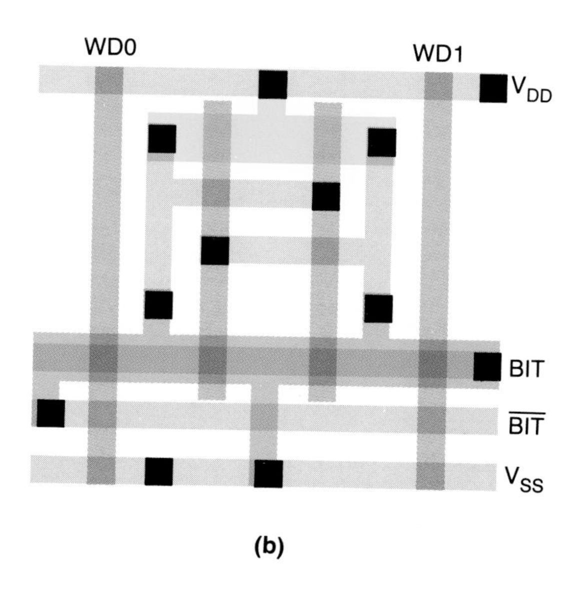

(b)

Plate 14

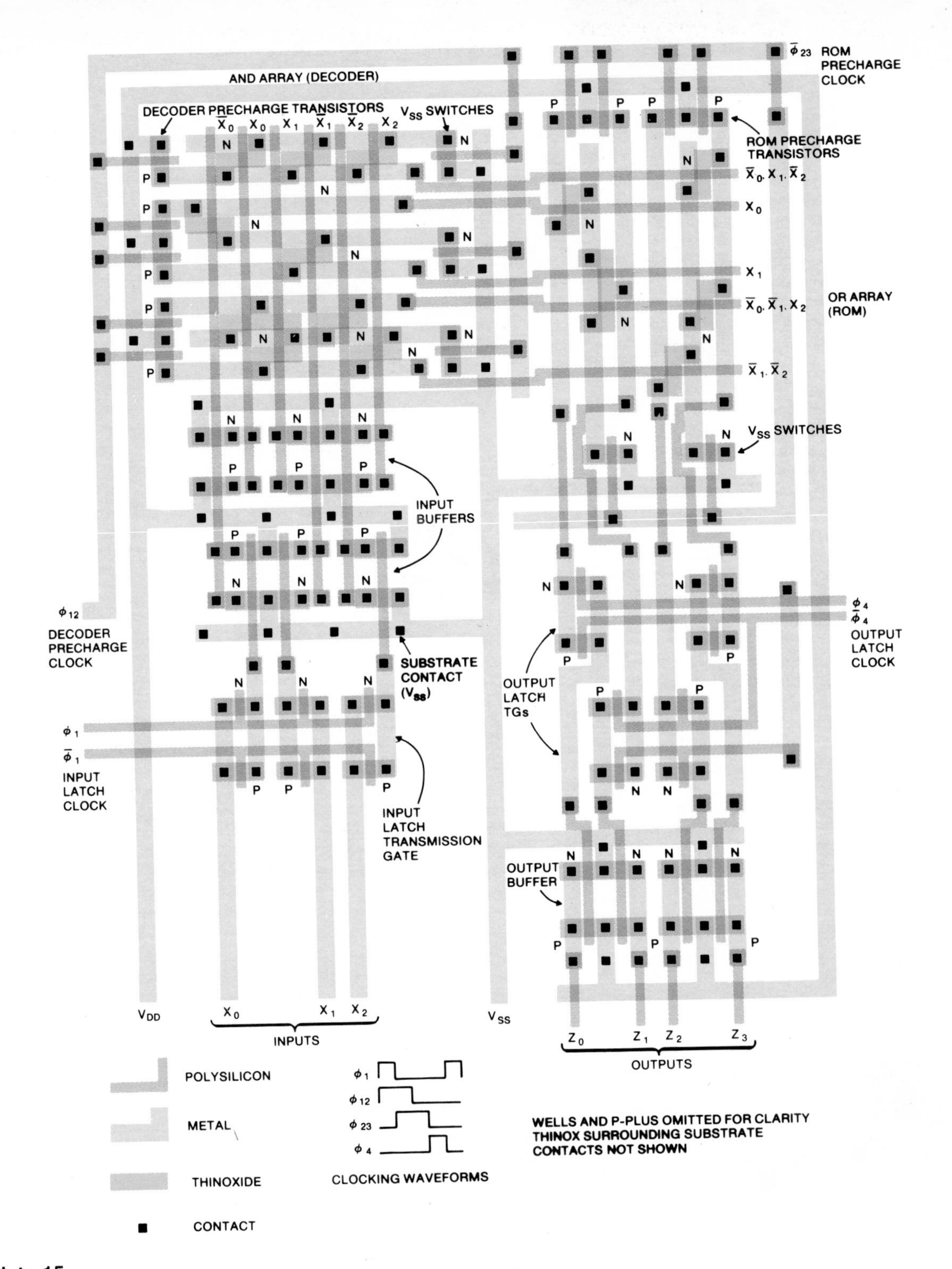
AND ARRAY (DECODER)
DECODER PRECHARGE TRANSISTORS
V_{SS} SWITCHES
$\bar{\phi}_{23}$ ROM PRECHARGE CLOCK
ROM PRECHARGE TRANSISTORS
$\bar{X}_0 \cdot X_1 \cdot \bar{X}_2$
X_0
X_1
$\bar{X}_0 \cdot \bar{X}_1 \cdot X_2$
OR ARRAY (ROM)
$\bar{X}_1 \cdot \bar{X}_2$
V_{SS} SWITCHES
INPUT BUFFERS
ϕ_{12}
DECODER PRECHARGE CLOCK
SUBSTRATE CONTACT (V_{SS})
ϕ_1
$\bar{\phi}_1$
INPUT LATCH CLOCK
INPUT LATCH TRANSMISSION GATE
OUTPUT LATCH TGs
ϕ_4
$\bar{\phi}_4$
OUTPUT LATCH CLOCK
OUTPUT BUFFER
V_{DD}
X_0 X_1 X_2
INPUTS
V_{SS}
Z_0 Z_1 Z_2 Z_3
OUTPUTS
POLYSILICON
METAL
THINOXIDE
CONTACT
CLOCKING WAVEFORMS
WELLS AND P-PLUS OMITTED FOR CLARITY
THINOX SURROUNDING SUBSTRATE CONTACTS NOT SHOWN

Plate 15

SYSTEMS DESIGN AND DESIGN METHODS

PART 2

This Part deals with general IC design methods in Chapter 6. The emphasis is on methods and associated tools that are suited for CMOS, although the principles are general enough to be applied to most IC technologies. Chapter 7 treats at length symbolic layout systems, since these are viewed as highly important to the assimilation of CMOS into the system design community. Finally, Chapter 8 covers the design of a range of important subsystems, including logic and memory structures.

STRUCTURED DESIGN AND TESTING

6.1 Introduction

In Chapter 1 we found that the design description for an integrated circuit may be described in terms of three domains, namely: 1) the behavioral domain, 2) the structural domain, and 3) the physical domain. In each of these domains there are a number of design options that may be selected to solve a particular problem. For instance, at the behavioral level, the freedom to choose say a sequential or parallel algorithm is available. In the structural domain, the decision about which particular logic family, clocking strategy, or circuit style to use is initially unbound. At the physical level, how the circuit is implemented in terms of chips, boards, and cabinets also provides many options to the designer. These domains may be hierarchically divided into levels of design abstraction. Classically these have included the following:

- architectural or functional level
- register transfer level
- logic level
- circuit level.

Again, various implementation options are available at each level of abstraction.

We will summarize some popular design styles in this chapter and focus on structured hierarchical design. Before embarking on an examination of design styles it will be instructional to identify a representative approach to design today and an ideal approach that we might desire in the future. Fig. 6.1a illustrates a typical design flow of a contemporary design system. Fig. 6.1b illustrates an ideal approach to design. Note that the existing approach relies on verification of the equivalence of successive hierarchies and domain descriptions, while the ideal approach completely synthesizes the design.

6.2 Design styles

6.2.1 Introduction

A good VLSI design system should provide for consistent descriptions in all three description domains and at all relevant levels of abstraction. The means by which this is accomplished may be measured

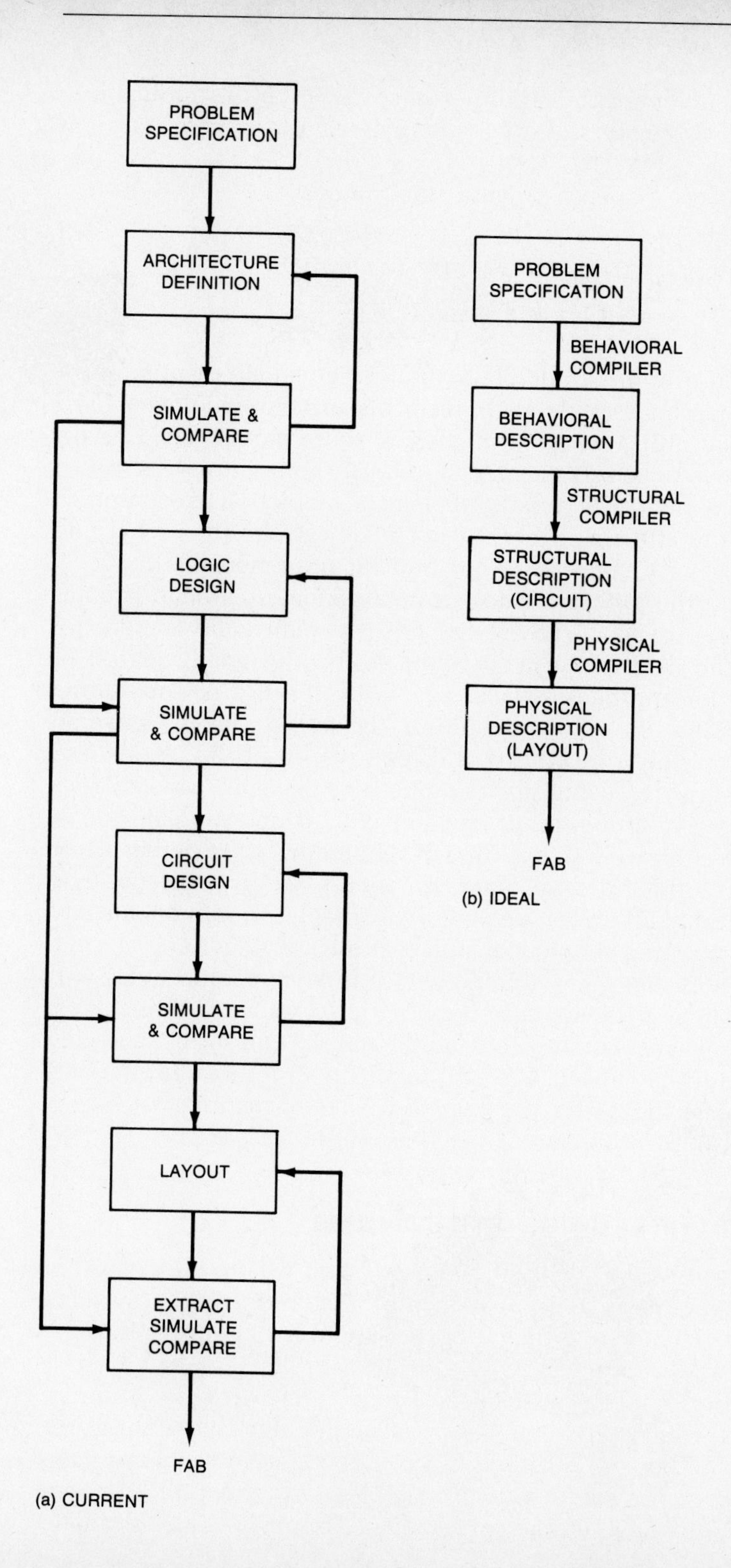

FIGURE 6.1. Current and ideal design approaches

in various terms that differ in importance based on the application. These design parameters may be summarized in terms of

- performance — speed, power, function
- size of die
- time to design — ease of use
- ease of test generation and testability.

Design is a continuous trade-off to achieve adequate results for all of the above parameters. As such, the tools and methodologies used for a particular chip will be a function of these parameters. Certain end results have to be met (i.e., the chip must conform to performance specifications), but other constraints may be a function of economics (i.e., size of die affecting yield) or even subjectivity (i.e., what one designer finds easy, another might find incomprehensible).

Given that the process of designing a system on silicon is complicated, the role of good VLSI design aids is to reduce this complexity and assure the designer of a working product. A good method of simplifying the approach to a design is by the use of constraints and abstractions. By using constraints the tool designer has some hope of automating procedures and taking a lot of the "leg-work" out of a design. By using abstractions, the designer can collapse details and arrive at a simpler concept with which to deal.

In this chapter we will examine design methodologies that allow a variation in the freedom available in the design strategy. The choice, assuming all styles are equally available, should be entirely economic. According to function, suitable design methods are selected. It may be found that due to inefficiencies in layout some styles will not be capable of implementing the function. Following these steps, the required die cost is estimated and the quickest means of achieving that die should be chosen. We will focus on structured approaches to design since they offer the best prospects of dealing with large and diverse VLSI problems of the present and future.

6.2.2 Structured design strategies

A primary aim of the Mead and Conway text was to allow system designers the option of implementing high performance systems directly in silicon. From this point of view, it is imperative that the complexity of designing an IC or complete system be reduced. After all, if teams of expert industrial designers take man-years to finish chip designs, why should one expect a team of nonexperts to perform any better? Methods of dealing with complex design problems have been developed for large software problems. By adapting (or readapting) these to the IC design environment, we can not only formulate methods to deal with the apparent complexity of the IC design process to a novice, but also propose methods by which experts can cope with the ever increasing complexity of designing circuits with millions of devices.

In the following sections some of the techniques for reducing the complexity of IC design will be summarized [Buch80].

6.2.2.1 Hierarchy

The use of hierarchy involves dividing a module into submodules and then repeating this operation on the submodules until the complexity of the submodules is at an appropriately comprehensible level of detail. This parallels the software case where large programs are split into smaller and smaller sections until simple subroutines, with well defined functions and interfaces, can be written. As we have seen, a design may be expressed in terms of three domains. We can employ a "parallel hierarchy" in each domain to document the design. For instance, an adder may have a subroutine that models the behavior, a gate connection diagram that specifies the structure, and a piece of layout that specifies the physical nature of the adder. Composing the adder into other structures can proceed in parallel for all three domains, with domain-to-domain comparisons ensuring that the representations are consistent.

6.2.2.2 Modularity

Hierarchy involves dividing a system into a set of submodules. If these modules are "well formed" the interaction with other modules can be well characterized. The notion of "well formed" may differ from situation to situation but a good starting point are those criteria placed on a "well formed" software subroutine. First of all, a well defined interface is required. This is an argument list with variable types in the software case. In the IC case this corresponds to a well defined physical interface that indicates the position, name, layer type, size, and signal type of external interconnections. For instance, connection points may indicate the power and ground, inputs and outputs to a module. The function must also be defined in an unambiguous manner. Modularity helps the designer to clarify and document an approach to a problem, and also allows a design system to be of more utility by checking attributes of a module as it is constructed. The ability to divide a task into a set of well-defined modules also aids in a team design where a number of designers have a portion of a complete chip to design.

In structured programming, proponents advise the use of only three basic constructs. These are concatenation, iteration, and conditional selection. In the IC design world these constructs have parallels. For instance, concatenation is mirrored by cell abutment, where IC cells (in the physical domain) are connected by placing them adjacent to each other and inter-cell connections are formed on the common boundary. Iteration is handled in the IC case by one- and two-dimensional arrays of identical cells, typified by a memory. The use of conditional selection is typified in a programmable logic array (PLA), the function of which is determined by the location of transistors in an array. When combined with the ability to parametize designs, these three programming notions can greatly aid the designer in modularizing a design.

6.2.2.3 Regularity

The use of iteration to form arrays of identical cells is an example of the use of regularity in an IC design. However, extended use may be made of regular structures to simplify the design process. For instance, if one was constructing a "data-path," the interface between modules (power, ground, clocks, busses) might be common but the internal details of modules may differ according to function. Regularity can exist at all levels of the design hierarchy. At the circuit level, uniform transistors might be used rather than the manual optimization of each device. At the logic module level, identical gate structures might be employed. At higher levels, one might construct architectures that use a number of identical processor structures. By using regularity in the ways mentioned, a design may be judged correct by construction. Methods for formally proving the correctness of a design may also be aided by regularity.

6.2.2.4 Locality

By defining well-characterized interfaces for a module, we are effectively stating that the other internals of the module are unimportant to any exterior interface. In this way we are performing a form of "information hiding" that reduces the apparent complexity of that module. In the software world this is paralleled by the reduction of global variables to a minimum (hopefully to zero). Using this model, for instance, we would not physically overlay connections to a physical module, as this may modify the internal structure and operation of a previously defined module.

Modules can also be located to minimize the "global wiring" that may be necessary to connect a number of modules in an unstructured system. A common theme in design systems today is use "wires first, then modules" — rather than the more common "place modules, then route them together."

6.2.3 Handcrafted mask layout

Handcrafted mask layout is the term applied to less constrained design techniques that involve, at some stage, the layout of functional subsystems at the mask level. Of course, this is the oldest form of chip design and still the most widely used by semiconductor vendors. Essentially it requires that a design be divided among designers with expertise in logic, circuit, and process details.

The method of completing such layouts has progressed from cutting RUBILITH®, to drawing on MYLAR® and digitizing, to interactive graphics entry and onward. The argument is that by attending to each transistor and optimizing layout and circuit parameters, the highest performance and smallest die size results. Of course, this can be a daunting task for VLSI circuits consisting of hundreds of thousands or millions of transistors. Note that as total freedom is allowed at the physical level, the structural specification and hence the behavioral description may differ from that required. This hampered mask level layout before the advent of adequate circuit extraction tools, which take as input a mask description and then, by various recognition techniques, present the designer with the circuit description that corresponds to the interrelationship of the mask shapes present.

6.2.4 Gate array design

Gate arrays are currently enjoying widespread popularity as an LSI-VLSI implementation medium. This arises through a combination of readily available vendors, design tools, and a compatibility with TTL design that makes it easy for the system designer to transfer a design to silicon with the minimum amount of effort. Above all, the cost of a gate array is potentially the lowest of the methods described in this chapter for certain classes of integrated circuits. The widespread availability of design tools is in some ways a result of the strictly constrained physical layout. Although this does not necessarily reduce the complexity of the tools required, it does give a bounded problem.

Gate arrays come in various flavors, but can be categorized by a design that uses a large number of identical "sites," each site consisting of a number of circuit elements. In CMOS, these sites consist of a number of n-transistors and p-transistors. Some typical sites are shown in Fig. 6.2. The arrangement shown in Fig. 6.2a is a typical six transistor site, composed of three n-transistors and three p-transistors. The gate signals are connected in common and the n- and p-transistors are connected as illustrated in the site schematic. A sketch of the layout is shown in Fig. 6.2b. Another site

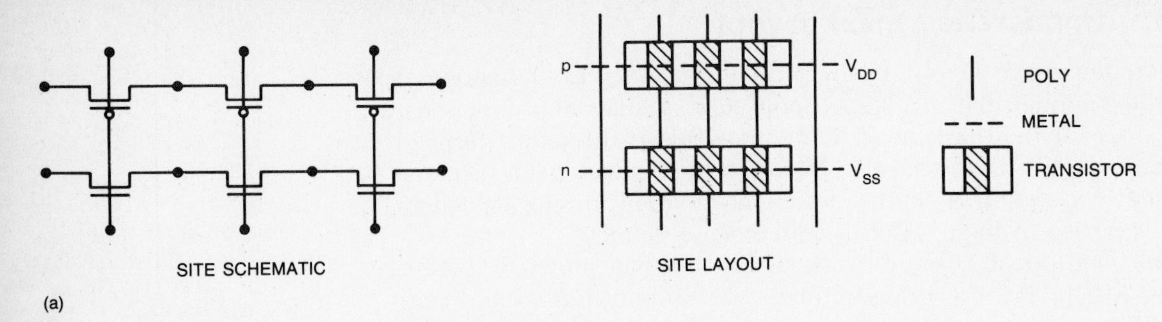

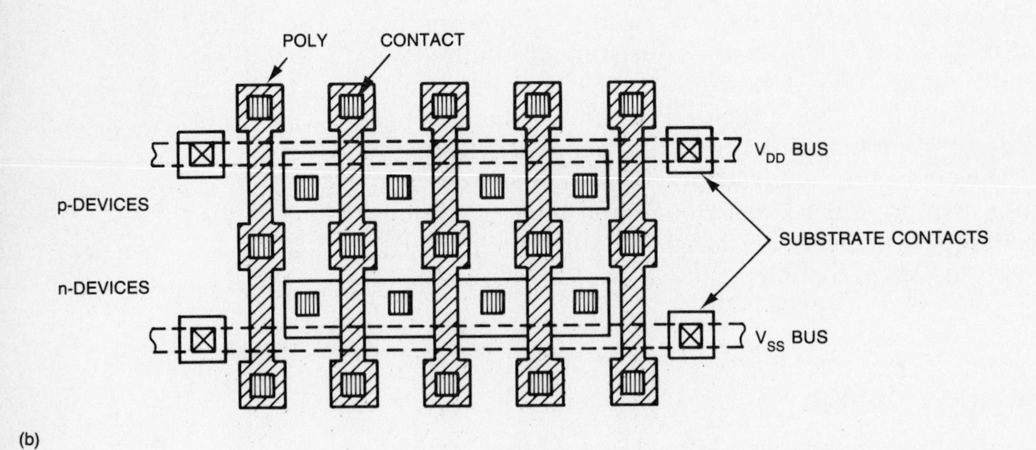

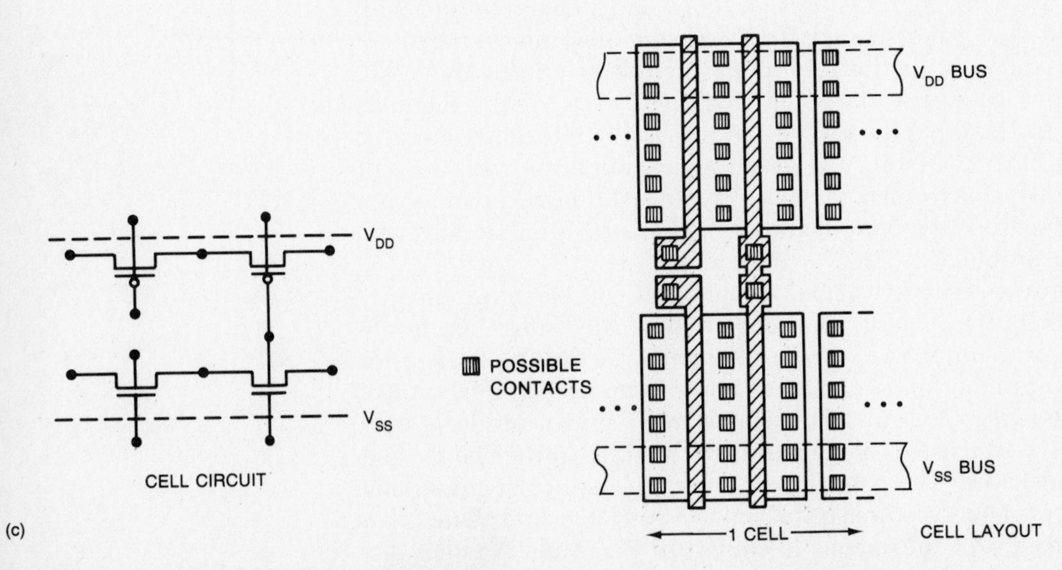

FIGURE 6.2. Gate array site configurations

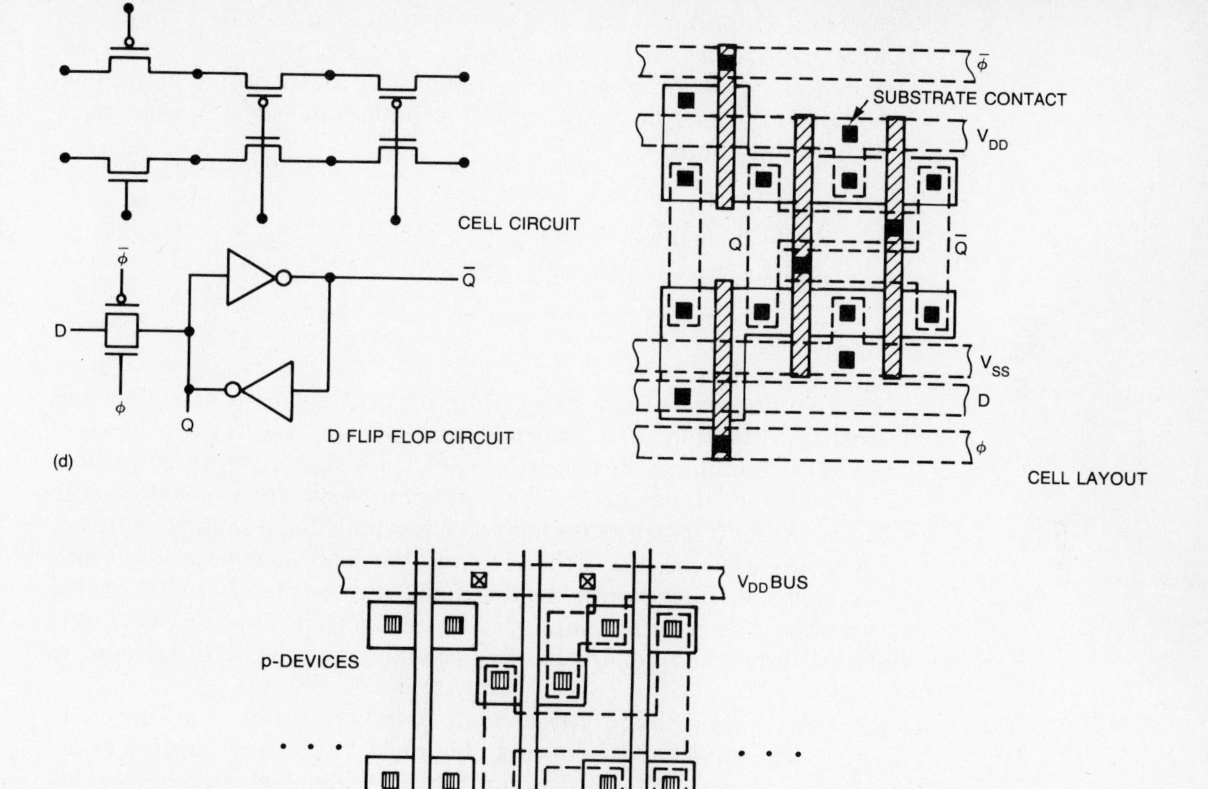

FIGURE 6.2. (continued)

structure is shown in Fig. 6.2c [Tana84]. This has four transistors. One n-p pair has a common gate connection, while the other pair has separate gate signals. This allows the easy implementation of a transmission gate and inverter as might be used in latches. A third structure is shown in Fig. 6.2d [TIIF84]. This structure has six transistors ratioed in a particular manner to allow the static latch connection shown. This is one of two types of cell used on this particular array, catering for memory structures. The logic cell is similar to that shown in Fig. 6.2a. A final cell is shown in Fig. 6.2e. Here a continuous array of transistors has been used with metal completely programming the cell[Wern84]. This may be compared with the CMOS cell array strategy discussed in Chapter 5. The predefined topology allows a vendor to stockpile wafers processed to a given fabrication step (usually metallization) and then "personalize" a set of wafers to implement a given design. This personalization may be completed in a number of ways. For instance, possible methods are:

- single layer metal
- single layer metal and contacts
- double layer metal and contacts and vias.

or in numerous other ways. Thus for any given design the mask cost is a fraction of the normal cost (1/8 to 1/4). This also saves time in reducing a design to practice — a strong selling point in a competitive environment. The tradeoff in a gate array is wasted chip area; all the transistors for a given array have to be in place whether they are used or not. Other factors that lead to suboptimum area usage arise from the fixed placement and the fixed circuit configurations that may be realized. For instance, if RAM or ROM is to be included, the vendor has to estimate a good ratio of memory to logic.

A typical gate array floor plan is shown in Fig. 6.3a. Arrays of sites are separated by routing channels. Usually strict directional control is maintained over routing (i.e., metal mostly horizontal, polysilicon vertical, or vice versa). Mask-programmable I/O cells surround the inner core.

A six transistor CMOS site is shown in circuit schematic form in Fig. 6.3b. With this structure we may build a variety of 3-, 2-, or 1-input gates. In the example, the site is programmed by metal internal to the cell (i.e., between the power rails). Connection to the cell would be via vertical metal and horizontal polysilicon runners. Typical design decisions in a gate array would include transistor sizing (see Chapter 4), selection of the width of the routing channel, and placement and nature of discretionary wiring.

A typical flowchart used in the IGC-20000D [GEIn83] gate array product is shown in Fig. 6.4. The customer is responsible for creating a logic schematic and a set of test vectors, which are used initially to verify the customer's logic. This logic schematic is then converted to CMOS gate-array macros of the type shown in Fig. 6.3b. After simulation, the CMOS cells are placed on the appropriate array and automatically routed. Any necessary revisions are communicated to the customer and this procedure is repeated until an acceptable CMOS implementation is found. Final placement and routing precedes a final simulation with all parasitics. The array is then manufactured and tested with a customer-generated test vector set.

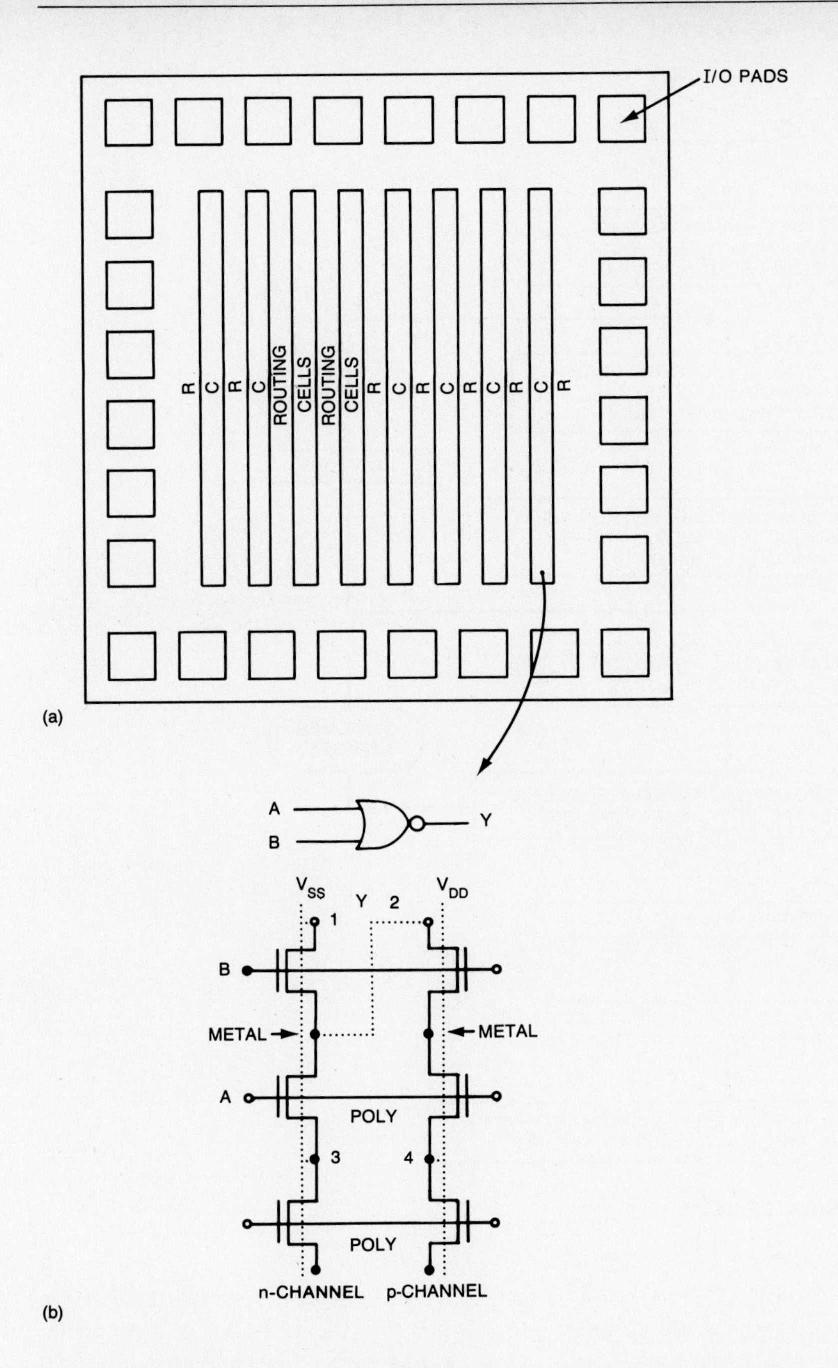

FIGURE 6.3. Typical gate array floor plan

6.2.5 Standard cell design

Standard cell systems rely on a set of predefined logic/circuit cells to complete a design. The attraction to the TTL designer is also present as in gate-array design, as the cells can be organized to

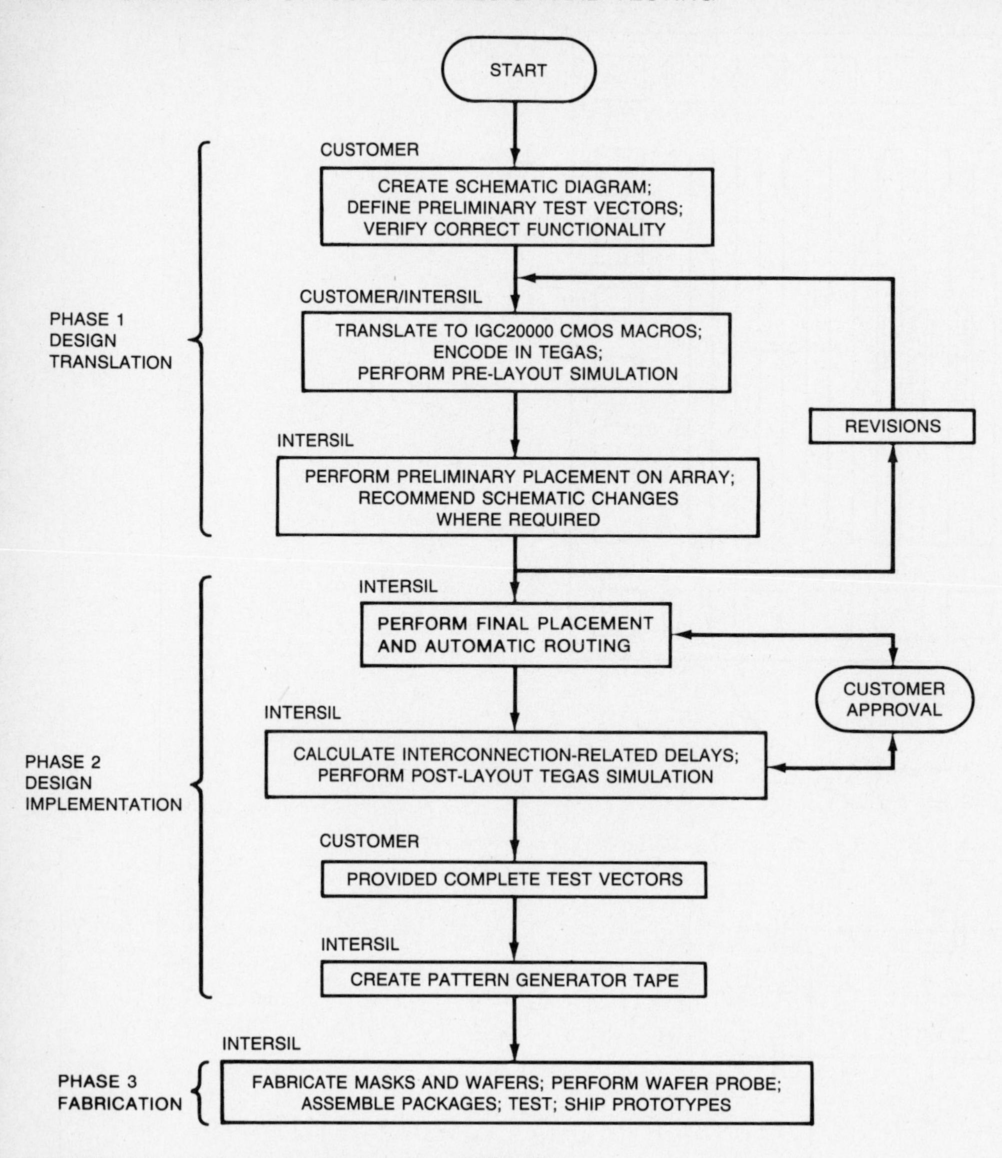

FIGURE 6.4. Gate array design flow (GEIn83)

follow a TTL data book. Complexity of cells can vary from SSI-type component, such as gates and latches, to MSI/LSI-type components, such as RAMs, ROMs, and PLAs. Comprehensive systems for standard cell design have long been available in large companies [PeDS77] and, recently, commercial offerings have appeared.

A full mask set is required for normal standard cell chips, although design tradeoffs may be made. For instance, one might design a chip that can be used for a number of applications by personalizing

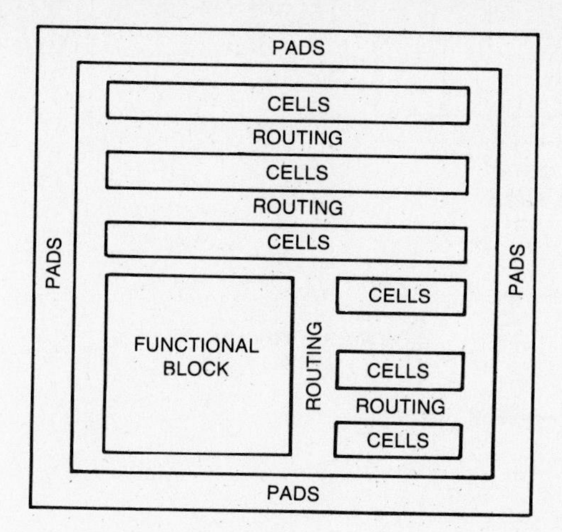

FIGURE 6.5. Typical standard cell floor plan

a control PLA or ROM that is part of a larger custom chip. In comparison to gate array design, standard cell design systems use cells with predefined layout and logic, which can be placed anywhere in the area defining the chip.

A typical floor plan for a chip designed with standard cells is shown in Fig. 6.5. MSI type cells are placed in rows separated by routing channels. These rows are then arranged in columns. In addition, large LSI functional blocks (e.g., a RAM) may be located to optimally connect with the random logic. MSI cells are often of fixed height, with variable length catering for circuits of differing complexity. This is shown in Fig. 6.6a. This layout strategy is representative of mature standard cell systems. Recently, systems have been proposed and implemented that hierarchically group primitive cells together to form larger blocks and then combine these on up to the complete chip description.

A skeleton cell layout is shown in Fig. 6.6b, with power rails running horizontally in first layer metal and I/O connections running vertically in polysilicon or second layer metal. The I/O access to cells may vary, with a variety of methods shown in Fig. 6.6c.

6.2.6 Symbolic layout methods

In an earlier section we discussed various methods to reduce the complexity of the IC design task. These included the use of hierarchy, regularity, modularity, and locality. There is current interest in design methods that simplify the lower level details of IC design by hiding process design rules and capturing structural and physical domain information in a loose format. These "symbolic layout" approaches yield a kind of "assembly" language representation format in the

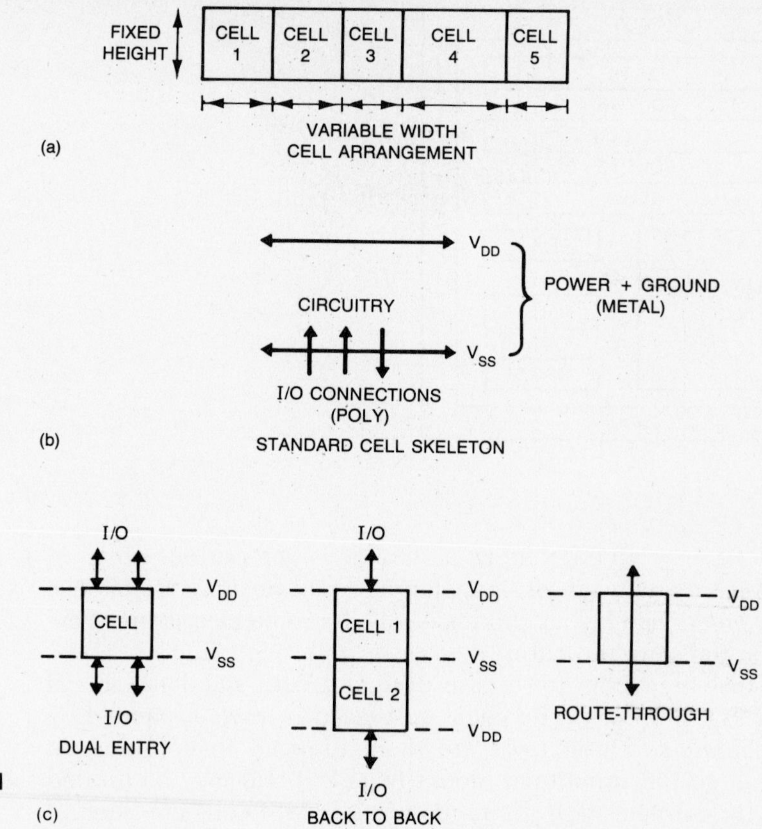

FIGURE 6.6. Standard cell formats

software sense. They allow simplified design at a very low level of the design hierarchy. In the next chapter we will examine symbolic layout systems with the notion that this is the lowest level that a system designer might become involved in in layout. The case history section provides good examples to substantiate this point.

6.3 Automated synthesis

Numerous approaches have been considered that map behavioral descriptions to intermediate forms such as boolean logic and then, by conventional techniques, to layout (mask level, standard cell, gate array, symbolic). The following is a list compiled by Shiva [Shiv83] listing the main approaches.

1 Register transfer level to logic level

2 Global design space "explorers"

3 Local design space synthesizers
4 Logic array synthesizers
5 Silicon compilers
6 Human creativity based systems.

The main problem with most of these approaches is that boolean logic is the accepted lower level description of most systems. This necessarily limits the types of CMOS circuits that may be realized. For the rest of this section we will concentrate on two related methods that allow effective high level design in CMOS.

6.3.1 Procedural module definition

Once a particular design style or architecture has been defined, a procedural modeling approach is advantageous to encapsulate the many design variations that a designer might require. This involves writing "generators" in a high level language that specify a lower level format — either symbolic or mask level. The advantage of specifying layouts at the symbolic level is the maintenance of the library over technology scaling and variations.

One of the features that is frequently absent from interactive-graphics-based design systems is the ability to parametize designs to allow repetition, conditionalization, or other forms of procedural definition. This kind of flexibility is frequently needed when composing designs using modules that may need slightly different functionality, depending on the context in which the particular design is used. For instance, quite often the first and last cell in an array of cells have to assume some additional functionality. The first bit of an adder might have the carry grounded. This requirement leads one to consider higher level languages to implement procedurally generated structures. A good example of this approach may be found in the DPL/Daedalus system [BMSS81]. This system allows the construction of objects called TYPEs, using a LISP based language called DPL. TYPEs call other TYPEs to build more complicated objects. When a TYPE is called with a set of parameters, it creates a PROTOTYPE. VIRTUAL COPYs are created for each parameter invocation of a TYPE. Combined with an interactive graphics editor (Daedalus), this system provided a powerful mask design capability. The graphics editor maintained knowledge about cell stretch points and objects that could be conditionally inserted. Using this approach, both interactively designed cells and generators for structures such as PLAs and data-paths were successfully built[Shro82].

If generators are built up hierarchically, then extremely powerful subsystems may be built, as illustrated by the PLEX project [BuMC83]. Here, low level functions such as multiplexers and decoders are specified in a "C" like language. Other components are then built up from these lower level functions with connections made by abutment, river routing, or channel routing.

6.3.2 Silicon compilers

A silicon compiler is ideally an automatic translation tool that converts a behavioral description to a mask level description. An early example of an attempt at this objective was "Bristle Blocks" [Joha79], which compiled a fixed floor plan layout of a fixed architecture microprocessor. This was a system that saw numerous offshoots, some of which are summarized in this section. Most "silicon compilers" today are still characterized by a notion of a fixed floor plan in which a designer has a limited amount of freedom to explore the design space. In addition, they may be broadly categorized as "specialist" standard-cell systems, which take account of a particular architecture and floor-plan style to automate the design process. No silicon compilers existing at this point in time have demonstrated any ability to deal with the complexity contained in contemporary full custom ICs. One valid use of current experimental silicon compilers today is to explore space/time tradeoffs of a particular design. Once this step is completed, an optimized design might be attempted using more conventional techniques. Hopefully, in the future, silicon compilers that "synthesize" floor plans and circuits in the same manner as a human designer will emerge.

6.3.2.1 The FIRST silicon compiler

An interesting example of a contemporary silicon compiler is FIRST, which was designed at the University of Edinburgh [DeRB82]. This was developed to build signal processing circuits for telecommunications environments. In the original implementation the technology was nMOS, and a bit-serial arithmetic approach was adopted.

A typical floor plan is shown in Fig. 6.7. Two ranks of circuitry surround a central "umbilical" routing channel. The circuit elements

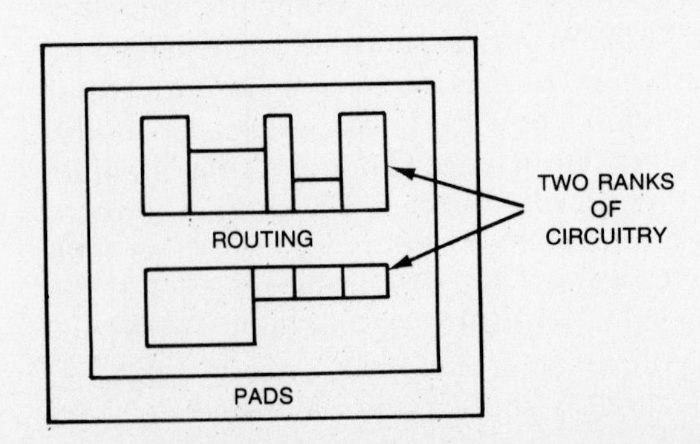

FIGURE 6.7. FIRST floor plan style

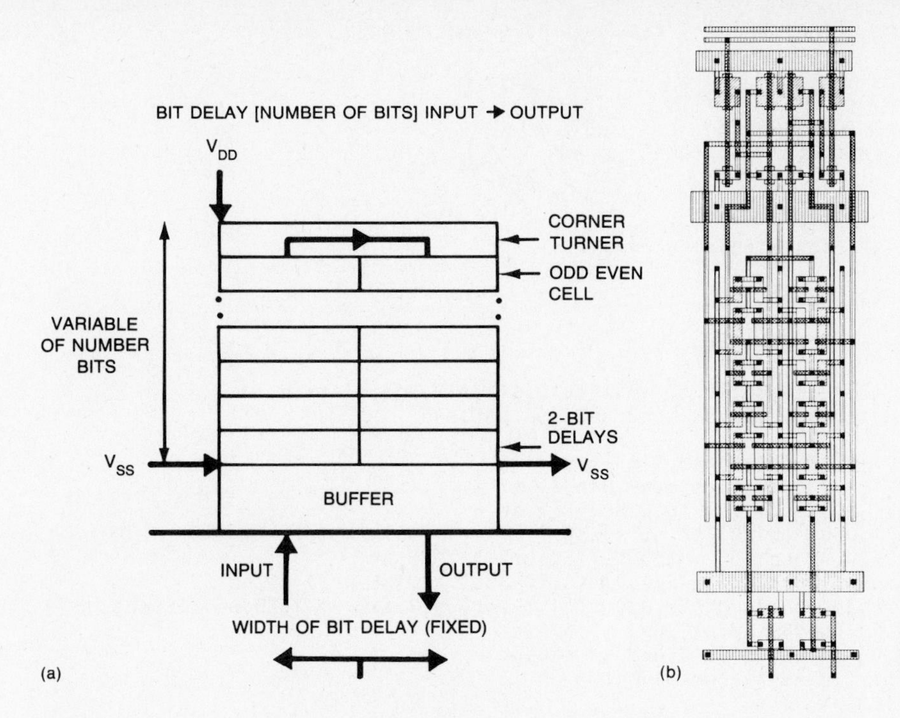

FIGURE 6.8. FIRST cell layout style

are composed of single bit and multiple bit delay elements, word delays, serial adders, subtractors, magnitude comparators, and multipliers. Fig. 6.8 shows a typical tessellated cell with clock, power routing, and I/O circuitry for a possible CMOS implementation of a delay line. As bit-serial circuitry is used throughout, the intercell routing is minimum. Clock buffers are used in each module to guarantee a clock frequency of 8 MHz into a specified loading and worst-case routing capacitance.

Design is initiated by writing a structural description for the subsystem desired. Note that behavior is not specified. The structural description is really a means of defining the interconnections between hand-optimized standard cells. An important difference between this system and other standard cell systems is the ability to customize a circuit module in terms of a set of parameters such as the length of a bit-level delay line. A typical description for an adaptive transversal filter is shown in Fig. 6.9. Each circuit module has a functional model that has been verified by extensive simulation. Thus simulation of the circuit may proceed at the functional level, which allows the designer to explore the design space in a more effective manner. Following successful simulation, the layout is generated completely automatically. The designer does not have to be aware of any MOS circuit or layout design, providing all modules that are needed are available. Assuming that the potential problem can be couched in

```
        FIRST COMPILER - Copyright Denyer, Renshaw, Bergmann - 1982

SOURCE FILE: LMSFIR1

!Adaptive (LMS) Transversal (FIR) Filter
!filter stage of multiplexed filter sections

CONSTANT round=1
CONTROL INPUT c0,c1,init
CONTROL INPUT c1t0,c1t1,c1t2,c1t3,c1t5,c1t6,c2t0
INPUT Xin,TMEIN,WMSBin,WLSBin
OUTPUT Xout,YMSBout,YISBout,YLSBout

        OPERATOR FIR[wordlength,idles,stages]

                SIGNAL s1,s2,s3,s4,s5,s6,s7,s8,s9,s10,s11,s12,s13,s14,s15
                   s16,s19,s20,s21,s22,ss1,ss2,ss3,ss4,ss5,ss6

                Multiplex[1,0,0] (c2t0) s3,Xin -> s1
                Multiplex[1,0,0] (INIT) s15,WMSBin -> s7
                Multiplex[1,0,0] (INIT) s16,WLSBin -> s8
                Multiplex[1,0,0] (c1t2) s6,s5 -> s6
                Multiplex[1,0,0] (c1t6) YMSBout,YISBout -> YMSBout
                Multiply[round,wordlength-2,0,0] (c1t0->nc) s2,TMEin -> s5,nc
                Multiply[round,wordlength-2,0,0] (c1t5->nc) s22,s15 -> YLSBout,YISBout
                Add[1,0,0,0] (c1t3) s6,s11,gnd -> s15,nc
                Add[1,0,0,0] (c1t3) s5,s12,gnd -> s16,nc
                constant p1=(idles+1)*(wordlength)
                Bitdelay[18] s1 -> ss1
                Bitdelay[18] ss1 -> ss2
                Bitdelay[18] ss2 -> ss3
                Bitdelay[18] ss3 -> ss4
                Bitdelay[18] ss4 -> ss5
                constant p2= p1-90, p3 = p2/2, p4 = p2-p3
                Bitdelay[p3] ss5 -> ss6
                Bitdelay[p4] ss6 -> Xout
                Worddelay[idles+stages-1,wordlength-2,0] (c1t1) s7 -> s19
                Worddelay[idles+stages-1,wordlength,0] (c1t1) s8 -> s10
                constant half = stages/2, rest = (stages-1)-half
                Worddelay[half,wordlength-2,0] (c1t0) Xout -> s4
                Worddelay[rest,wordlength-2,0] (c1t0) s4 -> s2
                Bitdelay[wordlength] s1 -> s19
                Bitdelay[wordlength] s19 -> s20
                Bitdelay[wordlength-1] s2 -> s3
                Bitdelay[wordlength-4] s9->s11
                Bitdelay[wordlength-4] s10 -> s12
                Bitdelay[wordlength/2 +1] s21 -> s22

        END

! ALLOCATE VALUES TO PARAMETERS : wordlength, idles, stages

FIR[14,10,47]

ENDOFPROGRAM
```

FIGURE 6.9. Typical FIRST listing (Courtesy of N. Bergmann, P. Denyer, and D. Renshaw, University of Edinburgh)

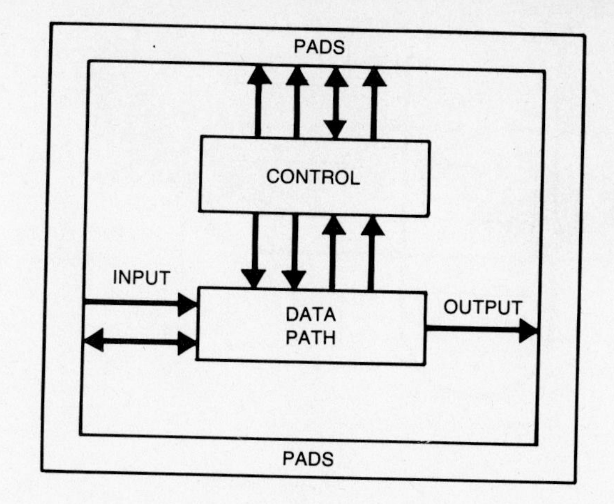

FIGURE 6.10. Generic MacPitts floor plan

bit-serial terms, the density of this approach is relatively good. However, the algorithms that may be implemented are characterized as "stateless," following a data-flow model of computation.

6.3.2.2 The MacPitts silicon compiler

Another example of a fixed floor plan silicon compiler, developed to deal with concurrent parallel data path architectures, is the MacPitts Silicon Compiler [SiSC82]. The generic floor plan is shown in Fig. 6.10. It consists of a data-path with input and output ports which is orchestrated by a control section which also has control inputs and outputs.

Design proceeds by writing a LISP-like behavioral specification of the intended function. This may be interpreted functionally to verify performance at a high level. A scanner is employed to find the parallelism and data operators needed in the data-path to be designed. These are then generated from small "organelles," as shown in Fig. 6.11. An organelle is one bit of an m-bit slice that performs a particular function. For instance, the organelle might invert the value of a particular bus signal. Control is generated by examining the functional description and is implemented as a Weinberger Array [Wein67]. This compiler was used to generate experimental designs, but the Weinberger Arrays generated for moderate examples are large and slow. For CMOS, the control might well be implemented as PLAs or random standard cell logic.

Fig. 6.12a shows a sample MacPitts data-path organelle that computes the absolute value on a bit slice basis. It uses a "polycell" or standard cell type of cell construct. With basic modules such as adders, inverters, comparators, and transmission gates, a wide range

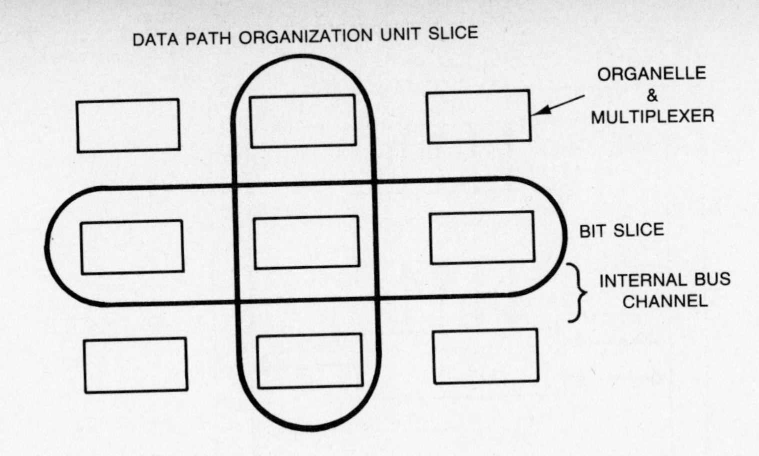

FIGURE 6.11. Style of MacPitts organelle

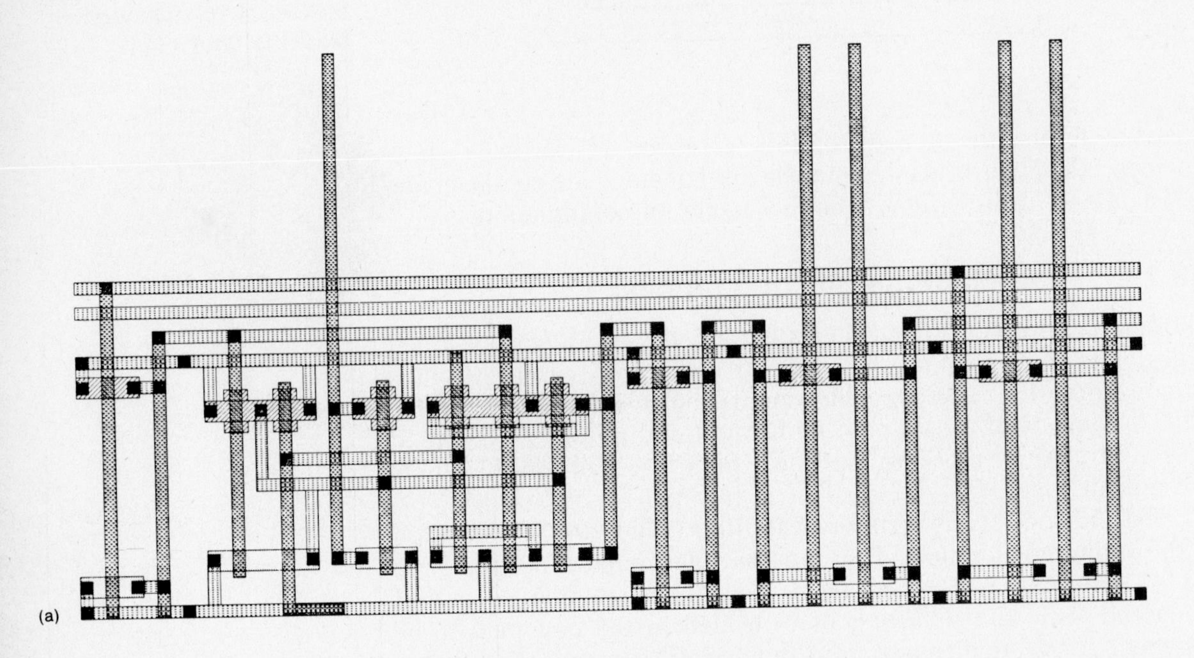

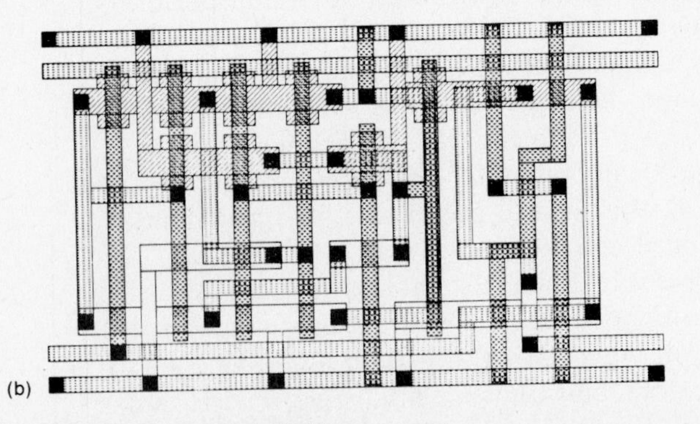

FIGURE 6.12. MacPitts style using "polycells"

of functions may be implemented. Economies can be made when custom modules for frequently used operations (e.g., the absolute value module) are designed (Fig. 6.12b). This seems to be a principal weakness not provided by the original implementation of MacPitts. The power of a system like MacPitts is to allow a novice designer to explore space/time tradeoffs as they impact a design. As systems similar to MacPitts mature, they will become capable of designing very large sections of VLSI chips with minimal low level input from the designer.

6.4 The custom design tool box

6.4.1 Introduction

Regardless of the design style chosen, there are a number of standard tools needed to ensure first-time successful silicon. Absence of a particular tool will not necessarily mean failure but will surely contribute to it. Conversely, all the tools in the world will not cover incompetence. This section will summarize a number of the typical tools required. These are relatively independent of CMOS technology. We will stress tools and algorithms that operate down to the transistor level and ignore tools that only deal at the logic level.

6.4.2 Circuit level simulation

Circuit analysis programs are typified by the SPICE program developed at the University of California at Berkeley [Nage75]. The basis for this type of program is the solution of the matrix equations relating the circuit voltages, currents, and resistances (or conductances). This type of simulator is characterized by high accuracy but long simulation times. Simulation time is typically proportional to n^m, where n is the number of nonlinear devices in the circuit, and m is the range from 1 to 2. This type of program is used to verify in detail small circuits or to verify the simulation results of more efficient but less accurate simulators, such as timing simulators. It is unrealistic to use this type of program for the verification of large VLSI chips.

6.4.3 Timing simulation

It is possible to simplify the general circuit analysis approach used above to allow simple nonmatrix calculations to be employed to solve for circuit behavior. This usually involves making some approximations about the circuit. Typical of a simulator using this approach is the MOTIS simulator [ChGK75]. The accuracy of such simulators is less than SPICE-type simulators, but the execution time is almost two orders of magnitude less. The simulators can not be used for all circuit configurations. In particular, circuits in which there are many internode capacitances, such as bootstrap circuits, are not handled easily. For most current CMOS circuits, these simulators do work well. A simple example of such a simulator is given in Chapter 7.

6.4.4 Logic level simulation

Many simulators have evolved to deal with simulation at the logic level. They use primitive models such as NOT, AND, OR, NAND, and NOR gates. Timing parameters may be assigned to the logic models based on prior circuit simulation and known circuit parasitics. As all logic circuits are rarely active simultaneously, logic events may be scheduled on a queue. This means that the state of the network is evaluated on an event driven basis, rather than on a timing substep basis, as are the two previous simulators.

Logic simulators are adequate for well-characterized CMOS circuits that have regular logic counterparts. They are relatively fast and are thus suitable for large circuits. This has been also aided by hardware engines that compute the simulation algorithm. However, the link between the logic models used and the physical structure is normally weak. Characterization often requires massive simulation of gates using a circuit analysis program combined with designer input. Some logic simulators developed for other forms of logic have problems with the bidirectional nature of transmission gates. These weaknesses have given rise to the next form of simulator, which uses many of the same principles but uses transistors as the lowest level logic primitive.

6.4.5 Switch level simulation

A good method of ensuring correspondence between physical layout and simulation is to use transistors as primitives in the simulator. MOSSIM is a simulator that does just this [Brya81]. The simulator is a three-state simulator using (1,0,X) states. For each set of input vectors the circuit is assigned a *steady-state level*. The network of nodes interconnecting transistors has a particular state. When the inputs are perturbed, the new states are evaluated by using a relaxation technique on the nodes in the circuit. Not all circuits may be simulated with this kind of simulator (for example, the transmission gate XOR discussed in Chapter 8). Recently, switch level simulators with timing modes have been implemented.

6.4.6 Timing verifiers

A timing verifier takes a different approach to circuit verification. Here, the delays through all paths in a circuit are evaluated and the user is provided with information about these delays. An example of this type of analyzer is the TV program [Joup83]. An MOS verifier works best at the transistor level. The circuit to be analyzed is first statically examined to determine the direction of signal flow in all transistors. This is necessary to evaluate only those delays that will be critical in actual circuit operation. Each transistor is examined and the direction of signal flow is calculated using nine rules. These rules may be determined from:

- circuit design methodology rules
- electrical rules
- user supplied rules.

The TV program calculates an RC delay for each node. These are then evaluated in a *breadth-first* manner. Delay paths are qualified by appropriate clocks.

A timing analyzer implemented at the transistor level can provide a designer with rapid feedback about critical paths. Combined with a switch-level simulator for rapid global functional simulation, a timing simulator for detailed module verification, and a circuit analysis program for critical path evaluation, the timing analyzer completes a set of powerful verification tools.

6.4.7 Schematic editors

The initial task in designing an IC often is to capture the circuit connectivity prior to completing a physical layout. This can be completed textually, but many designers prefer a schematic description. Recently, many commercial schematic capture systems have become available. These systems frequently have libraries of components based on TTL components or gate array or standard cell primitives. Frequently, in addition to graphical editing, as part of a complete ***c****omputer* ***a****ided* ***e****ngineering* (CAE) workstation, they allow simulation of circuits with direct feedback from the schematic (i.e., probing and setting signals).

Some parts of particular IC designs are ill-specified using graphics. A typical example is a PLA. State equations or high level language algorithm form a better specification.

6.4.8 Net list comparison

If a schematic or circuit description is entered to define an IC, at some stage a physical layout is generated. This may be completed automatically, as in the case of a gate array or place and route standard cell system. Alternatively, the physical layout may have a manual component. Ideally, the signal names between parallel representations would be the same, allowing easy comparison between desired and actual circuit. In reality, signal names are often omitted from internal nodes in a circuit and only applied to I/O ports. Thus there is the problem of comparing two graphs that are labeled in a limited manner. Programs that verify the equality or lack thereof of two graphs are thus needed.

Typical of a program that performs this function is a program called WOMBAT [SpNe83]. Signatures are calculated for each transistor in the test and reference circuit. Signatures include:

- fan-in
- fan-out
- transistor type
- bound nets connected to transistor.

Test and reference circuits are then repeatedly checked to correlate transistors. Another example of a net-list comparator may be found in [EbZa83].

6.4.9 Layout editors

Current commercial layout editors fall into two main categories. These are broadly classified as CAD and CAE approaches. The CAD approach is typified by systems that use layout editors to manipulate multilevel polygons with operations such as window, move, copy, and stretch. These are, in effect, clever drafting tools. The second approach typically limits mask manipulation to manhattan shapes and 45° figures, but concentrates more on overall system design, allowing the design to be integrated with other design tools such as simulators.

Recent research contributions include layout editors that monitor design rules as the designer works [Rubi83] and editors that monitor connectivity. In the next chapter we will examine a system that has no design rules at all. Another good application of layout editors is in the area of graphical floor-planning, where high level placement of modules is completed and wiring and communication strategies are formulated before detailed layout commences.

6.4.10 Design rule checkers

If mask design is completed manually, it is necessary to verify that the layout conforms to the geometric design rules. This is achieved with a design rule checker. Many variations exist, but typical approaches are found in [S2WV83][BaTe80][Bair77].

Hierarchical design rule checkers are necessary for large circuits [Whit80]. These design rule checkers use the hierarchical nature of a design to reduce the number of cells that have to be individually checked.

6.4.11 Circuit extractors

Allied to design rule checkers are circuit extractors that examine the interrelationship of mask layers to infer the existence of transistors and other components. Various approaches have been implemented to approach this problem [HoLa83][BaTe80]. Commonly, parasitic capacitances and resistances are reported in addition to transistor connectivity. Algorithms commonly use geometric shape intersections to recognize active devices. The need for such tools by the system designer will decrease as higher level design techniques provide "correct-by-construction" modules. High volume designs specified at the mask level will continue to require this type of tool.

6.5 Testing

6.5.1 Introduction

A critical factor in all LSI and VLSI design is the need to incorporate methods of testing circuits. This task should proceed concurrently with any architectural considerations and not be left until fabricated parts are available.

Fig. 6.13a shows a combinational circuit with n-inputs. To test this circuit exhaustively a sequence of 2^n inputs (or test vectors) must be applied and observed to fully exercise the circuit. This

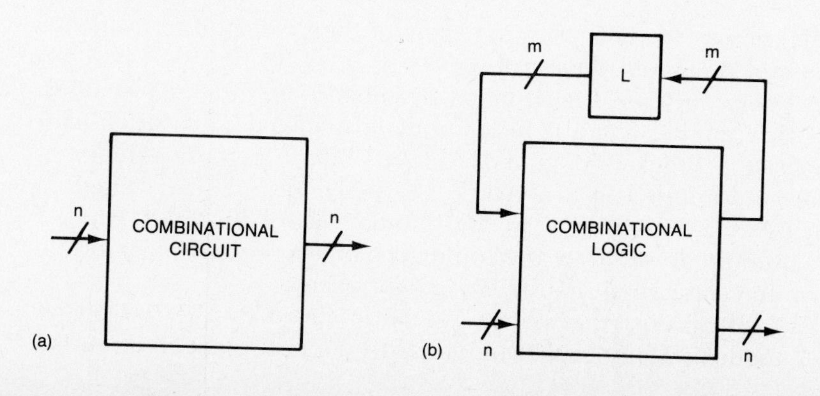

FIGURE 6.13. Combinational and sequential testing

combinational circuit is converted to a sequential circuit with addition of m-storage latches, as shown in Fig. 6.13b. The state of the circuit is determined by the inputs and the previous state. A minimum of 2^{n+m} test vectors must be applied to exhaustively test the circuit. To quote Williams [Will83],

> With LSI, this may be a network with N = 25 and M = 50, or 2^{75} patterns, which is approximately 3.8×10^{22}. Assuming one had the patterns and applied them at an application rate of 1 μs per pattern, the best time would be over a billion years (10^9).

Clearly this is an important area of design that has to be well understood. In the remainder of this section, a summary of design techniques to cope with testing will be given. Three main areas are of importance:

1. Test generation
2. Test verification
3. Design for test.

Test generation relates to the problems of the generation of a number (minimum) of tests to verify the behavior of a circuit and the "goodness" of a given percentage of internal nodes. The problem of test verification is concerned with finding measures of the effectiveness of a given set of tests. This is commonly gauged by performing "fault simulations." Design for test is the task of designing circuits from the outset, so that the previous two endeavors are limited in magnitude. In relation to test generation, test inputs to verify functionality are generally supplied by the designer. These could be, for instance, a variety of programs that run on a microprocessor, if the microprocessor was the device under test. The other form of test inputs are those applied by the manufacturer to verify a certain percentage of good internal circuit nodes prior to shipping parts.

6.5.2 Fault models

In this section we will examine some of the models used to model faults and their relevance to CMOS.

A commonly used fault model is called the "Stuck-At" model. With this model, a faulty gate input is modeled as a Stuck-At-0 (S-A-0) or a Stuck-At-1 (S-A-1) value. When a certain number of vectors are applied to a network, the percentage fault coverage is often quoted. This relates to the number of S-A-0 or S-A-1 faults that could be detected by the input sequence as a percentage of the total number of single faults that might occur.

Not all failures that occur can be modeled by the S-A-0 and S-A-1 models. Many faults are caused by short- or open-circuited

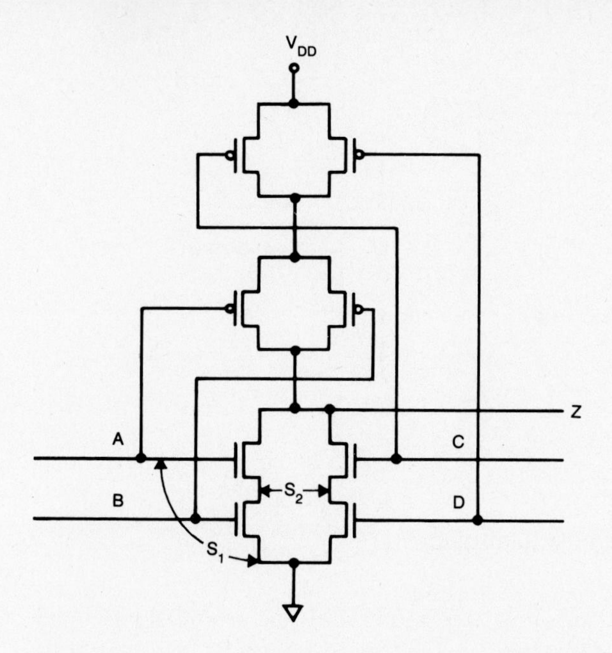

FIGURE 6.14. Faults in CMOS

networks. Considering the faults shown in Fig. 6.14, it can be seen that short S1 is modeled by a S-A-0 fault at input A, while short S2 modifies the function of the gate. What becomes evident is that to ensure good modeling, faults should be modeled at the transistor level, as it is only at this level that the complete circuit structure is known. For instance, in the case of a simple NAND gate, the intermediate node in the series n-pair is "hidden" by the schematic. What this implies is that test generation must be done in such a way as to take account of possible shorts and open circuits at the switch level [GaCV80]. Although the switch level may be the most appropriate level, many existing systems rely on boolean logic representations of circuits. Thus models that incorporate such logic must also be considered.

A particular problem that arises with CMOS is that it is possible for a fault to convert a combinational circuit into a sequential circuit. This is illustrated for the case of a 2-input NOR gate in which one of the transistors is rendered ineffective (stuck open or stuck closed). This might be due to a missing source, drain, or gate connection. If one of the n-transistors (A connected to gate) is stuck open, then the function displayed by the gate will be

$$F = \overline{A + B} + A \cdot \overline{B} \cdot F_n,$$

where F_n is the previous state of the gate. Similarly if the B n-transistor is missing, the function is

$$F = \overline{A + B} + \overline{A} \cdot B \cdot F_n.$$

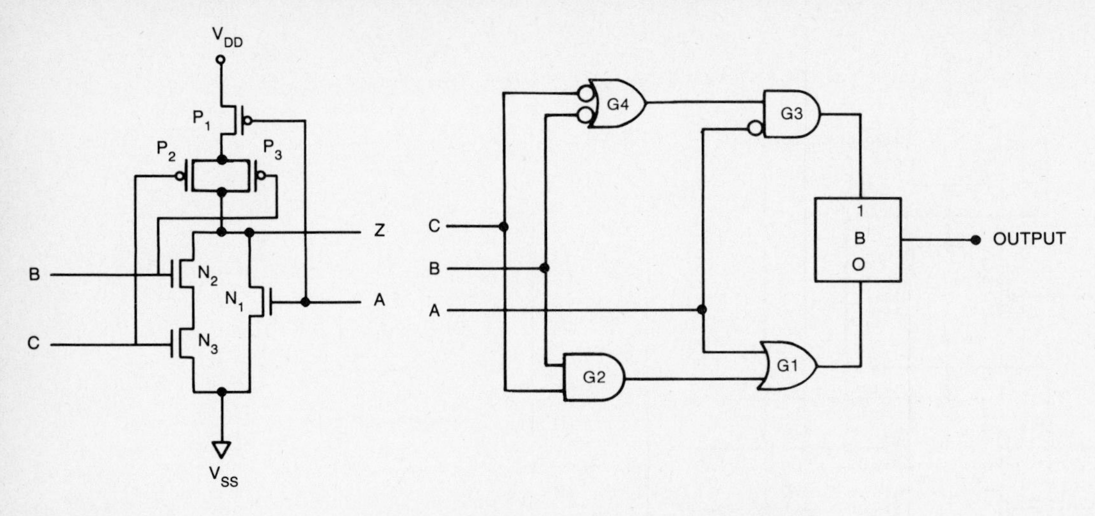

FIGURE 6.15. Logic fault model for CMOS © IEEE 1983 ([JaAg83])

If either p-transistor is missing, the node would be arbitrarily charged until one of the n-transistors discharged the node. Thereafter it would remain at zero, bar charge leakage effects. This problem has caused researchers to search for new methods of test generation to detect such behavior [ElCl81].

A model for CMOS circuits that allows test generation using methods such as the D-algorithm [RoBS67] is represented by Fig. 6.15. Fig. 6.15a shows the CMOS gate in circuit form, while Fig. 6.15b shows the model used for test generation [JaAg83]. The n-tree and p-tree are represented by the logic blocks shown. They are connected to a "B-block", which has the characteristics shown in Table 6.1. The *m*-state in this table indicates that the block retains the previous state. The *X*-state may be biased towards a 0 or 1, depending on the gate and technology (ratioed logic).

Considering the logic model shown in Fig. 6.15b, open and short faults at the circuit level may be mapped to equivalent faults in the logic representation. Some of these are summarized in Table 6.2.

TABLE 6.1.

S_1	S_0	Y
1	1	X
1	0	1
0	1	0
0	1	0
0	0	m

© IEEE 1983 ([JaAg83])

TABLE 6.2

EQUIVALENT FAULTS

Fault in MOS Circuit	Fault in Logic Circuit
A S-A-0/1	A S-A-0/1
B S-A-0/1	B S-A-0/1
C S-A-0/1	C S-A-0/1
Z S-A-0/1	Z S-A-0/1
P1 short	G3, A S-A-0
P1 open	G3, A S-A-1
N1 short	G1, A S-A-1
N2 open	G1, A S-A-0

© IEEE 1983 ([JaAg83])

6.5.3 Design for testability

The key to designing circuits that are testable are two concepts called controllability and observability. Simply stated, controllability is the ability to set and reset every node internal to the circuit. Observability is the ability to observe either directly or indirectly the state of any node in the circuit. Given the circuit structure, programs such as SCOAP [Gold79][GoTh80] are available that calculate the ability to control and observe internal circuit nodes.

There are three main approaches to what is commonly called "design for testability." These may be categorized as:

1 Ad hoc testing
2 Structured design for testability
3 Self-test and built-in testing.

6.5.4 Ad hoc testing

The techniques grouped under the ad hoc category are basically techniques to reduce the combinational explosion of testing. Common techniques involve partitioning large sequential circuits and adding test points. Long counters are good examples of circuits that can be partitioned into smaller counters that may be exercised with fewer test vectors. Another technique classified in this category is the use of the bus in a bus-oriented system for test purposes.

Included in this category are the strategies used to test bit-sliced systems [SrHa81]. The theory relating to testing of bit-sliced systems concerns a class of components known as iterative logic arrays (ILAs). An ILA is classed as C-testable if it can be tested with a constant number of input patterns independent of array size [Frie73]. An ILA is I-testable if the test responses from every cell in the ILA can be made identical. This allows the ILA to be tested with a minimum number of tests by using an equality circuit [SrHa81]. Furthermore, ILAs may possess both characteristics, rendering them CI-testable. An example of an ILA modified to allow I-testability is shown in Fig. 6.16. It consists of a cascaded 1-bit counter cell to which two gates have been added to allow the counter to be tested.

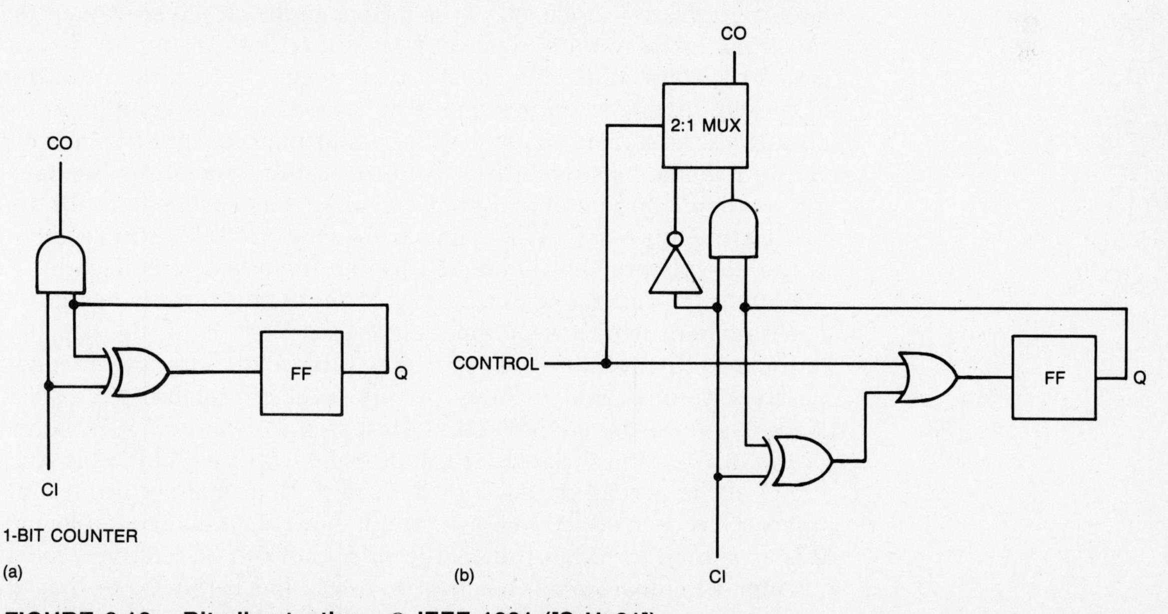

FIGURE 6.16. Bit-slice testing © IEEE 1981 ([SrHa81])

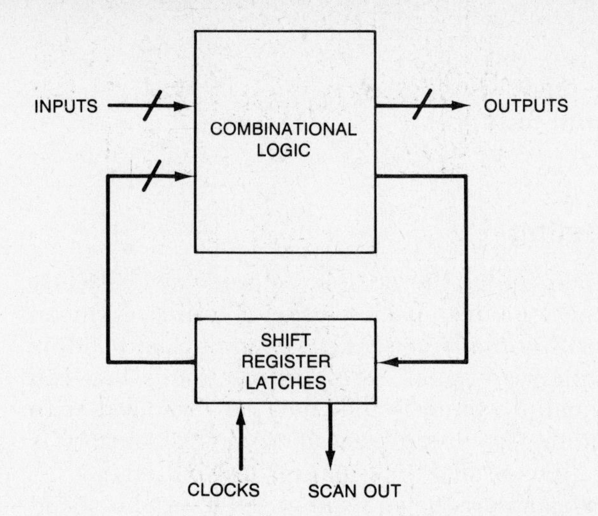

FIGURE 6.17. LSSD testing

6.5.5 Structured design for testability

A collection of approaches have evolved for testing that involves a structured approach to testability. The approaches stem from the basic tenets of controllability and observability outlined earlier in this section. A popular approach is called Level Sensitive Scan Design or the LSSD approach, introduced by IBM [EiWi78]. This is illustrated in Fig. 6.17. The latches in the circuit are all what are termed "shift register latches" or SRLs. In the normal mode of operation, the registers act as the regular storage latches in the circuit. In the test mode, all of the latches in the circuit are connected in series. In this mode, data may be shifted into or out of the cascaded registers. With this capability, testing is reduced to inputting a known sequence (controllability), exercising the combinational circuitry and storing the results, and shifting the stored values out of the register (observability). Automatic test generation programs are available for combinational circuits, thus further simplifying this testing approach. The primary objection to this testing method is the complexity that is created by the increased circuit count in the latch, the increased external pin count, and to some extent the need to chain widely separated latches together. Thus, the decision to include this testability approach would involve trading area and possibly some speed to achieve this level of testability. A static latch based on the static-D latch discussed in Section 5.4 is shown in Fig. 6.18a. D is the regular input to the latch, while I feeds from the Q of the preceding latch in the chain. Note that a 2-input multiplexer has been added to the circuit. It may be possible to implement this as shown in Fig. 6.18b, as the shift path can be relatively slow. Further versions are shown in Fig. 6.18c and 6.18d. Note that all

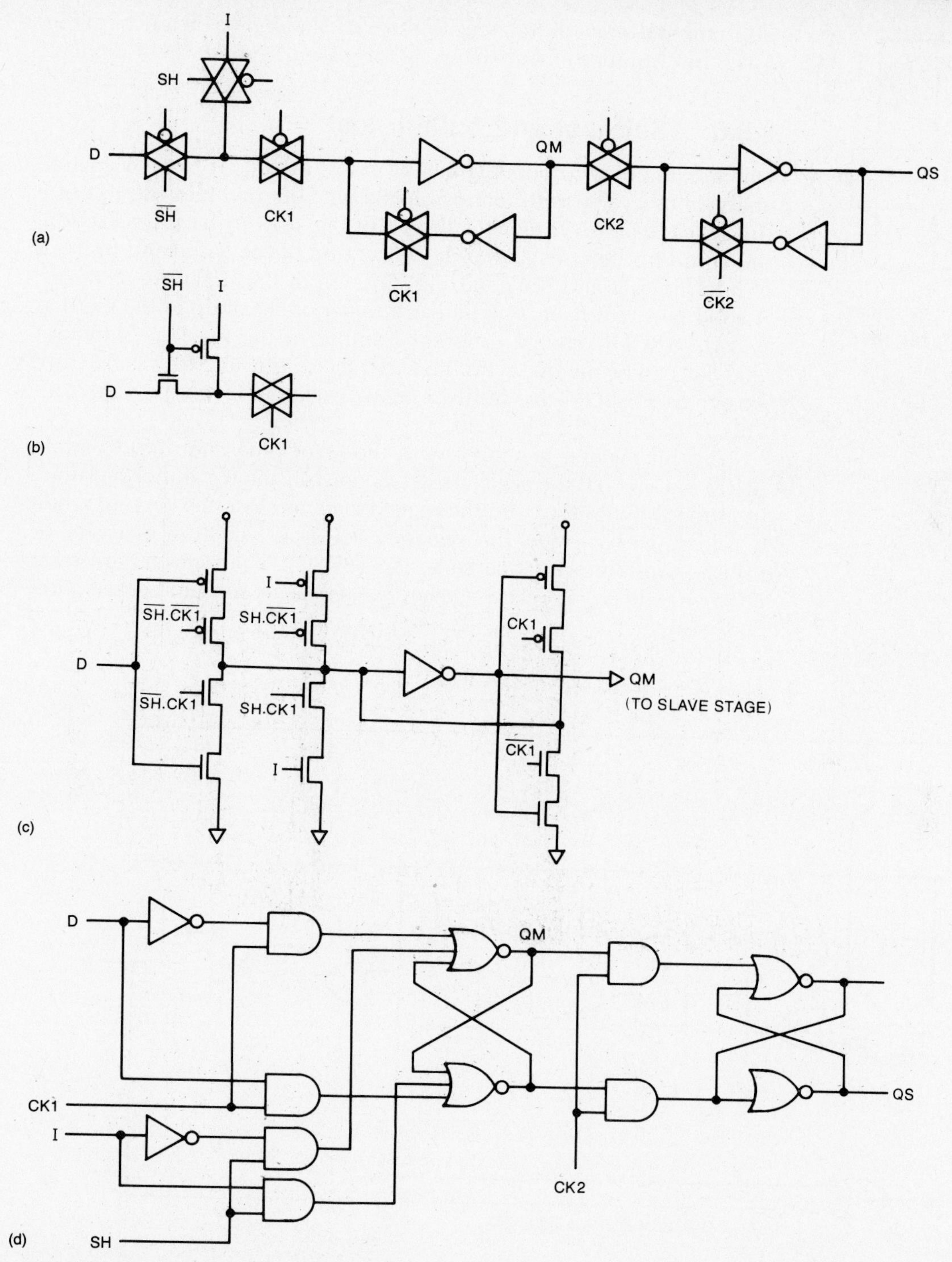

FIGURE 6.18. LSSD implementations

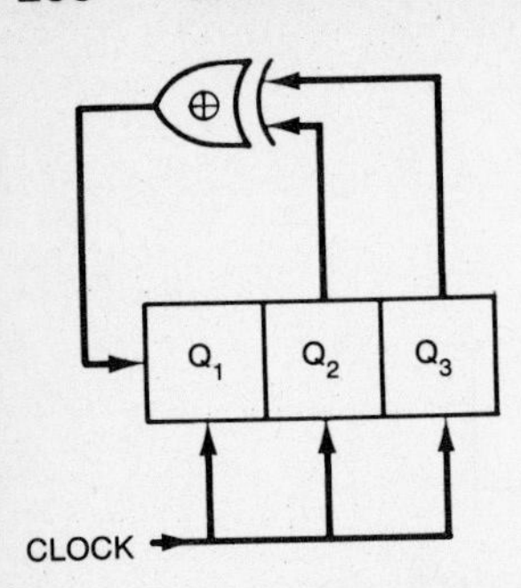

FIGURE 6.19. A linear feedback shift register

implementations add devices to the basic latch. For instance, in Fig. 6.18c the number of transistors is increased from 10 to 14 devices.

6.5.6 Self-test and built-in test

One method of incorporating a built-in test module is to use signature analysis or cyclic redundancy checking. This involves the use of a linear feedback shift register shown in Fig. 6.19. After initialization, the value in the register will be a function of the value and number of latch inputs and the counting function of the signature analyzer. A good part will have a particular number or signature in the register. A bad part will have a different number in the register. Signature analysis can be merged with the LSSD technique to create a structure known as BILBO — for Built-In Logic Block Observation [KoMZ79]. This is outlined in Fig. 6.20.

A 3-bit register is shown with the associated circuitry. In mode A ($C_0 = C_1 = 1$), the registers act as conventional parallel registers. In mode B ($C_0 = C_1 = 0$), the registers act as scan registers. In mode C ($C_0 = 1 \quad C_1 = 0$), the registers act as a signature analyzer or pseudo-random sequence generator (PRSG). The registers are reset if $C_0 = 0$ and $C_1 = 1$. Thus a complete test generation and observation

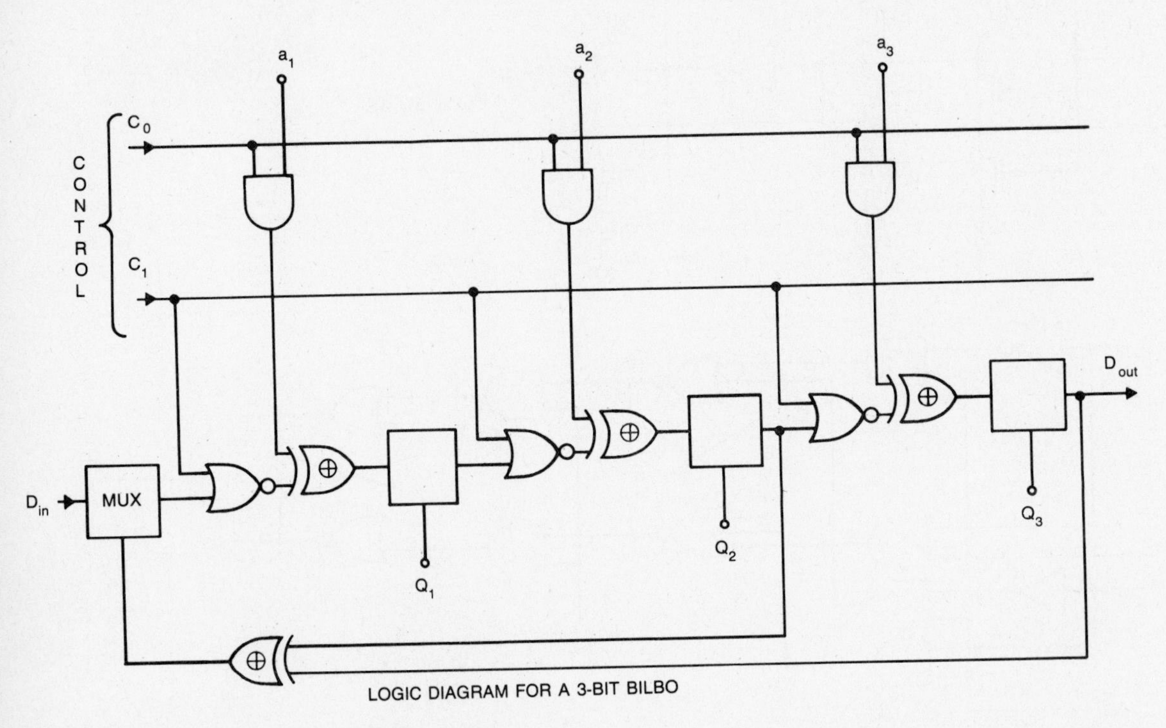

FIGURE 6.20. BILBO circuitry

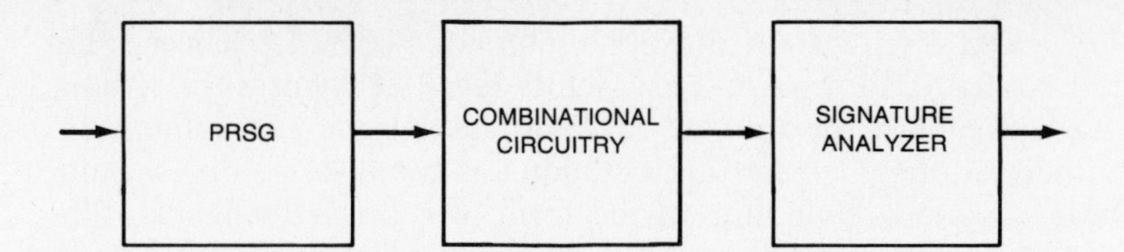

FIGURE 6.21. BILBO usage

arrangement can be implemented as shown in Fig. 6.21. In this case two sets of registers have been added in addition to some random logic to effect the test structure.

Another approach to built-in test is called Design for Autonomous Test [McCB81]. In this approach, modules are partitioned into small modules, which are then tested exhaustively. The main method for partitioning involves the use of multiplexers. Fig. 6.22a shows a

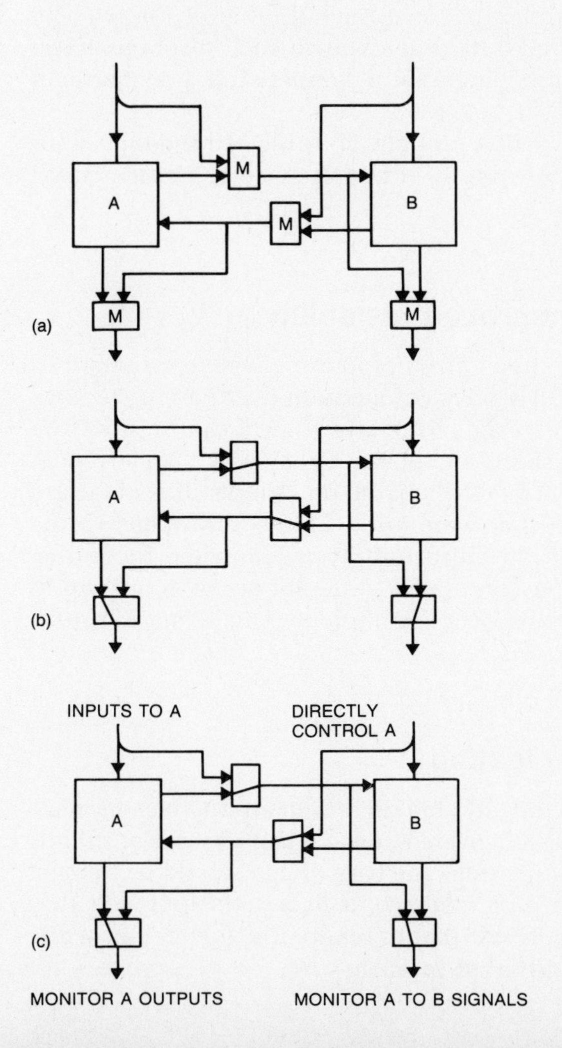

FIGURE 6.22. Multiplexer segmenting for autonomous test © IEEE 1981 ([McCB81])

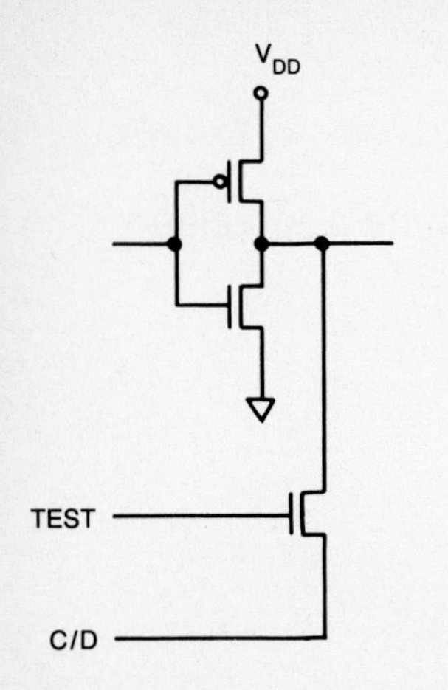

FIGURE 6.23. C/D circuitry © IEEE 1981 ([McCB81])

typical circuit with multiplexers included. Fig. 6.22b shows the circuit configured for normal use, while Fig. 6.22c shows the circuit configured to test module A. No fault models or test generation techniques are required for this technique. A complete module would include a pattern generator in the form of a linear feedback shift register, a signature analyzer, and test control circuitry. Exhaustive testing is not suitable in situations where stuck-open faults exist. For this reason the circuit shown in Fig. 6.23 was proposed [McCB81]. This provides charge and discharge control over the output of the logic gate. After each test, the charge-discharge circuit (C/D) is exercised by asserting TEST and strobing C/D high then low. Outputs that are not stuck open return to the correct value, while stuck open lines will remain charged or discharged.

Another test method requiring exhaustive testing is called Syndrome Testing [Savi80] [Savi81]. Here, all possible inputs are applied to the circuit and the number of 1's at the output are counted. The resultant value is compared to that of a known good machine. Extra circuitry includes a pattern generator, a counter, and a comparison circuit.

Further techniques involve double or triple redundancy with voting circuitry. On chip "stimuli" ROM are also sometimes used to generate tests.

6.5.7 Layout for improved testability

We have covered the methods of incorporating test structures into circuits. As faults occur in a physical medium over which we have control, some precautions taken at the layout level can improve the likelihood of undesirable shorts and opens. Galiay [GaCV80] advances some rules for improving testability based on observations of failure modes in nMOS circuits that were built. Shorts and opens in the metal layer and shorts in the diffusion layer dominated the faults. It is quite probable that a completely different set of rules would be needed for each different CMOS technology. This seems a fruitful area for further research.

6.5.8 Summary — testing

In this section we have concentrated on design alternatives to make testing easier. Each approach covered adds certain design constraints such as area and speed. If possible, the base design should be thought of with testing in mind. For instance, a data-path approach to a problem might be easier to test than a random logic version, while not being substantially different in area.

6.6 Summary

In this chapter a number of approaches to designing CMOS circuits have been summarized. Some of the tools required to effect these methods were then covered. Finally, some design approaches that take into account the need to test circuits were surveyed.

6.7 Exercises

6.1 A family of gate arrays requires an arbitrary amount of logic and small RAMs with up to sixteen 8-bit words located anywhere on the die. Design a gate array cell that would accommodate these requirements. Assume a two level metal CMOS process. Complete layouts for an 8-input multiplexer, a 1*8 RAM, and a 1 of 16 decoder.

6.2 Propose testing methods for the bit-serial silicon compiler covered in Section 6.3.2. Repeat this exercise for the data-path compiler.

6.3 Figure 6.18 shows some LSSD latches. Design an LSSD latch for use with an N-P pipelined logic circuit. Propose a testing strategy for circuitry composed of N-P logic.

SYMBOLIC LAYOUT SYSTEMS

7.1 Introduction

We have examined a number of strategies to design ICs. Gate array methods are best suited to quick, turnaround random logic functions, especially for small numbers of parts. Speed and density are sacrificed. Standard cell systems improve on the utilization of silicon and range of functions available to the designer, but still incur some area, power, and speed penalties. Usually both design methods encourage the designer to think in terms of low level TTL SSI functions or such well-known structures as RAMs, ROMs, and PLAs. In fact, this last point is the main reason that such systems are gaining popularity in the general electronics industry. Full custom layout at the mask level has been traditionally error prone, time consuming, and the domain of experts well-trained in the art. The adoption of simple geometric design rules and global system design rules, such as employing strict 2-phase clocking, has led to the methodologies that allow system designers to design chips at the mask level [MeCo80]. This allows system designers to use the full circuit potential that silicon provides. An improved methodology that is gaining acceptance is to design low level cells in a symbolic manner at the circuit level of abstraction, thus alleviating the designer from the burdensome task of dealing with geometric design rules. These cells are then combined by using a variety of well-defined composition techniques, which extend the capability of such systems to the chip level. A good mixed methodology might combine standard cell blocks (defined symbolically) that are automatically designed, a method of constructing regular repeated cells in the symbolic domain, and control structures such as programmable logic arrays. As tools improve, the symbolic level of description provides a good "assembly language" that may be targeted by silicon compilers. In the next sections we will examine some existing approaches to symbolic IC design. During the course of this discussion, we will describe two successful approaches developed at AT&T Bell Laboratories that demonstrate the power of these techniques.

7.2 Coarse grid symbolic layout

The idea behind coarse grid symbolic layout involves dividing the chip surface into a uniformly spaced grid in both the X and Y directions. The grid size represents the minimum feature or placement tolerance that is desired in a given process and is usually selected by close consultation between design tool developers and semicon-

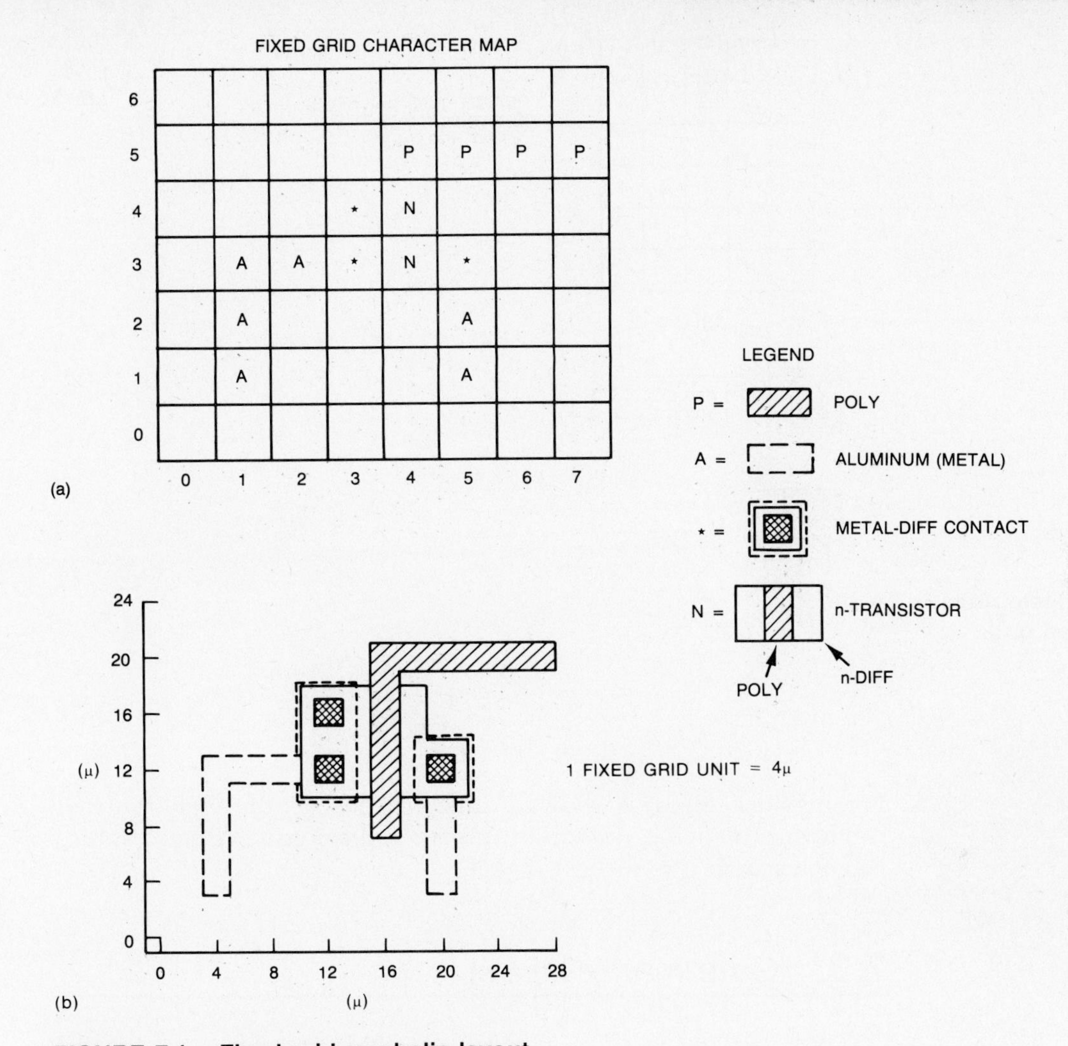

FIGURE 7.1. Fixed grid symbolic layout

ductor process engineers. For each combination of mask layers that exist at a grid location, a symbol is defined. Fig. 7.1 shows a typical symbol set and layout. Given a particular design system, these symbols are then placed on the grid to construct the desired circuit, much in the same way as one would tile a floor. Symbol sets may be defined as characters or perhaps graphical symbols, if a graphics display is used for design.

Rockwell International [Lars78] and American Microsystems International (AMI) [GiNa76], [ClKS80] have made use of character-based symbolic layout for some time. The design process consists of laying symbols on the coarse grid. The use of fixed-size symbols simplifies geometric design rules, but does not totally alleviate them.

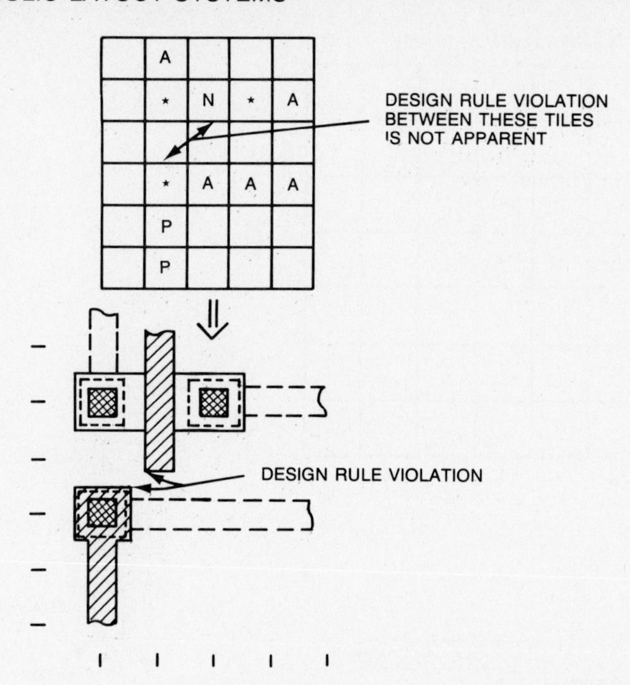

FIGURE 7.2. Design rule violation in fixed grid layout

For instance, Fig. 7.2 shows a diagonal design rule violation that can occur in such a system. In summary, fixed grid systems capture geometry with a reduced rule set.

7.3 Gate-matrix layout

A character-based symbolic layout style was developed at Bell Labs [LoLa80] specifically for custom CMOS circuitry. It improves on coarse grid symbolic layout by providing a regular layout style where a matrix of intersecting transistor diffusion rows and polysilicon columns is employed. The intersection of a row and column is a potential transistor site (poly crossing diffusion). A related style is featured in [PZSB84].

The evolution of this technique from a standard cell viewpoint is shown in Fig. 7.3. Fig. 7.3a shows a circuit implemented in terms of standard cells (four 2-input NANDS and one inverter). Note that inter-cell connections are in metal. Rather than running these connections in metal, we can run vertical polysilicon columns corresponding to each gate signal. The transistors may then be placed on the polysilicon signals and interconnected, as shown in Fig.

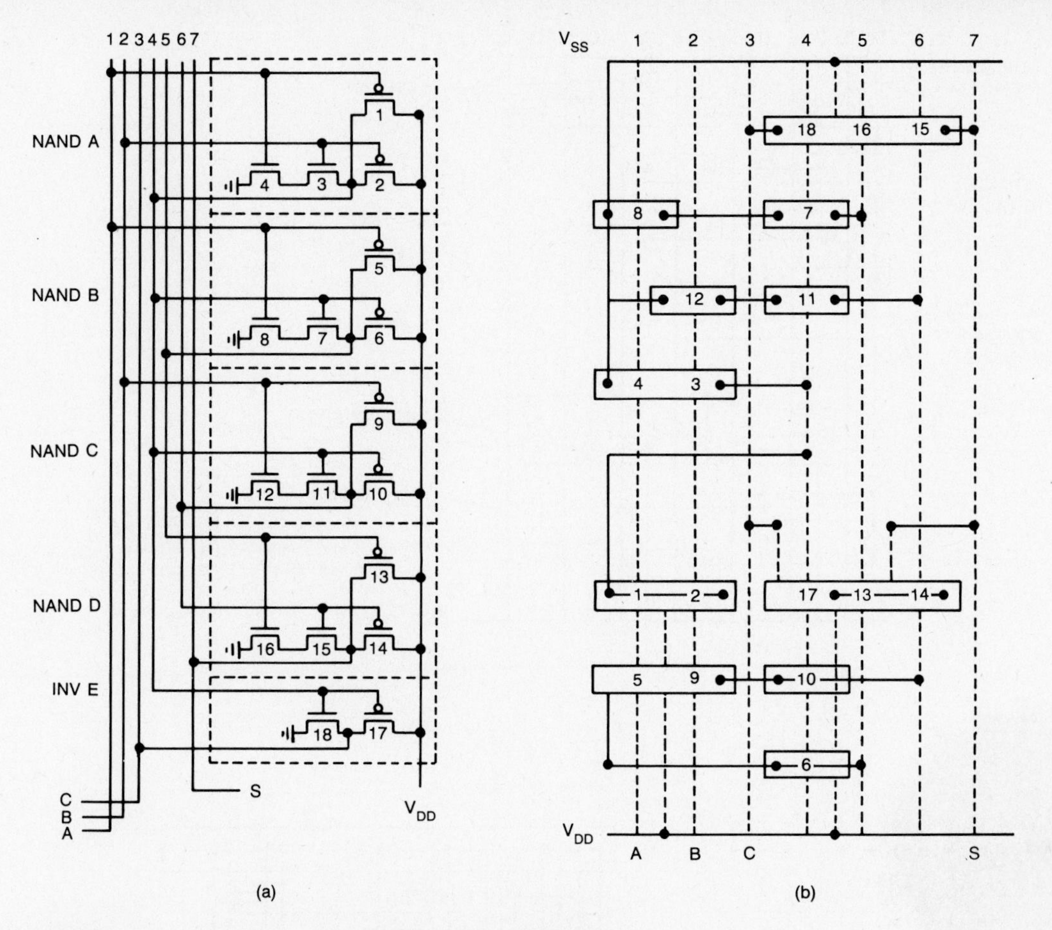

FIGURE 7.3. Evolution of gate-matrix design style

7.3b. Note that vertical columns may be either polysilicon (S) or diffusion (D) (Fig. 7.4a). Horizontal rows are transistors and/or metal routing tracks. Metal may also run vertically. A character symbolic layout for the layout shown in Fig. 7.4a is shown in Fig. 7.4b. The following table identifies the symbols:

```
N  n-channel transistor
P  p-channel transistor
+  metal-poly or metal-diffusion crossover
*  contact
|  polysilicon or n-diffusion wire
!  p-diffusion wire
:  vertical metal
-  horizontal metal
```

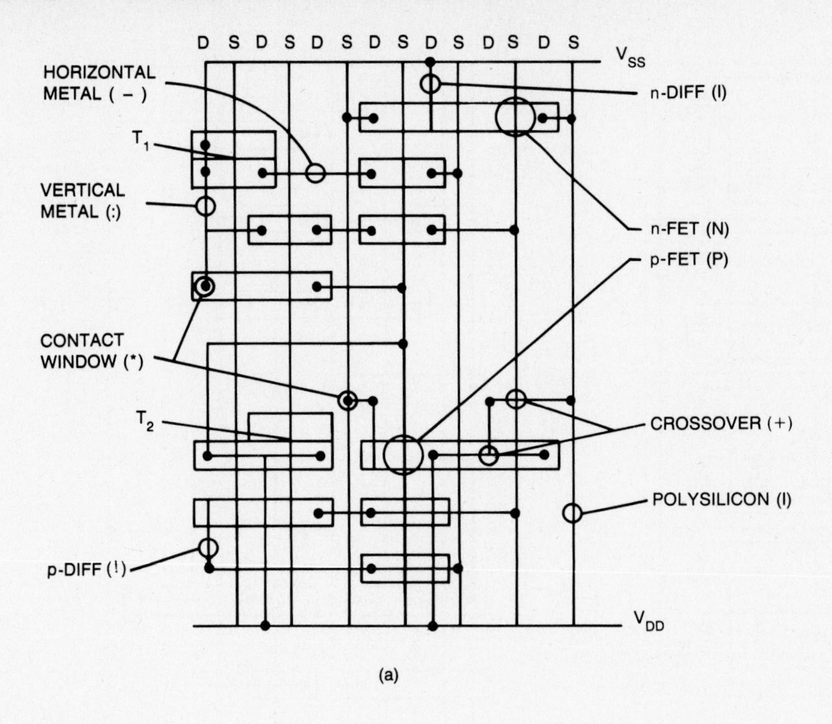

(a)

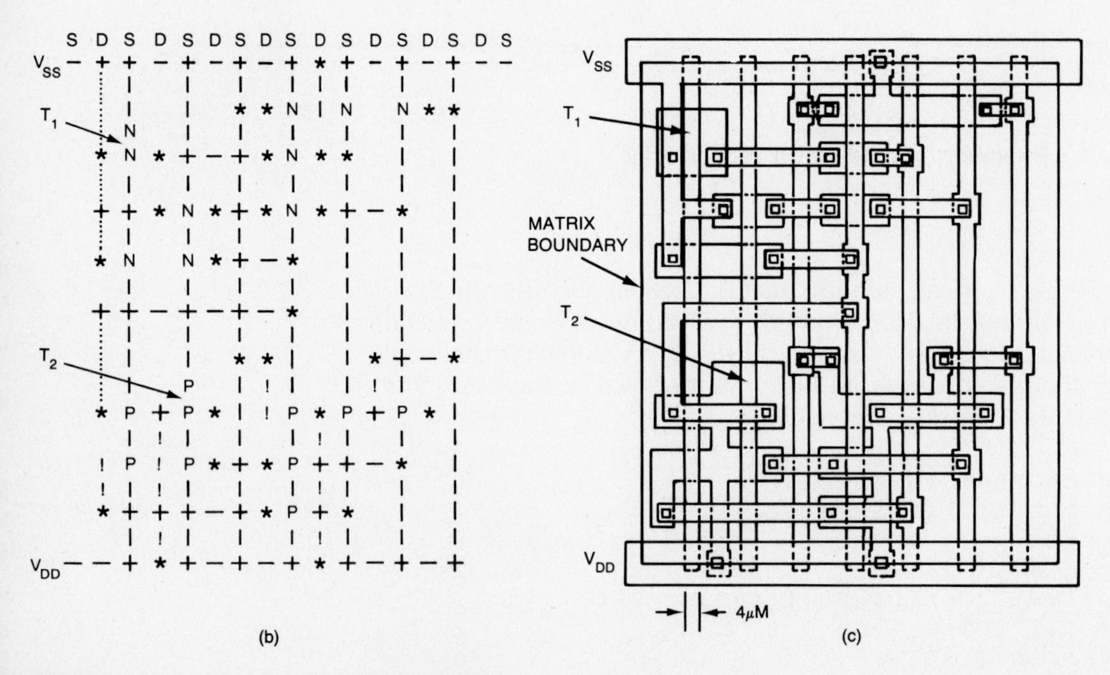

(b)

(c)

FIGURE 7.4. Typical gate-matrix layout

A mask layout is shown in Fig. 7.4c. The following rules summarize the gate-matrix technique:

1 Polysilicon runs only in one direction and is of constant width and pitch.
2 Diffusion wires (of constant width) may run vertically between polysilicon columns.
3 Metal may run horizontally and vertically. Any pitch departures from minimum (e.g., power rails) are manually specified.
4 Transistors can only exist on polysilicon columns.

Wide transistors may be specified by abutting two or more N or P symbols.

To convert from the character symbolic to mask artwork, the character matrix is examined and the symbols are expanded to their equivalent mask entities. Operations such as merging horizontal dashes into one metal wire and merging adjacent devices are performed during this phase. Fig. 7.5 shows typical grid spacings, in

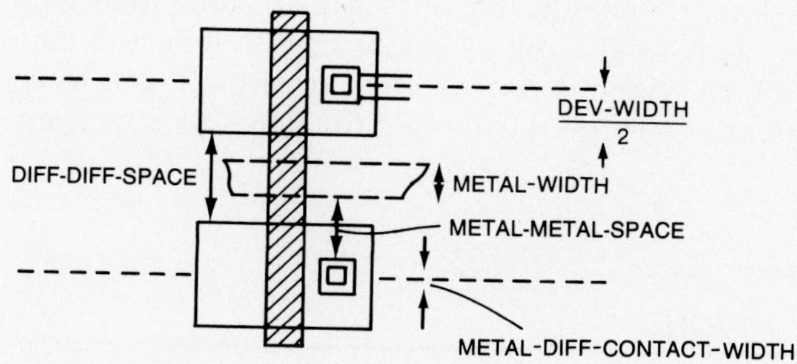

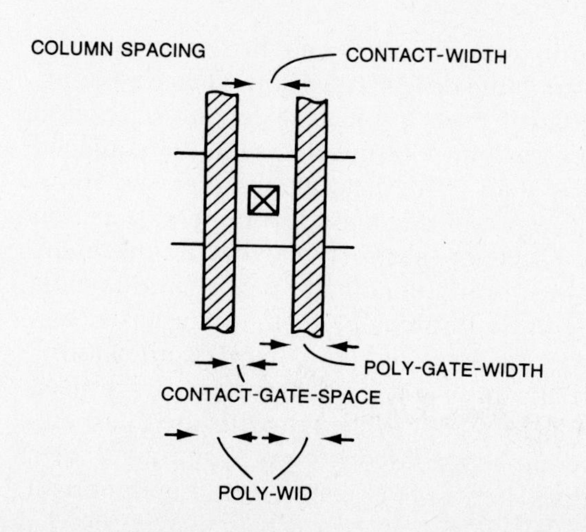

Figure 7.5. Gate-matrix row and column spacings

terms of the design rules presented in Chapter 3. The row pitch is determined by the minimum drain/source separation of two unconnected transistors with metal diffusion contacts. The column pitch is determined by the spacing of two polysilicon lines with a diffusion contact between the polysilicon runners.

Note that this design style is "technology updatable" as the design description is held in this symbolic form. This means that the essential mask information is encoded by the character format. Examples may be seen in Chapter 9, Section 9.5. Circuit extraction may be done at the symbolic level or at the mask level by conventional circuit extractors. The extraction at the symbolic level is considerably faster. An example of this form of extraction is given in Section 7.6.7.

Such a layout style still has layout rules to observe, similar to the example shown in Fig. 7.2. The use of this constrained style can aid in the design of large circuits where many design teams have to design large blocks concurrently. Here, regularity has been used in the layout style. Modularity is encouraged by the block nature and "through routing" aids in locality considerations. The basic format, that is, the character symbolic description, is not hierarchical. Modules must be assembled in their entirety and "pasted" together at the mask level. In addition, there is no freedom to locally optimize geometry, such as altering transistor sizes to reflect circuit loading. The character symbolic specification eventually gets very baroque when trying to express all the requirements of advanced layouts.

7.4 Sticks layout

The term "sticks" is a generic term given to symbolic design systems that do not necessarily constrain the designer to a grid when designing. Rather, a free form topological description of a layout is entered via an interactive graphics system. Graphical symbols are placed relative to each other rather than in an absolute manner and interconnected by colored sticks representing mask level interconnection layers. This technique is based on abstractions that IC designers have used for many years to simplify pencil and paper hand layouts prior to digitization. In currently implemented computer based systems, correct mask spacings are achieved by using a "compaction" process. Since 1978 [Will78], quite a few such systems have been reported [HsPe79] [Dunl80] [KeWa83] [Most81], with the main emphasis being placed on compaction algorithms. A summary of a typical compaction algorithm is given in Section 7.6.9. It is important to note that many of these systems use, as a basis for layout manipulation, a mask level description (i.e., boxes, lines). This is to be contrasted with the approach described in the following sections.

7.5 Virtual grid symbolic layout

Virtual grid symbolic layout [West81a] is a symbolic layout method that draws on the experience gained in coarse grid symbolic systems, gate matrix, "sticks"-type systems, and other approaches such as ICSYS [Buch80] developed at the University of Edinburgh and Caltech. In essence, the system approaches design at the layout level by manipulating circuit elements such as transistors and wires as opposed to any form of geometric mask description. These elements are placed on a grid to facilitate easy design capture and simplified tools, with the final geometric spacing between grid lines determined by the density and interference of circuit elements on neighboring grid locations. This leads to the notion of a "virtual" grid. This concept is best illustrated by a simple example as shown in Fig. 7.6a. Three vertical wires are shown centered on a grid. The result of using a fixed grid of 10 units and a wire width and separation of 10 units leads to the mask description shown in Fig. 7.6b. By using a grid in which the spacing varies according to topology, the mask description in Fig. 7.6c is constructed. The dashed shape may be possible in certain virtual grid compactors. The end result for the designer is that placement on the grid may be done without regard to any design rules. In addition to eliminating design rules, the grid is also used to define circuit connectivity in a manner similar to that employed in schematic capture systems. Here, the notion of a "coordinode," as introduced by Buchanan, is used to

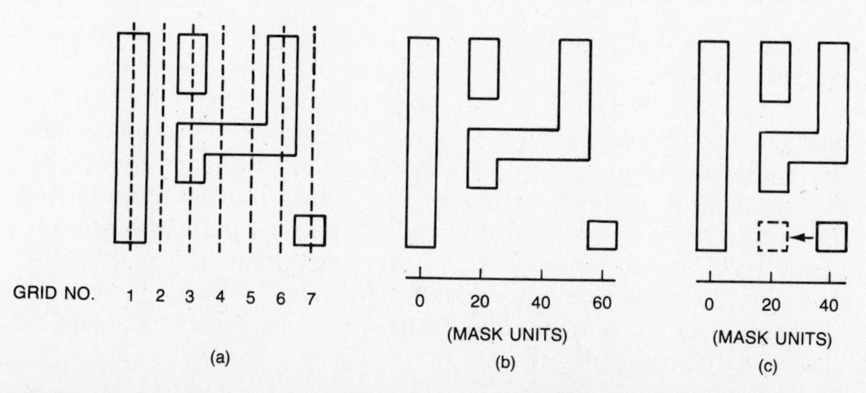

FIGURE 7.6. The virtual grid

capture physical location, structural connectivity, and behavioral state. As its name suggests, a coordinode has the properties of a coordinate, namely, some xy position that will eventually map to the silicon surface. In addition, it may possess the properties of a node in a circuit such as voltage or simulation state. Structurally, a coordinode defines the nodes in the network being designed. In the virtual grid context, a coordinode is mapped to a discrete set of grid points rather than a quasi-continuous set of X-Y coordinates. The grid coordinates form the *lines of action* in a circuit, defining the essential communication paths in and through a circuit. Local geometric perturbations are handled by software skilled in the art of manipulating geometry.

MULGA [West81b] [WeAc81] is an example of an integrated design system based on these principles. The VIVID system [RBDD83] is a system under development that employs the same principles. Further systems [ReIv83] have reported the use of similar techniques. A discussion of the underlying features of this system will form the basis for the discussion of the attributes of a contemporary symbolic layout system.

A virtual grid circuit capture system yields the following benefits:

- Design rule free topology capture.
- Rapid design capture through use of *point* interconnect.
- Fast grid-based algorithms for connectivity audit, compaction, and other processes.
- Ability to allow parametized cells with automatic geometry generation.
- Hierarchical module assembly.
- Natural target for higher level silicon compilers (geometry free).

In the following sections we will demonstrate these attributes.

7.5.1 Language

The language that forms the lowest level of design in the MULGA system describes circuit elements in terms of their topological relationship to each other and their electrical connectivity. This is based in some part on the language used in the ICSYS system [Buch80]. A central concept is that of the virtual grid, which unifies the description of layout and circuit. It does this by representing the circuit on a topological grid, which defines the electrical connectivity of the circuit in a straight-forward manner. The grid also describes in a relative manner the placement of components. The assignment of geometric information is delayed until a compaction process is completed.

The first language developed that embodied these concepts was called ICDL (for Intermediate Circuit Description Language). Subsequently, an extended version of this language, called ABCD, was developed by J. Rosenberg and N. Weste at the Microelectronics Center of North Carolina [RoWe82]. Both languages contain the following primitives:

- MOS transistors
- contacts
- wires
- pins
- instances
- points.

Further circuit level primitives, such as capacitors, resistors, and bipolar transistors, could be added. A language was chosen as the main data base representation (as opposed to a character symbolic format) for several reasons. First, a single parser may be written that interfaces the textual representation to an appropriate internal data format. This allows various utilities to be independently written by different tool designers. A readable data base also allows a definable interchange format and communication between designers. Furthermore, all the standard text utilities available on the operating system may be used to manipulate the design. A textual representation may be gracefully expanded to include new language features, especially high level constructs such as looping, conditionals, and parameter passing. Finally, during start-up, designs may be done textually if interactive tools are not available.

7.5.2 Devices

Devices or transistors may be textually specified with parameterized type, width, length, orientation, and location. A typical statement might be:

```
nt1: device ntype (3,2) or=east
```

This says that a transistor named `nt1` is an n-type device and is located at virtual grid point (3,2), with the drain facing east. It has a default minimum width. Fig. 7.7 shows a graphical specification of a transistor and illustrates the use of the virtual grid. The gate of `nt1` is assumed to be at (3,2). The drain is at (4,2) and the source at (2,2). One may also say,

```
pt1: device ptype (nt1,6) or=east w=2
```

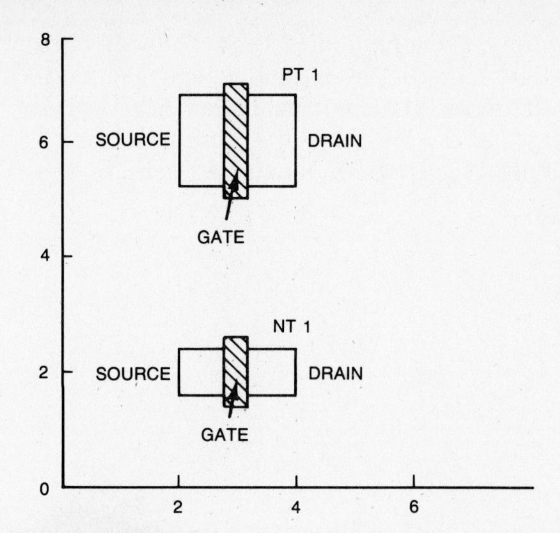

FIGURE 7.7. Symbolic layout-transistors

to place a p-transistor called `pt1` above `nt1`. The width is twice the default minimum width. Widths and lengths may be specified in terms of micron dimensions or default sizes (the normal case). Transistors are allowed multiple connection points as a function of their width or, alternatively, wide transistors may be built by stacking "unit" transistors as allowed in the gate matrix methodology.

Note that a transistor has a strict structural model — the virtual grid connection points that define the drain, gate, and source. In addition, based on the width, length, and orientation parameters, the transistor has a well-defined physical model. The behavior at the transistor level (i.e., "on" current, gain, etc.) could be well-characterized as the use of "unit" transistors is encouraged. Fig. 7.8a illustrates the manner in which an n-type transistor is constructed for two target processes. Note that in each case the procedure correctly constructs a valid transistor. The routine that produces the transistor might be used in a compactor, a capacitance evaluation program, or in the actual mask generation. In early virtual grid systems only rectangular transistors were allowed. There is no reason, however, why other shaped transistors could not be used (i.e., "el" shaped devices). Fig. 7.8b shows a *bent gate* transistor with its connection strategy. However, to maintain regularity, the "unit" regular transistors are preferred.

7.5.3 Contacts

Contacts may be named and parameterized by type and position. Typical statements would be:

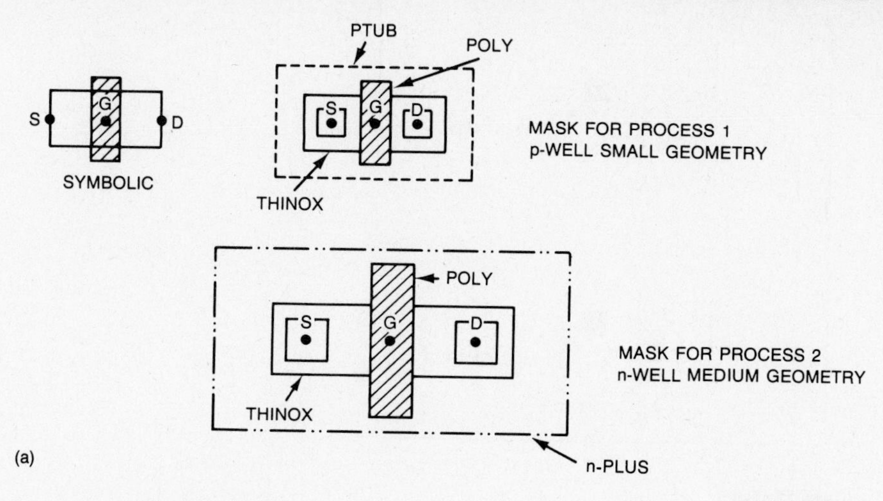

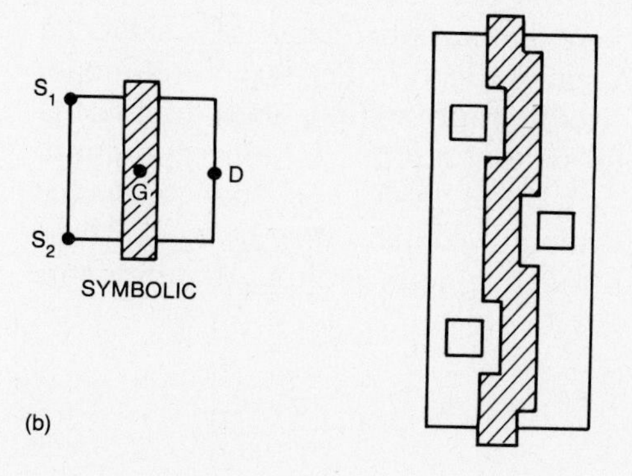

FIGURE 7.8. Physical construction of transistors

```
contact vss (nt1.d,0)
contact md nt1.d
contact md pt1.d
```

These place a substrate contact (Vss) and two metal diffusion (md) contacts, as shown in Fig. 7.9. Contacts join wires to wires or wires to transistors. The second two statements make use of a concept called symbolic bonding (used in ABCD), which allows components to reference the structural interconnection points by name rather than coordinate value.

The physical mask model of a contact is a set of overlapping rectangles on appropriate mask levels. This can be a fixed geometry,

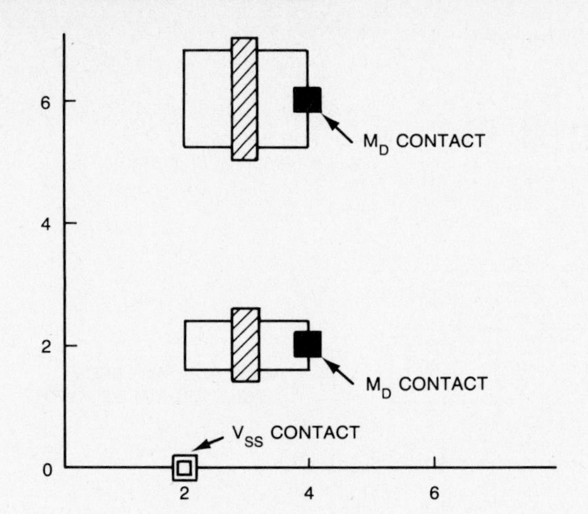

FIGURE 7.9. Symbolic layout—contacts

as shown in Fig. 7.10a, or a variable geometry, as may be required in the situation in Fig. 7.10b, where the polysilicon has to be extended in the direction that aluminum leaves the join. The structural model corresponds to a node in the circuit graph, which, according to the accuracy desired, can be a resistance or simply a ''join'' operation. Minimum sized contacts are used. For large area, low resistance contacts, two-dimensional replication of contacts may be used. These constraints are consistent with fine geometry etching requirements for precise contact definition.

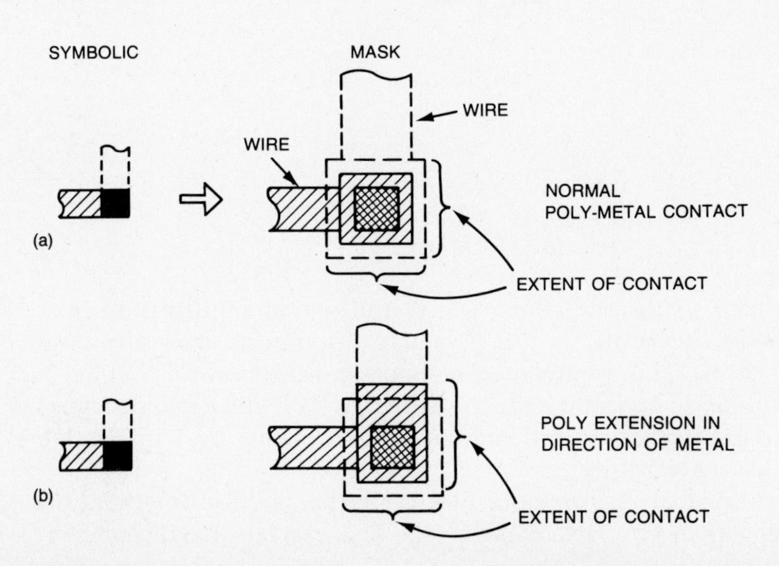

FIGURE 7.10. Physical construction of contacts

7.5.4 Wires

Wires may be parameterized in type and width. A typical statement might be

```
wire poly (3,2)(3,6)
```

or

```
wire alum nt1.d pt1.d
```

The first statement wires the gates of the two transistors together, while the second statement places a wire between the drains of the transistors `nt1` and `pt1`. Note that the first statement may also be specified as

```
wire poly nt1.g pt1.g
```

Again notice the use of symbolic bonding. Fig. 7.11 shows the partial cell with these wires in place. Note that as we have placed the appropriate contact at the drains of the two transistors, they are now connected.

Wires serve to connect transistors together. Physically, wires consist of paths of a certain width on prescribed layers. Fig. 7.12 illustrates the mask generation of an n-wire. Behaviorally, wires can have attributes of resistance and capacitance. The capacitance can exist between the wire and substrate or to other wires. The wire may have a capacitance that may be dependent on voltage.

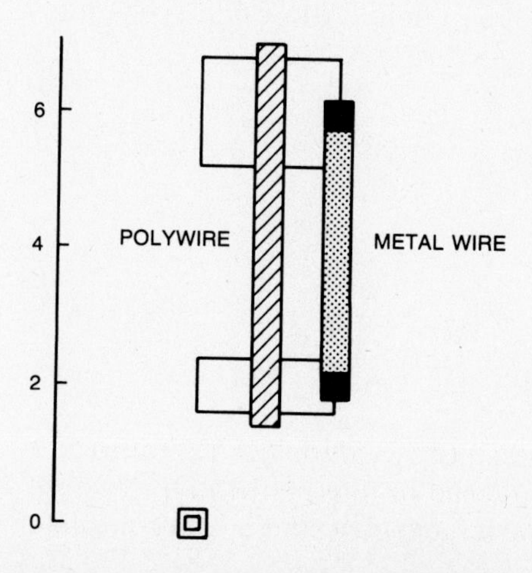

FIGURE 7.11. Symbolic layout—wires

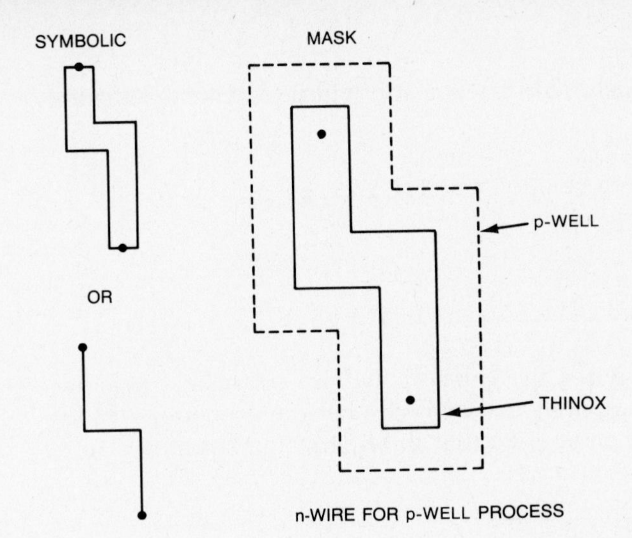

FIGURE 7.12. Physical construction of wires

7.5.5 Pins

Pins are very important because they link the structural, physical, and behavioral domains. In other words, we can isolate the state of any particular node (in time and space) in a circuit by careful naming. Pins, of course, have names and a layer associated with them. Thus the statements

```
vss: pin alum (0,0) vss
A:   pin poly (nt1,4) A
vdd: pin alum (0,8) vdd
```

denote points in the circuit that we desire to carry these node names. Pins may also have an optional "type" field. Allowable types are:

- vdd
- vss
- clock
- input
- output
- ioput
- generic.

These types are used in module assembly (Section 7.6.12) to further ensure correct construction. Pins are used in interpreting the structure of a cell and also during assembling larger subsystems. The complete

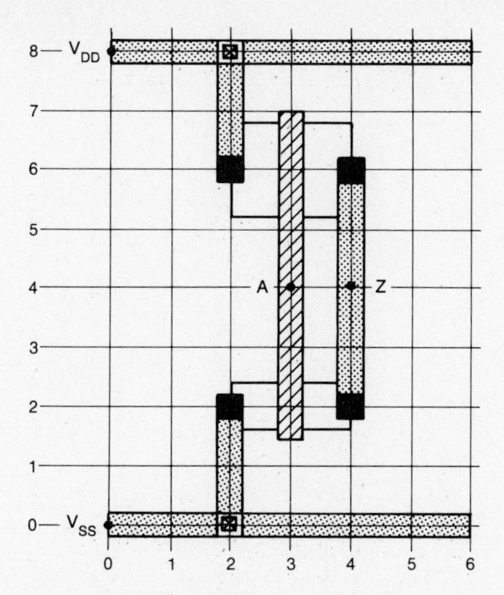

FIGURE 7.13. Completed virtual grid symbolic layout for inverter

inverter with the pins is shown in Fig. 7.13, with the complete ICDL textual description shown in Table 7.1. A mask layout is shown in Fig. 7.14. The main emphasis in this figure is to illustrate all the shapes generated.

TABLE 7.1. ICDL listing for CMOS inverter

```
BB 0 0 6 8
device n (3,2) or=east
device p (3,6) or=east
wire alum (0,0) (6,0)
wire alum (0,8) (6,8)
wire poly (3,2) (3,6)
wire alum (2,0) (2,2)
wire alum (2,6) (2,8)
wire alum (4,2) (4,6)
contact md (2,2)
contact md (4,2)
contact md (2,6)
contact md (4,6)
contact vss (2,0)
contact vdd (2,8)
pin alum (0,0) vss
pin alum (0,8) vdd
pin poly (3,4) a
pin poly (4,4) z
```

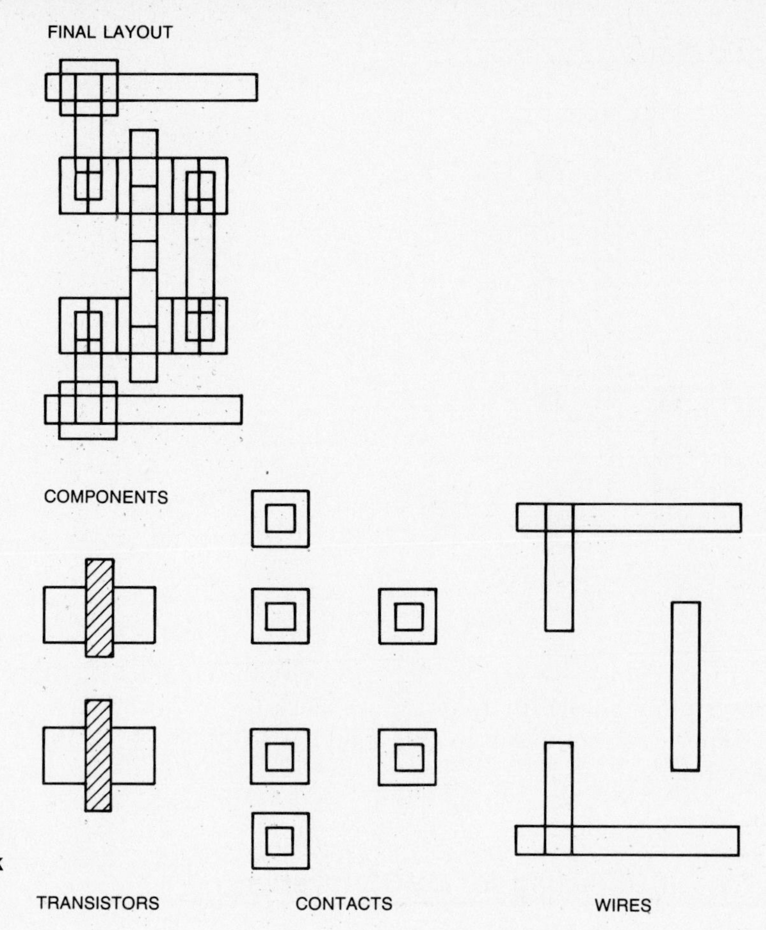

FIGURE 7.14. Conversion of symbolic layout to mask layout

As stated previously, pins serve to link the physical, structural, and behavioral attributes of a design. Physically, they denote a geometric point in a circuit with a name. Structurally, they can denote a net name associated by connectivity at the wire, contact, and transistor level. Behaviorally, pins can be assigned a voltage level or logic state during verification, thus completing the loop. If two pins name the same net, they will have the same "pin-name." However, they will have a different physical position. If we wish to connect to a cell, it may be desirable to connect to a particular pin, say, on the "north-west" corner. This is achieved by using a second name field and is achieved in the cases above in the following fashion. In ICDL

```
nw: pin alum (0,0) vss
```

In ABCD

```
vss_nw: pin alum (0,0)
```

7.5.6 Instances

The instance construct is used to create hierarchical circuits. Instances may be named, positioned relatively, replicated, rotated, and reflected. Fig. 7.15 shows a typical circuit layout that would be specified by the following ABCD fragment:

```
c1: instance gate1 ll=(0,0) num = 4 dir =
    vertical
c2: instance gate2 ll=c1[0][lr]
```

Here, four instances of a "gate1" cell are arrayed vertically, with the lower left of the group placed at virtual grid coordinate (0,0)(ll = (0,0)). This instance of gate1 is called c1. The second statement refers to an instance of gate2 called c2, whose lower left corner is placed at the lower right of the zero'th instance of c1 (ll = c1[0][lr]) [Rose83]. Instances may be represented structurally by nodes of arbitrary complexity in the graph representing the IC design. The ICDL/ABCD description for the cell, combined with the environment that the cell is used, provides all the physical and structural information required to generate the various models for the verification of the design. Instances may also be rotated and reflected. The orientation of a cell is defined in terms of a "major-

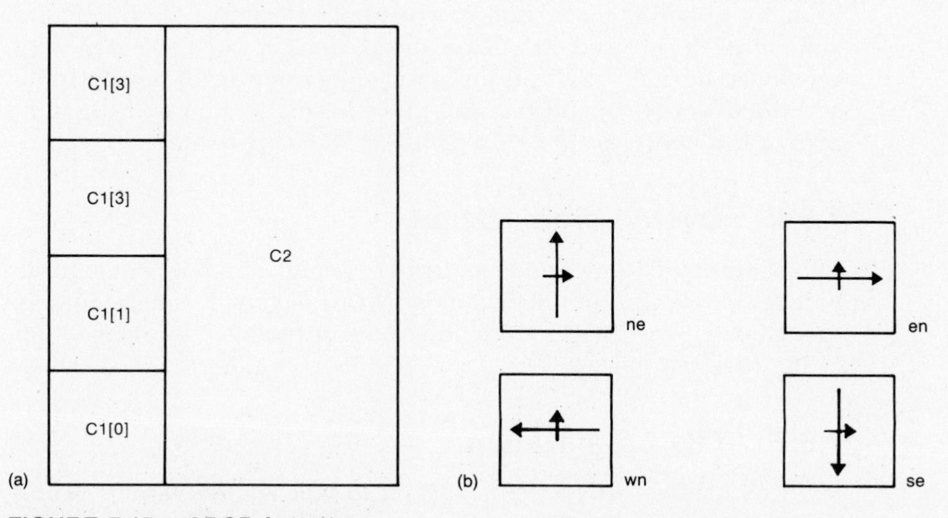

FIGURE 7.15. ABCD instances

minor" axis direction. The default for a cell is ne, with the major axis pointing north and the minor pointing east. Other variations are shown. There are eight in all.

7.5.7 Representations

All ICDL elements have well-defined structural and physical models. During the course of a design, one often needs different physical models to aid interpretation or checking of a design. This is easily achieved because the design description has been caught at a level that is able to accommodate these different views of the design. For instance, graphical representations of ICDL depend on the type of output device that is being used. A text description may be used or a two-dimensional character map may be used to display an ICDL description on a text terminal (with say * standing for a contact). On a color graphic display, we employ a notation that we have termed "logs" or "thick sticks." Instead of employing lines, wires and transistors are represented by boxes that reflect the relative width at the mask level. This enables designers to gauge the effect of oversize elements on topology. Note that due to the simple connection defined by the virtual grid, transistors may in fact be drawn in schematic form. On a line drawing peripheral (i.e., plotter) wires are collapsed to single lines, but transistors are still drawn with relative sized rectangles.

7.6 Symbolic design tools

In many ways the tools used in a symbolic design environment are similar to those used in custom mask design, but the absence of geometric design rules and design at the circuit level results in an additional set of capabilities and tools. In this section we will summarize the nature of tools used in the MULGA system.

7.6.1 Overall organization

The operation of the various tools that operate on the ICDL language will be summarized in the remainder of this section. Before beginning, though, it is worthwhile to review some of the overall organization of the MULGA system.

7.6.2 File organization

MULGA uses the UNIX™ file system to manage the various representations of a design. The design data is managed by using a number

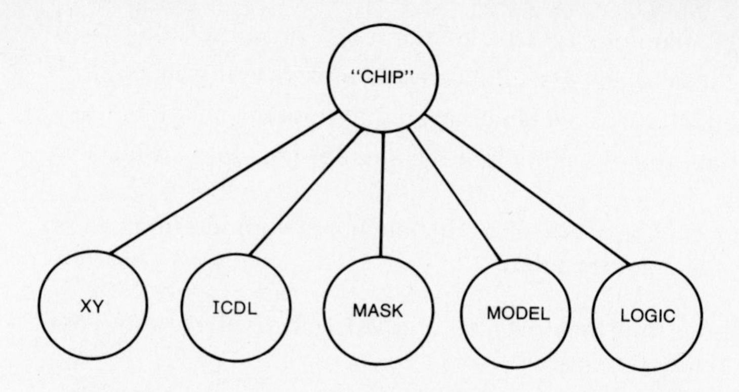

FIGURE 7.16. MULGA file organization

of functionally distinct subdirectories, as shown in Fig. 7.16. All files in a given directory are similar representations of different cells. For instance, the original ICDL cell description for a cell named "fulladder" is held in file icdl/fulladder. The mask description for this cell is held in file mask/fulladder, and so on. This allows arbitrary tools to be added with new representations occurring in a newly assigned subdirectory. We use this structure to maintain a set of parallel hierarchies for different design descriptions.

7.6.3 Software organization

As a large amount of new software was required, maximum use was made of library functions. The most common piece of software is the parser, which reads an ICDL description into memory and creates an internal data structure. This consists of a doubly-linked list with attributes identifying element type, subclass, and attendant parameters.

A common parser reports any ICDL syntax errors and insures a common data structure is presented to all tools. Accompanying software builds and maintains the internal data base.

The next most common subroutines are those that print the internal data structure either to a terminal or a disk file. Similar routines are needed to plot representations of ICDL on various types of graphic displays or hard copy devices. Note that although device independent graphics may be used to some extent, the type of graphic representation depends on the type of output device and the particular view of the layout required. Broad categories are raster versus line drawing devices and symbolic versus circuit drawing mode.

A range of general purpose routines may be identified, which include the following:

1 Bounding box calculation (recursive) (in terms of virtual grid coordinates).
2 Move a rectangular area.

3 Delete the elements in a rectangular area or point.
4 Identify those elements in an area or at a virtual grid point.
5 Copy elements in a rectangular area to another position.
6 Encapsulate the elements in a rectangular area into an instance [Rubi83].
7 Yank the elements out of an instance and include them in an existing instance [Rubi83].

All of the above routines may be applied selectively to any combination of symbolic entities.

7.6.4 Chip design process

The chip design process in the MULGA system begins with a hierarchical floor plan. This floor plan is required to conform to a "restricted hierarchy," similar in principle to the "separated hierarchy" of Rowson [Rows81]. This restricted hierarchy classifies cells as modules, composition cells, and leaf cells, as shown in Fig. 7.17. Leaf cells contain symbolic layout primitives; that is, they define the physical layout of the circuit. They are symbolically connected (by abutment) through a hierarchy of composition cells. Composition cells contain only instances of other cells (leaf and composition). Different modules contain cells that are structurally unrelated. For example, a PLA controller and a data path would normally be designed

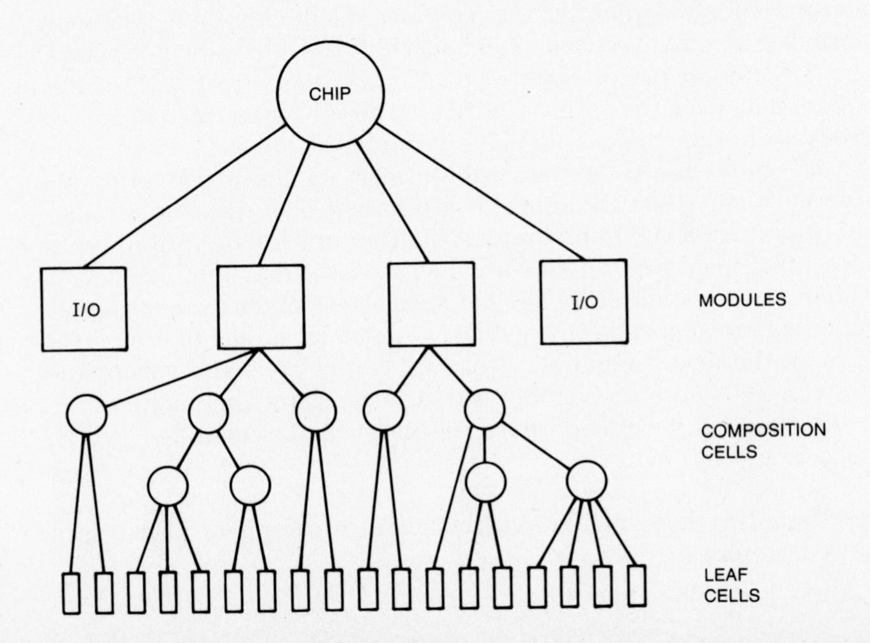

FIGURE 7.17. Restricted hierarchy

as two separate modules. These modules are then connected to each other and to the I/O modules (pads) using conventional place and route techniques.

It should be noted that within modules, cells are always connected across abutting boundaries. Modules thus form the highly structured, pitch-matched portions of the layout. This does not mean that cells within a module cannot be routed together. Intramodule routing is accomplished using symbolic routing cells that also pitch match to their leaf cell neighbors.

Through the use of this restricted hierarchy and some composition design rules that are outlined in Section 7.6.7 and Section 7.6.8, the floor plan can be used as a complete structural specification for the chip. This is similar to the technique used in the CHAS system [Mudg80], in which the floor plan is used to subdivide the logic, circuit, and layout representations into a set of consistent hierarchies. Because composition cells define only cell placement and connectivity, they are representation independent and can be used to generate (for example, functional) descriptions of the chip. Composition cells are, therefore, not only process, but also (potentially) technology independent.

7.6.5 Cell design process

The implementation of a floor plan begins with symbolic leaf cell design, either by using an interactive graphics editor or procedurally in terms of a higher level language specifying ICDL. Once a leaf cell has been symbolically designed, an audit program checks for circuit inconsistencies. A compaction program then examines the ICDL cell description and assigns valid mask dimensions to a file called the "design grid" file. Depending on the internals of cells, the final size of symbolically matched cells may vary, as may the locations of common interconnection points on a cell boundary. Thus abutting cells may have to be pitch matched to regain structural integrity. This is achieved by using a pitch match program. A circuit interpretation program examines the ICDL file, compaction data, and output of the audit program and produces structural descriptions for various styles of circuit simulator.

7.6.6 Interactive graphics editor

An interactive graphics editor is used for the initial capture and exploration of cell designs. Once this task has been completed, the prototype layout may be converted to a procedure and parameterized to generalize the cell. The editor is also used to create and modify floor plans.

Key attributes of an interactive editor include the speed of response, ease of user interface, clear presentation of design information, and absence of catastrophic failure. The notation of the virtual grid and the use of symbology aid in a number of these requirements. For instance, most operations to modify the topology of elements involve the definition of a "point"-based data structure. Thus although elements appear as proportionally weighted rectangular shapes (called "logs," for thick sticks), the selection criteria rely on a single point or group of points. This contrasts with geometric editors, which manipulate rectangular or polygonal areas at some finely resolved detail. In these systems, one always has to mentally keep track of corners or similarly ill-defined points of reference. The above impressions are best borne out by experiencing both types of systems. Users who have done this strongly favor the virtual grid capture method. It also has the added benefit that it meshes nicely with schematic capture systems, requiring little modification to such a system to design using the virtual grid. Concurrent textual and graphical references to the ICDL language are maintained throughout edit sessions.

The use of color is very important when designing at the leaf cell level, as it efficiently codifies the layered nature of the IC interconnect. A monochrome raster display with stipples may be used but is much less effective than a comparable color display.

The editor used in the MULGA system is relatively simple and provides the following mechanisms:

Selection methods:

- point (virtual grid point)
- area
- element type (i.e., device, wire)
- element layer
- node name.

Operations:

- identify
- move
- delete
- copy
- append
- change.

Viewpoint management:

- windowing
- panning.

Viewing modes:

- mask symbolic
- circuit symbolic
- mask
- schematic.

In addition, stroke and raster, color, or monochrome hardcopy devices are supported.

One of the interesting attributes implemented in the original MULGA editor is the ability to pan in real time across the ICDL data base. This allows a designer to scan wires or cell boundaries without losing visual continuity. This is achieved by keeping a bounding box (in virtual grid coordinates) for all modules in the editor. When the screen is panned to the left for instance, the image in the viewpoint is physically moved by a small amount to the left in a "wrap-around" mode. Thus a small portion of what was on the left of the screen view now appears on the right. This strip is erased and the ICDL data base is searched, culling out bounding boxes that do not correspond to the screen viewpoint. When cells are found that will be in the erased portion of the screen, they are entered and the resultant graphics entities that describe the elements in the window are transmitted to the graphics display. Due to efficiencies in tree culling, almost continuous motion can be achieved for modest hardware configurations.

7.6.7 Circuit interpreter

The circuit interpreter is responsible for interpreting the implicit connectivity in an ICDL module and creating explicit connectivity in the form of node names and numbers. We use the term circuit interpretation, rather than circuit extraction, because the circuit information is held implicitly by the symbolic description.

The algorithm for determining connectivity is relatively straightforward and in many ways parallels those used by schematic capture systems. A virtual grid matrix corresponding to the size of the cell is allocated. ICDL elements are then plotted into this matrix. As this proceeds, an algorithm merges node numbers of connected

LAYOUT

METAL WIRE

5

(a)

ALIAS MAP

NODE	CONNECTED TO	
0	0	V_{SS}
1	1	V_{DD}
2	2	
3	3	
4	4	
5	5	
6	6	
7	7	

6

5

METAL-POLY CONTACT

(b)

NODE	CONNECTED TO	
0	0	V_{SS}
1	1	V_{DD}
2	2	
3	3	
4	4	
5	5	
6	5	
7	7	

6

5

7

POLY WIRE

(c)

NODE	CONNECTED TO	
0	0	V_{SS}
1	1	V_{DD}
2	2	
3	3	
4	4	
5	5	
6	5	
7	6	

5

5

5

(d)

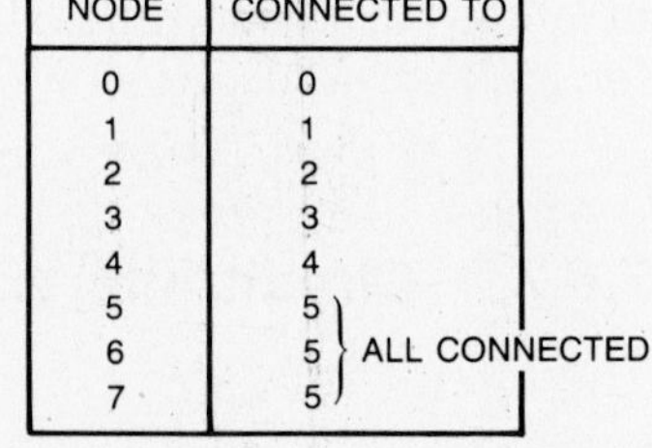

NODE	CONNECTED TO	
0	0	
1	1	
2	2	
3	3	
4	4	
5	5	ALL CONNECTED
6	5	
7	5	

FIGURE 7.18. Circuit interpreter algorithm

regions. A linear table keeps a map of node numbers and temporary aliases. For instance, imagine an aluminum wire (node number 5) is plotted into a blank grid. This is illustrated in Fig. 7.18a, along with a map of aliases. Nodes 0 and 1 are reserved for V_{SS} and V_{DD}. Next a metal-poly contact (node number 6) is plotted on top of the aluminum wire. At this stage, it is noted that the contact connects to the underlying wire, so an entry in the alias table is changed, as shown in Fig. 7.18b. Next a polysilicon wire is plotted (node number 7). When it intersects the aluminum wire and contact, a further alias entry is made, as illustrated in Fig. 7.18c. When this process is completed for all elements in the cell, the alias table is recursively reduced to illustrate a *true* alias mapping. This final table is shown in Fig. 7.18d. Node naming is incorporated by noting pin names. Unnamed nodes are given internally generated names.

In the course of this process, several simple circuit/composition rules are checked. These include the following:

- Polysilicon crossing diffusion as an *implicit* (and thus illegal) device.
- Nonmanhattan wires.
- Illegal contact use, i.e., metal-poly trying to connect metal-diffusion.
- Nodes electrically connected but with conflicting node names.
- Nodes named identically but not connected electrically.
- Floating inputs — gate signals not available on the periphery of a cell.
- Floating transistor connections.
- Declared *input* nodes not on boundary.
- Unattached pins.

Rules, in general, are built to reinforce a composition methodology. The rules are divided into *warnings* and *fatal* errors. The former allow further processes to act on cells in, say, the module assembly phase, while the latter require corrective action before the cell can be used further.

The circuit interpreter is used in all phases of design including the interactive editor, compactor, and chip assembler. In addition, it is used with appropriate formatting to generate descriptions for various types of simulators. This includes the MODEL language for a timing simulator called EMU, and the ADVICE [Nage80] (SPICE) language for the ADVICE circuit simulator. In this case, further routines summarize the capacitance of wires, transistors, and contacts using compaction results or statistically averaged grid spacings.

7.6.8 Virtual grid compaction

The compactor is the key element in any symbolic layout system where mask dimensions have to be determined automatically. A number of approaches have been proposed to deal with this problem. We will examine one simple approach in this section and summarize a graph-based approach in the next section to show the relationship to the underlying virtual grid notation.

The compaction approach originally used in the MULGA system was chosen for three reasons:

1 It was easy to implement.
2 It was fast.
3 It was extensible to deal with symbolic layout at the chip level.

As we will see, most of the attributes of other compactors may be embedded in the virtual grid compactor.

Virtual grid compaction is completed in only two passes — one an X-compaction step and one a Y-compaction step. It commences by plotting the virtual grid structure into a matrix, which is comprised of a sequence of pointers to ICDL elements that exist at that virtual grid location. This is illustrated in Fig. 7.19. In summary, the compaction process proceeds as follows. For the X-compaction phase, adjacent grid columns are examined for interference between elements

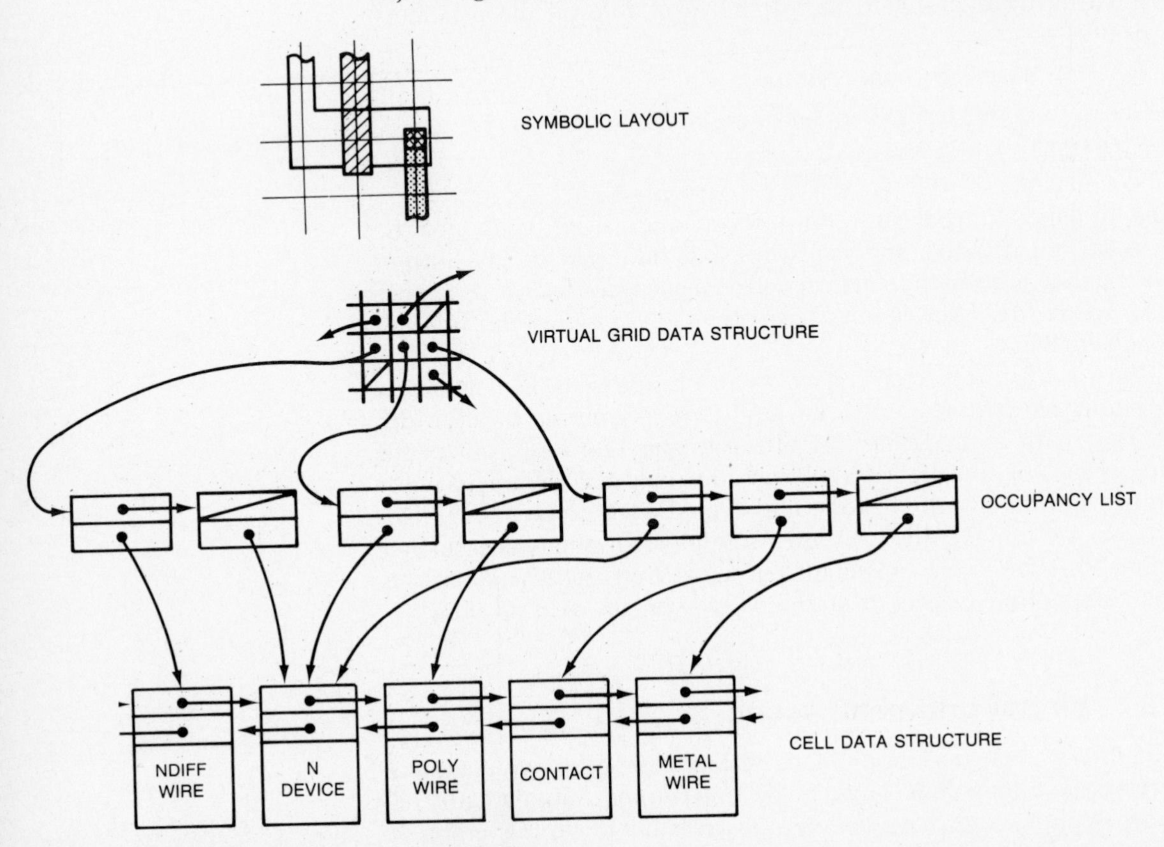

FIGURE 7.19. Compaction data structure

TABLE 7.2. Design grid table

DIRECTION	VIRTUAL GRID COORD.	MASK COORD. nm
X	0	12.5
X	1	16.25
X	2	21.25
X	3	26.50
X	4	31.25
X	5	36.25
X	6	41.25
Y	0	0.0
Y	1	8.75
Y	2	15.00
Y	3	23.75
Y	4	31.25
Y	5	37.50
Y	6	47.50
Y	7	56.50
Y	8	62.75

on the same row. Actually, rather than compare against the previous column, a "picket fence" [BoWe83] is maintained that keeps a record of the location of the last element in each layer at each row position (i.e., a composite point). This takes account of both "white space" (column locations with no entries) and circuit elements that need to be spaced from elements that are not on the adjacent columns. The worst-case column spacing — the spacing required by one of the elements to meet design rules in the column being placed — dictates the actual column placement. This is repeated for all columns and a table such as shown in Table 7.2 results, which tabulates the virtual grid location and the minimum mask location for this column. The Y-compaction is completed similarly, with the exception that diagonal clearances must be checked. This is shown in Fig. 7.20.

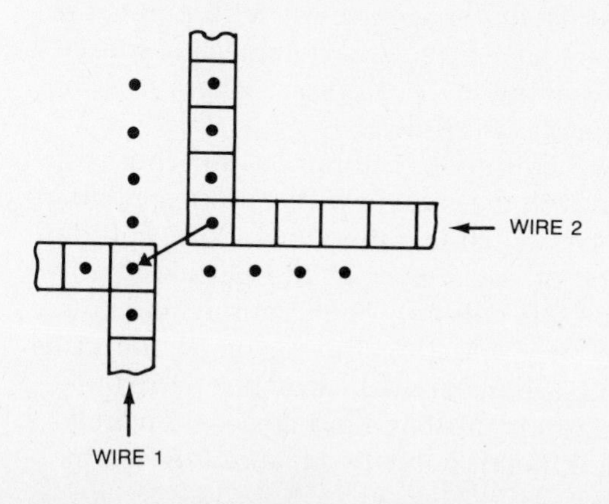

FIGURE 7.20. Y diagonal checking

The result of the process is a table of both X and Y locations that yield a minimum dimension cell. A cell may be stretched in a trivial manner by adding to the row or column mask location a desired value and propagating this amount to the end of the array. Thus pitch matching may be achieved without again examining the cell itself. Only the grid file needs to be manipulated. This is a major advantage of this approach, allowing hierarchical compaction to be implemented in a relatively straight-forward fashion.

In more detail, the compaction algorithm may be stated as follows. Central to the algorithm is the construction of *regions*, which are composed of rectangles placed over a virtual grid location. The rectangles have the dimensions of the particular mask layer that they represent. These include actual mask layers and *composite* layers such as gate polysilicon. Typical layers might be metal1, metal2, poly1, poly2, n-diffusion, p-diffusion, window, n-plus, p-plus, buried-contact, Vss-contact, Vdd-contact, implant (for depletion loads), p-well, gate-polysilicon, and active (which defines the active gate region less any required overlaps). A final layer, called the peripheral zone, ensures all internal structures are spaced half a design rule from the boundary. This layer enforces a composition rule that dictates how cells may be abutted. A rectangle has a center mask coordinate and a north, south, east, and west dimension. These may be zero, positive, or negative, allowing the rectangle to be skewed around the actual mask coordinate corresponding to the virtual grid coordinate.

Prior to compaction proper, the ICDL data structure is scanned to resolve context dependent entities. For instance, during specification a designer need only specify a metal-diffusion contact (or a generic contact, for that matter). The program resolves whether the contact is n-diffusion or p-diffusion, wire or transistor, and any further context information. This is the key to dealing with more complex design rules. Compaction is commenced by plotting the virtual grid data structure into a matrix, as shown in Fig. 7.19. For instance, contacts are plotted as single points. All wire points are plotted once. Simple transistors have three grid points (gate, source, drain), with wide transistors having $2n + 1$ points, where n is the number of source/drain connections allowed.

The matrix is then scanned column by column, comparing each composite rectangle in each layer in the column with its corresponding interference member in a *fence*, which is trailed behind the column recording the last occurrence of each layer at the particular row value under consideration for this column. The scanning and comparison process is achieved by evaluating the *region* at the grid points in question. To do this the matrix data structure is scanned noting the primitive ICDL elements present. For instance, if a metal-polysilicon contact is found, a metal, polysilicon, and window rectangle are added to the region description. If the contact is offset, the rectangles are appropriately skewed. If the contact had some further context information (such as "extend in direction of metal"), this would be included. A transistor drain or source is resolved in terms of the mask layers needed to build the device in the technology to be utilized. For instance, an n-transistor in a p-well process might have a n-diffusion layer and a p-well, while a p-transistor might have a p-diffusion layer and a p-plus layer. The gate of a transistor has a gate-polysilicon and active layer. Wire points are classified by their position on a wire. They can be ends, corners, mid-points, or dots. Composite regions are compared by a design rule checker that uses a table of spacing values to arrive at a spacing. The table also includes *skip* tags that tell the design rule checker to skip a check (i.e., metal to p-well). An auxiliary table defines which layers can touch other layers, regardless of node number, and which layers can overlap other layers if they have different node numbers. For X, the routine checks the west edge of the point in the column being checked against the east edge of the fence composite grid point. If layers are allowed to overlap, the test is skipped. The worst-case space is noted as the minimum distance the two grid points may be spaced to insure all possible design rules are obeyed. For diagonal tolerance checks, complete rectangles are checked against each other using correct euclidean distances.

Improvements to the basic virtual grid compactor allow the insertion of arbitrary grid lines and the storage of a spacing array for each layer rather than a fixed column or row value.

7.6.9 Graph-based compaction

Historically, in graph-based compaction systems, symbolic circuit elements were placed on a *fine grained* physical grid. The grid lines might have micron or lambda values initially. The compactor pushes apart groups that are too close together and reduces the spacing between overly-spaced groups. Generally, a group is defined as a set of circuit elements that share the same center line in a layout, are connected geometrically, and/or connected electrically. It is possible to have two or more groups on the same column or row.

For X-compaction, all circuit elements on the same vertical center line in the symbolic layout that are geometrically connected, are held together as a group. A directed constraint graph is built. The nodes of the graph are the groups, and the branches are used to connect groups that have potential design rule violations. The weights of the branches are the minimum separations necessary between two nodes. An example of the mapping of a symbolic circuit to a graph is depicted in Fig. 7.21. If there is no spacing necessary between two groups, an edge will not be created between the two

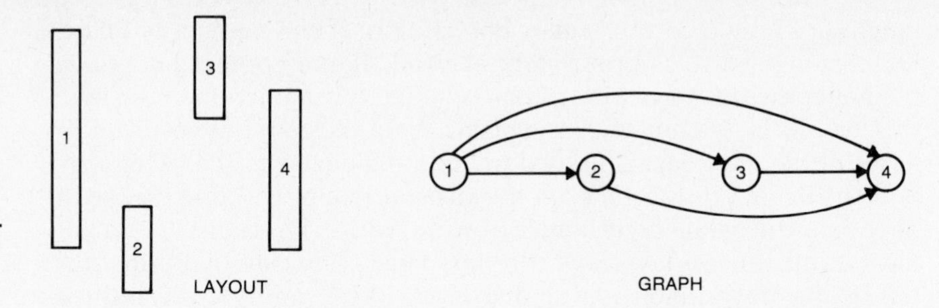

FIGURE 7.21. Graph generation for graph-based compactor

groups. Once the graph has been established, the critical path (i.e., the path with the greatest spacing requirement) through the graph can be determined. The nodes in the path can then be placed sequentially. For any given node, there may be a number of paths to it. The critical path to a given node will determine its minimum placement consistent with all design rules. Time consuming parts of a graph-based compactor include the generation of the graph itself and the selection of the minimum path through the graph. The link between virtual grid compaction and graph-based compaction is as follows. If one completes virtual grid compaction as normal, but assigns mask coordinates to *all* virtual grid coordinates, then the same result can be generated. The advantage gained by assigning one mask value to a column or a row is that pitch matching may be achieved by a simple manipulation of the design grid file. To achieve the same result with the multiple column version it becomes necessary to store more information about element interactions. This is, in essence, the graph described above. Thus this information has to be generated and stored. As systems evolve, graph-based compaction should yield the best results when combined with pitch matching additions. The virtual grid capture method can serve as the basis for both methods of compaction.

7.6.10 Mask generation

Mask generation proceeds after compaction and pitch matching have been completed. Virtual grid spacings for each mask cell are used in conjunction with the original ICDL description to generate the appropriate mask detail for each symbolic element. As we have seen, it is necessary to enclose n-transistors in a p-well or p-transistors in an n-well depending on the process. As diffusion wires, contacts, and transistors are built, a running tally of the n- and p-areas is kept. In addition, the location of all V_{DD} and V_{SS} substrate contacts is stored. At the conclusion of converting all the ICDL elements, an automatic procedure inserts the appropriate well. If for some topological reason this is impossible, an error message is generated.

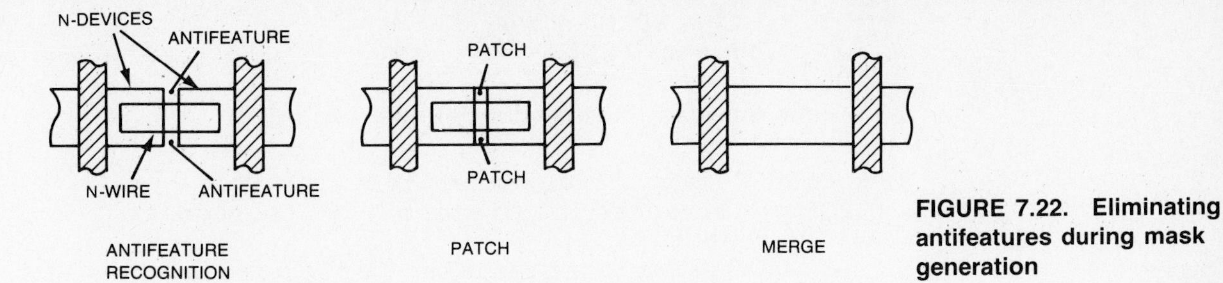

FIGURE 7.22. Eliminating antifeatures during mask generation

As the mask geometries are automatically generated, small antifeatures can result. An example of this is shown in Fig. 7.22. A small rectangle is added to "patch" the undersized feature. Although electrically these features do not affect performance, small photoresist features are likely to peel off during processing, creating problems elsewhere on the mask. To avoid this, an "antifeature" remover is passed over the final set of masks. Finally, a "rectangle merger" merges all rectangles into the largest possible polygons. Note that the mask generator has to be tightly coupled with the compactor to alleviate any confusion over symbolic mask mapping.

7.6.11 Cell verification

The first level of verification is provided by the circuit interpreter. Verification of the behavior of the cell must be carried out before performing a simulation on the cell. The circuit interpreter can translate to various forms to suit various simulators.

At the cell level in MULGA, first level simulation uses the EMU timing simulator, which is closely modeled on the MOTIS [ChGK75] timing simulator but allows arbitrary connections of transistors to facilitate simulation of dynamic logic and distributed gates (i.e., a precharged bus). The simulator gives reasonable results, which model charge redistribution and rise and fall times without incurring the time penalty of a SPICE-like simulation.

The simulator uses a simple "forward euler" method of integrating the current into a capacitive node. As this is potentially unstable, an automatic time-step control is implemented. Furthermore, voltages are clamped at V_{DD} and V_{SS}. The main loop may be summarized as follows:

```
PROCEDURE simulate(timing_period)
  WHILE ( tick  <=  timing_period )
    calculate_new_states(tick)
    output_voltages(tick)
    tick = tick + 1
```

```
        END_WHILE
      END_PROCEDURE
```

The new states are calculated as follows:

```
PROCEDURE calculate_new_states(tick)
  FOR ( all_nodes_in_circuit ) /* unordered
       search */
       integration_time_step = min_time_step
    FOR(sub_step = 0; sub_step <= min_time_step;
         sub_step += integration_time_step;
    FOR ( all_gates_connected_to_this_node )
         isum = sum_current_into_node
    END_FOR
     node_voltage = calculate_node_voltage(isum)
    IF ( node_voltage > max_allowed_step )
         subdivide_sub_step
      enter_new_voltage
    END_FOR
  END_FOR

  FOR ( all_nodes_in_circuit )
    old_node_voltage = new_node_voltage
  END_FOR
END_PROCEDURE
```

This time-step control is on a *per-node* basis. A *per-circuit* time-step control can be similarly implemented by keeping the maximum voltage change during a simulation period and repeating the iteration appropriately. The new voltage is calculated from the equation

$$new\_voltage = old\_voltage + \frac{sub\_step * isum}{node\_capacitance}.$$

Each node has the following structure:

```
        NODE_ATTRIBUTES
           old_node_voltage
           new_node_voltage
           current_into_node
           capacitance_on_node
           flags
           pointer_to_list_of_transistors
        END_ATTRIBUTES
```

Each transistor has the following structure:

```
TRANSISTOR_ATTRIBUTES
   pointer_to_next_transistor
```

```
    transistor_opcode
    transistor_scale
    pointer_to_other_(source/drain)
    _node_in_node_attribute_table
    pointer_to_gate_node_in_node_attribute_table
END_ATTRIBUTES
```

The transistor list can be modified to include compound gate structures to increase the computation rate.

It should be pointed out that this simulator deals with an idealized model of the circuit nodes sinking and sourcing current with capacitance to ground. Simulation accuracy is not as good as SPICE. It becomes time consuming to deal with large numbers of devices (>10000) as the simulator is not event-driven. This type of simulator is best suited for local workstation use to verify custom cells in a highly interactive environment. This type of simulator should be used in conjunction with higher level logic/switch simulators and more accurate circuit analysis programs.

7.6.12 Module assembly

The task of converting a virtual grid symbolic layout to a mask layout is more than just compaction. It consists of a number of steps, all of which must be automated to alleviate the possibility of human error. This is the task of the module assembler [AcWe83].

Typical steps in module assembly are as follows. Cells are first classified as leaf cells or composition cells. Cells with primitive elements are classified as leaf cells. Following this, the cells are further classified by noting the cells on the boundaries of each cell. This information is used during compaction. Nonleaf cells with primitive components in them are smashed (reduced to a flat nonhierarchical description) to yield only primitive components. The ports on each cell to be compacted are recognized and matched against neighboring cells. These ports define pitch-match points. Dummy cells are composed of the target compaction cell and compaction components included from abutting cells. Each particular cell is then compacted. Note that most of the time each prototypical cell is compacted only once. However, it may generate a number of mask versions through different pitch-match constraints. Using the pitch-match points, the module assembler then attempts to reconcile all the pitch matching constraints. A heuristic is used for this process. Following this, the mask level representation is generated for the pitch-matched cells and a hierarchical mask description matching the symbolic hierarchical description.

7.7 Future directions

7.7.1 Flexi-cells

With the symbolic approach described in the preceding sections, it becomes possible to adjust cells at the circuit level with automatic geometry generation and chip composition. The basic approach is to design the complete chip at the symbolic level. Critical path approaches may then be applied to optimally sized transistors according to the methods discussed in Chapter 4 and Chapter 5. This initial sizing uses some estimate of virtual grid spacing. The layout is compacted and pitch matched and the procedure may be repeated. Note that optimizations such as yield enhancement may be built into the compaction process.

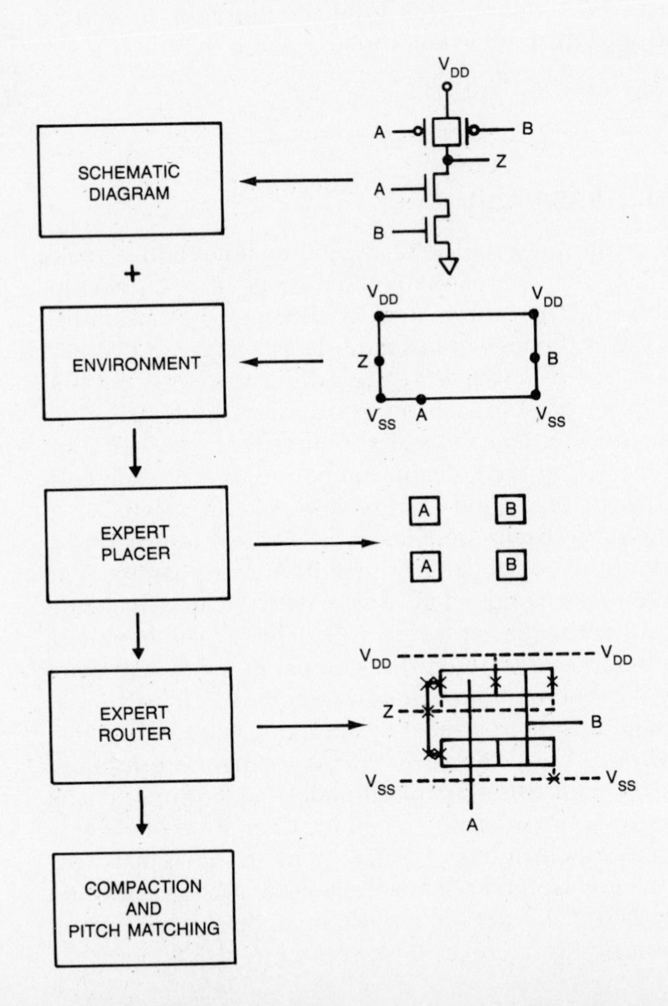

Figure 7.23. Expert system cell generator

Automatic geometry generation yields a powerful tool. Quite often topologies may have to be changed — a task for an *expert* system.

7.7.2 Expert systems

Recently, authors have attacked the automatic generation of layout and higher level design tasks by the use of rule-based expert systems [KiMc83] [KoTh83] [KoWe84] [Pall82]. These techniques provide an alternative programming approach in situations where algorithms are ill-defined.

Fig. 7.23 shows an outline of one such system for converting schematic diagrams to symbolic layouts like those included in this book [KoWe84]. The designer enters a schematic that describes transistor interconnections. In addition, an "environment" is defined by the designer (this may be generated by a floor plan expert). An expert placer then applies rules similar to those in Chapter 5 to a virtual grid array of transistors. The output of the expert placer is passed to a pre-router, which routes in whatever layer it can to complete a structurally correct layout. A rule-based router then applies a set of routing rules to the pre-routed cell, thereby improving the cell. The completed cell is passed to compaction and pitch matching. Although in its infancy, this program can deal with simple cells and has produced layouts approximately 10 percent larger than those designed by hand. This definitely seems an area for further fruitful research.

7.8 Summary

In this chapter we have zeroed in on symbolic design methods for CMOS. One symbolic design system was dealt with, at an excruciating level of detail, to hopefully show that the implementation of such a system is not difficult. The benefits of such a system in CMOS design are well worth the effort. Designs no longer become obsolete when the process design rules change nonlinearly. A number of different CMOS families can be targeted from the one symbolic source description. Even changes such as inclusion of second metal may be dealt with in a systematic manner. Finally, for those who have to deal at low levels, the design task is considerably simplified.

The chapter concluded with some ideas on where symbolic design methods might lead. Expert systems, combined with procedural design techniques, will surely be a mainstay of future design systems.

7.9 Exercises

7.1 Using the lambda-based rules in Chapter 3, design a fixed grid symbolic character set for a p-well CMOS process with one layer of metal (allow arbitrary layout style). Characters should include metal and polysilicon wires, contacts, and transistors (see Fig. 7.1). What design rules must be obeyed? What do you estimate the loss in area?

7.2 Outline a connectivity checking program to be used with the fixed grid system designed in Exercise 7.1.

CMOS SUBSYSTEM DESIGN

8.1 Introduction

In previous chapters the groundwork for CMOS circuit design was developed. In this chapter, we examine various design options available to the CMOS designer when designing at the subsystem level. A large design space is available where one may design for circuit simplicity, time of design, low power, or high speed, or combinations thereof.

We first examine adders as an example of the wide range of circuit options that are available to the designer when dealing at the transistor level. Of course, in time, we hope that advanced design systems can encapsulate such knowledge and automatically generate the optimum subcircuit for a given system design situation. Following adders, binary counters and multipliers are surveyed. Memory design is then treated. Finally, PLA-type structure styles are summarized.

8.2 Adders and related functions

Adders form important components in many systems. The truth table for a binary full adder is shown below in Table 8.1, along with some functions that will be of use during the discussion of adders.

A and *B* are the adder inputs, *C* the carry input, *SUM* is the sum output, and *CARRY* is the carry output. The generate signal *G* (*A.B*) occurs when a carry out (CARRY) is internally generated within the adder. When the propagate signal, P (A + B), is true, the carry in signal (C) is passed to the carry output (CARRY) when C is true. (In some adders $A \oplus B$ is used as the P term as it may be reused to generate the sum term.)

TABLE 8.1. Adder truth table

C	A	B	A.B(G)	A + B(P)	$A \oplus B$	SUM	CARRY
0	0	0	0	0	0	0	0
0	0	1	0	1	1	1	0
0	1	0	0	1	1	1	0
0	1	1	1	1	0	0	1
1	0	0	0	0	0	1	0
1	0	1	0	1	1	0	1
1	1	0	0	1	1	0	1
1	1	1	1	1	0	1	1

8.2.1 Combinational adder

Probably the simplest approach to designing an adder is to implement gates to yield the required majority logic functions. From the truth table these are

$$SUM = ABC + A\overline{B}\overline{C} + \overline{A}\overline{B}C + \overline{A}B\overline{C} \quad (8.1)$$

$$\begin{aligned} CARRY &= AB + AC + BC \\ &= AB + C(A + B). \end{aligned} \quad (8.2)$$

The gate schematic is shown in Fig. 8.1a on page 312, while the transistor schematic is shown in Fig. 8.1b. As the carry out signal ($\overline{CARRY}$) is used in the generation of SUM, SUM will be delayed with respect to CARRY. This is consistent with the use of such a circuit in an n-bit parallel adder. Here, the CARRY signal is allowed to "ripple" through the stages, as shown in Fig. 8.2a on page 313. In this case, the carry delay has to be minimized, as the delay associated with the adder is $T_a = n.T_c$, where T_a is the total add time, n is the number of stages, and T_c is the delay of one carry stage. To optimize the carry delay, the inverter at the output of the carry gate can be omitted. In this case, every other stage operates on complement data, as shown in Fig. 8.2b. This results in a significant decrease in carry delay. Any delay in inverting the adder inputs or sum outputs is finessed out of the critical path.

Rather than construct a ripple carry adder, a serial adder, shown in Fig. 8.3 on page 313, may be constructed. At time t, the SUM is calculated and the CARRY stored in the flip-flop. At time $t + 1$, the sum uses CARRY(t) to calculate a new SUM.

$$\begin{aligned} CARRY[t + 1] &= A[t + 1].B[t + 1] \\ &\quad + C[t].(A[t + 1] + B[t + 1]) \\ SUM[t + 1] &= \overline{CARRY}[t + 1].(A[t + 1] + B[t + 1] \\ &\quad + C[t]) + A[t + 1].B[t + 1].C[t] \end{aligned} \quad (8.3)$$

In this application, equal SUM and CARRY delays are advantageous, as this determines the fastest clock frequency at which the adder can operate.

Considering the combinational adder schematic in more detail, it may be seen that by optimizing the carry gate, we can reduce the ripple carry delay. This is of special significance for a parallel adder. The transistor schematic for the carry stage is shown redrawn in Fig. 8.4 on page 314. This more clearly shows the effect of the P and G terms outlined previously. Note that the p-chain is not the exact dual of the n-chain. It is left to the reader to verify the equivalence. The SUM stage is also presented in a similar form. We might start the physical design by using unit-sized n-transistors and p-

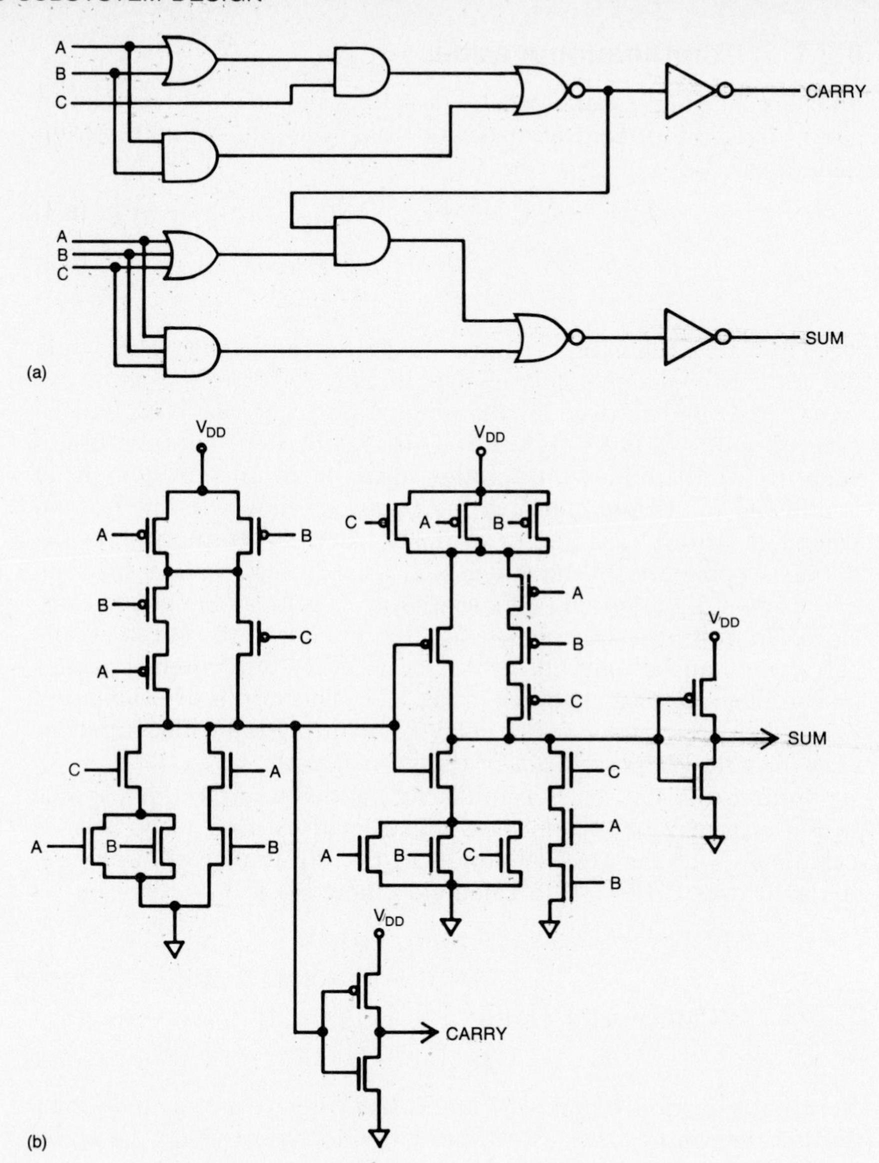

FIGURE 8.1. Combinational adder schematic

transistors. Using the styles of layout presented so far, two possible layouts for the combinational adder are depicted in Fig. 8.5 on page 315. The choice of aspect ratio would depend very much on the environment. For instance, in a data path, where the height of the data path had to be minimized, the layout in Fig. 8.5a (minimum transistor stacking) would be preferred. The layout in Fig. 8.5b illustrates a "minimum width" design. Fig. 8.5c shows a design

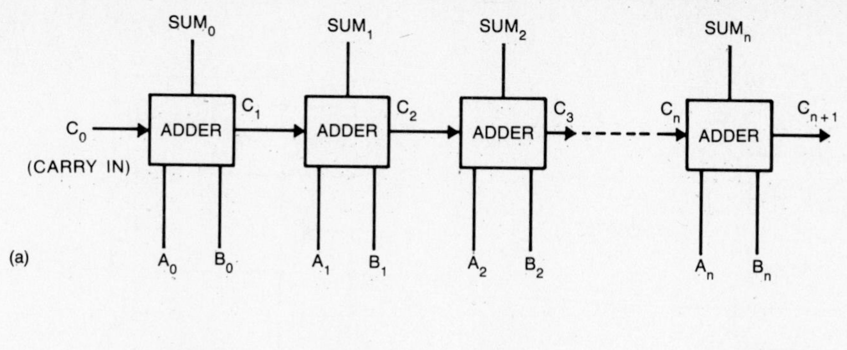

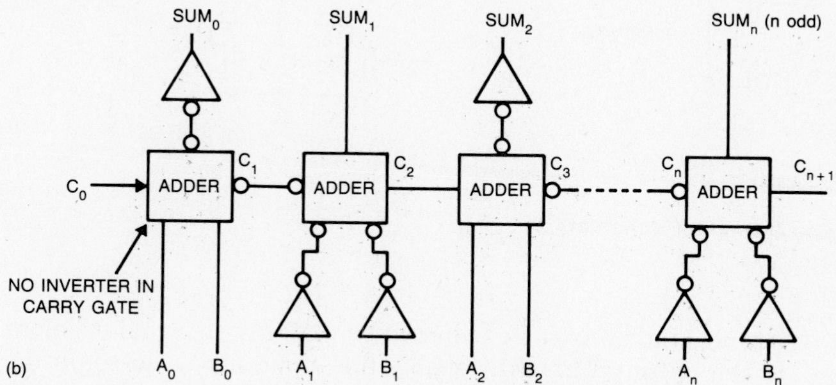

FIGURE 8.2. n-bit ripple carry adder

part way between Fig. 8.5a and Fig. 8.5b. Once the basic layout has been determined, some optimization of transistor sizing may take place. This is only necessary if after simulation the adder is found to be lacking in speed. Remember, the static CMOS gate, if implemented correctly, will always function correctly. The following optimizations may be made to the combinational adder (Fig. 8.4):

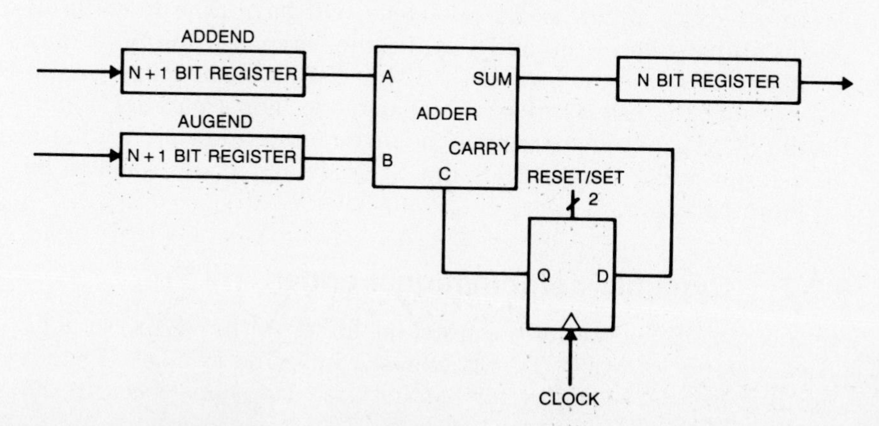

FIGURE 8.3. Serial adder schematic

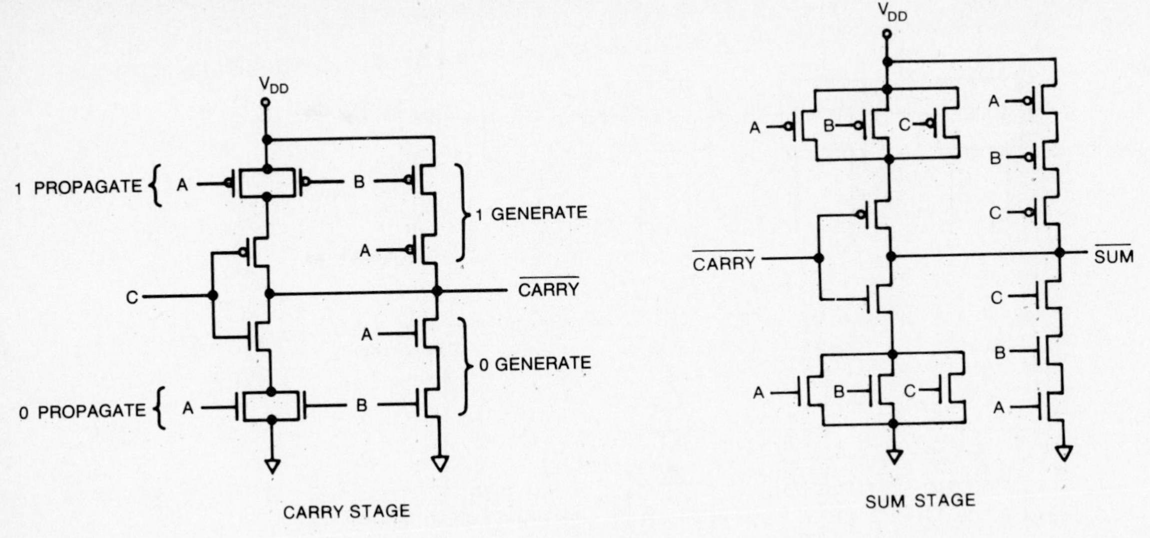

FIGURE 8.4. Carry and sum gate transistor schematics

1 Arrange the transistors switched by the carry in signal (*C*) close to the output. This will enable the input signals to settle the gate such that the *C* transistors are least influenced by body effect.

2 Make all transistors in the sum gate whose gate signals are connected to $\overline{CARRY}$ minimum size. This minimizes the capacitive load on this signal. Keep routing on this signal to a minimum and minimize the use of diffusion as a routing layer.

3 Sizing of series transistors can be determined by simulation. It may or may not pay to increase the size of the series n-transistors and p-transistors. For instance, it may pay to increase the size of the transistors connected to *A* and *B* in the carry gate in a ripple carry adder, as these signals will have time to settle in the upper bits of the adder while the carry is rippling. It may pay to increase the size of the C transistors in the carry gate to override the effects of stray capacitance. For a parallel adder, the SUM gate transistors may be made minimum size, while for a serial adder the CARRY and SUM delays would have to be more balanced.

8.2.2 Dynamic combinational adder

An N-P CMOS version of the previous adder is shown in Fig. 8.6a on page 316 [GoDM83]. This has been configured as a serial adder and thus has a 1-bit delay for feeding the carry signal back to the

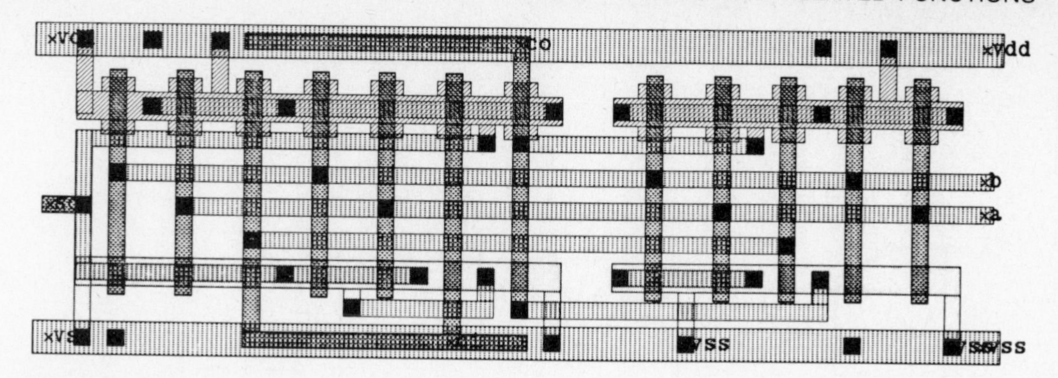

(a)

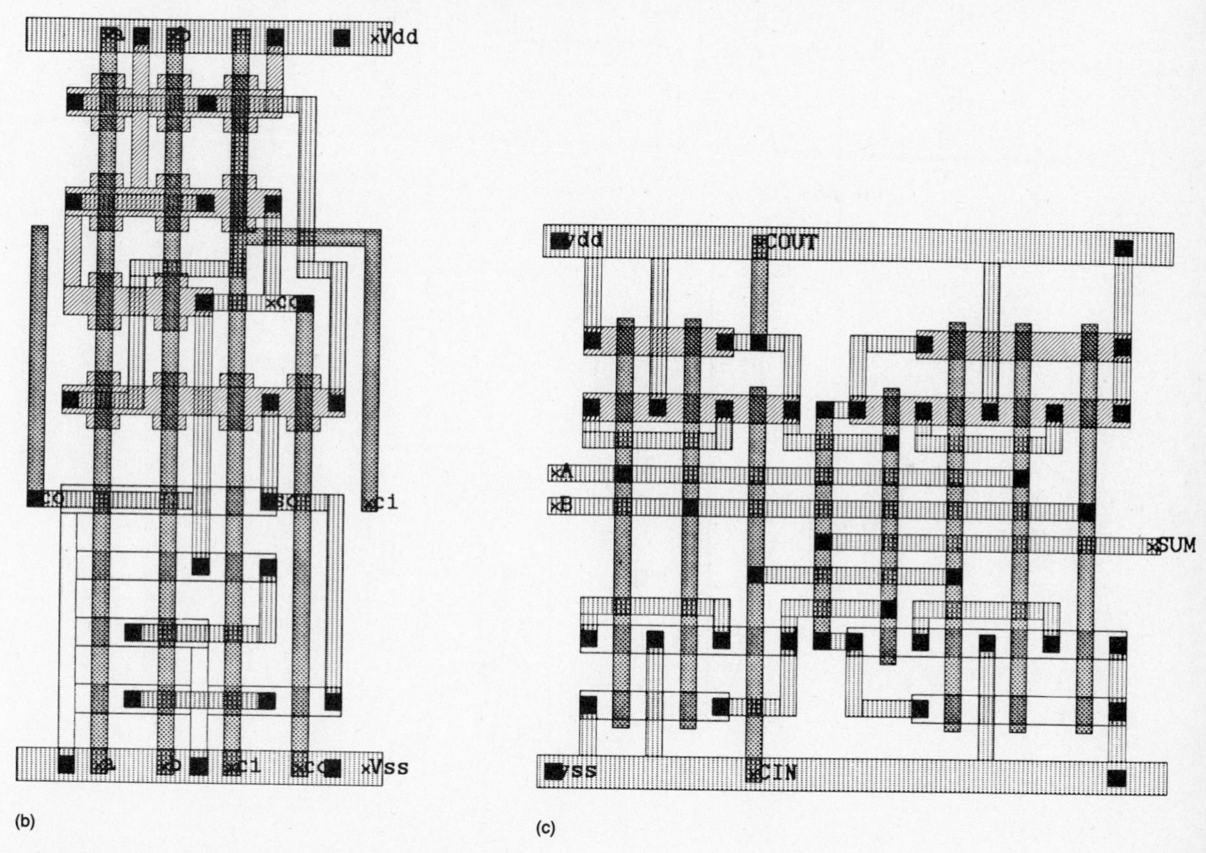

(b)

(c)

FIGURE 8.5. Combinational adder layouts

adder, and a carry reset and set signal for initialization. This allows the adder to act as a subtractor by setting carry on the 0th cycle and logically inverting the subtrahend. A possible layout is shown in Fig. 8.6b on page 317. One possible problem that might arise in

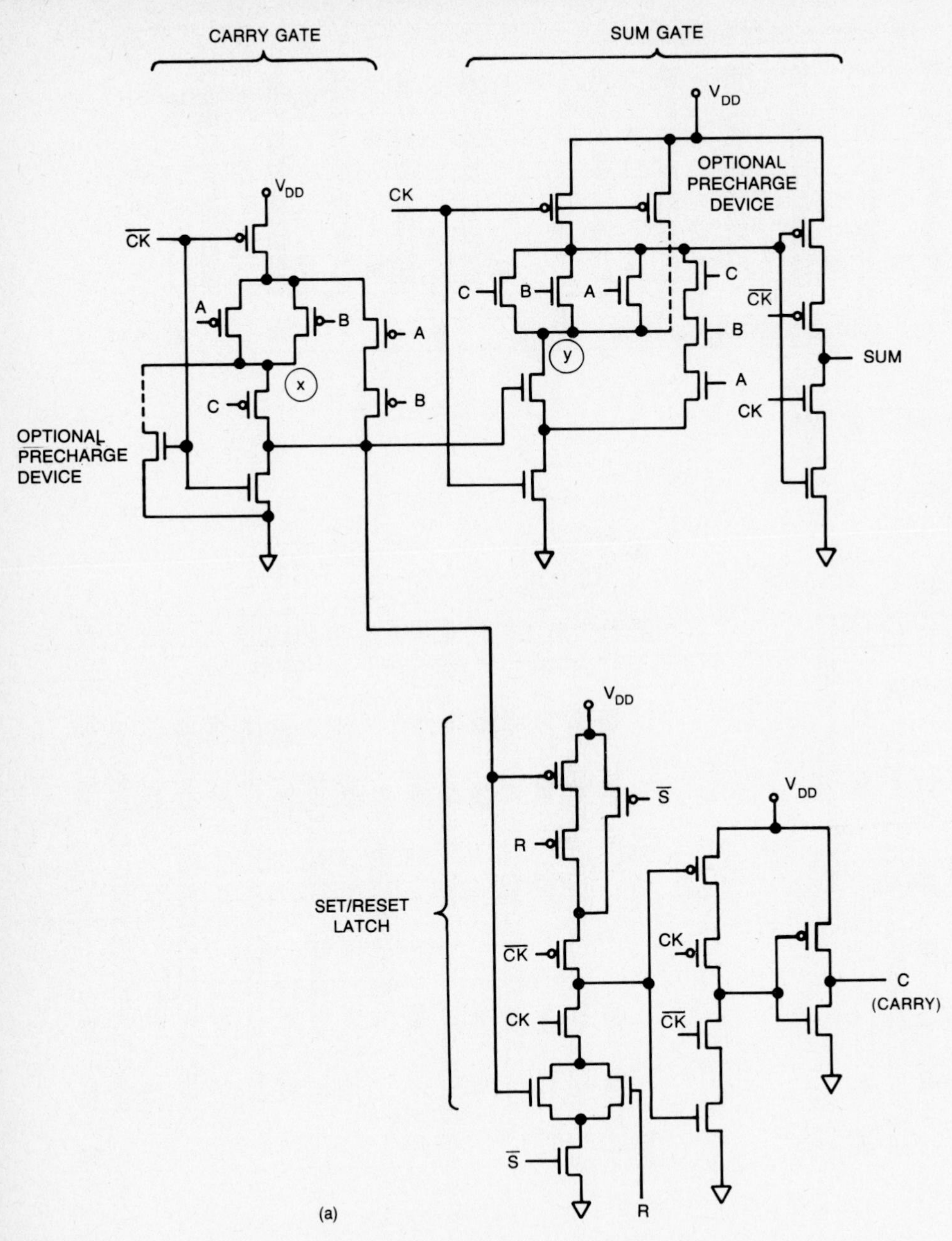

FIGURE 8.6. Dynamic serial adder schematic and layout

this gate is charge redistribution onto uncharged nodes. For instance, node X in the carry gate may have to be independently precharged. A similar requirement might be necessary for node Y in the SUM gate.

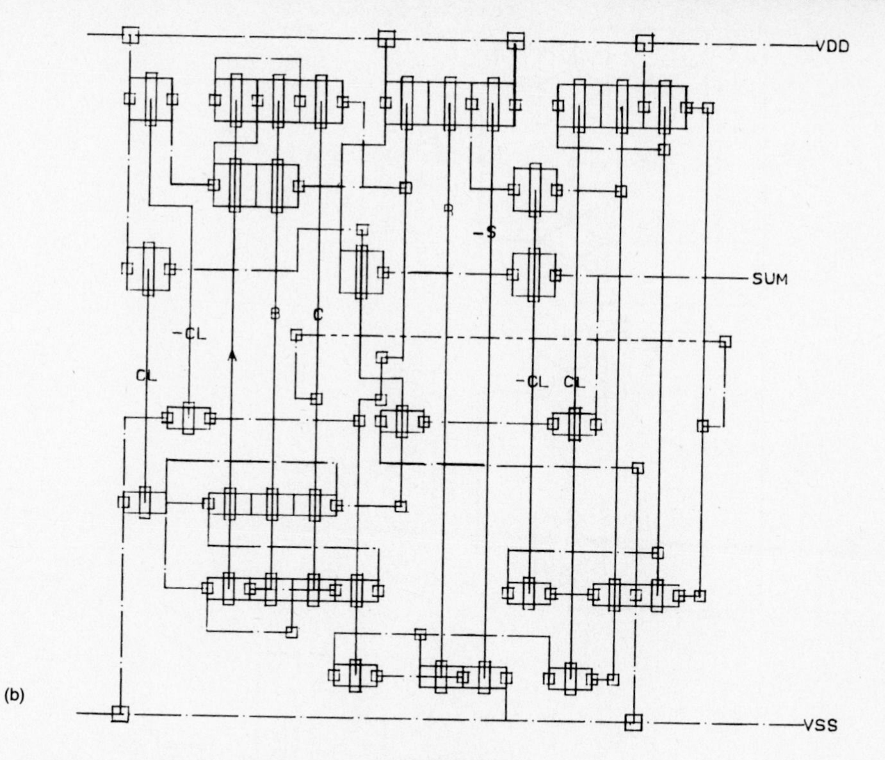

FIGURE 8.6. (continued)
a) single metal version
b) two metal version

8.2.3 Transmission gate adder

A rather different implementation of an adder uses a novel exclusive-or (XOR) gate. The schematic for this XOR gate is shown in Fig. 8.7. As a point to note, switch level simulators have problems with this gate. The operation of the gate is explained as follows:

1 When signal A is high, $\overline{A}$ is low. Transistor pair 1 and 2 thus act as an inverter, with $\overline{B}$ appearing at the output. The transmission gate formed by transistor pair 3 and 4 is open.

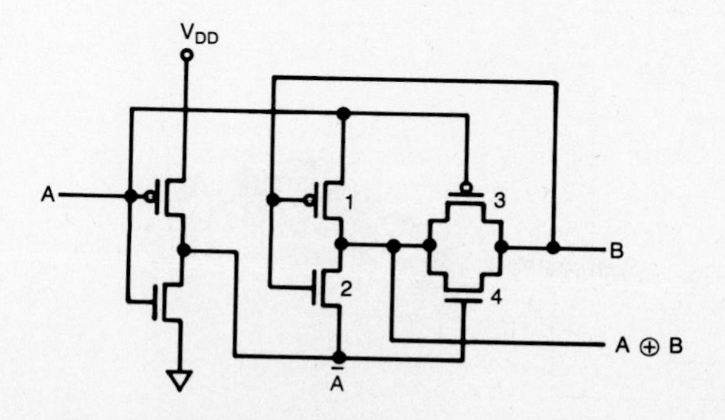

FIGURE 8.7. Transmission gate exclusive-or gate

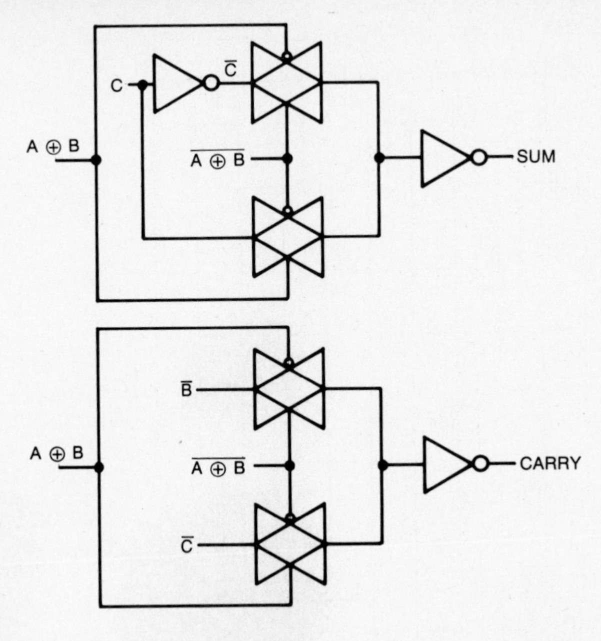

FIGURE 8.8. Transmission gate adder

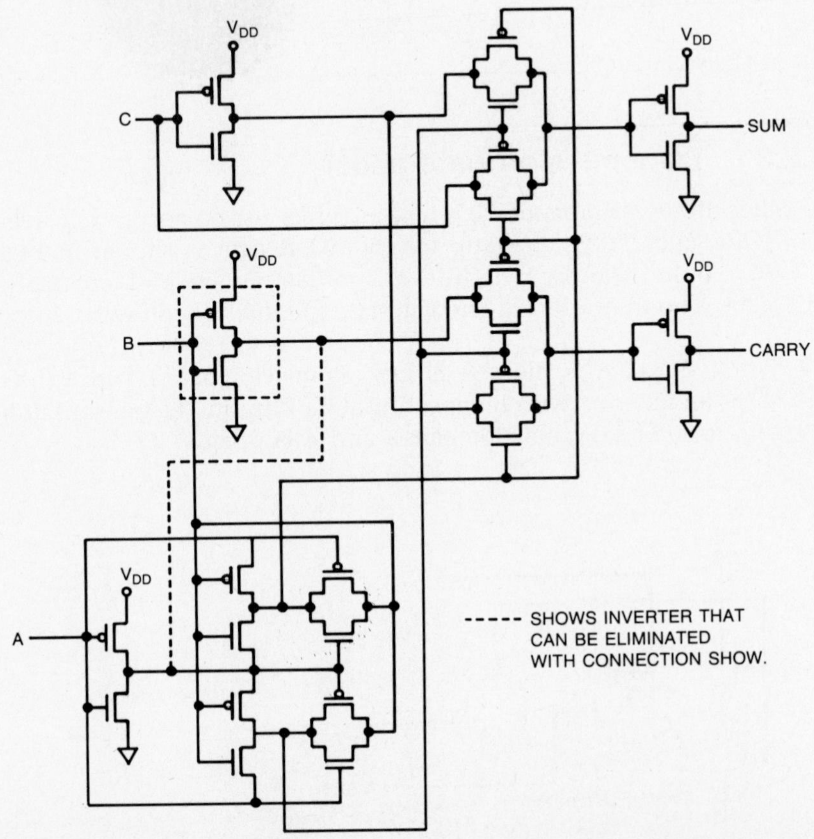

FIGURE 8.9. Complete TG adder

2 When signal A is low, $\overline{A}$ is high. The transmission gate (3 + 4) is now closed, passing B to the output. The inverter pair (1 + 2) are disabled.

Thus this transistor configuration forms a 6 (or 4) transistor XOR gate. By reversing the connections of A and $\overline{A}$, an exclusive-nor (XNOR) gate is constructed.

By using four transmission gates, four inverters, and two XOR gates, an adder may be constructed according to Fig. 8.8 [SuOA73]. From the truth table for the adder, it may be seen that when $A \oplus B$ is true, $SUM = \overline{C}$. When $A \oplus B$ is false, $SUM = C$. Similarly, when $A \oplus B$ is true, $CARRY = C$. When $A \oplus B$ is false, $CARRY = A$ (or B). This adder has 24 transistors, the same as the combinational adder, but has the advantage of having equal SUM and $CARRY$ delay times. In addition, the SUM and $CARRY$ signals are non-inverted. The completed schematic is shown in Fig. 8.9. Fig. 8.10 outlines a representative layout for this adder. The layout style

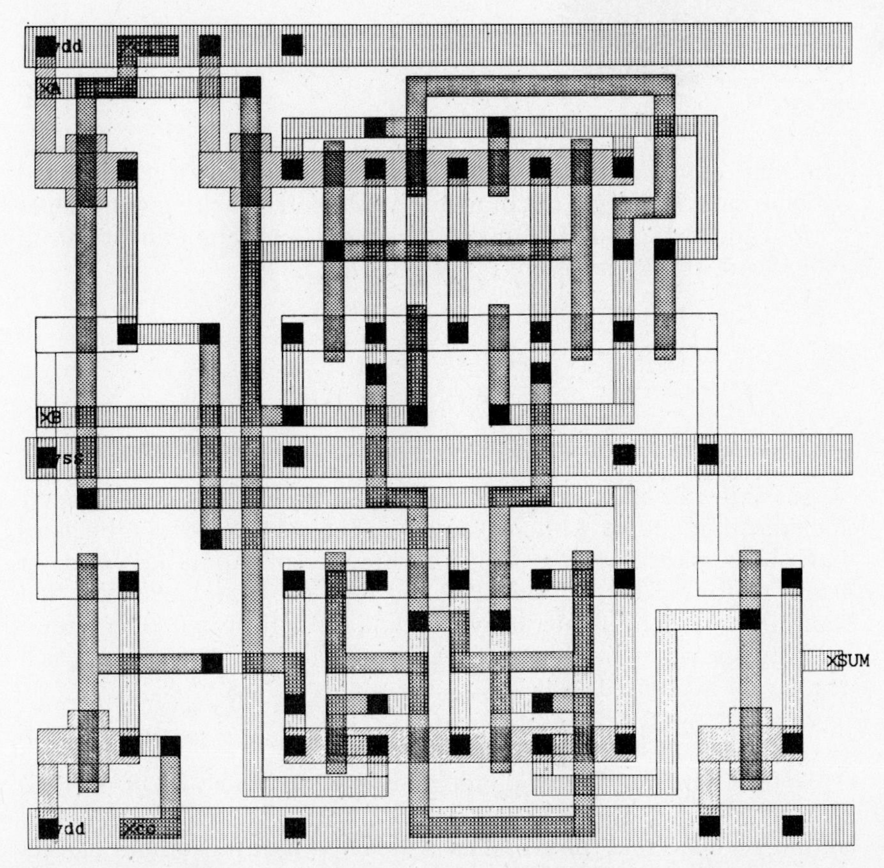

FIGURE 8.10. TG adder layout

is considerably different from the complementary gate as high usage is made of transmission gates. The generic two-way multiplexer structure mentioned in Chapter 5 may be seen in this implementation.

8.2.4 Carry lookahead adders

The linear growth of adder carry delay with the size of the input word may be improved by calculating the carries to each stage in parallel. The carry of the *ith* stage, C_i, may be expressed as

$$C_i = G_i + P_i \, . \, C_{i-1}, \tag{8.4}$$

where

$$G_i = A_i \, . \, B_i \qquad \textit{generate signal} \tag{8.5}$$

$$P_i = A_i + B_i \qquad \textit{propagate signal.} \tag{8.6}$$

Expanding this yields

$$C_i = G_i + P_iG_{i-1} + P_iP_{i-1}G_{i-2} + \cdots + P_i \cdots P_1C_0. \tag{8.7}$$

The sum S_i is generated by

$$S_i = C_{i-1} \oplus A_i \oplus B_i \quad (\text{if } P_i = A_i \oplus B_i). \tag{8.8}$$

The size of the gates needed to implement this carry lookahead scheme can clearly get out of hand. As a result, the number of stages of lookahead is usually limited to about four. For four stages of lookahead, the appropriate terms are

$$C_1 = G_1 + P_1C_0$$

$$C_2 = G_2 + P_2G_1 + P_2P_1C_0$$

$$C_3 = G_3 + P_3G_2 + P_3P_2G_1 + P_3P_2P_1C_0$$

$$C_4 = G_4 + P_4G_3 + P_4P_3G_2 + P_4P_3P_2G_1 + P_4P_3P_2P_1C_0.$$

A possible implementation of the carry gate for this kind of carry lookahead adder for 4 bits is shown in Fig. 8.11. Note that the gates have been partitioned to keep the number of inputs less than or equal to four. This is typical of the type of carry lookahead that would be used in a gate array or standard cell design. The circuit and layout are quite irregular. Taking the term of C_4, we note that it may be expressed as

$$C_4 = G_4 + P_4 . (G_3 + P_3 .(G_2 + P_2 .(G_1 + P_1 .C_0))). \tag{8.9}$$

This function may be implemented as a domino CMOS (nMOS) gate, as shown in Fig. 8.12 on page 322. Carry $C_1 - C_3$ are generated similarly. Note that the worst-case delay path in this circuit has six

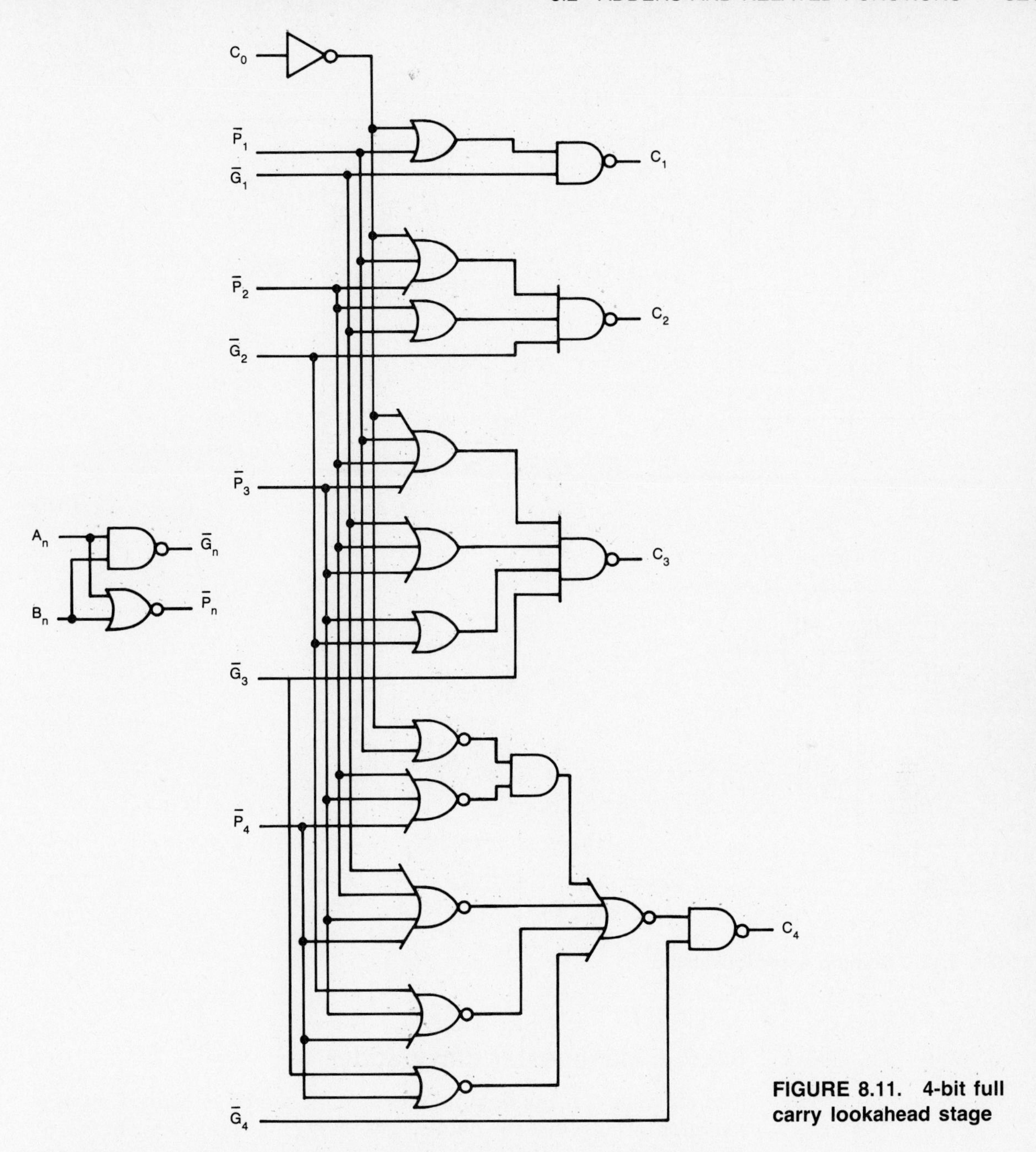

FIGURE 8.11. 4-bit full carry lookahead stage

n-transistors in series. A static version of the C_4 gate is shown in Fig. 8.13a on page 323. A symbolic layout of a gate to implement this function is shown in Fig. 8.13b. The circuit has been rearranged to allow the simple layout shown [UeVC81].

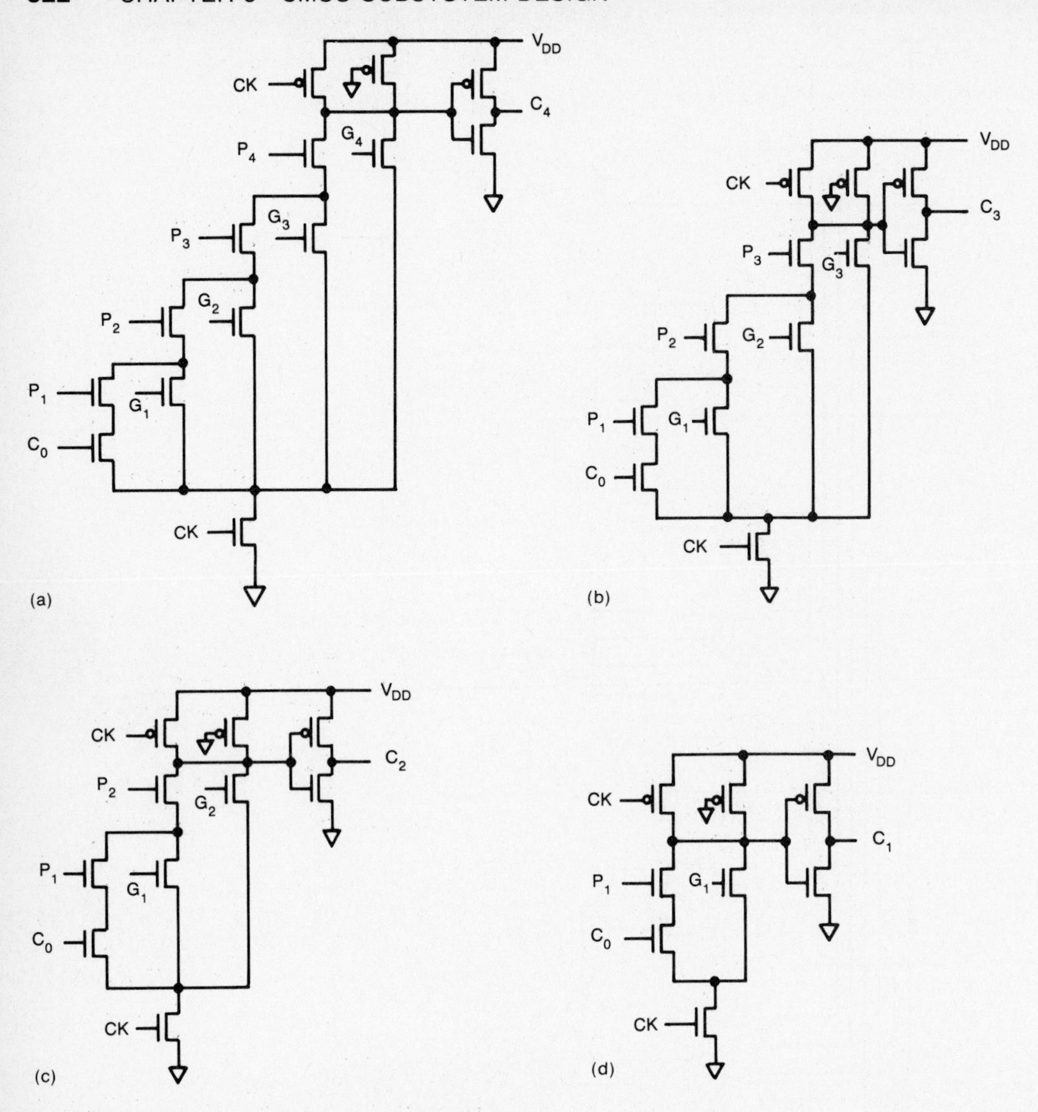

FIGURE 8.12. Domino carry lookahead

8.2.5 Manchester carry adder

The efficiency of the domino carry chain can be enhanced by precharging at appropriate points. The elemental circuit is shown in Fig. 8.14a on page 324. Operation proceeds as follows. When CLOCK is low, the output node is precharged by the p pull-up transistor. When CLOCK goes high, the n pull-down transistor turns on. If carry generate (A.B) is true, then the output node discharges. If carry

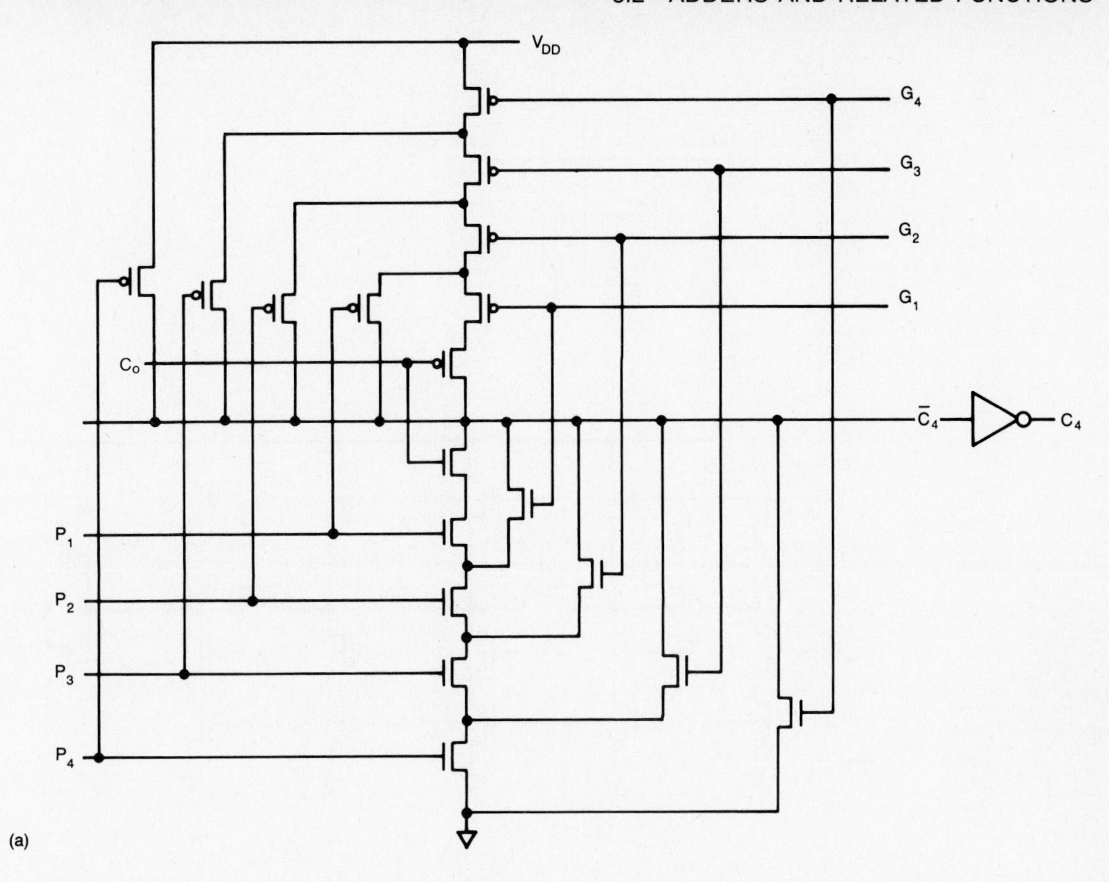

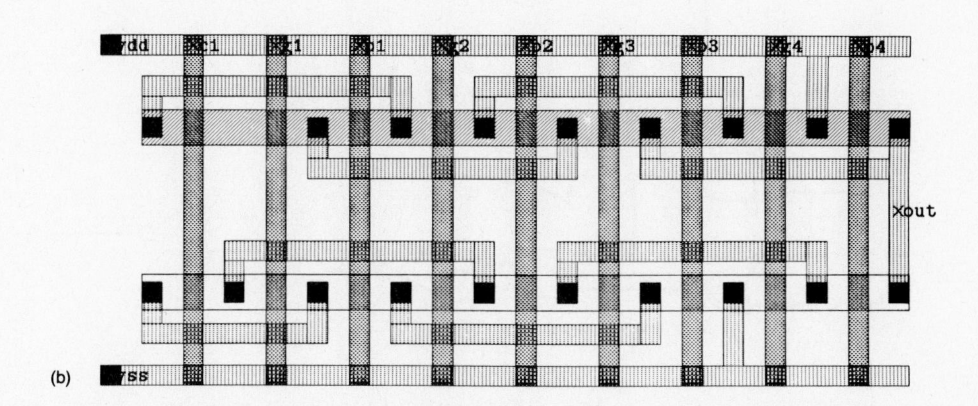

FIGURE 8.13. Static carry lookahead gate

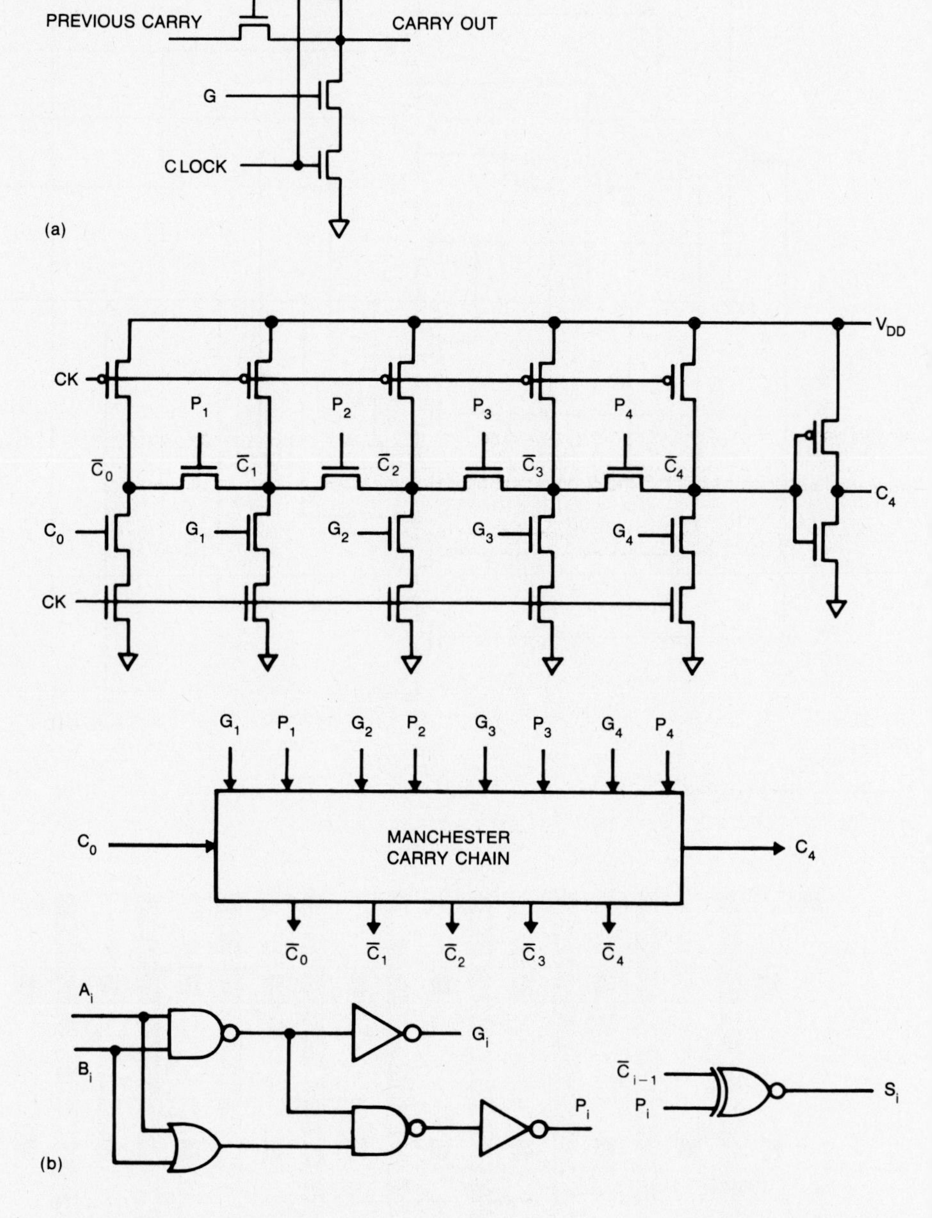

FIGURE 8.14. Manchester carry chain

propagate $(A + B)$ is true, then a previous carry may be coupled to the output node, conditionally discharging it. Note that in this circuit $\overline{CARRY}$ is actually propagated.

We can construct a 4-bit adder by cascading four such stages and constructing the circuitry to supply the appropriate signals.

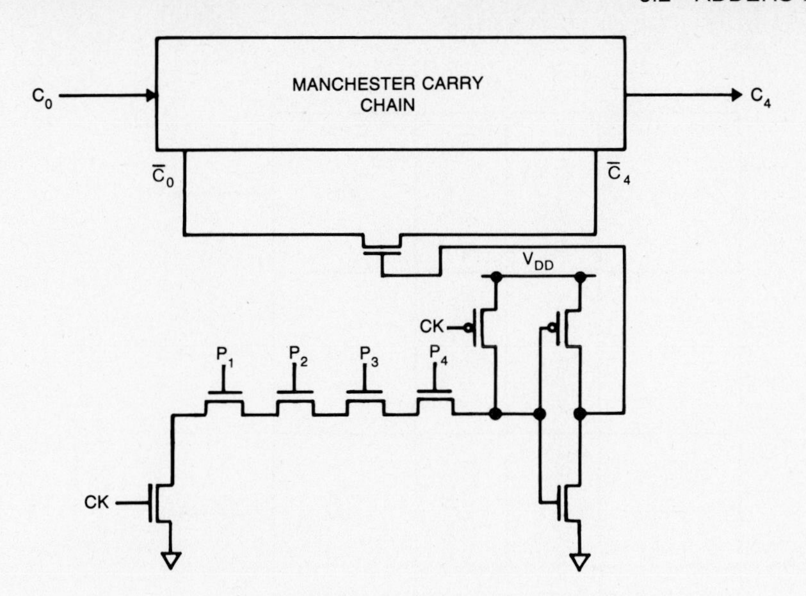

FIGURE 8.15. Manchester lookahead circuitry

This is commonly called a Manchester carry adder. Thus a 4-bit adder would be constructed as shown in Fig. 8.14b. Notice the similarity to the domino carry circuit. However, we no longer need the intermediate carry gates, as the carry values are available in a distributed fashion in this arrangement. We have chosen a 4-bit adder to reduce the number of series propagate transistors, which reduces the influence of body effect. Note that if all propagate signals are true, and C is high, six series n-transistors pull the output node low. This worst-case propagation time can be improved by bypassing the four stages if all carry propagate signals are true [PBHK82]. The additional circuitry needed to achieve this is shown in Fig. 8.15. It consists of a dynamic AND gate, which turns on a carry bypass signal if all carry propagates are true. Note that although the circuits have similar circuitry the node capacitance at intermediate nodes in the lookahead gate is approximately 1/2 that of the Manchester chain. Thus this arrangement should improve the overall speed of the adder. The optimum number of cascaded stages may be calculated for a given technology by simulation.

Fig. 8.16 demonstrates a layout plan for this adder. Two parallel carry chains are used, one to propagate the carry and the other to provide local carries for SUM generation. The latter signal is more heavily loaded and thus would slow the carry chain if this was used as the sole carry chain. A horizontal 1-bit strip consists of the carry propagate and generate block, the two carry chains, and the sum block. The end cells in the carry chain differ slightly to include the gates needed.

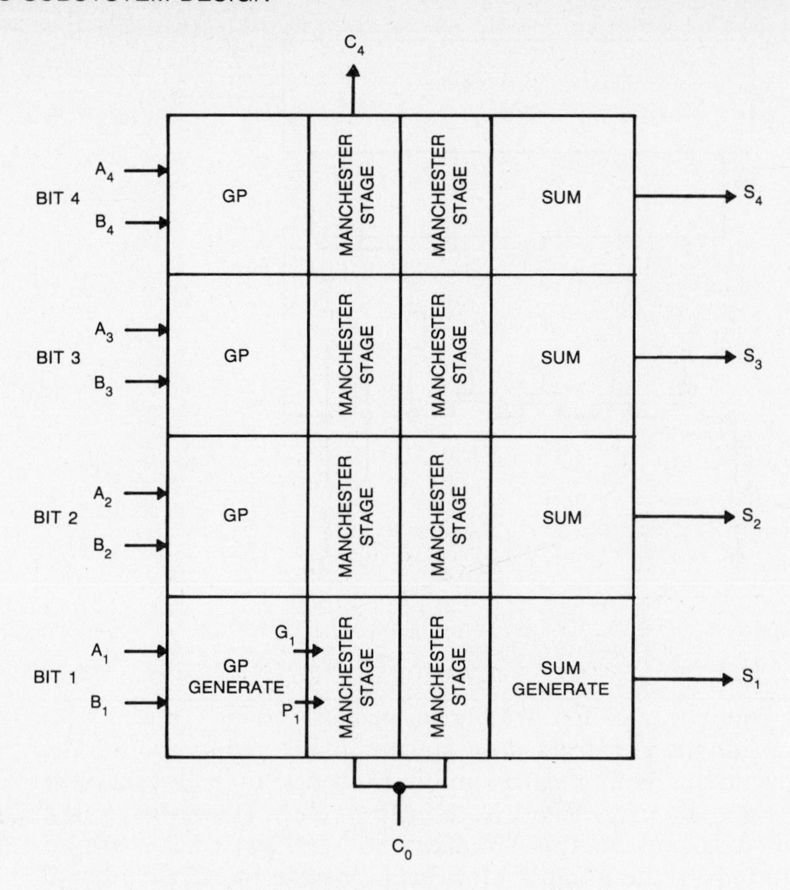

FIGURE 8.16. Manchester adder floor plan

8.2.6 Binary lookahead carry adder

Reviewing the equations for the binary adder we have

$$C_i = G_i + P_iC_{i-1}$$

$$P_i = A_i + B_i \text{ or } A_i \oplus B_i$$

$$G_i = A_i \,.\, B_i$$

$$S_i = C_{i-1} \oplus P_i \quad (\text{if } P_i = A_i \oplus B_i).$$

Both G_i and P_i can be determined in constant time, so C_i is the only time critical term that needs to be calculated. We can define a new operator o, which has the following function:

$$(g, p) \text{ o } (g', p') = (g + (p.g'), p.p'), \tag{8.10}$$

where g, p, g′, p′ are boolean variables. It can be shown that this new operator is associative [BrKu80], and the carry signals can be determined by

$$C_i = G_i,$$

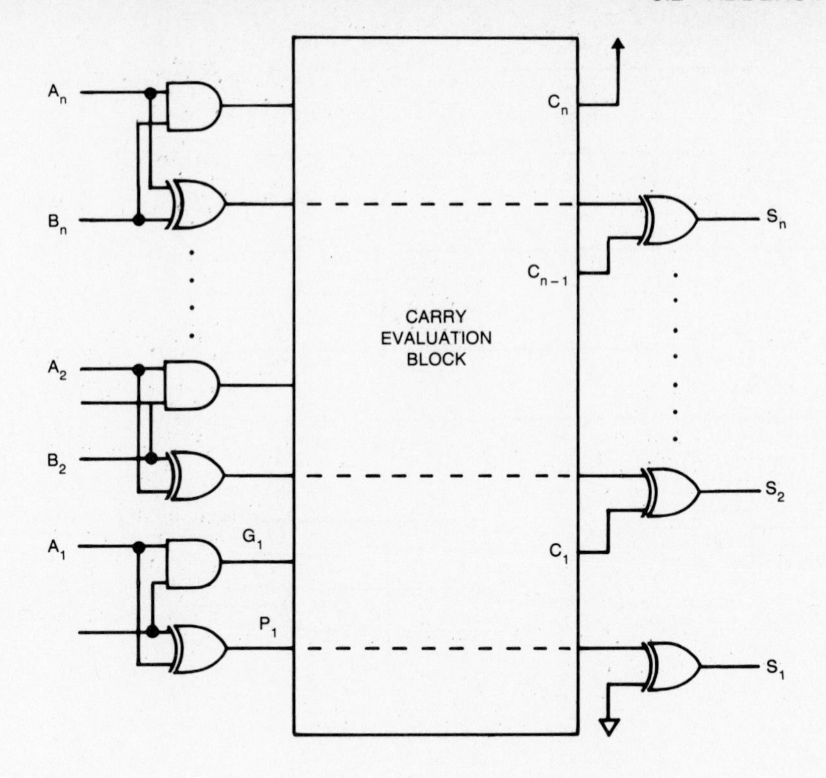

FIGURE 8.17. Carry look-ahead adder

where

$$(G_i, P_i) = \begin{cases} (g_1, p_1) & \text{if } i = 1 \\ (g_i, p_i) \cdots o \cdots (G_{i-1}, P_{i-1}) & \text{if } 2 \leq i \leq n \\ = (g_i, p_i)\ o\ (g_{i-1}, p_{i-1}) \cdots o \cdots (g_1, p_1). & \end{cases} \quad \textbf{(8.11)}$$

The associative property of the o operator allows the processing elements to be embedded in a binary tree structure of depth O(log n). A generalized carry lookahead adder is shown in Fig. 8.17. It is composed of a *G* and *P* term generator, the carry block, and a sum block.

The carry block is shown in more detail in Fig. 8.18 for a 4-bit adder. Note that the lookahead structure is implemented as a binary tree followed by an inverse binary tree. The carry propagation time in this structure is proportional to $\log_2$ of the size of the adder. A suitable floor plan for a 4-bit version of this adder is shown in Fig. 8.19, with the schematics shown in Fig. 8.20. If we use complementary gates, then the *o* function can be implemented as $(g, p)\ o\ (G, P) = (\overline{g + (p.G)}, \overline{p.P})$. Alternate columns use the inverse function, $(g, p)\ o\ (G, P) = (\overline{g.(p + G)}, \overline{p + P})$. In addition, signals are buffered in the locations where the *o* processors are absent.

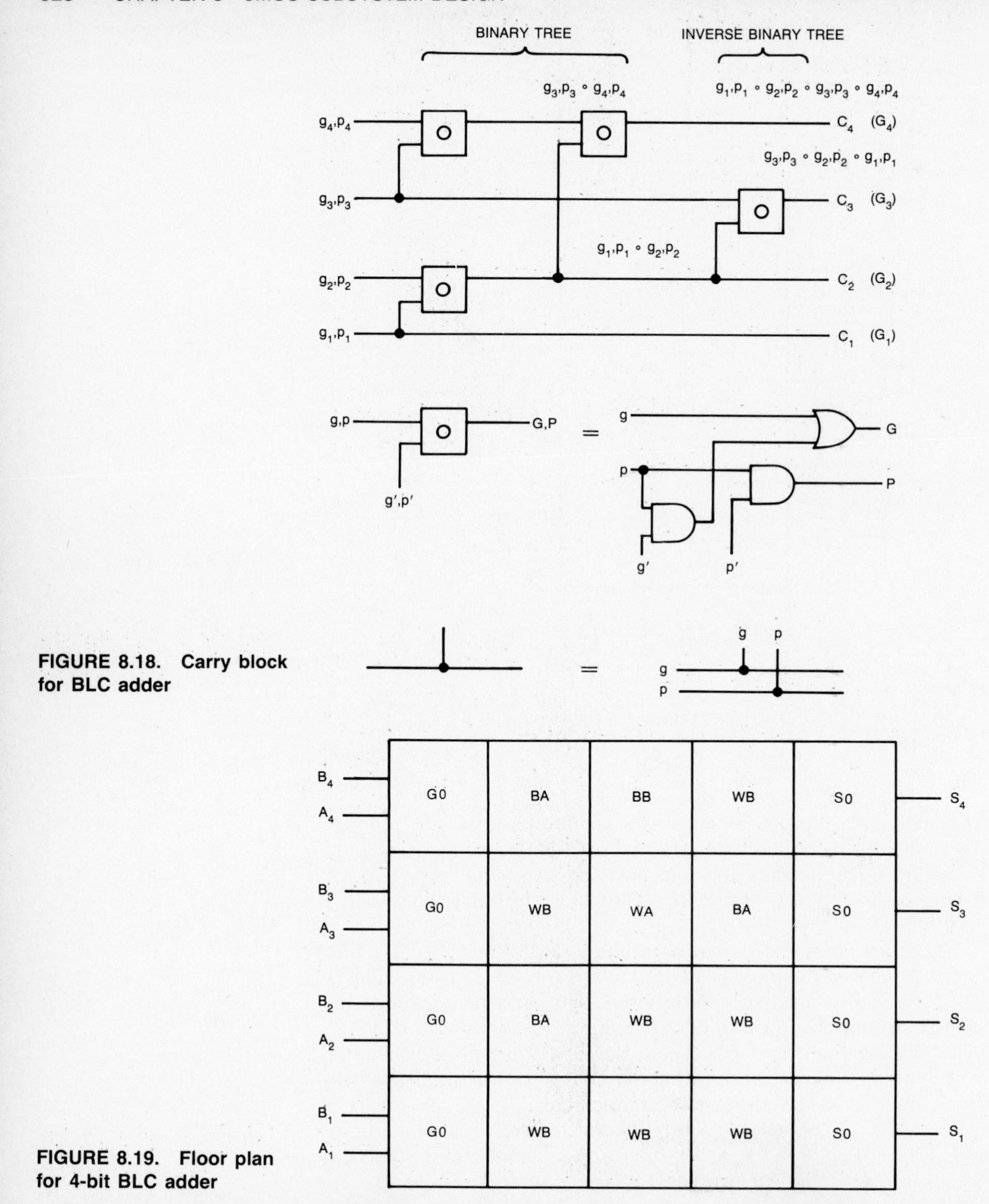

FIGURE 8.18. Carry block for BLC adder

FIGURE 8.19. Floor plan for 4-bit BLC adder

'o' OPERATOR

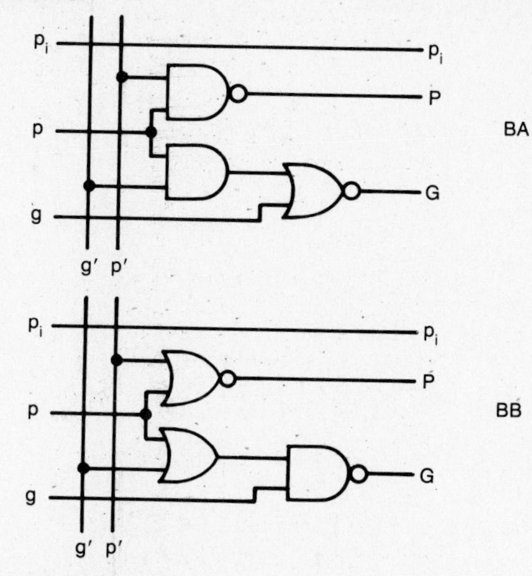

g_i p_i GENERATION-G0

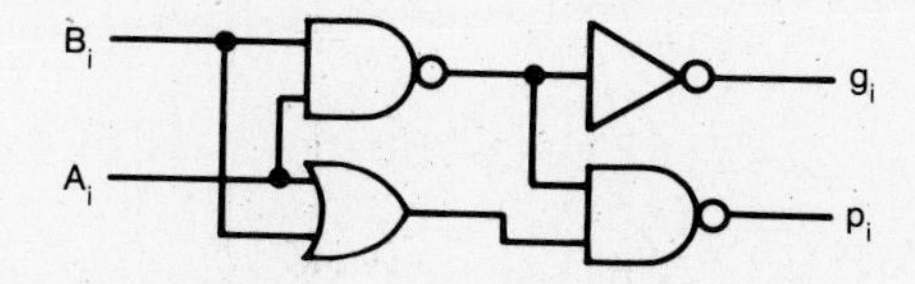

SUM GENERATION-S0

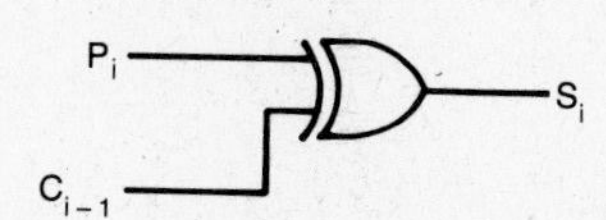

ROUTING CIRCUITS

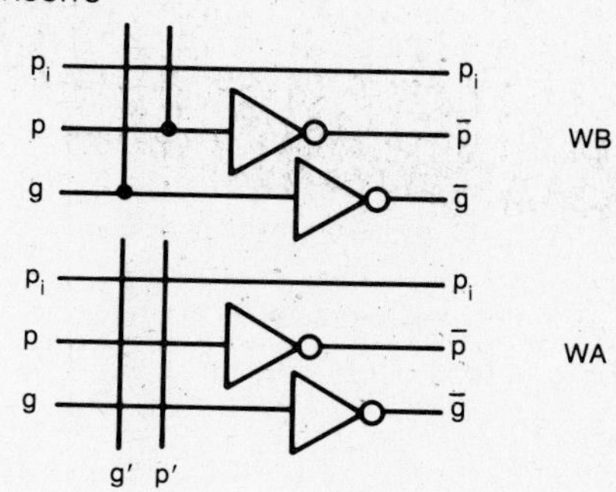

FIGURE 8.20. BLC adder schematics

This adder is best generated using a procedure. The following 'C' procedure, based on a procedure in [BrEw82], generates an n-bit adder. (An alternative floor-plan consists of a vertical carry block with a tree routing channel to connect the appropriate gates.)

Adder Procedure (n bit adder)

```
d = 2;
a = TRUE;
do {    /*normal tree portion*/
     for ( i = 1; i <= n; i ++ ) /*one
     column*/
     {

          if ((i%d) == 0)
          {
               if (a) plot cell BA
               else plot cell BB
          }
          else if ((i%d) <= (d/2))
          {
               if (i%2) plot cell WB
               else plot cell WA
          }
          else plot WA
     move cell position up
     }
     reset cell position to bottom and move
     right one position
     d *= 2;
     a =   ~a;
}
while (d <= n);
d / = 4;
if ((3*d) > n) d / = 2;
do { /* inverse tree portion */
     for (i = 1; i <= n;i++)
     {
          imod = i%(2*d);
          if (i < (2*d))
          {
               if (i%2) plot cell WB
               else plot cell WA
          }
          else if (imod == d)
          {
               if (a) plot cell BA
               else plot cell BB
          }
          else if (imod == 0 )
          {
               if (i%2) plot cell WB
               else plot cell WA
```

```
                }
                else if (imod < d) plot WA
                else
                {
                        if (i%2) plot WB
                        else plot WA
                }
        move cell position up
        }
        reset cell position to bottom and
        right one position
        d / = 2;
        a =   ~ a;
    } while (d >= 1 );
```

Note that a similar floor plan could be used if dynamic logic was used to implement the functions. If domino logic is used, the *o* function can be directly implemented. However, careful consideration would have to be given to buffering the long vertical lines that occur in the carry generation block. A possible circuit strategy is shown in Fig. 8.21. Note that the basic floor plan can remain the same for the static version, a domino version, or an N-P CMOS version. This adder seems to be of greatest use for adders that are larger than 16 bits.

8.2.7 Carry select adder

Another approach to fast adders that expends area in favor of speed is to use a carry select adder. The basic scheme is shown in Fig. 8.22a [UyKY84]. Two ripple carry adder structures are built, one with a zero carry-in and the other with a one carry-in. This is repeated for a certain sized adder, say, 4-bits. The previous carry then selects the appropriate sum using a multiplexer or tri-state adder gates. The stage carries and the previous carry are gated to form the carry for the succeeding stage. As a further optimization, each succeeding ripple adder may be extended by one stage to account for the delay in the carry lookahead gate. Thus for a 32-bit adder, the stage numbers are 4-4-5-6-7-6, as shown in Fig. 8.22b. This yields an adder with approximately (4 + 1 + 1 + 1 + 1 + 1), or 9 gate delays for a 32-bit addition. The ripple adders may be designed statically, dynamically, or by using a combination of these approaches.

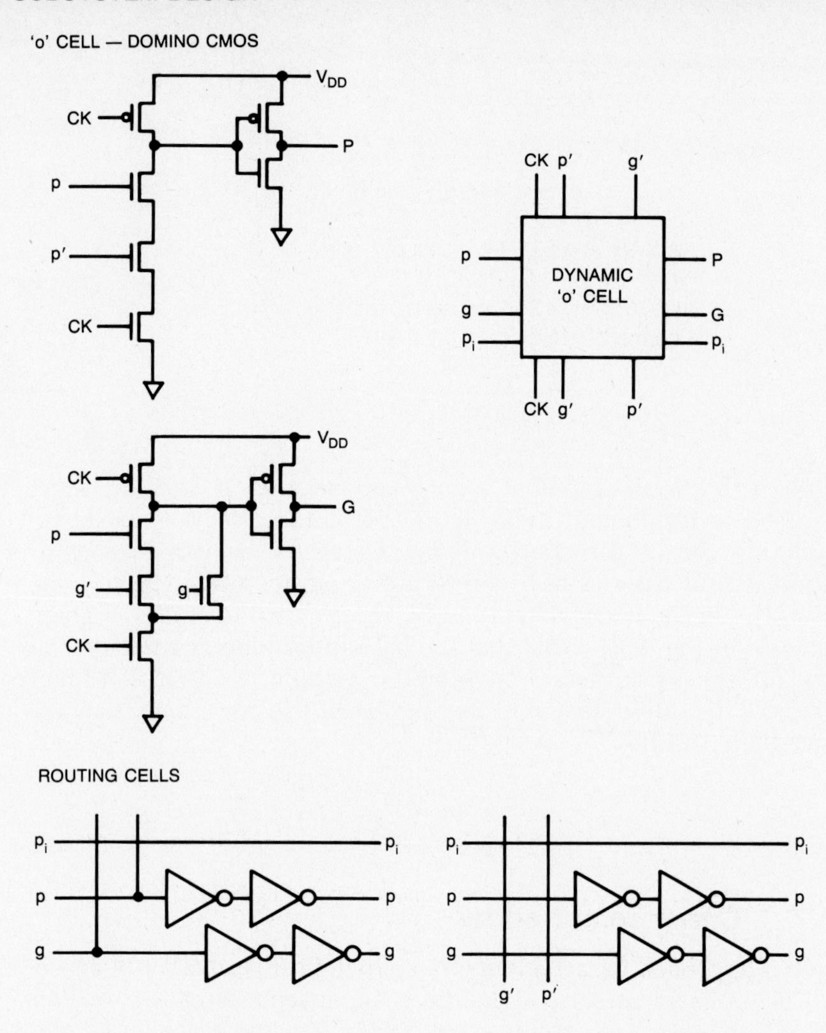

FIGURE 8.21. Dynamic BLC adder

8.2.8 Parity generators

A function related to binary addition is parity generation. Frequently it is necessary to generate the parity of, say, a 16- or 32-bit word. The function is

$$PARITY = A_0 \oplus A_1 \oplus A_2 \oplus A_3 \cdots \oplus A_n. \quad \textbf{(8.12)}$$

Fig. 8.23a shows a conventional implementation. Fig. 5.11d shows a schematic of a 4-input parity generator that employs cascade logic. A number of these may be cascaded to perform a 32-bit parity function (Fig. 8.23b) [Tsai83]. A static 4-input XOR that could be used is shown in Fig. 8.23c [GrHi83]. In a data path, Fig. 8.23a may

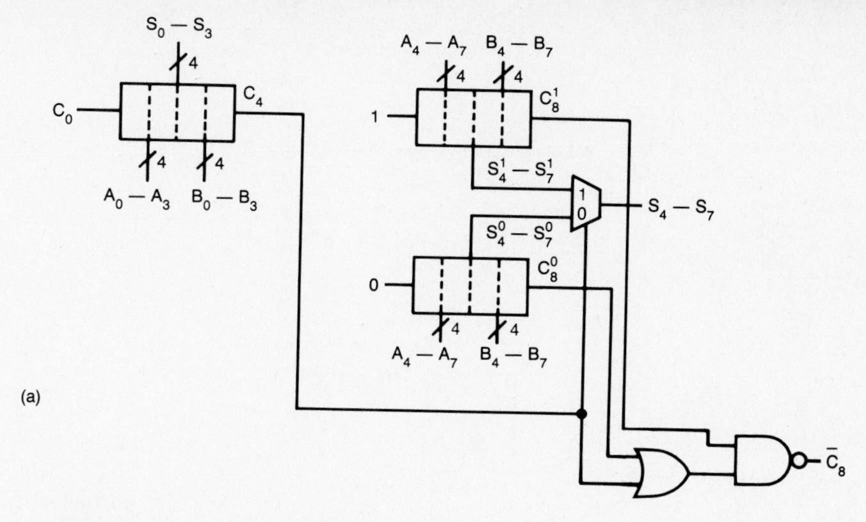

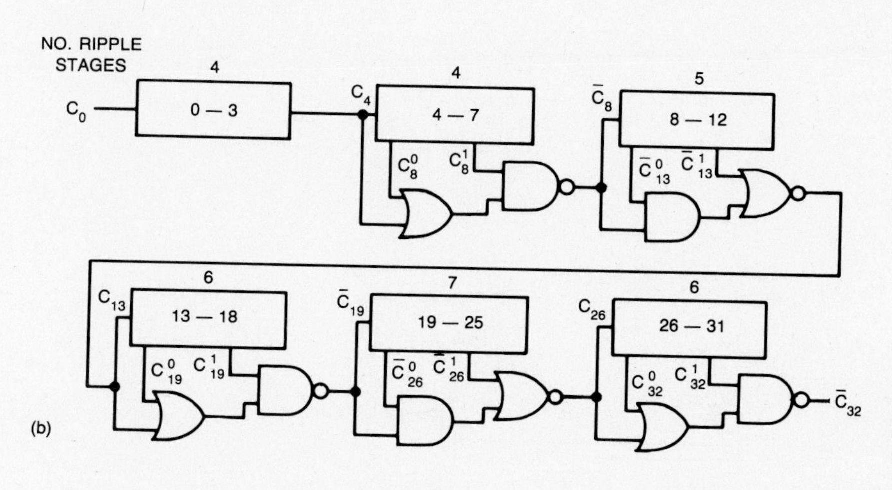

FIGURE 8.22. Carry select adder

be implemented as a linear column with a tree routing channel connecting the XOR gates.

8.2.9 Comparators

A magnitude comparator is useful to compare the magnitude of two binary numbers. One can build a comparator from an adder and a complementer, as shown in Fig. 8.24a. Another approach is to use a pass logic function, as shown in Fig. 8.24b [Whit83]. This may be single-ended or complementary, as discussed in Chapter 5 and is used as shown in Fig. 8.24.

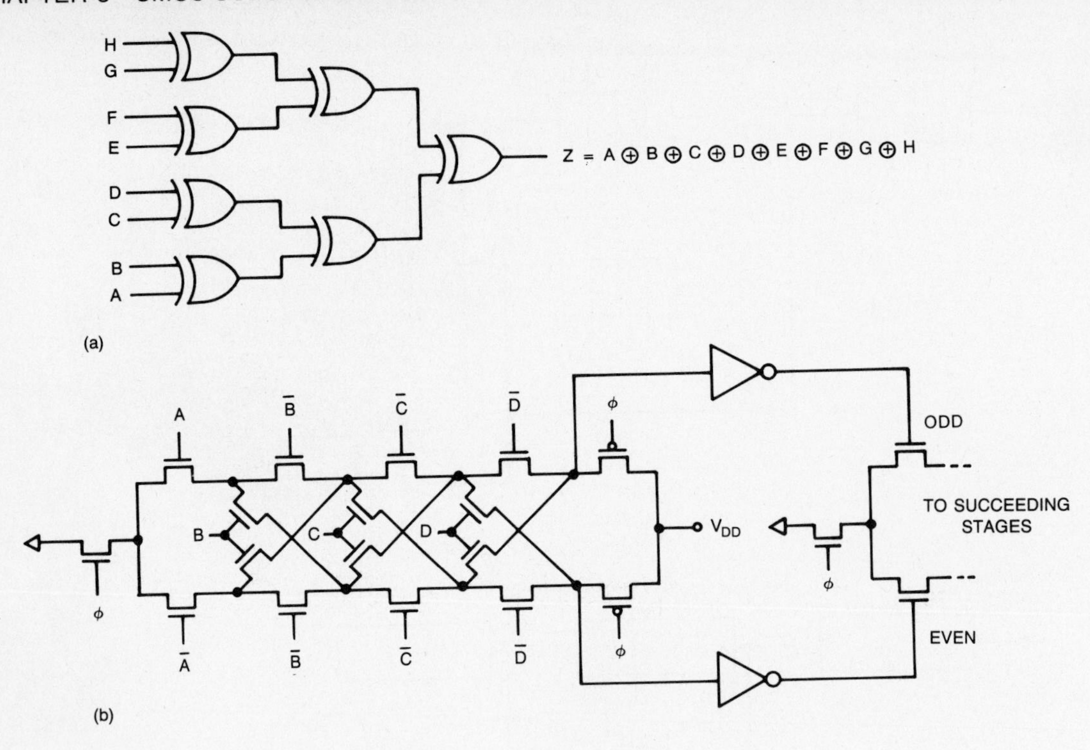

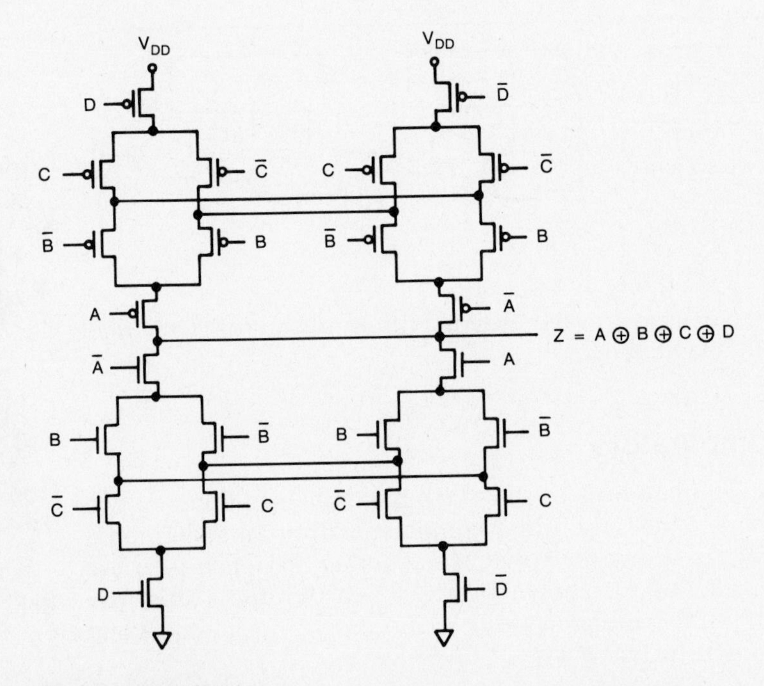

FIGURE 8.23. Parity generators

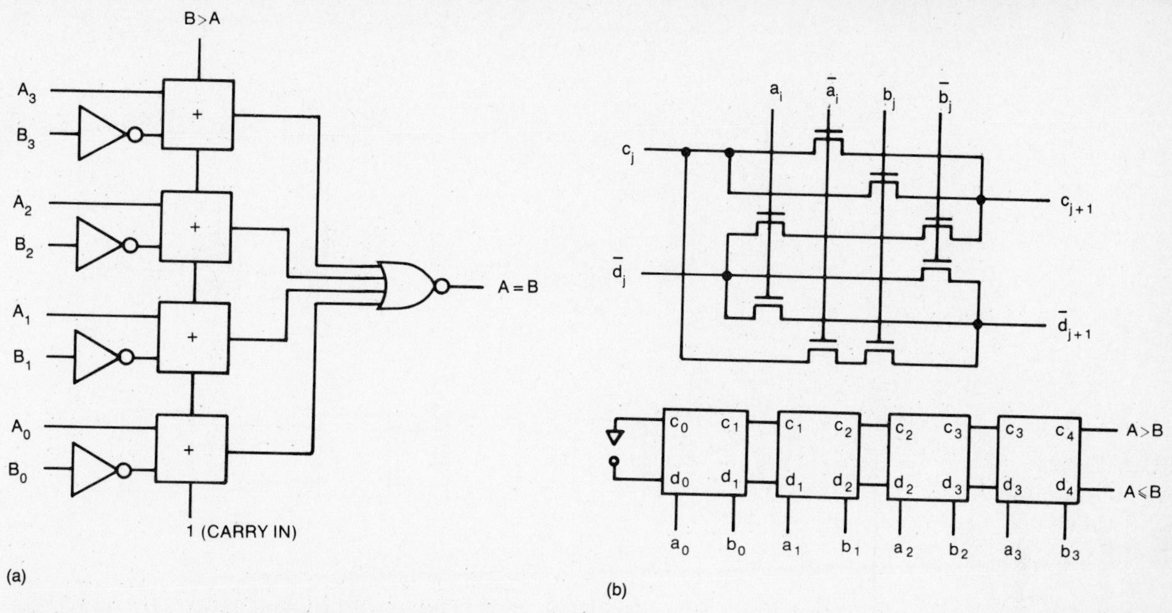

FIGURE 8.24. CMOS comparator structures. Reprinted from *ELECTRONICS*, September 22, 1983.

8.3 Binary counters

Binary counters are used to cycle through a sequence of binary numbers. One can consider asynchronous or synchronous counters.

8.3.1 Asynchronous counters

A "ripple-carry" binary counter is shown in Fig. 8.25. This is based on the 2-phase static D flip-flop introduced in Chapter 5. A counter stage is shown in Fig. 8.25a. This counter stage may be cascaded, as shown in Fig. 8.25b. Note that the clocking of each stage is carried out by the previous counter stage, and thus the time it takes the last counter stage to settle can be quite large for a long counter chain. This is an asynchronous counter as the counter outputs change at different time instances.

8.3.2 Synchronous counters

Synchronous counters generally require a "lookahead" signal to be generated, similar to that used in the design of fast adders. These counters tend to be more complex than simple asynchronous counters

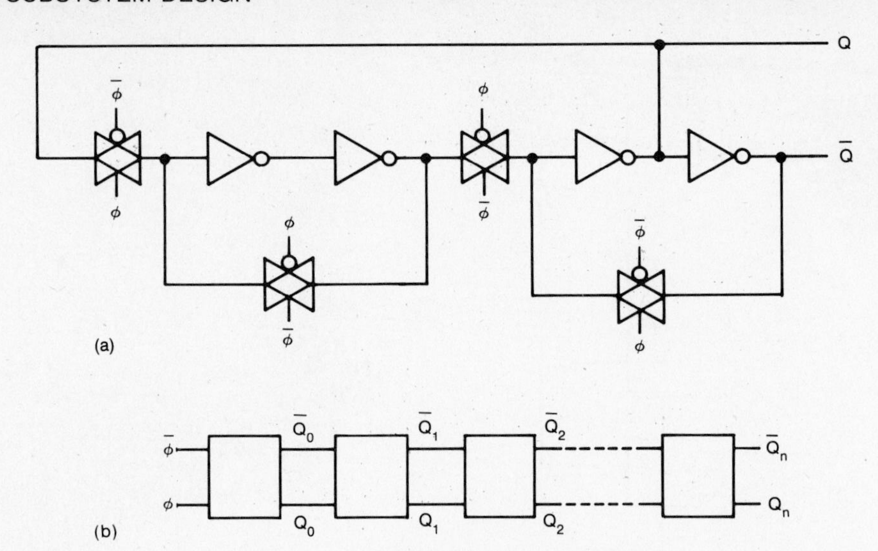

FIGURE 8.25. Asynchronous counter schematics

but have the advantage that stages are clocked simultaneously and the outputs change in a synchronous manner. One approach is to use D flip-flops and steering circuits to provide the D input signals. Fig. 8.26a shows such a structure [KrLi83]. In this circuit, the first stage operates as a simple divide by two stage, with $\overline{Q}$ fed back to the *D* input. Subsequent stages drive *Q* or $\overline{Q}$ back to *D* via a multiplexer. This switching is enabled when the two previous stage *Q* signals are true. This style of counter does not really lend itself to a regular layout but a possible floor plan using standard cells is shown in Fig. 8.27. The counter is representative of a counter that might be used in a gate array or standard-cell design system.

One can also design a counter by using an adder and a register. The overall scheme for this is shown in Fig. 8.28a. Basically, the

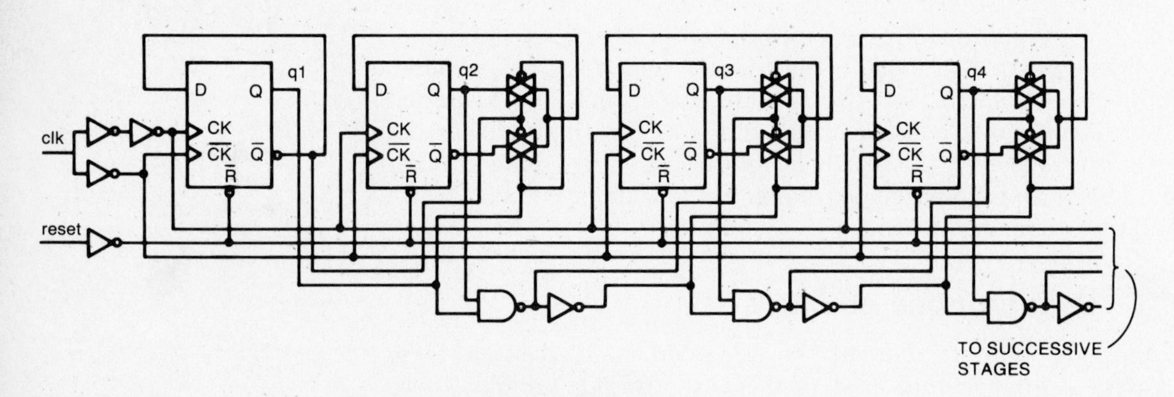

FIGURE 8.26. Synchronous counter schematic

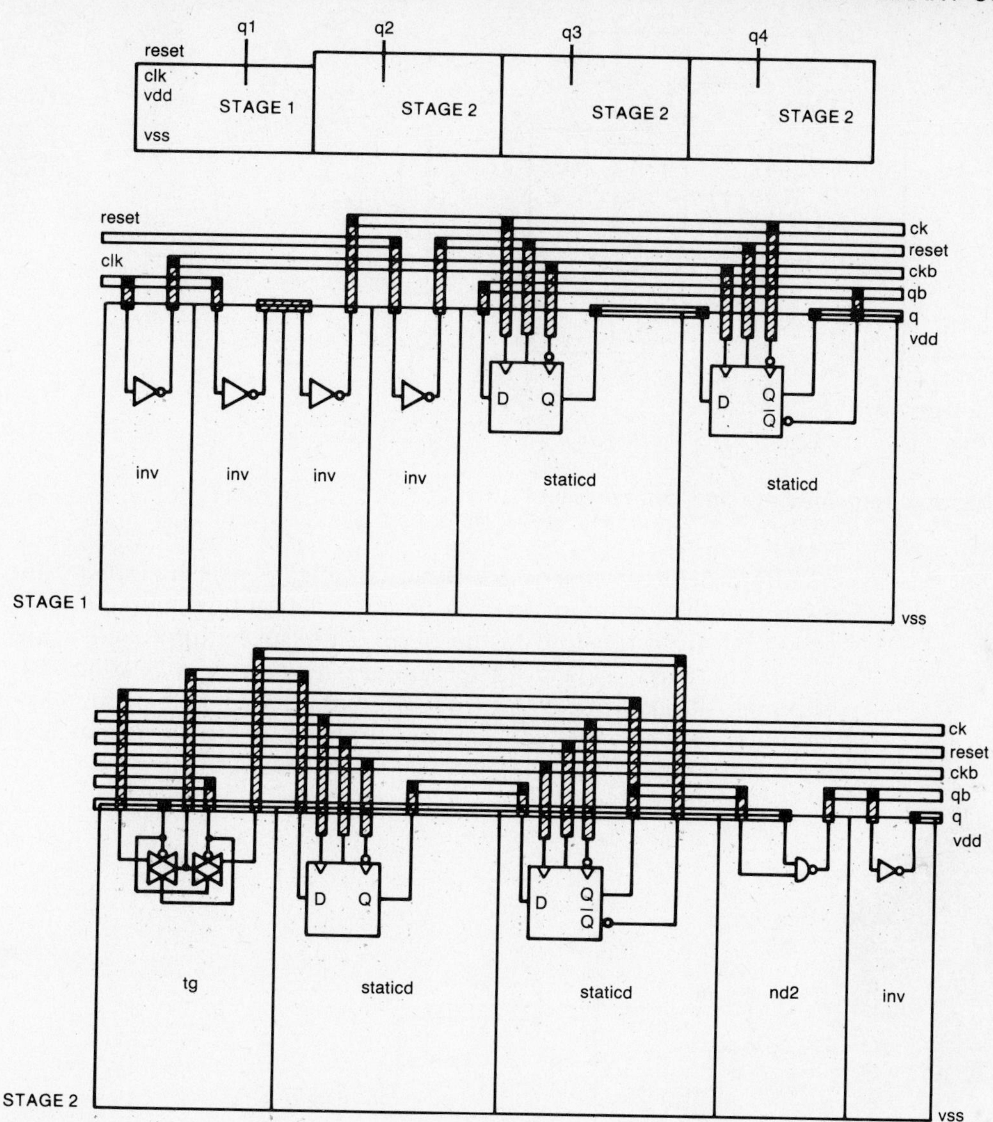

FIGURE 8.27. Synchronous counter floor plan using standard cells

adder is arranged as an accumulator such that it increments (in the same manner as a counter). Note that a decrementer may be constructed by changing the inputs as shown in Fig. 8.28b. Where only an incrementer (or decrementer) is required, the adder may be considerably simplified as one of the input terms is permanently zero (or one). A further circuit using an XOR gate, an AND gate, and a D flip-flop is shown in Fig. 8.28c. A circuit for the combinational incrementer is shown in Fig. 8.29. This is derived from the adder

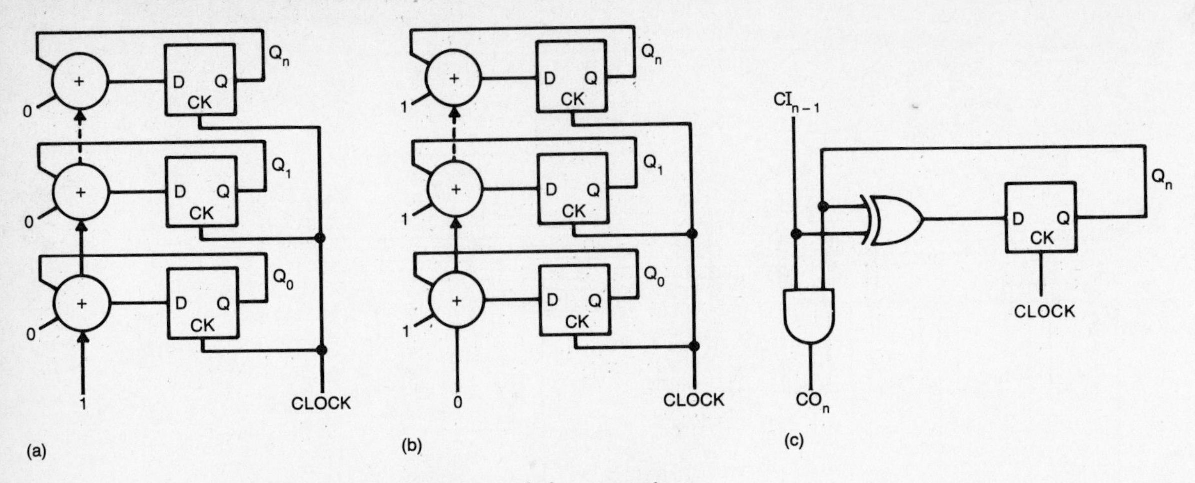

FIGURE 8.28. Adders as incrementers and decrementers

circuits previously discussed. Reset circuitry is also included. The speed that this incrementer can operate is determined by the ripple-carry time from the LSB to the MSB. This can be improved using any of the carry lookahead techniques discussed in Section 8.2. Although the complexity of this structure is greater than the conventional synchronous counter, the structure is highly regular and can form the basis for a simple counter building block.

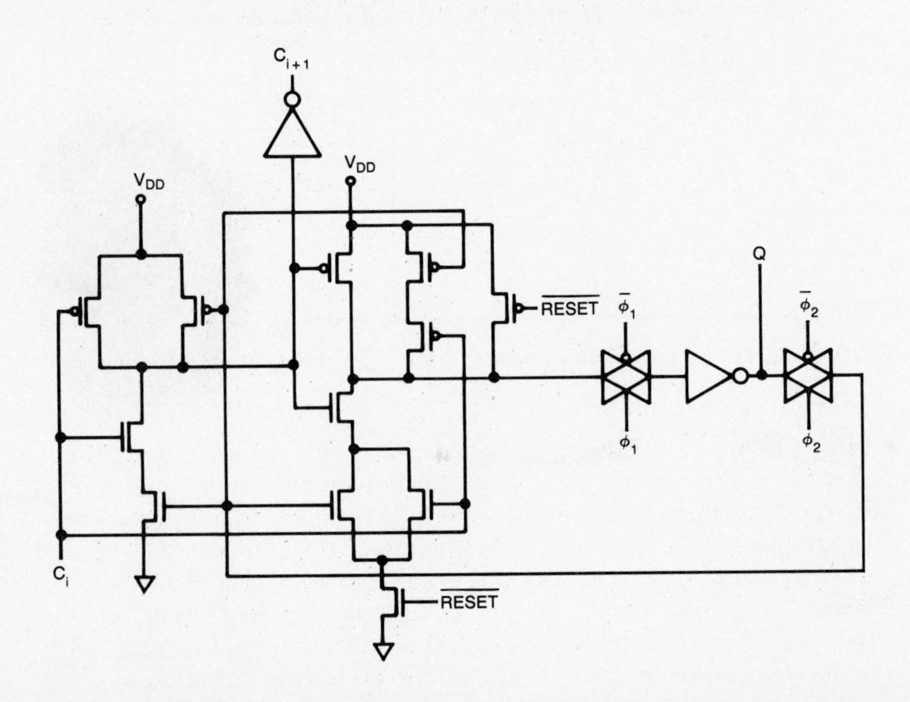

FIGURE 8.29. An adder counter circuit

8.4 Multipliers

In many signal processing operations, such as correlations, convolution, filtering, and frequency analysis, one needs to perform multiplication. We will use the multiplication algorithms to illustrate methods of designing different cells so that they fit into a larger structure. In order to introduce these designs, we will briefly introduce simple serial and parallel multipliers. The appropriate texts should be consulted for more definitive system architectures. The most basic form of multiplication consists of forming the product of two positive binary numbers. This may be accomplished through the traditional technique of successive additions and shifts in which each addition is conditional on one of the multiplier bits. For example, the multiplication of two positive binary integers, 12_{10} and 5_{10}, may proceed in the following manner:

```
multiplicand;    1100   :   12₁₀
multiplier  ;    0101   :    5₁₀
                 ----
                 1100
                0000
               1100
              0000
              -------
              0111100   :   60₁₀
```

Therefore, multiplication process may be viewed to consist of the following two steps:

1. Evaluation of partial product.
2. Accumulation of the shifted partial product.

It should be noted that binary multiplication is equivalent to a logical AND operation. Thus evaluation of partial products consists of the logical ANDing of the multiplicand and the relevant multiplier bit. There are a number of techniques that may be used to perform multiplication. In general, the choice is based upon factors such as speed, throughput, numerical accuracy, and area. As a rule, multipliers may be classified by the format in which data words are accessed, namely,

- serial form
- serial/parallel form
- parallel form.

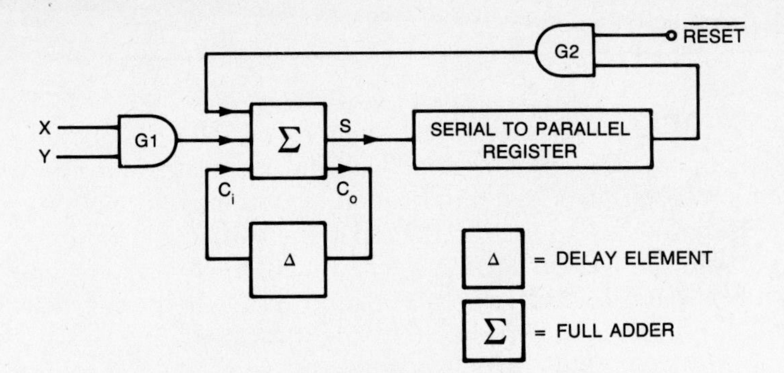

FIGURE 8.30. A basic serial multiplier

8.4.1 Serial multiplier

The simplest form of serial multiplier shown in Fig. 8.30 uses the successive addition algorithm and is implemented using a full adder, a logical AND circuit, a delay element (i.e., either static or dynamic flip-flop), and a serial to parallel register.

The two numbers X and Y are presented serially to the circuit (at different rates to account for multiplier and multiplicand word lengths). The partial product is evaluated for every bit of the multiplier and a serial addition is performed with the partial additions already stored in the register. The AND gate G2 between the input to the adder and the output of the register is used to reset the partial sum at the beginning of the multiplication cycle. If the register is made of N-1 stages, then the 1-bit shift required for each partial product is obtained automatically. As far as the speed of operation is concerned, the complete product of $M + N$ bits can be obtained in $M * N$ intervals of the multiplicand clock.

8.4.2 Serial/parallel multipliers

Using the general approach discussed previously, it is possible to realize a serial/parallel multiplier with a very modular structure that can easily be modified to obtain a pipelined system. The basic implementation is illustrated by Fig. 8.31. In this structure, the multiplication is performed by means of successive additions of columns of the shifted partial products matrix. As left-shifting by one bit in serial systems is obtained by a 1-bit delay element, the multiplier is successively shifted and gates the appropriate bit of the multiplicand. The delayed, gated instances of the multiplicand must all be in the same column of the shifted partial product matrix. They are then added to form the required product bit for the particular column.

This structure requires $M + N$ clock cycles to produce a product. The main limitation is that the maximum frequency is limited by

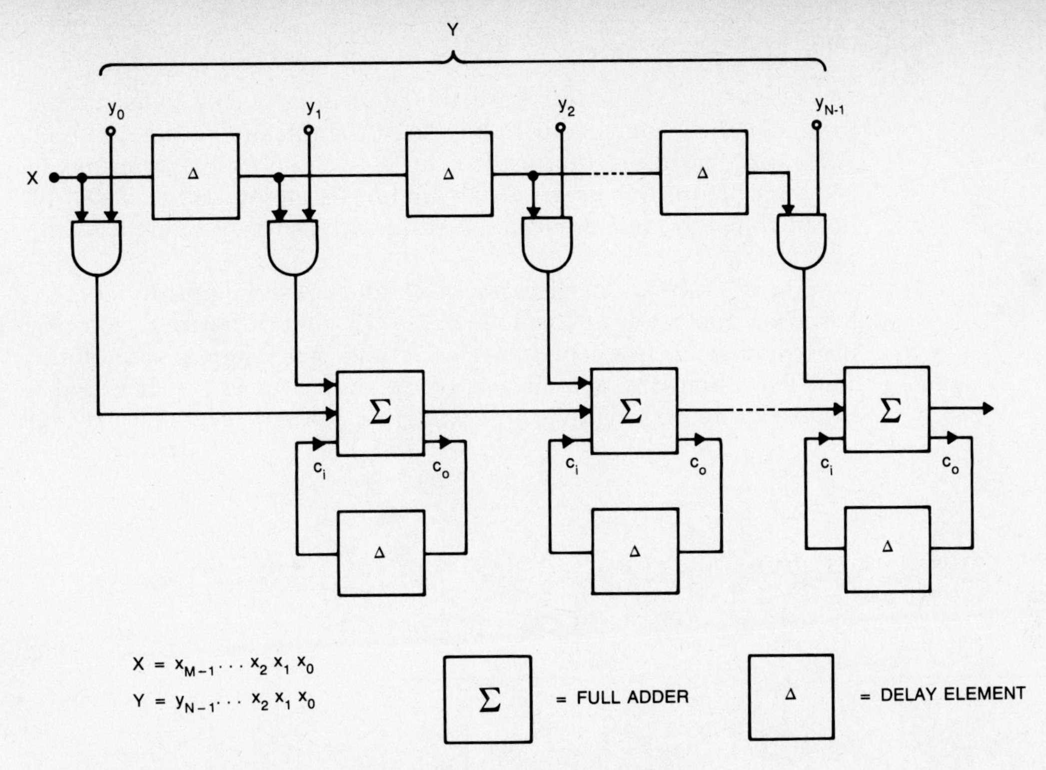

FIGURE 8.31. Basic structure for serial/parallel multiplier

the propagation through the array of adders. The structure of Fig. 8.31 can be modified into pipelined systems by the introduction of two delay elements in each cell, as shown in Fig. 8.32. If rounding or truncation of the product term to the same word length as the input is tolerated, then the time necessary to produce a product is

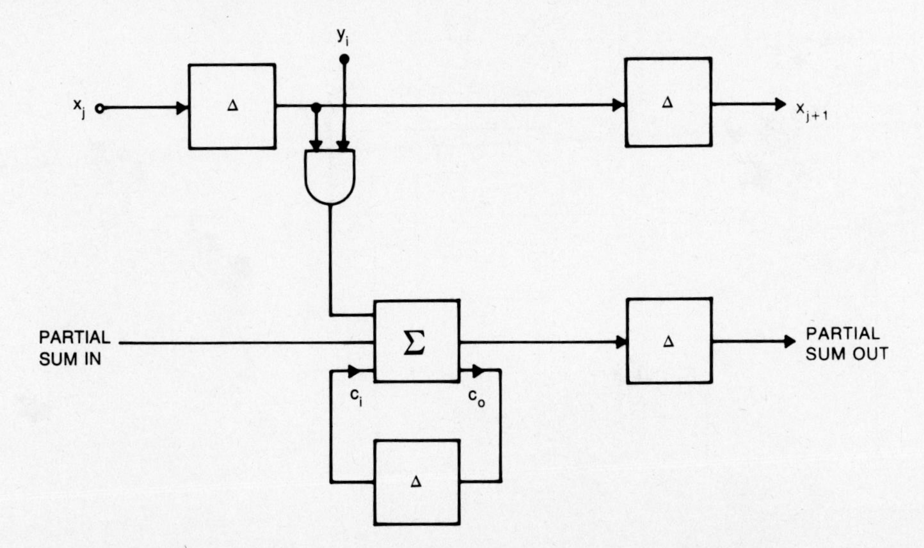

FIGURE 8.32. Pipeline multiplier structure

2M clock cycles. In this case the multiplier accumulates partial product sums, starting with the least significant partial product. After each addition, the result is an N-bit number that shortens to N-1 bits before the next partial product is added. Here, it can be noted that the chip area increases linearly with the length of the multiplier.

Fig. 8.33 shows a schematic of a two-stage serial multiplier stage based on the work of Lyon [Lyon76], in which the basic solution described so far has been modified so both words are in serial form. The floor plan for a 2-bit section is shown in Fig. 8.34a, and a symbolic layout is shown in Fig. 8.34b. (Designed by C. Durwood

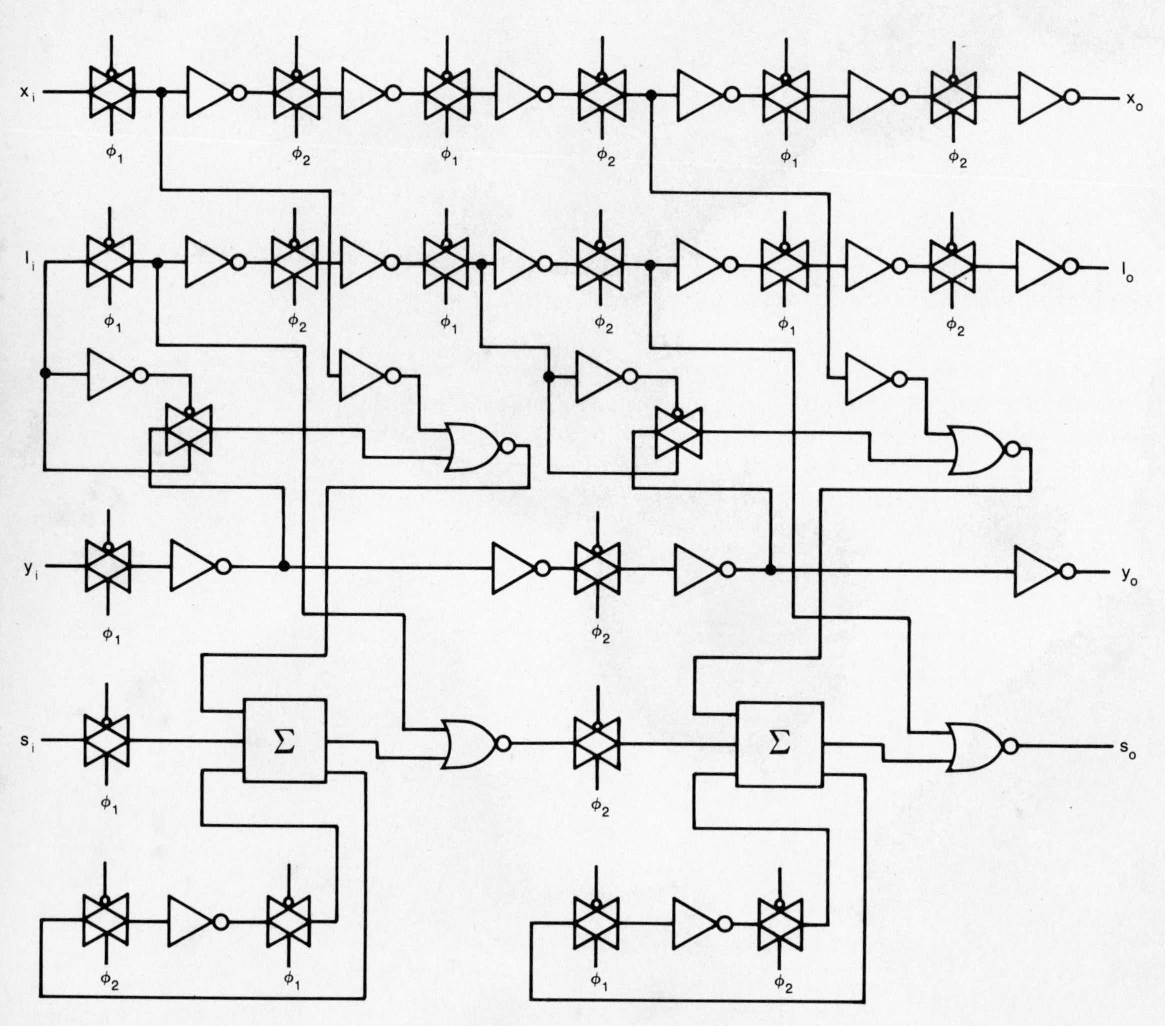

FIGURE 8.33. Two-stage serial multiplier

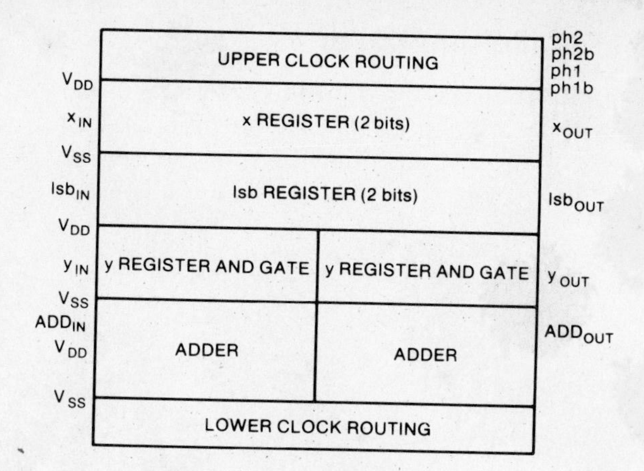

(a)

(b)

FIGURE 8.34. Serial multiplier cell layouts

Rogers and S. W. Daniel, MCNC) Note that although the designers used a 2-phase clocking scheme, the clock routing is not particularly difficult. The design illustrates the modularity of this type of multiplier. Plate 11 shows this design in color.

8.4.3 Parallel multiplier

A parallel multiplier is based on the observation that partial products in the multiplication process can be independently computed in parallel. For example, consider the unsigned binary integers X and Y:

$$X = \sum_{i=0}^{m-1} X_i 2^i$$

$$Y = \sum_{j=0}^{n-1} Y_j 2^j.$$

The product is found by

$$P_r = X_y Y_r = \sum_{i=0}^{m-1} X_i 2^i \cdot \sum_{j=0}^{n-1} Y_j 2^j$$

$$= \sum_{i=0}^{m-1} \sum_{j=0}^{n-1} (X_i Y_j) 2^{i+j} \tag{8.13}$$

$$= \sum_{k=0}^{m+n-1} P_k 2^k.$$

Thus P_k are the partial product terms called summands. There are mn summands, which are produced in parallel by a set of mn AND gates. For 4-bit numbers, the expression above may be expanded as in Table 8.2.

An $n \times n$ multiplier requires $n(n\text{-}2)$ full adders, n half adders, and n^2 AND gates. The worst-case delay associated with such a multiplier is $(2n + 1)\tau_g$, where τ_g is the worst-case adder delay. Fig. 8.35 shows a cell that may be used to construct a parallel multiplier. The X_i term is propagated vertically, while the Y_j term is propagated horizontally. Incoming partial products enter at the top left. Incoming CARRY IN values enter at the top of the cell. The bit-wise AND is performed in the cell, and the SUM is passed to the next cell at the lower right. The CARRY OUT is passed to

TABLE 8.2. 4-bit multiplier partial products

				X3	X2	X1	X0	Multiplicand
				Y3	Y2	Y1	Y0	Multiplier
				X3Y0	X2Y0	X1Y0	X0Y0	
			X3Y1	X2Y1	X1Y1	X0Y1		
		X3Y2	X2Y2	X1Y2	X0Y2			
	X3Y3	X2Y3	X1Y3	X0Y3				
P7	P6	P5	P4	P3	P2	P1	P0	Product

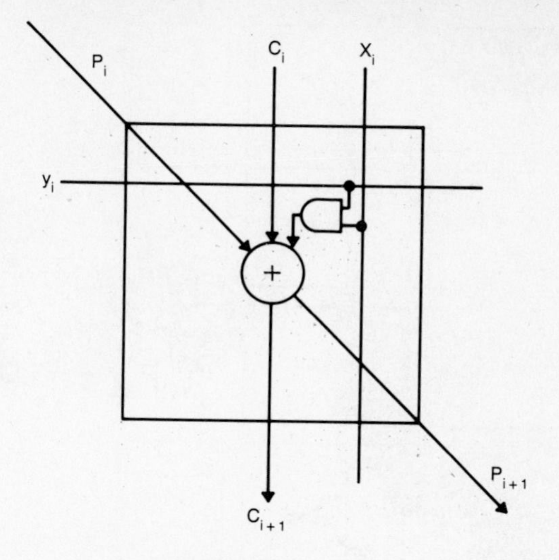

FIGURE 8.35. Parallel multiplier cell

the bottom of the cell. Fig. 8.36a shows the multiplier array with the partial products enumerated. This arrangement may be drawn as a square array, as shown in Fig. 8.36b, which is the most convenient for implementation.

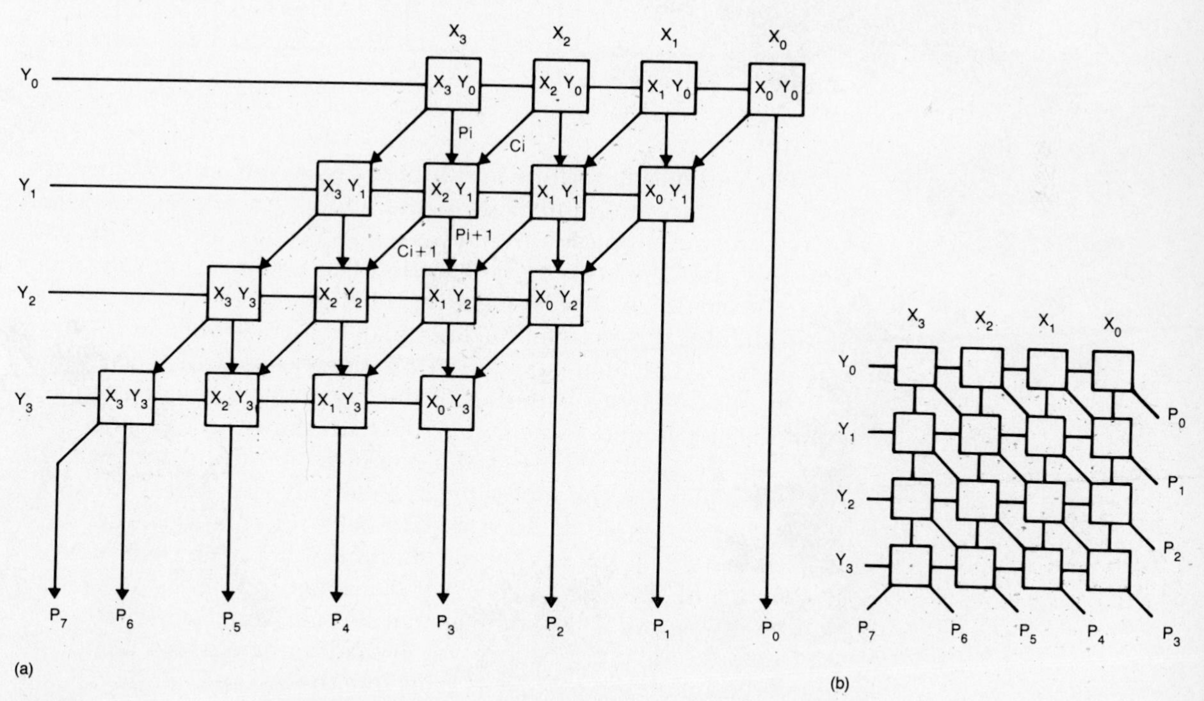

FIGURE 8.36. Parallel multiplier array

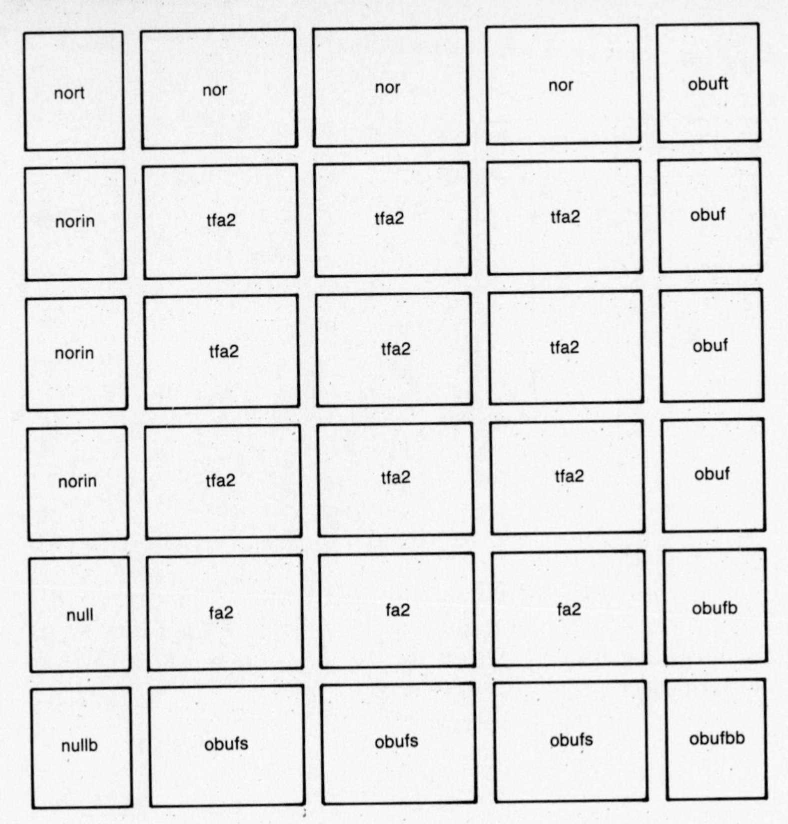

FIGURE 8.37. Parallel multiplier floor plan (4*4)

The cell design for this multiplier is relatively straight-forward, with the main attention paid to the adder. An adder with equal carry and sum propagation times is advantageous, as the worst-case multiply time depends on both paths. The transmission gate adder was chosen in an implementation of a 16 × 16 version of this multiplier for this reason. The floor plan of a 4 × 4 version is shown in Fig. 8.37. The final summation can use a carry lookahead adder, but in this version a variation of the TG adder cell was used as a rippled-carry adder. Some cell layouts are shown in Fig. 8.38 along with the detailed layout for the complete multiplier in Plate 13. The floor plan for the multiplier is relatively simple but it may be used as an example of how to plan cells for a layout. The adder cells (Plate 12) form the core of the multiplier as a 3 * 3 array in the center of the design. The various boundary cells are arrayed around these cells. Buffers line the bottom and right edges (Fig. 8.38b) of the layout. V_{DD} and V_{SS} are fed vertically and tapped off horizontally to each cell. AND gates line the top and left (Fig. 8.38a) border of the array. The final adder cell is shown in Fig. 8.38c. Note

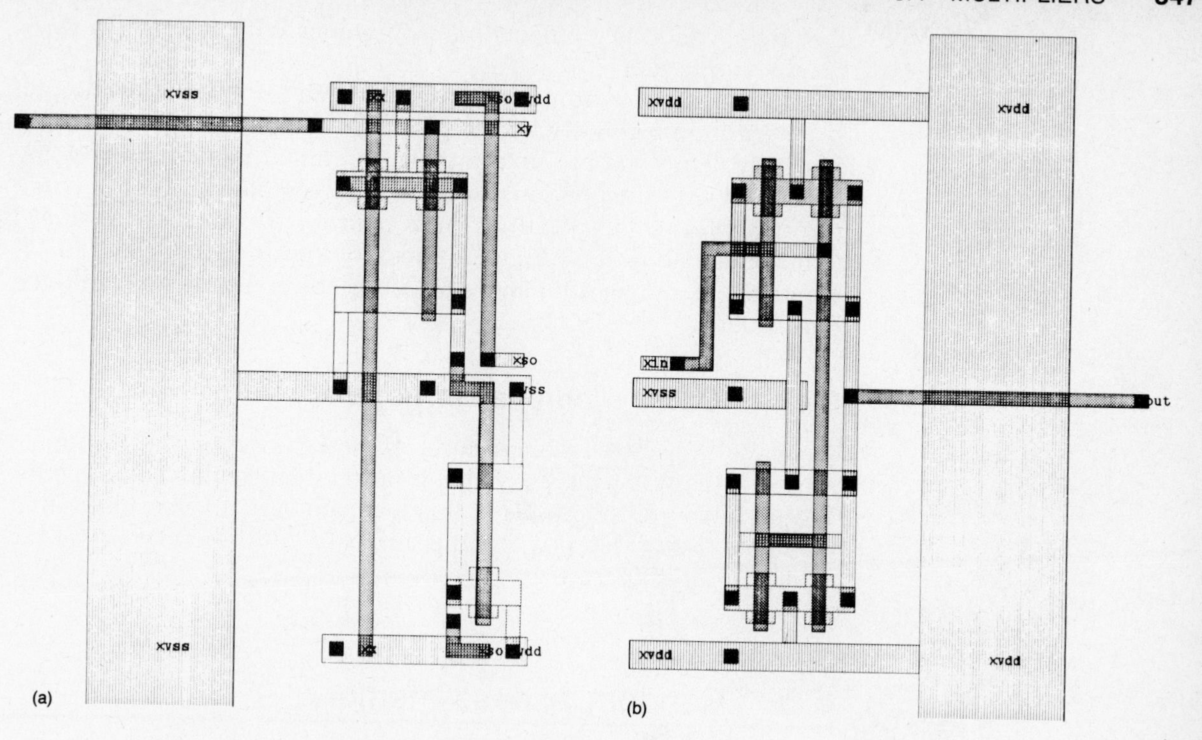

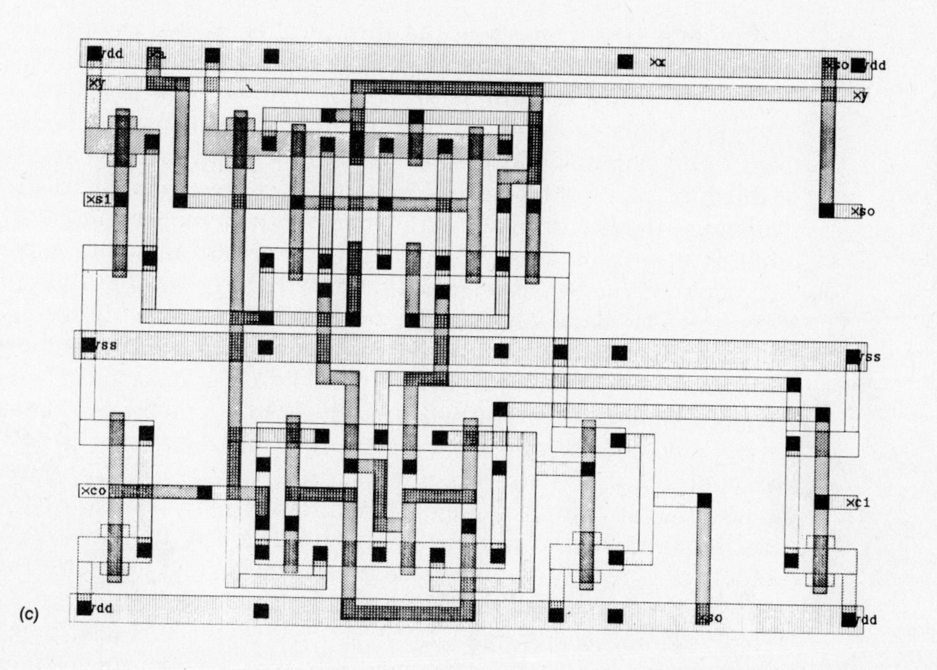

FIGURE 8.38. Multiplier cell layouts

that it is a modification of the array adder with rearranged ports for the ripple carry.

As can be seen from the figure, the time for product evaluation is determined primarily by the propagation delay of signals within the array. Thus the product is produced much more quickly when compared with the previous serial or serial/parallel multiplier structures. However, the penalty that is paid for this increase in speed is that the amount of chip area now required is proportional to N^2, where N is the word length for both the multiplicand and the multiplier.

8.4.4 Other multiplier structures

We have presented some simple approaches to multiplication to illustrate the way a larger system is built out of smaller components. The architectural trade-offs at this level are very diverse, and rather than deal in any detail with other forms of multipliers, we refer the reader to the literature.

8.5 Random access memory

Random access memory at the chip level is classed as memory that has an access time independent of the physical location of the data. This is contrasted with serial access memories, which have some latency associated with the reading or writing of a particular datum, and with content addressable memories. Within the general classification of random access memory, we can consider read only memory (ROM) or read/write memory (commonly called RAM). ROMs usually have a write time much greater than the read time (programmable ROMs have write times of the order of milliseconds), while RAMs have very similar read and write times. Both types of memory may be further divided into synchronous and asynchronous categories. Synchronous RAMs or ROMs require a clock edge to enable memory operation. The address to a synchronous memory only needs to be valid for a certain setup time after the clock edge. Asynchronous RAMs recognize address changes and output new data after any such change. Synchronous memories are easier to design and usually form the best choice for a system level building block, as they can generally be clocked by the system clock.

The memory cells used in RAMs can further be divided into static structures and dynamic structures. Static cells use some form of latched storage, while dynamic cells use dynamic storage of charge on a capacitor. We will concentrate on static RAMs as they are

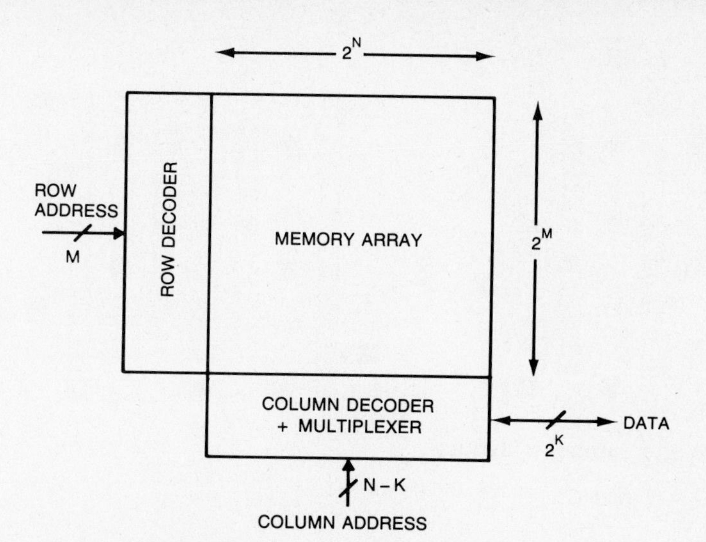

FIGURE 8.39. Random access memory structure

easier to design and are potentially less troublesome than dynamic RAMs. Static RAMs tend to be faster (but much larger) than dynamic RAMs.

A typical random access memory chip architecture is shown in Fig. 8.39. Central to the design is a memory array consisting of 2^n by 2^m bits of storage. A row (or word) decoder addresses one word of 2^m bits out of 2^n words. The column (or bit) decoder addresses 2^k of 2^m bits of the accessed row. This column decoder accesses a multiplexer, which routes the addressed data to and from interfaces to the external world.

8.5.1 Static RAM cells

Fig. 8.40 illustrates a generic static MOS RAM cell. The circuit consists of a crosscoupled inverter connected by pass transistors to

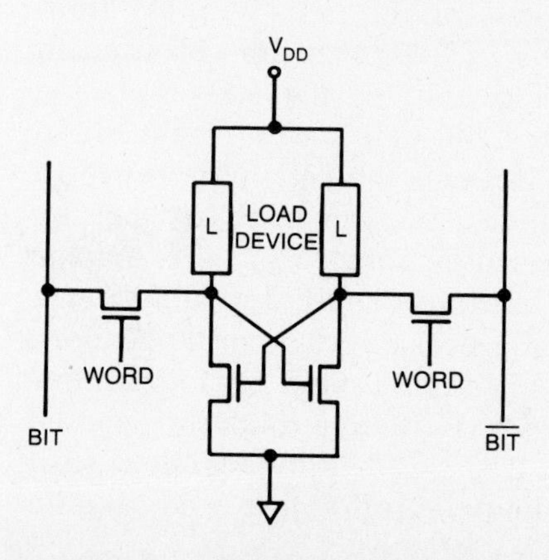

FIGURE 8.40. Static MOS RAM circuit

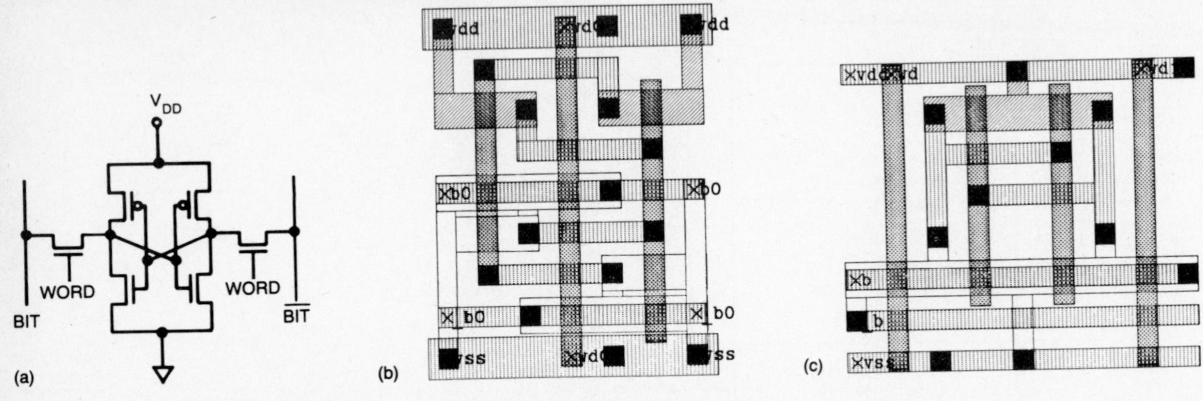

FIGURE 8.41. Static CMOS RAM cell circuit and symbolic layouts

a BIT and $\overline{\text{BIT}}$ line. The load device may be a depletion or enhancement transistor, a p-transistor, or an undoped polysilicon resistor. The purpose of the load is to counteract the effect of charge leakage at the drains of the pull-down and pass transistors.

To write the cell, DATA is placed on the BIT line, and $\overline{\text{DATA}}$ is placed on the $\overline{\text{BIT}}$ line. Then the WORD line is asserted. A read operation commences by precharging the BIT and $\overline{\text{BIT}}$ lines. The word line is asserted and either the BIT or $\overline{\text{BIT}}$ line will be discharged by one of the pull-down transistors in the cell. An alternative to using precharge is to use static pull-ups on the BIT lines.

Typical static CMOS cell layouts with p-transistor loads are shown in Fig. 8.41a. As the WORD line controls the two pass transistors, this is routed in polysilicon through the center of the cell (Fig. 8.41b). The BIT and $\overline{\text{BIT}}$ lines are routed horizontally in aluminum. An alternative cell is shown in Fig. 8.41c. If second level metal is available, the cell shown in Fig. 8.41c benefits by having the BIT lines run in this layer. Plate 14 shows the layouts in color.

Variations on the six-transistor cell include cells with five transistors and two-ported memory cells. A five-transistor cell used in a 16K CMOS/SOS RAM [DiSt79] is shown in Fig. 8.42a. By replacing metal contacts with buried contacts the cell size is reduced. In addition, only one bit line is run through the cell. This causes a problem when writing the cell, which is solved in this design by raising the WORD line voltage to a value above V_{DD}. This voltage is generated on-chip. Note that the butting N and P diffusions (legal in SOS) create the diodes shown in Fig. 8.42a. A two-ported memory cell is shown in Fig. 8.42b. This cell is often used in microprocessor structures. If only one port needs to be written, a transistor may be deleted. The layout shown in Fig. 8.41c allows independent read ports for dual-ported reads, while the two bit lines are used to write

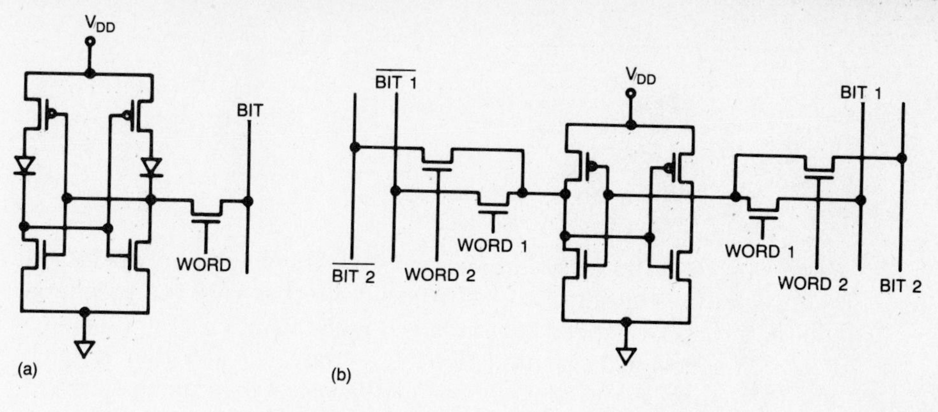

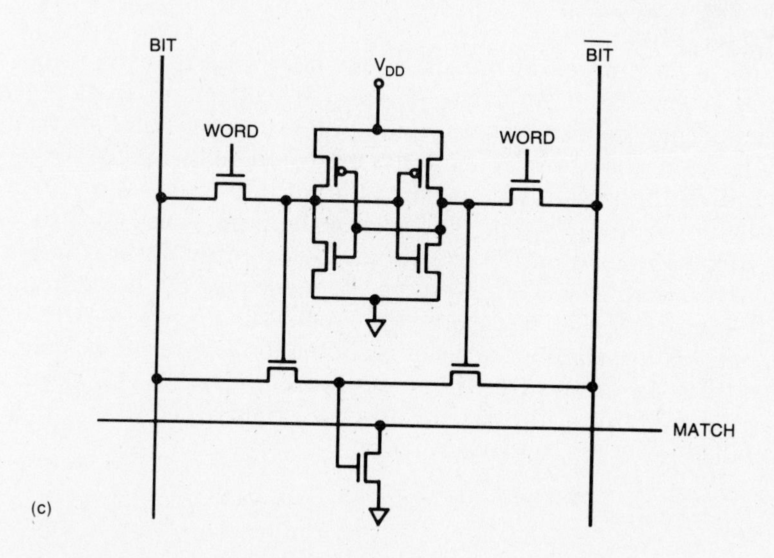

FIGURE 8.42. CMOS RAM variations a) 5-transistor RAM; b) 2-port RAM; c) content-addressable memory (CAM)

the cell. This is of particular use in a three-address architecture microprocessor (A op B → C).

Finally, in Fig. 8.42c, a content addressable memory (CAM) cell is shown [Hou83]. In normal operation, reads and writes are performed in an identical manner to the static cell. For a comparison operation, DATA is placed on $\overline{\text{BIT}}$ and $\overline{\text{DATA}}$ is placed on the BIT lines (as opposed to DATA on BIT and $\overline{\text{DATA}}$ on $\overline{\text{BIT}}$ during a normal read or write). If the data matches that in the cell, then the match transistor will remain turned off. If any cell has data that does not match, the match transistor pulls a previously precharged MATCH line low.

8.5.2 CMOS static RAM cell design

In order for the RAM cell to combine reliability with small size, some design criteria will be reviewed. In the following design, a

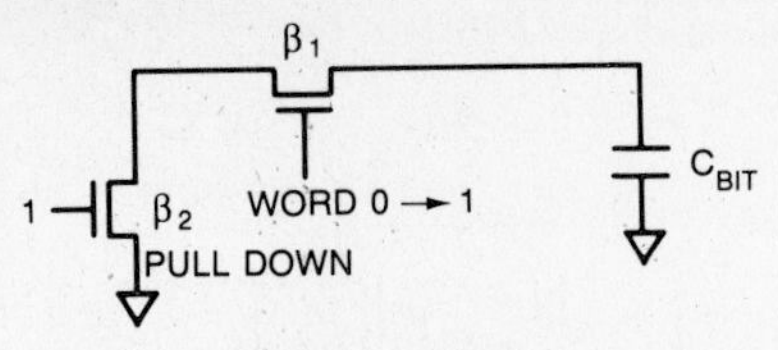

FIGURE 8.43. Read process in a static RAM

RAM design with fully precharged BIT lines will be discussed. There are other variations on this structure that vary the precharge voltage to improve speed performance. Some RAMs use no precharge at all. The design presented here is a "safe" design that is fairly tolerant to process variations but is not the fastest that can be designed. The following explanation is relatively brief but will serve to illustrate some of the design points.

Starting with the p-load transistor, we have seen that this can be minimum size, as it only has to offset the effects of leakage. Next, considering the pass transistor and pull-down transistor sizes, it should be noted that the effective series ratio of these two transistors will determine the pull-down speed of the bit line. This ratio will also affect the write operation. The size of the pass transistor also directly affects the load on the word line, thus altering the rise and fall time of the word line. This effect may be moderated by increasing the word-line driver. Let us assume the model shown in Fig. 8.43. A "unit" (minimum)-sized n-transistor is chosen for the pass transistor to reduce the load on the word line and to minimize the cell size. Consider the effect of a unit-sized pull-down transistor.

The fall time is $2\tau_F$, where

$$\tau_F = \frac{4}{\beta_n}\frac{C_{BIT}}{VDD}$$

β_n gain of unit-sized n-transistor.

If the pull-down is a double "unit"-sized transistor, the fall time will be approximately $3/2\ \tau_F$. Apart from speed considerations, the relative sizes of the pass transistor to the cell pull-down also have to be ratioed to prevent spurious write operations occurring while the cell is being read. For instance, if the pass transistors were infinitely large, the cell storage nodes would both be pulled to $V_{DD} - V_{t_n}$. This effectively erases any state held in the cell. Thus the pass to pull-down has to be ratioed such that the pull-down is able to clamp the stored "low" to well below the inverter switching point. A conservative rule would be to ratio the transistors so that the node voltage is not raised above the n-transistor threshold. Note that the dynamic operation of the cell will be affected by the ratio of the BIT line capacitances to internal memory cell capacitances due to charge sharing.

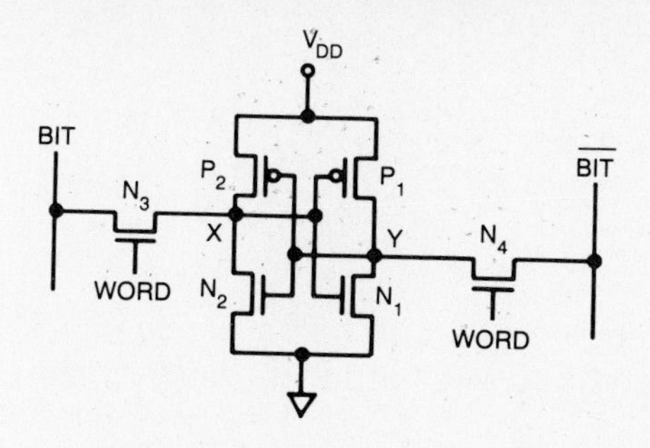

FIGURE 8.44. Write process in a static RAM

Considering the write operation, the ratios have to be selected in such a way as to switch the cell. Fig. 8.44 shows a model of the write process. We assume a 0 is stored at node X and a 1 stored at Y, and we desire to set X to 1 and Y to 0. When the WORD line is asserted BIT is 1 (V_{DD}) and $\overline{\text{BIT}}$ is 0. Node Y falls towards V_{SS}. The voltage that Y falls to has to be sufficient to turn transistor P_2 "on." Thus the ratio of N_4 to P_1 has to be such that this voltage is approximately 2/3 V_{DD} (greater than the threshold of P_2). Node X rises to a voltage determined by the ratio of transistors N_2 and N_3. In this manner the crosscoupled inverter pair is unbalanced and positive feedback then causes the cell to switch states.

8.5.3 Dynamic RAM cells

A four-transistor dynamic RAM cell may be achieved by deleting the p loads of the static cell, as shown in Fig. 8.45a. This cell and

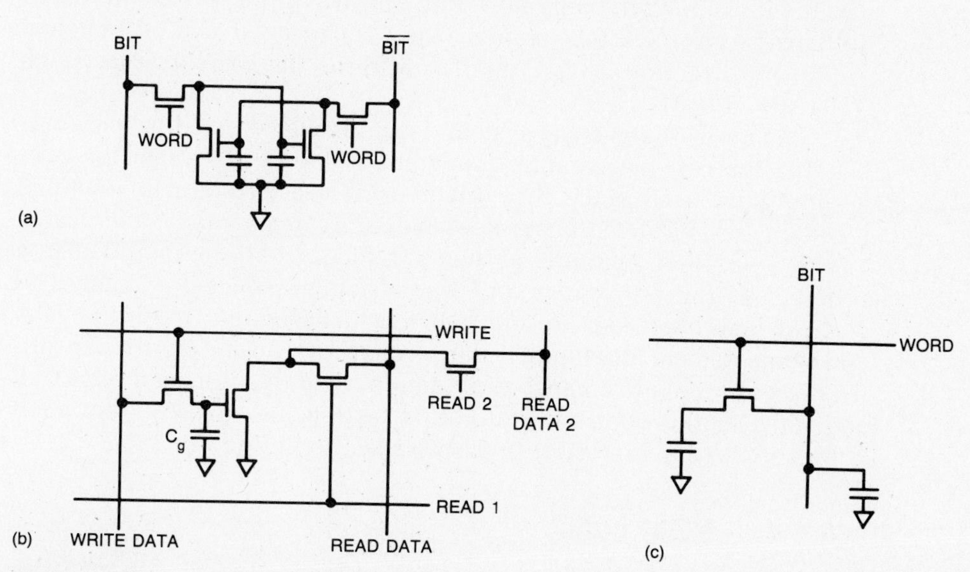

FIGURE 8.45. Dynamic RAM circuits

READ DECODER	MEMORY ARRAY	WRITE DECODER

FIGURE 8.46. Multiple read/write RAM floor plan

the other dynamic cells have to be refreshed to retain the contents of the memory. Design considerations are similar to those of the six-transistor static RAM.

A three-transistor cell is shown in Fig. 8.45b. The cell stores data on the gate of the storage transistor. Separate read and write control lines are used. Multiple read ports may be added quite easily, as shown in the figure. In addition, separate or merged read and write data busses may be used. A typical floor plan is shown in Fig. 8.46, where the read decoder is placed on one side of the memory array and the write decoder is placed on the other.

For the cell shown in Fig. 8.45b, the write operation proceeds by placing DATA on WRITE DATA and asserting the WRITE line. A read operation proceeds by precharging READ DATA and then asserting the READ control line. Design considerations in the three-transistor cell may include sizing the capacitance C_g to ensure adequate data storage time. All transistors are usually made minimum size. The write transistor can be made a little longer than minimum to reduce any subthreshold leakage problems. This cell needs to be refreshed to retain the memory contents.

A one-transistor cell is shown in Fig. 8.45c [Ride79]. The memory value is again stored on a capacitor. Sense amplifiers sense the small change in voltage that results when a particular cell is switched onto the BIT line. This type of cell forms the basis for most high density DRAMs.

As far as the average system design is concerned, the static six-transistor cell or dynamic four-transistor or three-transistor cells should be used since they involve the least amount of detailed circuit design and process knowledge. As a general system design principle, large amounts of memory should only be included in a design if the performance of the system is affected. Commercial RAM manufacturers are much better at designing RAMs than the average system designer. If dense memory can be partitioned off-chip with no performance degradation, then this is a good approach to take. Quite often one finds that fast caches or CAMs are good candidates for inclusion in a user specific design.

8.5.4 ROM cells

Read only memory cells may be implemented with only one transistor per bit of storage. A ROM is a static memory structure in that the

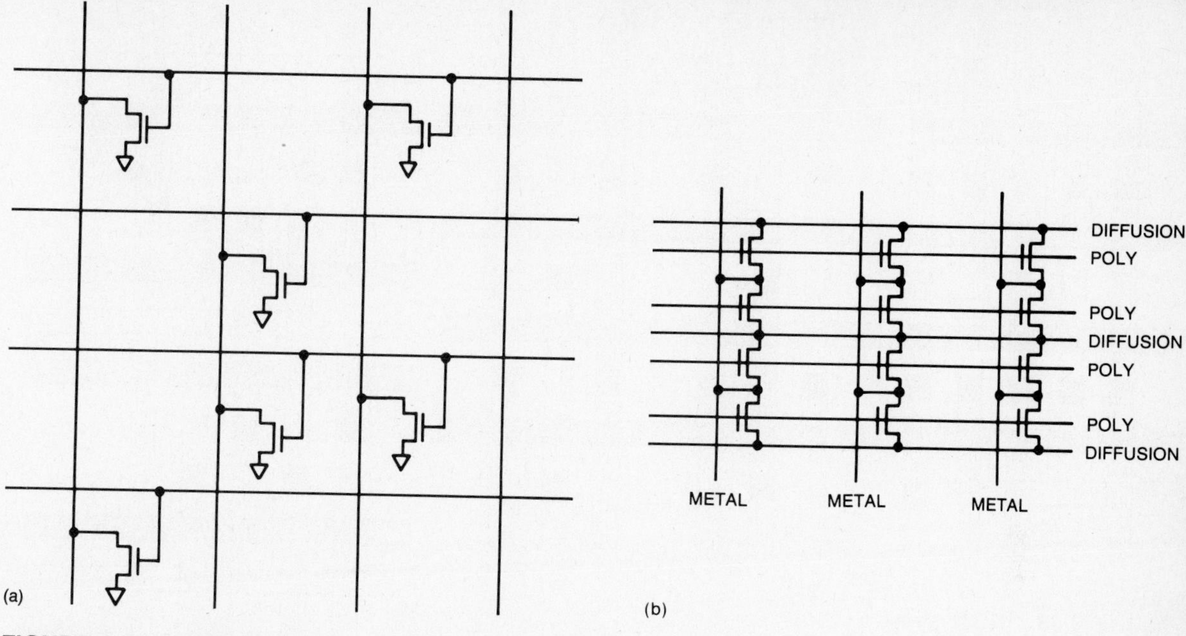

FIGURE 8.47. ROM arrays

state is retained indefinitely — even without power. A ROM array is usually implemented as a NOR array, as shown in Fig. 8.47a. Fig. 8.47b shows a possible layer assignment for a ROM. Note that a NAND array may be used if ultra-small ROMs are required, but as discussed elsewhere, these implementations will be quite slow.

Mask programmability may be achieved via contact programming, presence or absence of a transistor, or an implant to turn a transistor permanently off or on. Other technology options may be possible such as electrically erasable random access memories, but discussion on these types of ROM is beyond the scope of this text.

Several symbolic layouts for ROM cells are shown in Fig. 8.48, along with some programming techniques. The programming technique would depend on the amount of programming that is required. In a microcode ROM in a microprocessor, transistor programming would be preferable, as this would minimize the dynamic power dissipation (less capacitance on word lines). It can also affect speed if the load on word lines can be balanced in a sparse ROM. In a generic circuit that is mask programmable, metal programming may be desirable.

8.5.5 Row decoders

The row decoder in a random access memory is required to select 1 of 2^m rows. For instance, for a four-word memory, the truth table

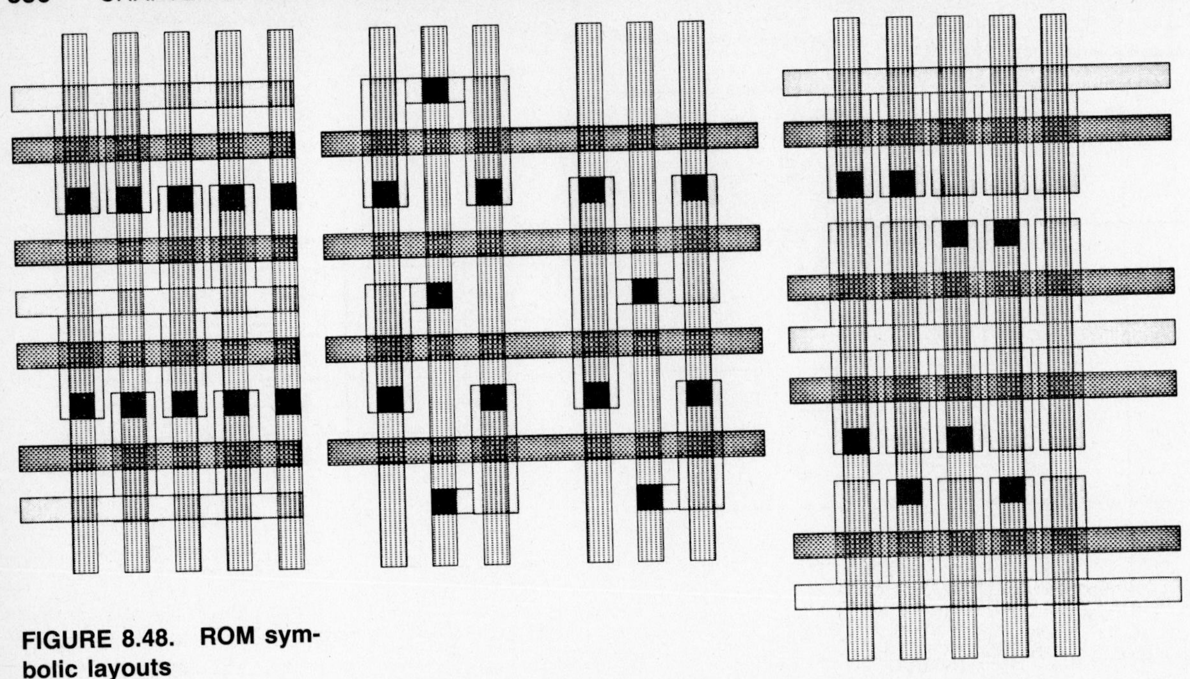

FIGURE 8.48. ROM symbolic layouts

is shown in Table 8.3. This may be implemented as a set of NOR gates, as shown in Fig. 8.49a, or in AND gates, according to Fig. 8.49b. As we will see, the NOR implementation is preferred.

When considering a memory cell (RAM or ROM) with precharged BIT lines, the simple NOR structure may be used. However, usually buffering is required to drive the WORD line. In this situation, we then have an OR structure in which all decoders except one are activated. This leads to multiple WORD lines being asserted and hence to corrupted data in the cells addressed. An additional problem in CMOS occurs if a static NOR gate is used. The series p-transistor structure slows the WORD line rise time and also increases the size of the decoder.

If a static decoder is required, then the gate input reduction techniques previously mentioned in Chapter 5 may be used. Fig.

TABLE 8.3. 2-bit row decoder function

INPUT		OUTPUT			
A_0	A_1	R_0	R_1	R_2	R_3
0	0	1	0	0	0
0	1	0	1	0	0
1	0	0	0	1	0
1	1	0	0	0	1

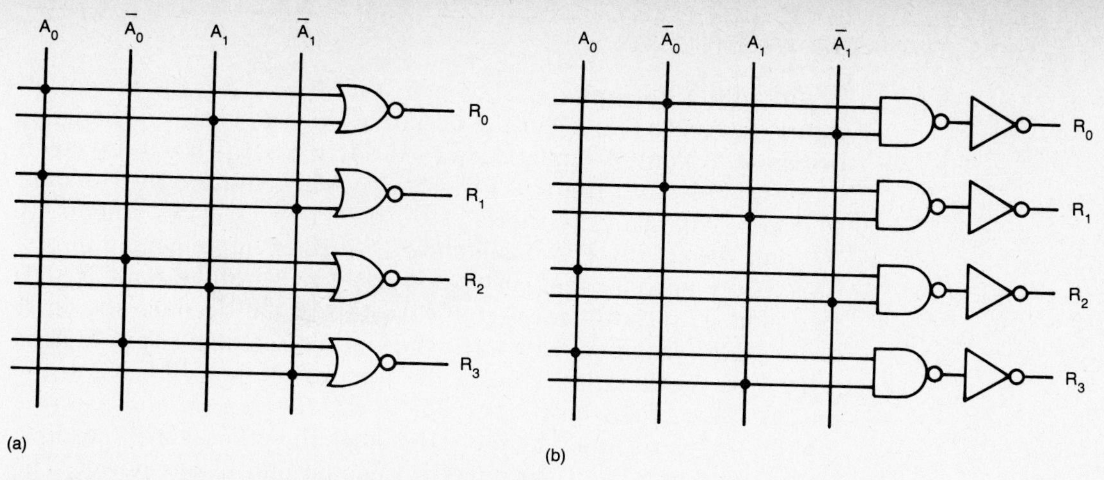

FIGURE 8.49. Row decoder logic schematics

8.50a shows one approach, while Fig. 8.50b shows an alternative implementation. A 2-input NOR gate is preferred for the last stage to allow fast rise time. The p-transistors in the gate should be designed to achieve the required rise time on the WORD line. A further

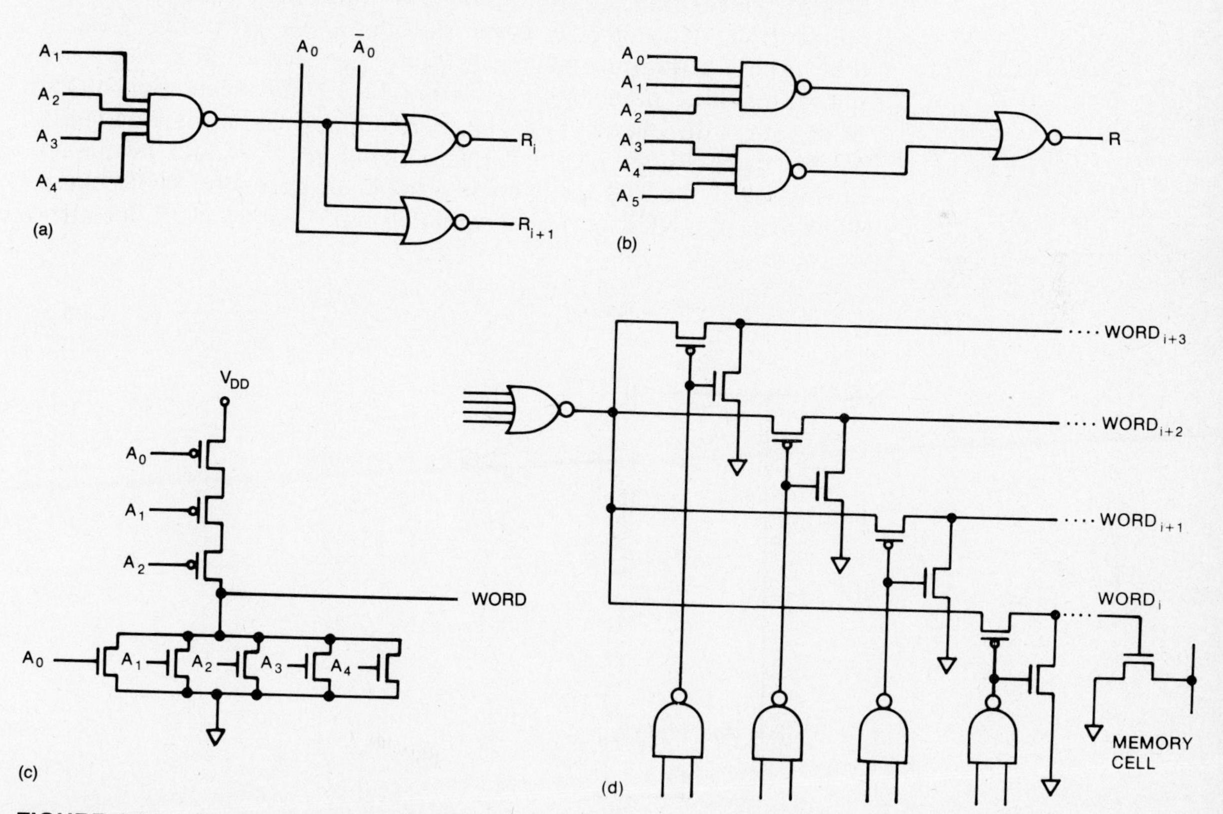

FIGURE 8.50. Static row decoder circuits

pseudo-nMOS decoder is shown in Fig. 8.50c. In this design, all inputs have to be "off" for the output of the gate to rise. Transistors driven by A3 and A4 must be ratioed relative to the p pull-up chain to adequately pull the word line down when those transistors are selected. This gate dissipates static power in all gates where A0, A1, and A2 are '0'. Fig. 8.50d shows a further implementation of a CMOS row decoder that employs a static NOR gate, a static NAND gate, and a specially connected inverter. In the decoder shown, 1 of 16 NOR decoders will have the output set high. This high is passed to the inverter. One of the four inverters will have a low output. This turns the p-transistor in the inverter on and passes a high to the word line. The n-devices keep the other word lines low. Note that 1 of 64 row decoders is enabled and draws power. The circuit could be slow due to the series p devices, but is small in size.

Dynamic row decoders are small, fast, and relatively safe to design. A very simple row decoder is shown in Fig. 8.51. This uses a pseudo-domino AND gate. This was successfully used for the 1 of 32 decoder in the RAM discussed in Section 9.1. The speed was carefully measured via simulation and found to be adequate for the application. A faster NOR based decoder is shown in Fig. 8.52. In this circuit a domino NOR gate does the decode. For pseudo 2-phase, clocking precharge is during $\phi 1 = 1$ and evaluation of the NOR gate is during $\phi 1 = 0$. The second clocked stage is a dynamic NAND gate, which is precharged during $\phi 1 = 1$ and evaluated during $\phi 2 = 1$. This feeds an inverter that drives the WORD line. Thus after the NOR gate has evaluated only 1 of 2^m NOR decoders

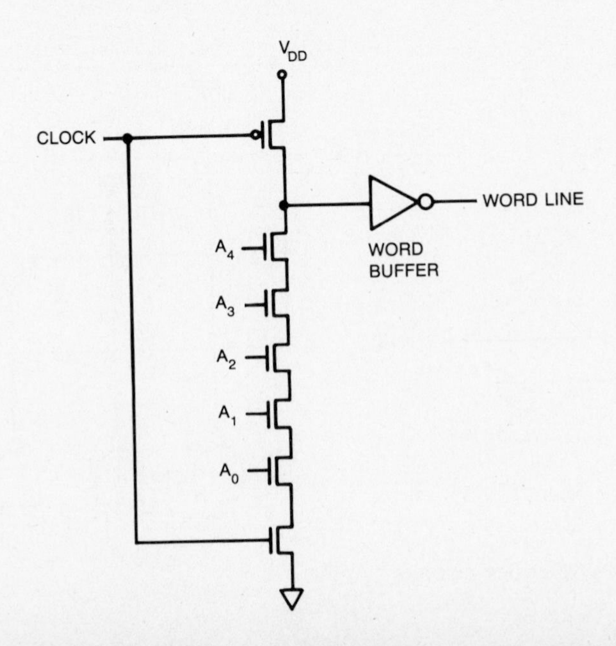

FIGURE 8.51. Dynamic AND decoder

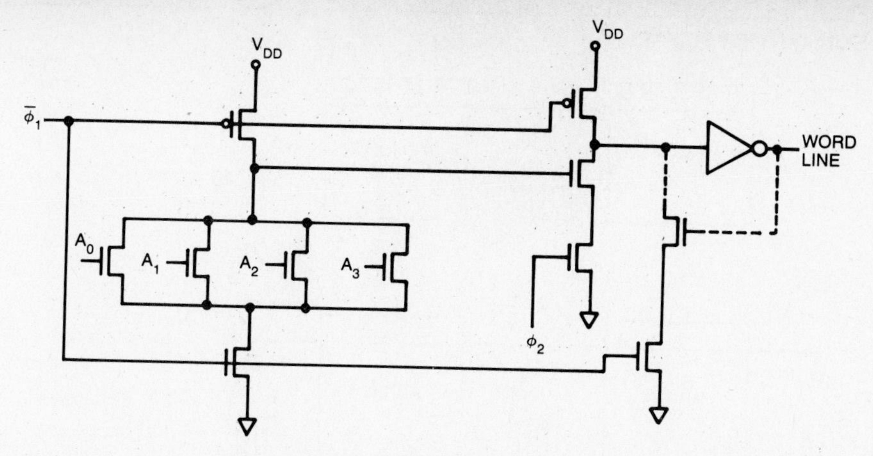

FIGURE 8.52. Dynamic NOR decoder

will remain precharged. Hence, only 1 of 2^m NAND gates will discharge and only 1 WORD line will be asserted through the inverter. Note that further decoding may be incorporated into the NAND gate. If required, this value can be latched by adding the circuitry included in the dotted outline. Concentrating on the last circuit, a circuit and representative symbolic physical implementation is shown in Fig. 8.53. The NOR transistors are usually made minimum size. The precharge transistors are sized to allow appropriate pull-up times.

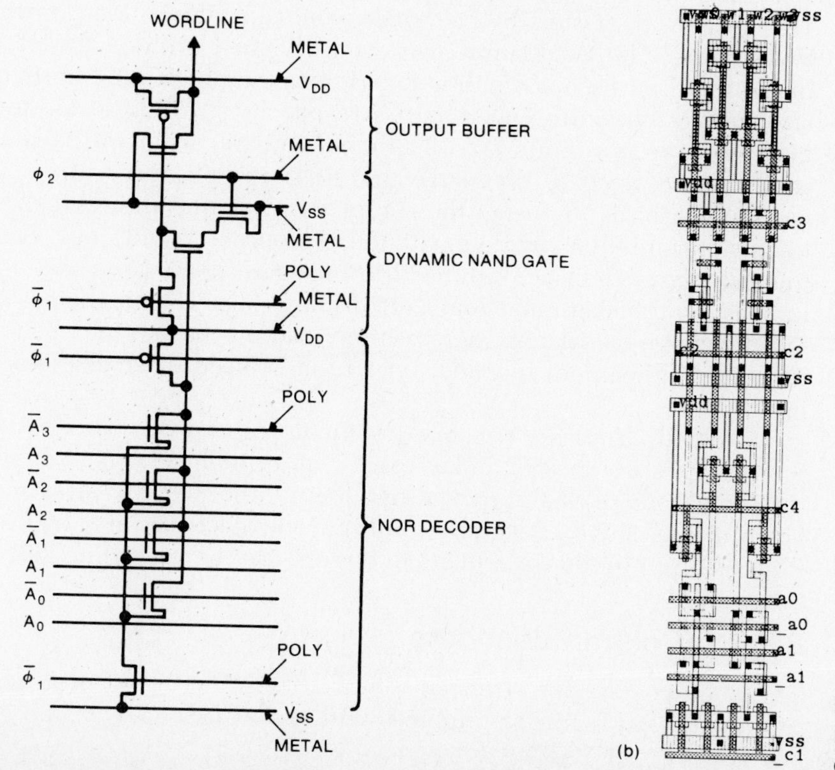

FIGURE 8.53. Symbolic layout for dynamic NOR decoder

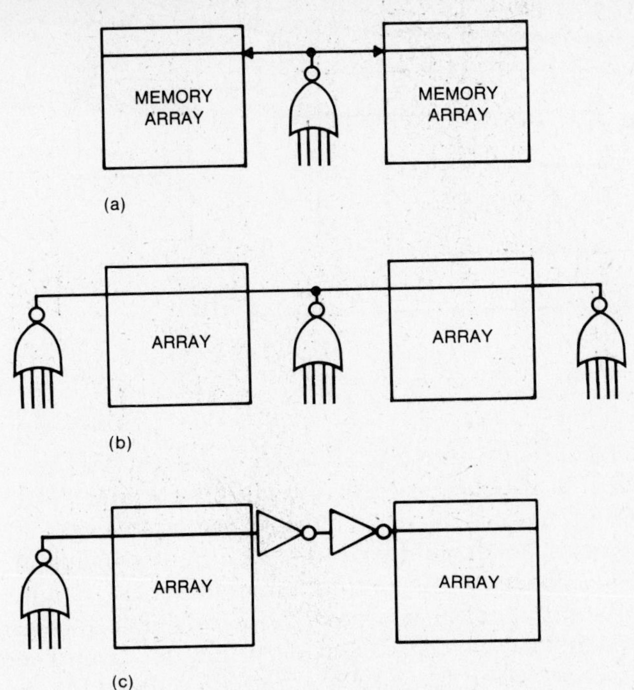

FIGURE 8.54. Methods of reducing poly word line delay in a RAM

The design might start by calculating the capacitance load on the WORD line. The WORD buffer is then designed to drive this load in the required time. This then determines the precharge transistor size in the dynamic NAND gate. At first cut all other transistors may be made minimum size. The precharge transistor in the NOR gate may be sized to precharge this node in appropriate time. A good approach is to design the complete gate with minimum sized transistors and then simulate with the appropriate WORD load. Each transistor may then be optimized to improve speed. Note that any increase in transistor size loads the driving stage. Hence, increasing the precharge transistors increases the load on the $\overline{\phi}_1$ clock driver. Thus any adjustment in transistor size must keep these changes in mind.

If WORD lines are run in polysilicon (as opposed to silicide), then the distributed RC delay can be quite high. Fig. 8.54 shows three methods of dealing with this. Firstly, the memory array may be split (Fig. 8.54a). Secondly, multiple row decoders may be used (Fig. 8.54b). Finally, intermediate buffers may be used (Fig. 8.54c).

8.5.6 Column decoders

The column decoder is responsible for selecting 2^k out of 2^m bits of the accessed row. Two methods will be considered. The simplest

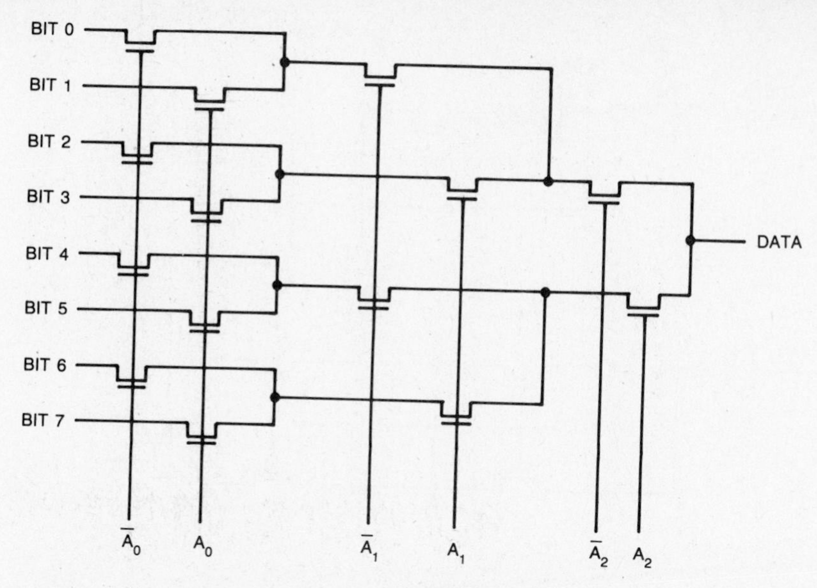

FIGURE 8.55. Column tree decoder

is the tree decoder shown in Fig. 8.55. Note that in the case of a memory with precharged BIT lines, care must be taken to precharge the sense amplifier (Section 8.5.7). This decoder is also potentially slow due to the series pass gates. It is a unilateral circuit, quite well suited for a ROM. An alternative column decoder is shown in Fig. 8.56. In this design, a NOR gate switches a single transmission gate.

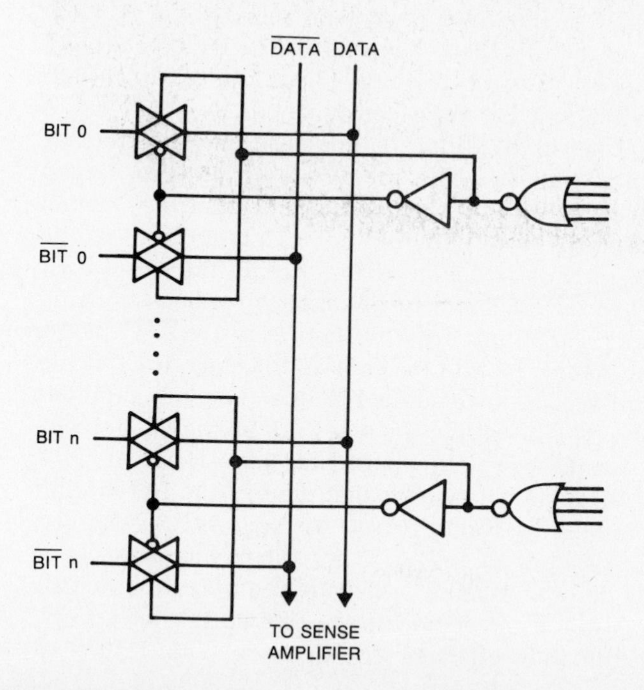

FIGURE 8.56. Logic column decoder

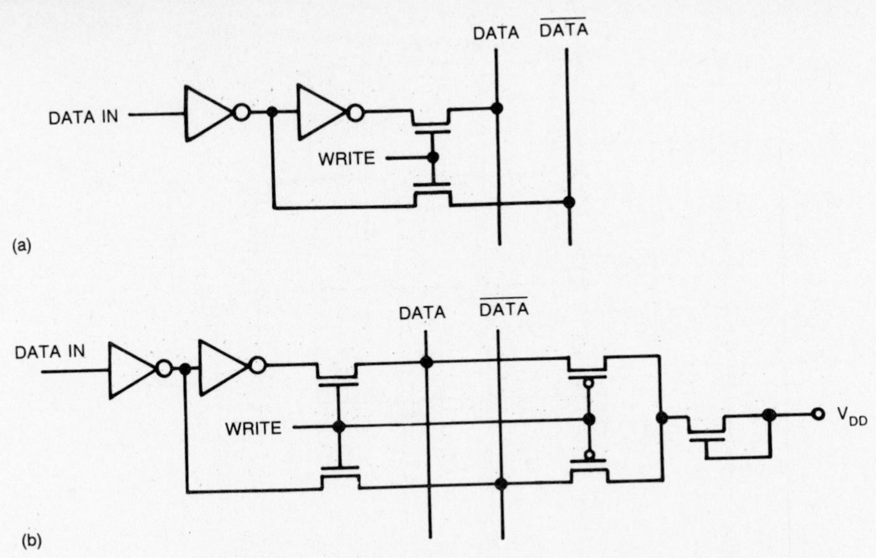

FIGURE 8.57. RAM write circuitry

As the column decoder has a long setup time ($T_{PRECHARGE}$ + T_{WORD} + T_{BIT}), static gates may be suitable in this role. The transmission gate may be a single transistor or a complementary transmission gate.

8.5.7 Read/write circuitry

Once a particular pair of BIT lines have been selected, circuitry is required to write and read the cell. The write circuitry is relatively straight-forward. Fig. 8.57 shows a variety of techniques [MMSN80].

The read circuitry can be more complicated. One can use a regular inverter as a sense amplifier. In this case, the inverter has to detect a low going signal (where the BIT lines are precharged high). The problem with this is that the fall time can be quite large. The speed can be improved by moving the switching point of the inverter towards V_{DD}. This is achieved by altering the ratio of β_n to β_p. A suitable inverter is shown in Fig. 8.58 (in this case, using unit-sized transistors, a 2-input NAND structure is used). We have in effect reduced the noise margin to increase the sensitivity of the amplifier. Note that if leakage lowered the BIT line, the sense amplifier could trigger falsely. An alternative sense technique that also trades noise margin for sensitivity is shown in Fig. 8.59. Here, the BIT line is precharged through a transistor that clamps the BIT line to a voltage a threshold below a clamp voltage. Capacitors C_1 is charged to this voltage and C_2 is charged to V_{DD}. When the memory transistor pulls down the BIT line, the clamp transistor attempts to replenish the charge. As C_1 is much larger than C_2, a charge transfer occurs, which has an amplification effect on the C_2 node. This results in

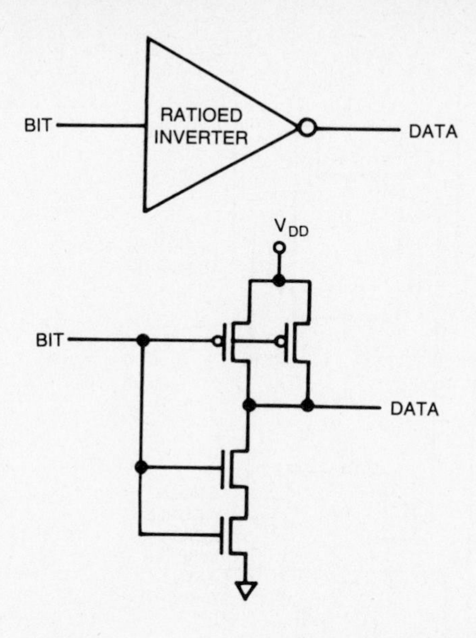

FIGURE 8.58. Simple sense amplifier

very fast switching. This technique can be used in RAMs, ROMs, and even PLAs to improve the sense speed. Note that when precharge ceases, charge may leak towards C_1 if the BIT line is not fully charged (as could result if the series n-device was too small to charge C_1 in the allowed precharge time). If the clamp voltage is such that it allows the voltage at C_2 to fall below the inverter threshold, then the output can switch erroneously. The bottom line with this and the previous technique is that these are not design options that can be "eye-balled." One has to carefully simulate the proposed circuit for the actual loading and typical cycle timings.

A differential sense amplifier senses a small difference between levels on the BIT lines and amplifies this to provide very fast sensing. Thus the BIT lines have to only change slightly in level to detect

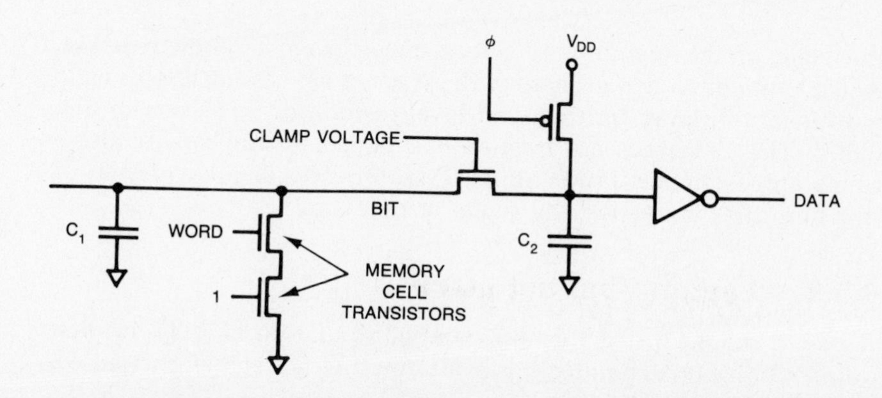

FIGURE 8.59. Charge balance sense amplifier

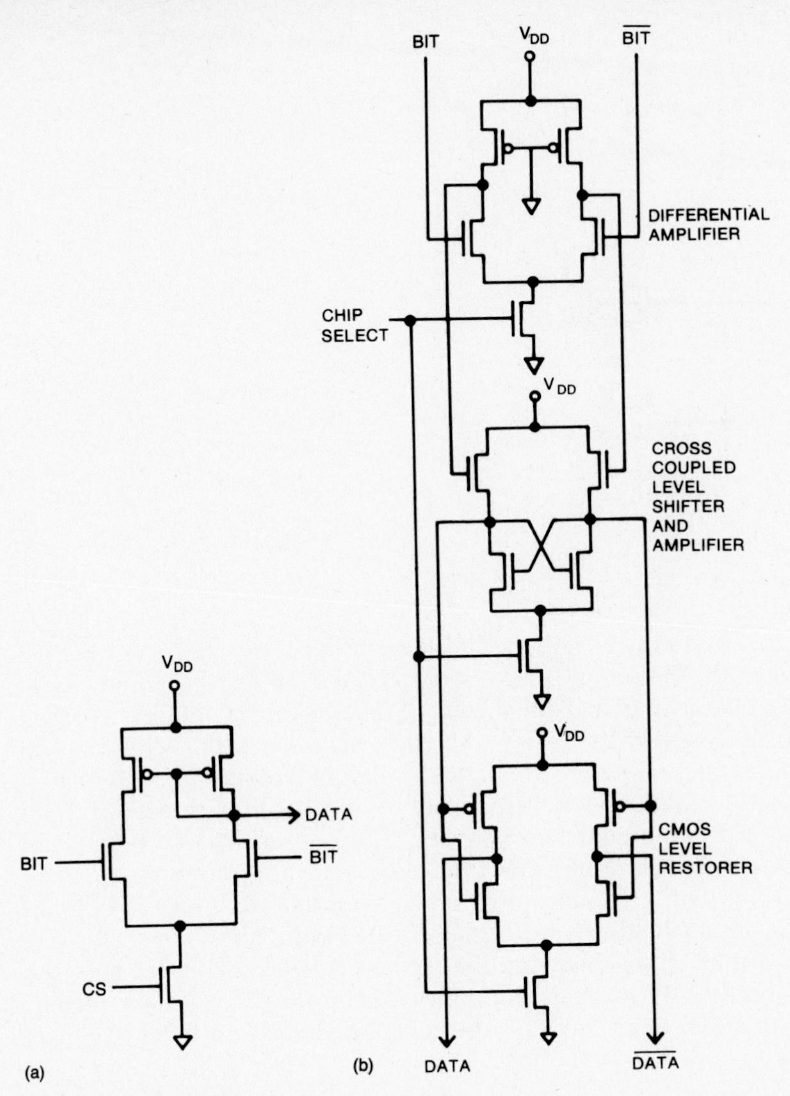

FIGURE 8.60. Differential sense amplifiers

the state of the memory. A single stage design is shown in Fig. 8.60a. Quite often two or more of these stages are cascaded to provide amplification, level shifting, and level restoration as shown in Fig. 8.60b. The BIT lines are frequently charged to mid-rail for these sense amp schemes [MMSN80] [OYIH80]. The detailed design of this amplifier is beyond the scope of this text.

8.5.8 Last in, first out stack

A useful memory structure for data manipulation is a last in, first out stack (LIFO). The block diagram for such an element is shown

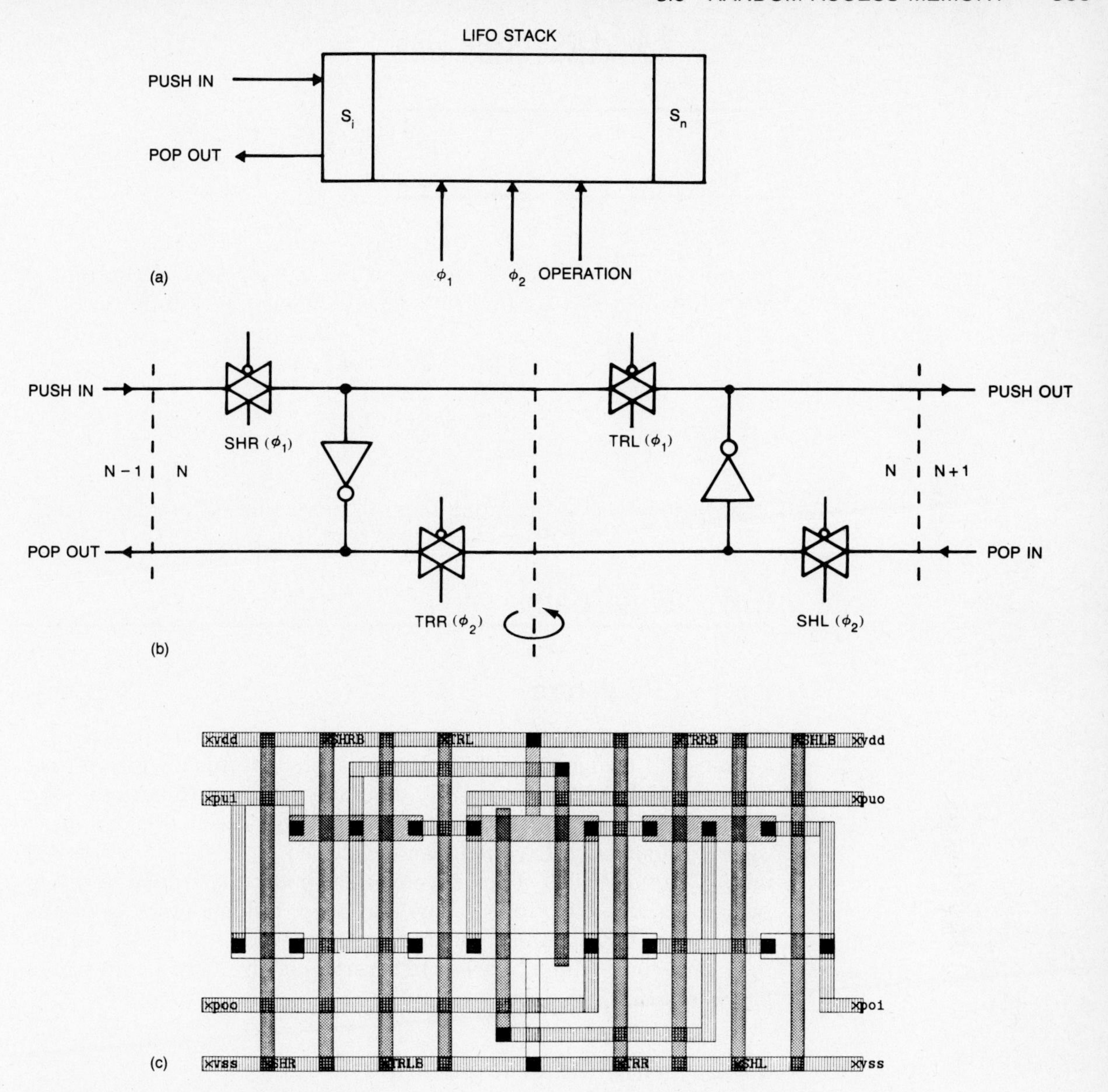

FIGURE 8.61. LIFO implementation

in Fig. 8.61a. The terminals consist of an input data port, output data port, stack controls, and clocks.

A PUSH operation pushes data onto the top of the stack at the input port. All internal locations are pushed, with the last location being pushed off the end. A POP operation pops the top location to the output port and all internal locations. The circuit diagram

TABLE 8.4. LIFO operations

OPERATION	ϕ_1	ϕ_2
PUSH	SHR	TRR
POP	TRL	SHL
NO-OP	TRL	TRR

for one bit of the stack is shown in Fig. 8.61b. A symbolic plot is also shown in Fig. 8.61c. The control signals are as follows:

TRL = Transfer Left
TRR = Transfer Right
SHL = Shift Left
SHR = Shift Right

Table 8.4 shows the operation in terms of a 2-phase clocking scheme.

8.6 Data paths

8.6.1 Registers

Any of the register designs covered in Chapter 5 may be used for a data path design. The static RAM covered in this chapter is also very useful. In CMOS, a good layout policy is to route busses through the middle of a cell between the n- and p-devices as this can mask the n to p spacing. Layout styles developed in Chapter 5 are useful here. Chapter 9 illustrates several data paths that illustrate these principles. Bus clocking strategy may vary. Using a precharged bus allows wire-ORing, and some memory structures may in fact require it. A static bus driven by tri-state inverters is also very useful where precharge clocks are not available.

8.6.2 Arithmetic logic units

The adder section provides the data path designer with a wide range of area/performance adder structures. Boolean operations are easily catered for by use of function blocks such as shown in Chapter 5. Chapter 9 also shows an implementation of an arithmetic unit with multiply hardware.

8.6.3 Barrel shifters

Barrel shifters are important elements in many microprocessor designs. Fig. 8.62 shows an implementation that is easy to implement in

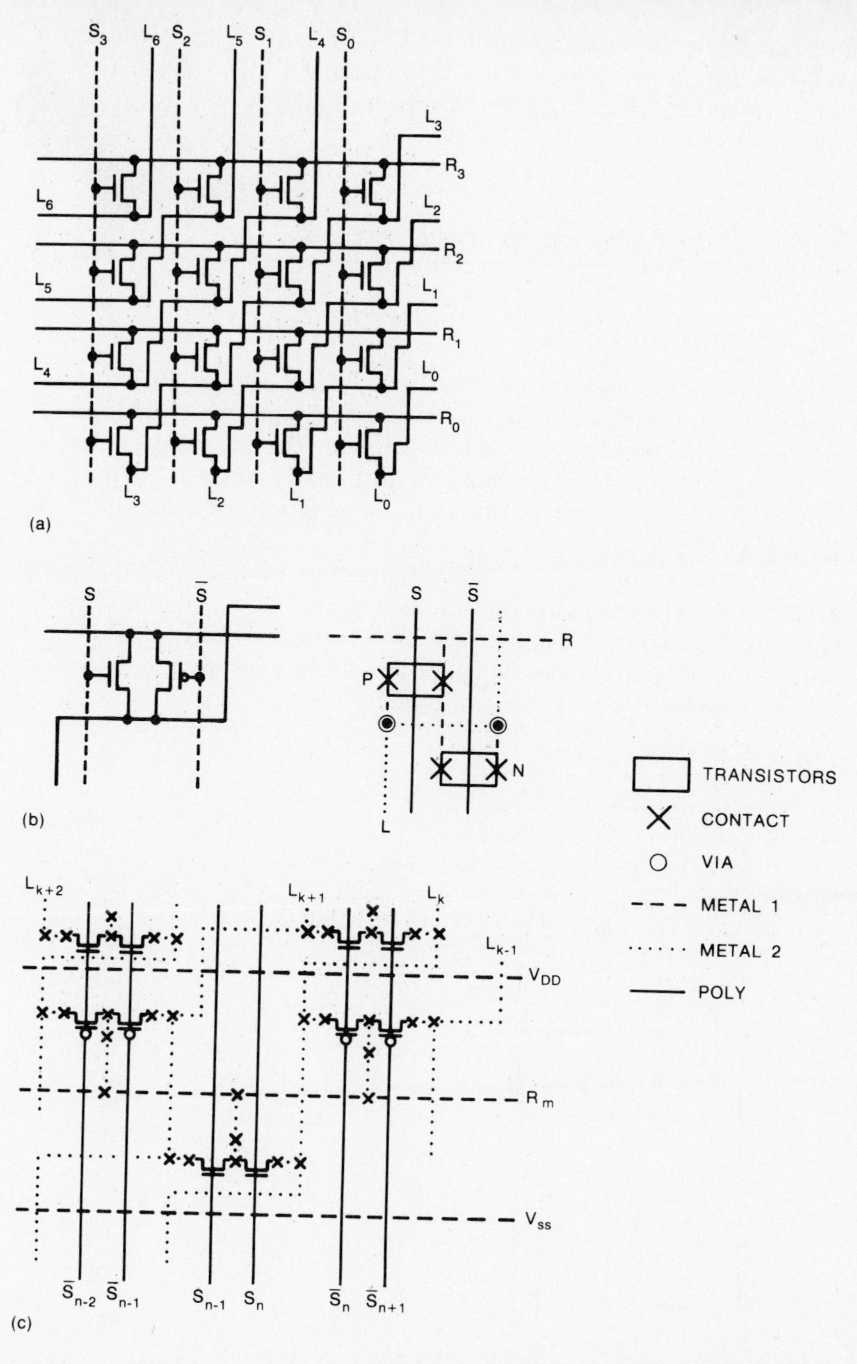

FIGURE 8.62. Barrel shifter implementation

CMOS. Fig. 8.62a shows a regular design based on a design in [SMKP82]. Layouts in a two level metal process for a cell is shown in Fig. 8.62b and Fig. 8.62c. This was simply arrived at by replacing a full transmission gate for each n-device in the original design. A precharged version may be implemented by omitting the p-transistor and inserting a precharge transistor per output line. Literal lines Li are switched by shift lines Sk to appear on result lines Rn to effect shift operations.

8.7 Programmable logic arrays

8.7.1 Introduction

A programmable logic array (PLA) provides a regular structure for implementing combinational and sequential logic functions. A PLA may be used to take inputs and perform some combinational function of these inputs to yield outputs, or additionally some of the outputs may be fed back to the inputs, thus forming a finite state machine shown in Fig. 8.63.

A typical PLA uses an AND-OR structure similar to that shown in Fig. 8.64. This implementation also shows clocks to latch inputs and outputs. The basis for a PLA is sum of products form of representation of binary expressions. For example, consider the following expressions that have to be evaluated:

$$Z_0 = X_0$$
$$Z_1 = X_1 + \overline{X_0} \,.\, \overline{X_1} \,.\, X_2$$

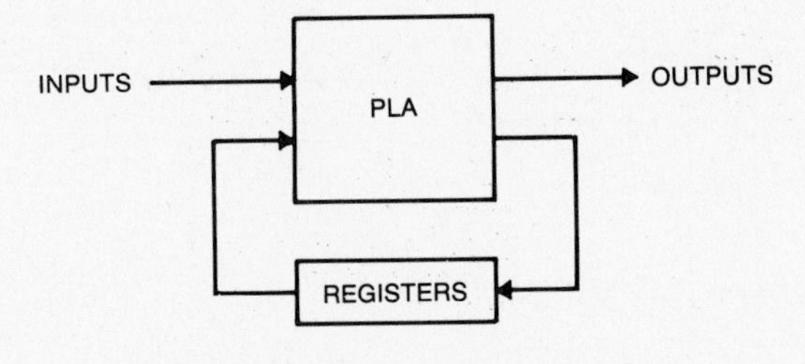

FIGURE 8.63. A finite state machine

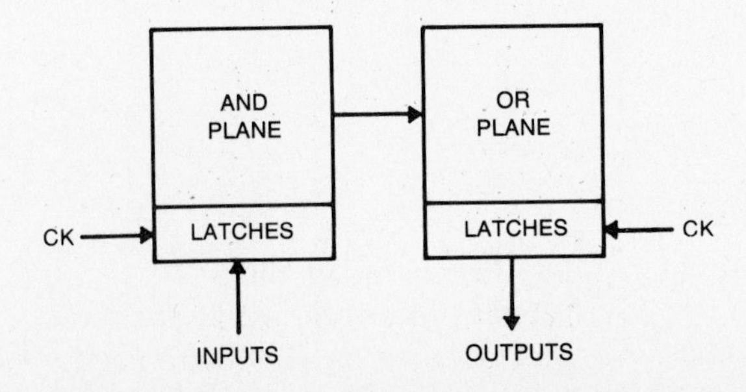

FIGURE 8.64. AND-OR PLA

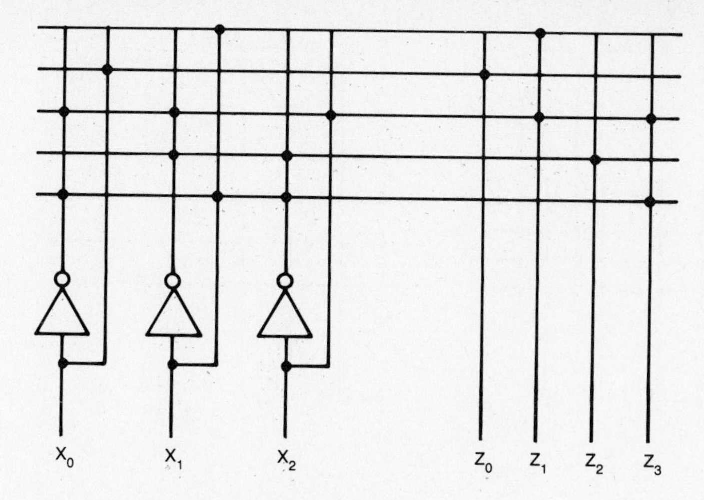

FIGURE 8.65. PLA example

$$Z_2 = \overline{X_1} . \overline{X_2}$$

$$Z_3 = \overline{X_0} . \overline{X_1} . X_2 + \overline{X_0} . X_1 . \overline{X_2},$$

where Z_0, Z_1, Z_2, Z_3 are the four output terms and X_0, X_1, X_2 are the input variables. There are five product terms, namely X_0, X_1, $\overline{X_0} . \overline{X_1} . X_2$, $\overline{X_1} . \overline{X_2}$, and $\overline{X_0} . X_1 . \overline{X_2}$. Thus these terms would be formed in the AND array of the PLA, as shown in Fig. 8.65. The four outputs are formed by ORing the appropriate product terms. Normally, high speed PLAs are implemented as two NOR arrays, as shown in Fig. 8.66. (Although NAND arrays may be used for slow applications.) By using inverting inputs and outputs, the AND-OR structure is maintained.

8.7.2 Electrical and physical design of CMOS PLAs

The electrical design of a CMOS PLA depends on the generic style of PLA. A straight-forward physical implementation for a PLA is

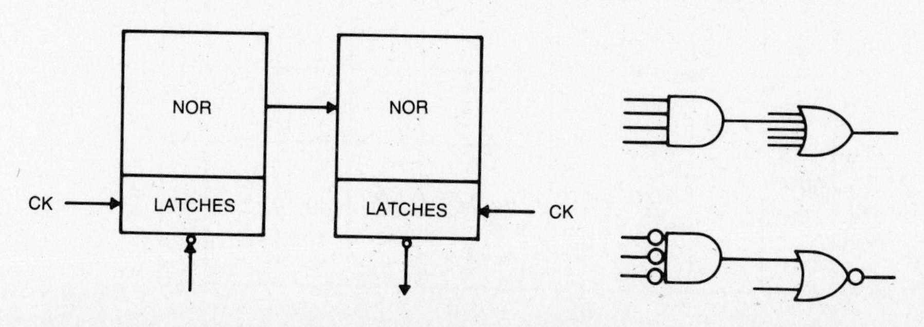

FIGURE 8.66. NOR-NOR PLA

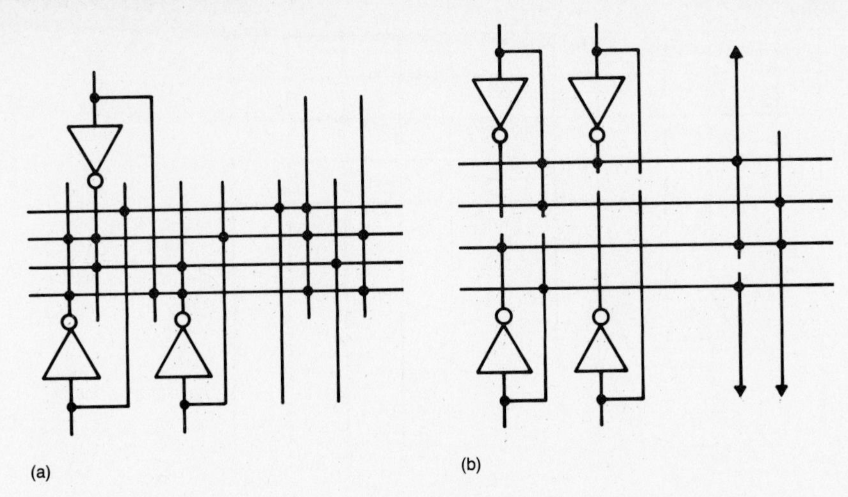

FIGURE 8.67. PLA variations

represented by Fig. 8.65. Variations to this involve multiple-sided access (Fig. 8.67a) and simple folding (Fig. 8.67b).

A generic floor plan for a "simple" PLA is shown in Fig. 8.68. This has been designed as a set of tiles, which are designated by letters. In the treatment of various circuit options this naming convention will be used to designate particular cells. Brief descriptions for the cells are as follows:

A	AND plane programming cell
O	OR plane programming cell
AO	AND-OR communication cell
IN	AND plane Input cell

TL	TA	TA	TA	TA	TM	TO	TO	TO	TR
LA	A	A	A	A	AO	O	O	O	RO
LA	A	A	A	A	AO	O	O	O	RO
LA	A	A	A	A	AO	O	O	O	RO
LA	A	A	A	A	AO	O	O	O	RO
LA	A	A	A	A	AO	O	O	O	RO
BL	IN	IN	IN	IN	BM	OUT	OUT	OUT	BR

FIGURE 8.68. PLA generic floor plan

OUT	OR plane Output cell
LA	Left AND plane cell
RO	Right OR plane cell
BL	Bottom left cell
BM	Bottom middle cell
BR	Bottom right cell
TL	Top left cell
TA	Top AND cell
TM	Top middle cell
TO	Top OR cell
TR	Top right cell

8.7.3 Pseudo-nMOS NOR gate

The most straight-forward PLA design uses a pseudo-nMOS NOR gate. Fig. 8.69 shows the circuit diagram with the key cell positions identified. Cell AO can either be a layer-change cell or can be used to buffer the AND array outputs. Design of the pseudo-nMOS NOR gates would follow the guidelines given in previous chapters. Advantages of this PLA include simplicity and small size. Disadvantages occur due to the static power dissipation of the NOR gates and possible speed problems. Any convenient latch may be used, with a simple dynamic 2-phase latch shown. This PLA could be fairly independent of the overall system clocking strategy. Cells TL, TA,

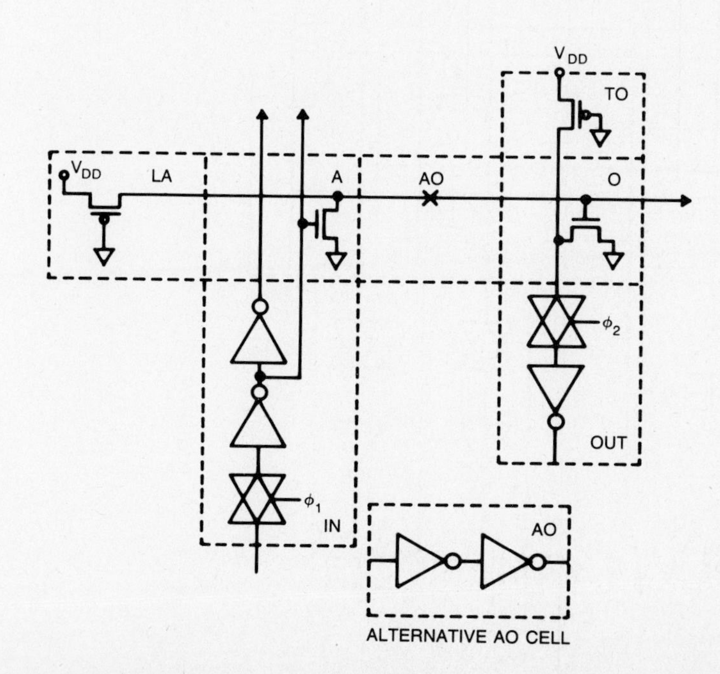

FIGURE 8.69. Pseudo-nMOS PLA circuits

BL, BM, TM, TR, RO, and BR are used to route power and clocks as necessary.

8.7.4 Dynamic CMOS — 2-phase clocking

By using dynamic CMOS, the circuit shown in Fig. 8.70 may be used. This assumes a strict 2-phase clocking strategy. In this clocking strategy the inputs are latched during $\phi 1$ and the AND stage is precharged during $\phi 1$. If $\phi 2$ is used to latch the second stage, an intermediate clock is required to precharge the OR plane. This is generated in this implementation by the portion of circuitry in cells TL, TA, and TM in Fig. 8.70. This uses a dummy product row that discharges at the worst-case rate according to the loading of the AND array. This self-timed clock provides a precharge/evaluate clock for the OR array.

An alternate implementation is shown in Fig. 8.71. Here, a pipeline latch is placed between the AND and OR planes. The first stage is latched and precharged during $\phi 1$. It is evaluated during $\overline{\phi 1}$. The

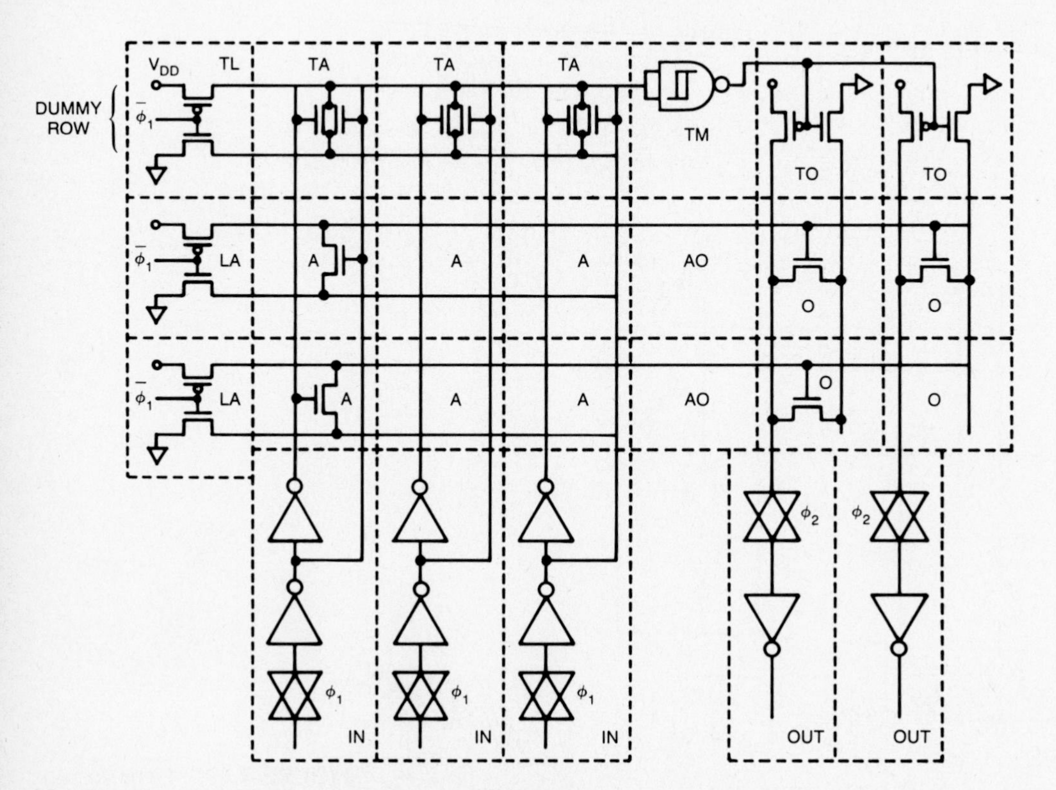

FIGURE 8.70. Dynamic CMOS 2ϕ PLA circuits—1

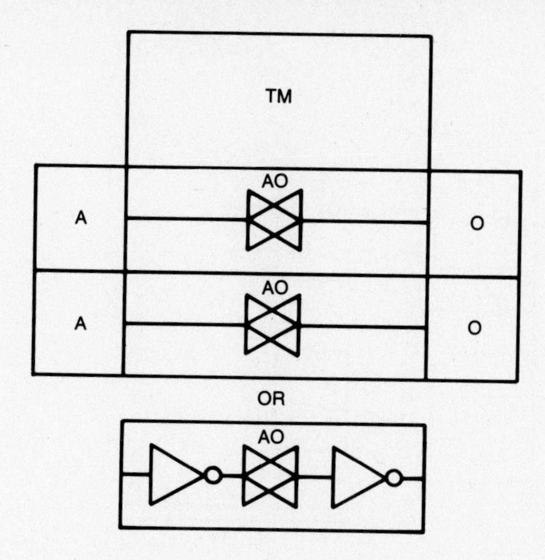

FIGURE 8.71. Dynamic CMOS 2ϕ PLA circuits—2

pipeline latches the AND output during $\phi 2$. The OR state is precharged during $\phi 2$ and evaluated during $\overline{\phi 2}$. The output latch is a master-slave device to provide the correct relationship to external logic. This PLA is described in more detail in Section 9.2.

A further PLA using 2-phase clocking is shown in Fig. 8.72. This uses a dynamic NOR gate without a ground switch. The input to the AND plane is gated by a set of NOR gates to ensure valid precharge. This structure generates less power and ground noise than the two other structures at the expense of area.

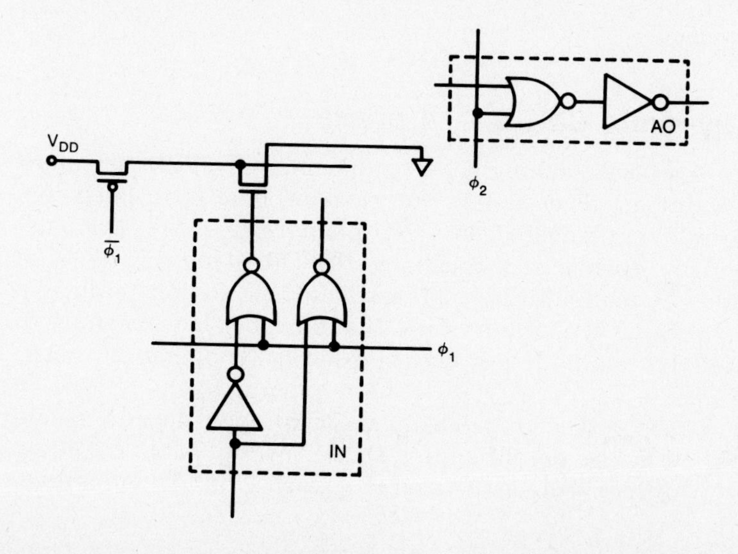

FIGURE 8.72. 2-phase PLA

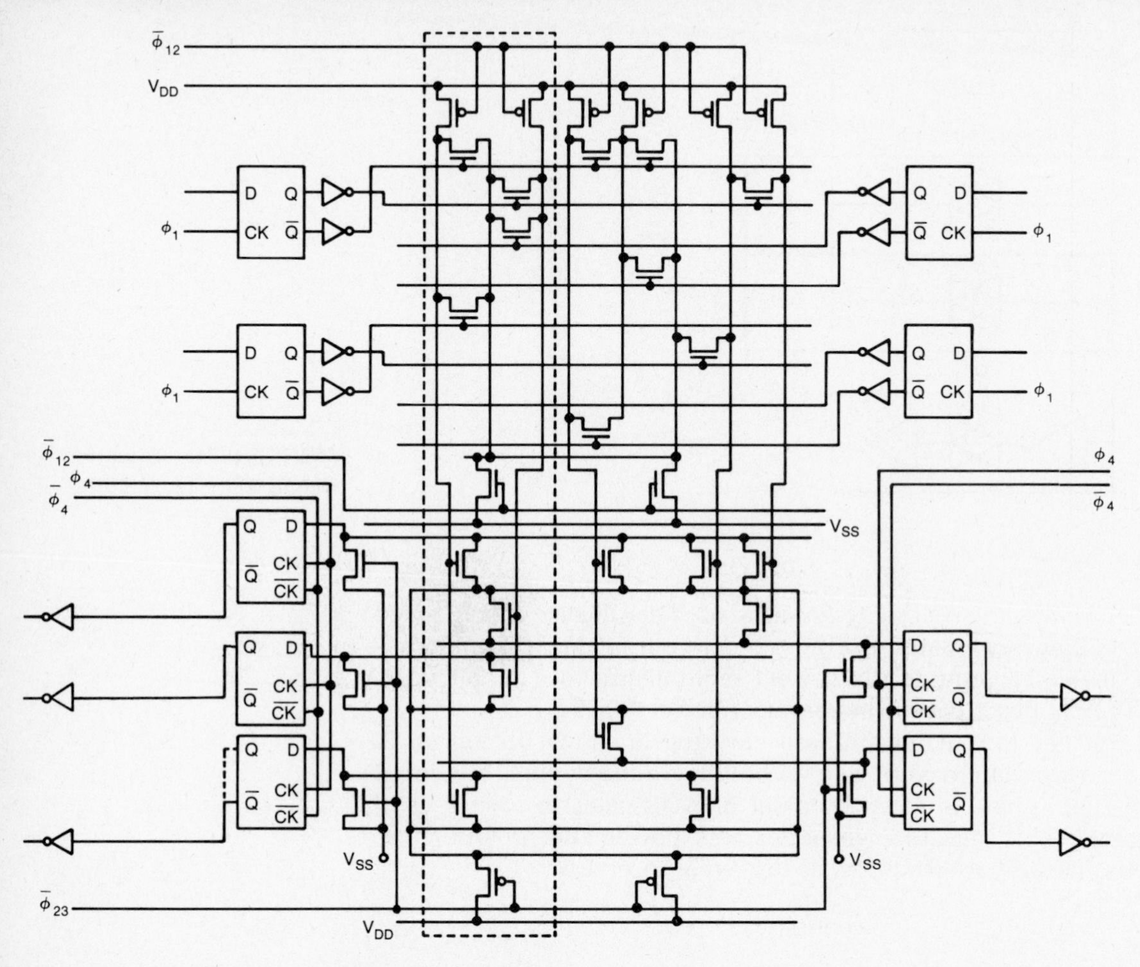

FIGURE 8.73. Dynamic CMOS 4ϕ PLA circuits

8.7.5 Dynamic CMOS — 4 phase

By using a 4-phase clocking scheme the internal clock generator may be eliminated. The strategy for clocking here is to latch the inputs during $\phi 1$, precharge the AND plane during $\phi 12$, evaluate the AND plane during $\overline{\phi 12}$, precharge the OR plane during $\phi 23$, evaluate the OR plane during $\overline{\phi 23}$, and latch the OR plane during $\phi 4$. Fig. 8.73 shows the circuits used. Static latches have been used in this PLA. (Dynamic latches may be used if DC operation is not required.)

Some increase in packing density and extra logic decoding may be obtained by using the structures shown in Fig. 8.74 [LaSh82]. Here four "wordlines" are associated with each ground-switch line.

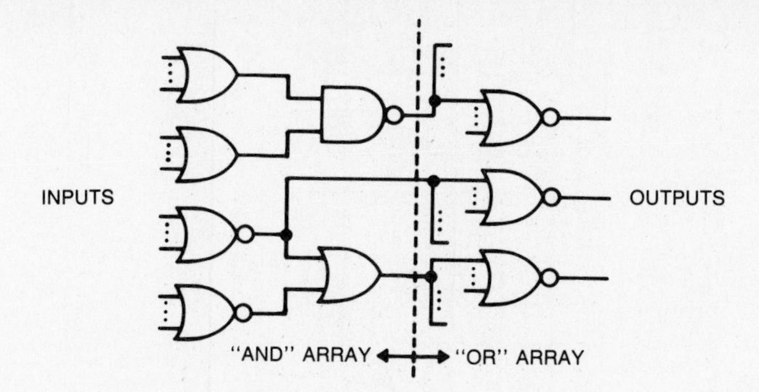

FIGURE 8.74. Decoded PLAs

This leads to an OR-NAND structure in the AND plane of the PLA. This may be represented as a NOR-OR structure as shown. This style of layout is represented by the PLA in Fig. 8.73.

8.7.6 Detailed PLA layout

A detailed layout for the PLA shown in Fig. 8.65 is shown in Fig. 8.75 for a PLA which uses a four phase clocking scheme. Dynamic latches are included in the input and output stages. Two wordlines have been used with each ground-switch line in the AND plane. Plate 15 shows the layout in color.

8.7.7 PLA else clause implementation

Quite often it is desirable to implement an *IF-THEN-ELSE* conditional statement in the control flow of an algorithm. Fig. 8.76 shows the basic circuitry that can be disposed between the AND and OR planes of the PLA to implement this function [KMSW83]. Thus *IF* the AND array has a particular pattern, *THEN* the signal *I* is activated, *ELSE* signal *E* is activated.

8.7.8 PLA design points

Dynamic PLAs can generate substantial noise as a result of the size and regular clocking applied to the circuits. Fig. 8.77 shows an equivalent circuit for a dynamic PLA plane. When the n-transistor ground switch turns on, the discharge current follows the path shown. This causes a transient in the power supply bus. By locally grounding the PLA, this noise can be substantially reduced [LaSh82].

Note that in the dynamic PLAs the ground-switch line is run in metal. The effect of running this in diffusion would be to present a much larger capacitance to the n-transistor ground switch, thus slowing the circuit.

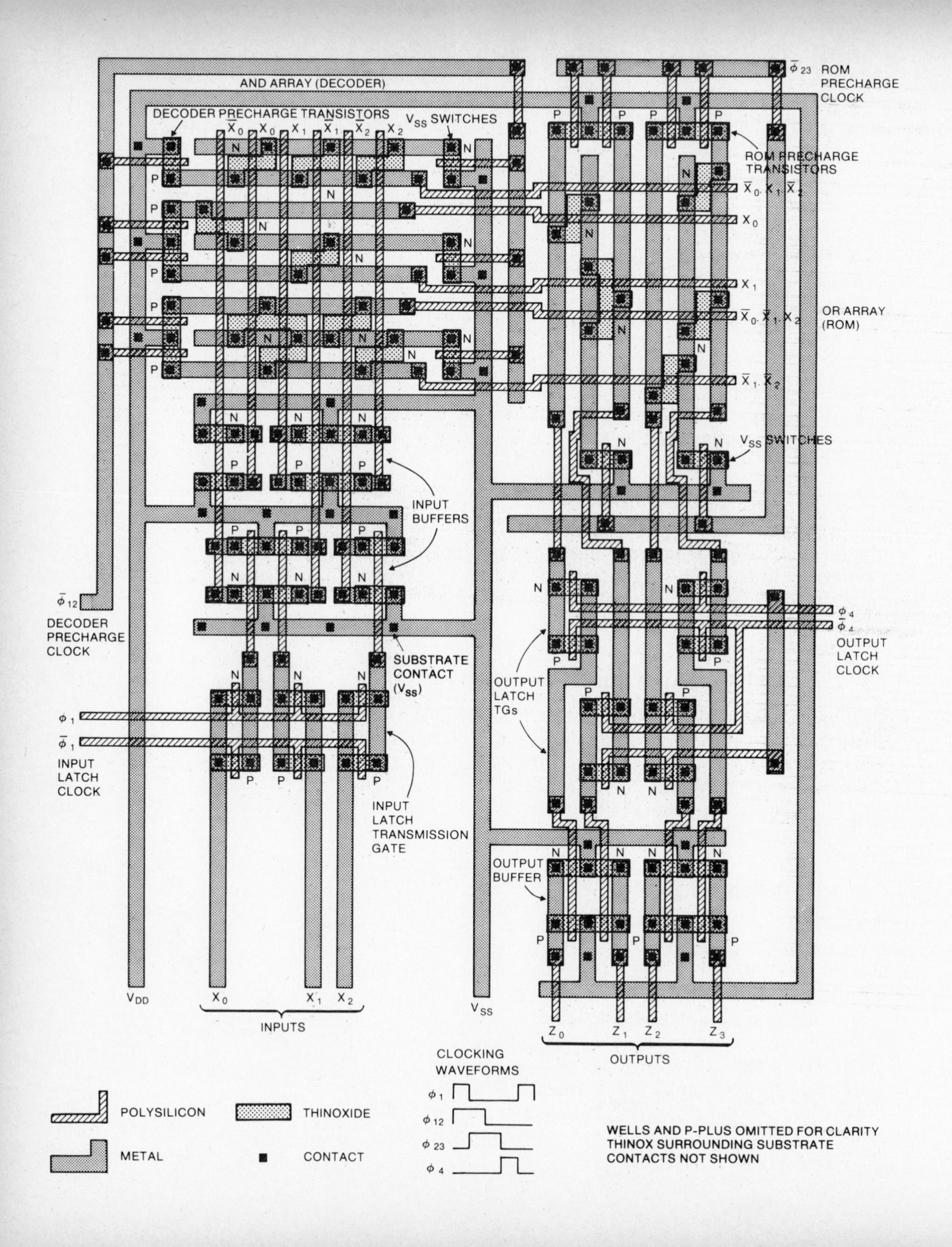

AND ARRAY (DECODER)
DECODER PRECHARGE TRANSISTORS
V_{SS} SWITCHES
$\bar{\phi}_{23}$ ROM PRECHARGE CLOCK
ROM PRECHARGE TRANSISTORS
$\bar{X}_0$ X_0 X_1 $\bar{X}_1$ $\bar{X}_2$ X_2
$\bar{X}_0 \cdot X_1 \cdot \bar{X}_2$
X_0
X_1
$\bar{X}_0 \cdot \bar{X}_1 \cdot X_2$
$\bar{X}_1 \cdot \bar{X}_2$
OR ARRAY (ROM)
V_{SS} SWITCHES
INPUT BUFFERS
$\bar{\phi}_{12}$
DECODER PRECHARGE CLOCK
SUBSTRATE CONTACT (V_{SS})
ϕ_4
$\bar{\phi}_4$
OUTPUT LATCH CLOCK
OUTPUT LATCH TGs
ϕ_1
$\bar{\phi}_1$
INPUT LATCH CLOCK
INPUT LATCH TRANSMISSION GATE
OUTPUT BUFFER
V_{DD}
X_0 X_1 X_2
INPUTS
V_{SS}
Z_0 Z_1 Z_2 Z_3
OUTPUTS
CLOCKING WAVEFORMS
ϕ_1
ϕ_{12}
ϕ_{23}
ϕ_4
POLYSILICON
THINOXIDE
METAL
CONTACT
WELLS AND P-PLUS OMITTED FOR CLARITY
THINOX SURROUNDING SUBSTRATE CONTACTS NOT SHOWN

◀ **FIGURE 8.75. Detailed PLA layout**

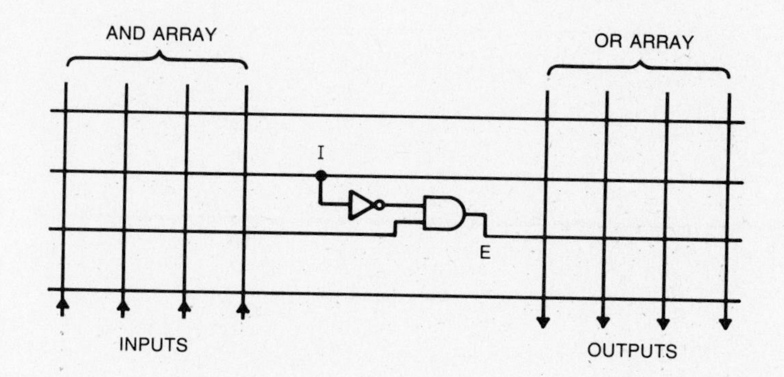

FIGURE 8.76. PLA else clause

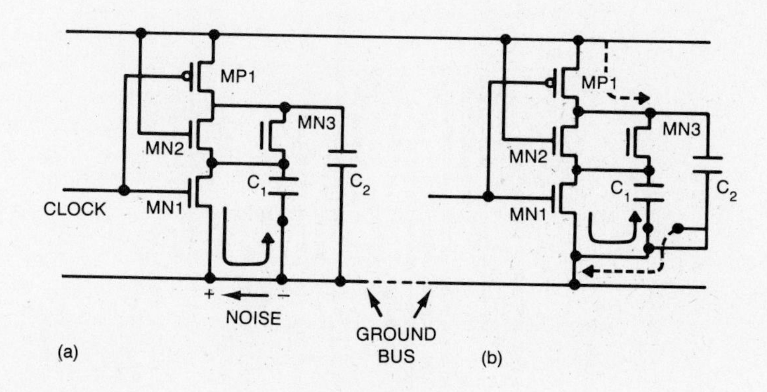

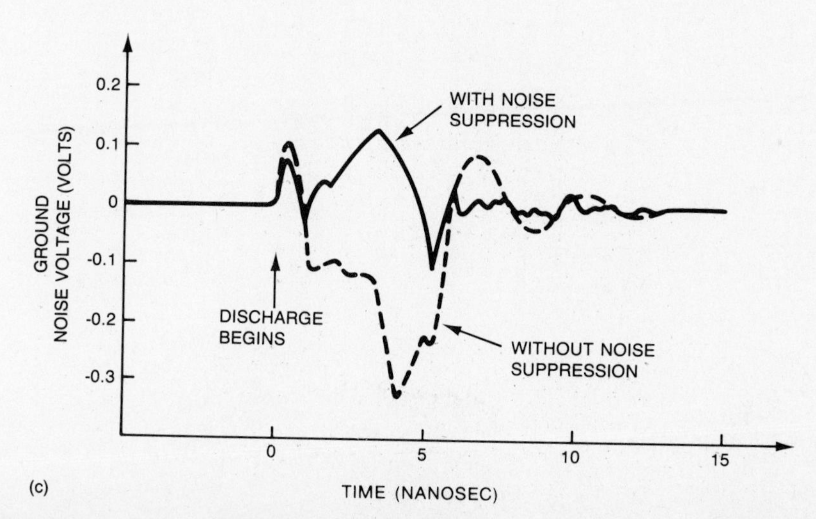

FIGURE 8.77. Noise in PLAs

TABLE 8.5. Truth table for 2-bit PPL counter

	K	C	A1	A0	D1	D0
	0	0	0	0	0	0
Count	0	0	0	1	0	1
Hold	0	0	1	0	1	0
	0	0	1	1	1	1
	0	1	0	0	0	1
Count	0	1	0	1	1	0
	0	1	1	0	1	1
	0	1	1	1	0	0
Clear	1	X	X	X	0	0

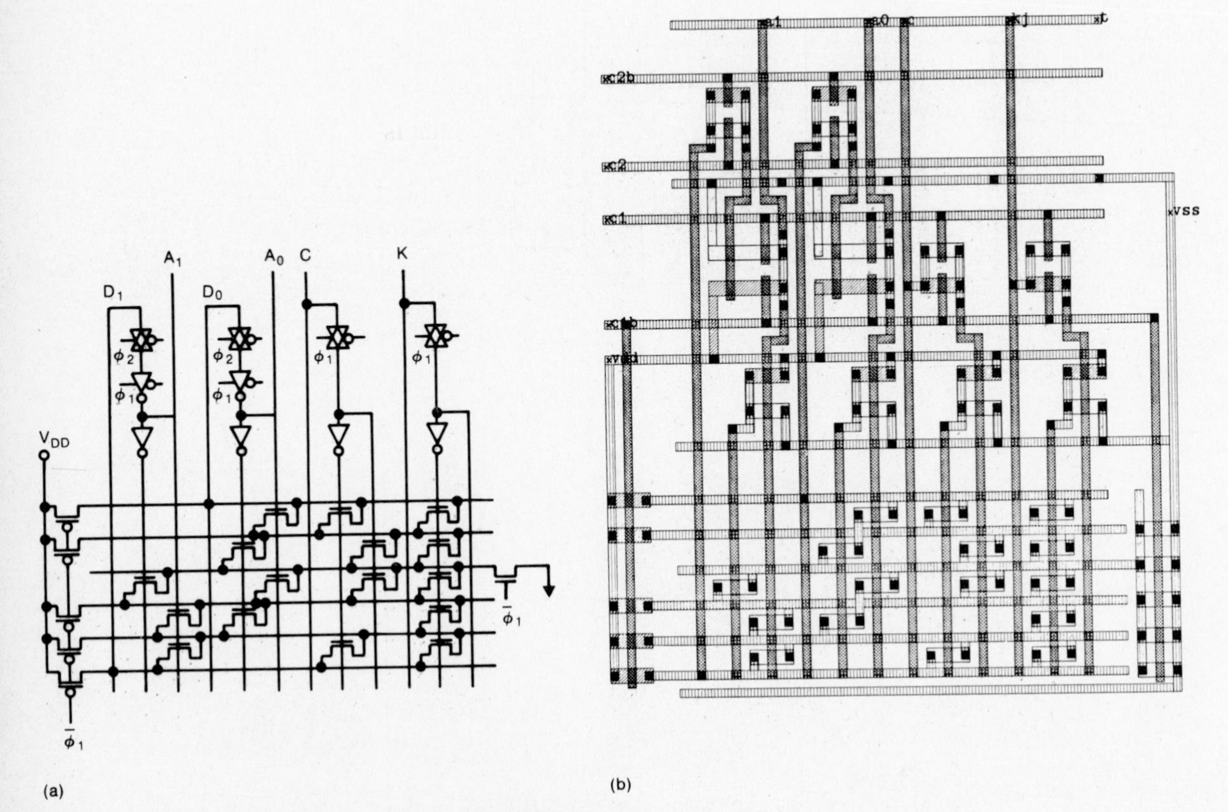

FIGURE 8.78. PPL design © IEEE 1983 ([Smit83])

With reference to the PLAs described in this section, the following recommendations may be made:

- Pseudo-nMOS — use where single clocking is available, for physically small PLAs (because they are small and to minimize static power dissipation).
- 2-phase — use with 2-phase clocking. Use switched planes for a small PLA size. Revert to input switching (Fig. 8.72) if noise is a problem and space is available.
- 4-phase — as above.
- 2-phase with intermediate latch — use if pipeline delay can be tolerated.

8.7.9 Programmable path logic (PPLs)

A PPL [Smit83] is similar to a PLA, except that the OR and AND planes are merged. This can result in a much smaller design for simple functions. An added advantage of the PPL and the SLA (storage logic array) structure that it is derived from, is the ability to symbolically represent the logic structure and have this directly map to geometry. To illustrate the PPL concept, an example of a 2-bit counter will be treated. The truth table for the 2-bit counter is presented in Table 8.5. The circuit is shown in Fig. 8.78a. A symbolic layout is shown in Fig. 8.78b. More symbolic views of the circuit may be used which map logic symbols onto a macro-grid representing the IC.

8.8 Exercises

8.1 Design a 4-phase dynamic serial adder. Show the style of input and output registers that would be used. Design a static CMOS version for the same function. Estimate the maximum clock frequency for both circuits.

8.2 Compare the speed of the static carry lookahead gate shown in Fig. 8.13 with the Manchester lookahead stage shown in Fig. 8.15. Estimate the speed for a 32-bit adder employing eight 4-bit stages.

8.3 Estimate the performance of a 32-bit BLC adder and a 32-bit carry select adder. Compare with the adders in Exercise 8.2. What adder would you use for an ALU in a 32-bit microprocessor? What adder would you use for a 512-bit adder?

8.4 Design a symbolic layout for a 16-bit parity generator using the circuit shown in Fig. 8.23b.

8.5 Design the circuit for a minimum area 12-bit comparator.

8.6 Design the circuit and layout for a minimum-sized incrementer using the circuit shown in Fig. 8.28c. Assume a pseudo 2-phase clocking strategy.

8.7 Characterize the charge pump sense amplifier for different clamp voltages. To what conclusions do you come?

PART 3

CMOS SYSTEM CASE STUDIES

Part 3 comprises five case studies of chips that have been designed in CMOS. They are included as examples to indicate architectural choices and design methodology as applied to CMOS chips. Four of the case studies feature small team designs, while one represents a large team design.

Section 9.1 deals with a programmable systolic processor designed for speech recognition. Section 9.3 summarizes work carried out on a self-routing switch network. Section 9.2 discusses a chip used in robotic vision. Section 9.5 covers the methodology used in the design of a 32 bit microprocessor. Section 9.4 treats the design and architecture of an advanced parallel graphics machine.

SYSTEM CASE STUDIES

9.1 Introduction

In this chapter, a number of CMOS chip designs will be studied. They vary in complexity from a wide band switch architecture to a full 32-bit microprocessor. In general, the aim of this chapter is to reinforce the circuit, system, and design principles outlined in this text. Wherever possible some of the thoughts behind a choice in architecture will be given.

The first three examples were designed on the MULGA system in the Computer Systems Research Laboratory at AT&T-Bell Laboratories, Holmdel, New Jersey. The pixel-planes chip was designed at the University of North Carolina (Chapel Hill), using the VIVID design system. Finally, the most complex chip, the WE32000, was designed at AT&T-Bell Laboratories, using an expansive array of production CAD tools.

9.2 Dynamic time warp processor

B. D. ACKLAND*

9.2.1 Introduction

In this section the design of a special purpose processing element used to solve a class of dynamic time warp (DTW) pattern matching operations in speech recognition is summarized. Sections trace through the design of the IC, beginning with a description of the problem and an algorithmic solution. The algorithm leads to a functional specification for the processing element. From this functional specification, a detailed structural description (including floor plan) can be derived. This, in turn, leads to a symbolic physical description, which can be translated into real physical mask data. Testing of the IC, both during the design to verify correctness and after fabrication to identify good chips, is also described.

The real design process was, of course, not as straight-forward as is described here. There was constant interaction between the bottom-up problems of physical design and the top-down requirements imposed by the functional specification. Functional, structural, and physical descriptions were often modified to maintain consistency as the design evolved. The final design described here represents an engineering compromise between what would have been functionally ideal and what was physically realizable.

* AT&T Bell Laboratories, Holmdel, N.J.

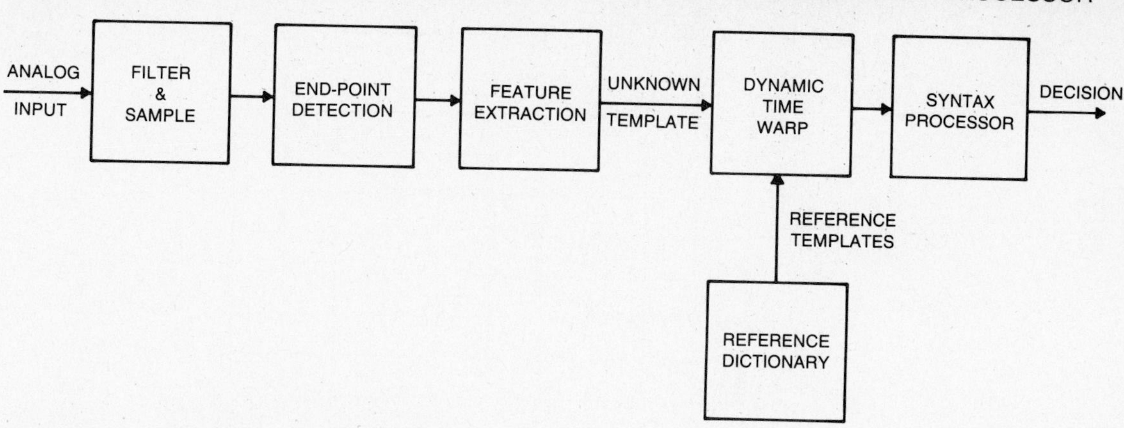

FIGURE 9.1. Isolated word recognizer

9.2.2 The problem

Speech recognition systems rely heavily on pattern matching for time alignment and identification of words or phrases in a speech sample. A block diagram of a typical isolated word recognizer is shown in Fig. 9.1. The input speech waveform is first split into isolated words. These words are then analyzed at regular intervals to yield a set of feature vectors. These features may be band energies, format frequencies, linear predictor coefficients, or others. The set of all feature vectors, from the beginning to the end of a word, is known as the feature vector template.

This unknown template is compared to a dictionary of reference templates to identify the word. In simple systems, the "closest" match identifies the unknown word. More sophisticated systems use a syntax or context postprocessor to accurately choose between close matches. Nonlinear pattern matching is normally used to allow for natural timing variations between the unknown and reference templates. This means that template matching tends to be a computationally intensive task, which limits the size of the reference dictionary in real-time applications.

Dynamic time warping (DTW) [Bell57] [SaCh71] is one such pattern matching technique and is illustrated in Fig. 9.2. The horizontal axis represents the reference template R, which has been divided into m time intervals. Associated with each interval is a feature vector, R_i. The vertical axis represents the unknown template U, characterized by n feature vectors, U_j. Each grid intersection (i,j) represents a possible match between interval i of the reference and

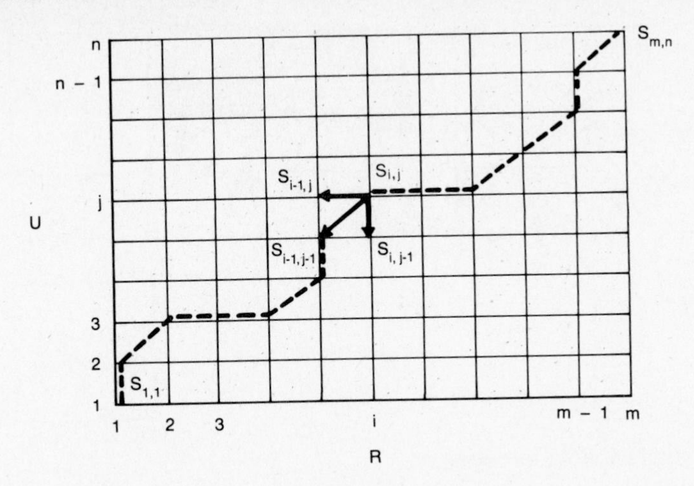

FIGURE 9.2. Dynamic time warping

interval j of the unknown. A distance measure $D_{i,j}$ describes the acoustic distance between (i.e., the dissimilarity of) these two intervals.

Any path from (1,1) to (m,n) represents a possible mapping or warp of the unknown onto the reference. For a given unknown/reference pair, the object is to find the warp function that minimizes the accumulated distance between these endpoints. This accumulated distance gives a measure of how well the two patterns match when one is allowed to warp to optimally match the other.

Because of the number of samples involved, it is not possible to exhaustively search all possible warp paths. Dynamic programming allows us to determine the optimal warp path iteratively. A partial sum term, $S_{i,j}$, is assigned to each grid intersection. This partial sum is defined by a recursion of the form

$$S_{i,j} = D_{i,j} + min(S_{i-1,j-1}, S_{i-1,j}, S_{i,j-1}). \tag{9.1}$$

$S_{m,n}$, the value that measures the overall dissimilarity between the two templates, is found by recursive evaluation of $S_{i,j}$ for all points in the lattice. These calculations are conventionally done by processing the warp matrix shown in Fig. 9.2 in serial raster scan fashion from lower left to upper right. At each grid intersection, a difference term $D_{i,j}$ is evaluated, followed by a partial sum $S_{i,j}$ calculation. This clearly involves huge amounts of computation, which must be repeated for each template in the reference dictionary. It is therefore attractive to look at alternative schemes that take advantage of the intrinsic parallelism in the DTW approach.

9.2.3 The algorithm

Suppose, rather than using a single processor to perform these calculations, we place a single processing element (PE) at each grid

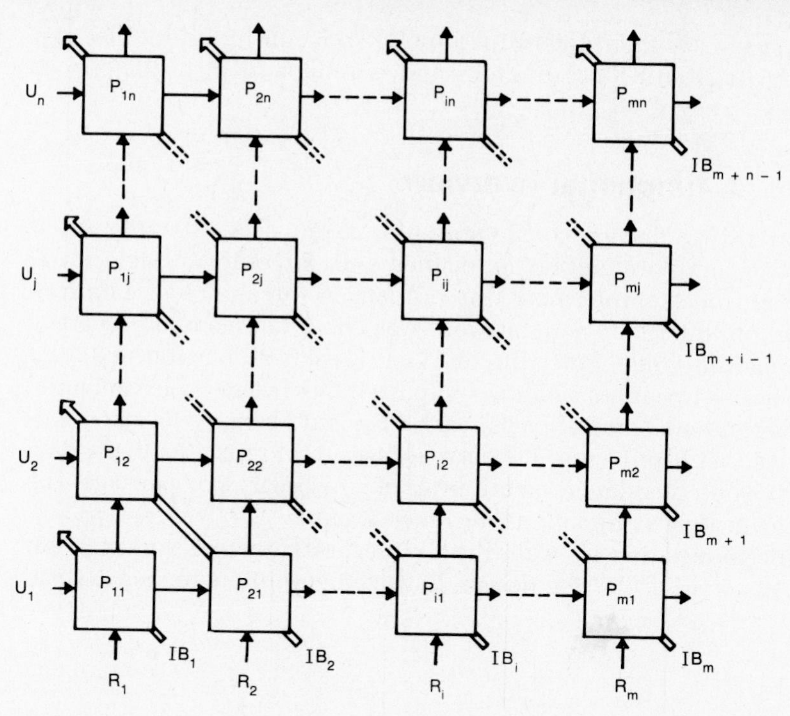

FIGURE 9.3. Orthogonal array of processing elements

intersection in the warp lattice, as shown in Fig. 9.3 [WeBA83] [BuAW84]. Each PE is then responsible for calculating the D and S terms corresponding to that intersection. To calculate the distance term $D_{i,j}$, the PE must be able to read in the appropriate feature vectors R_i and U_j and then pass them on to other PEs requiring this data. Partial sums are calculated according to Eq. (9.1). This requires previous partial sum data that is read in from neighboring PEs below and to the left. New data is then transferred to PEs above and to the right. A simple unidirectional orthogonal data network, as shown in Fig. 9.3, can be used to achieve these data transfers.

Because partial sum data is calculated iteratively, the operations that occur in any one PE must be appropriately sequenced with respect to operations occurring in neighboring PEs. For the recursion defined in Eq. (9.1), we must know $S_{i-1,j}$, $S_{i-1,j-1}$, and $S_{i,j-1}$ before we can calculate $S_{i,j}$. This suggests a control structure in which calculations are done on a diagonal by diagonal basis, as shown in Fig. 9.3. All PEs on the same diagonal simultaneously execute the same instruction. Parallel computation then proceeds in systolic waves from lower left to upper right; each diagonal taking over from the previous one until the final sum $S_{m,n}$ appears as output from the top right PE.

Maximum PE utilization is achieved by pipelining a number of template matches simultaneously through the array. The throughput of the system is then determined by how fast an individual PE can perform its D and S calculations and is independent of the size of the array.

9.2.4 A functional overview

An overriding design consideration was the need to integrate at least one, and preferably more than one, PE onto a single chip. Consider, for example, how this influenced the choice of data representation. Internally a 16-bit word was chosen to maintain accuracy and dynamic range. Referring to Fig. 9.3, each PE has four external data ports — two inputs and two outputs. If these were to be configured as independent parallel ports, each PE would require 64 data connections. With only one PE per package, this translates to 64 data pins. If one considers a small array of processors per package, the number of data connections becomes absurd.

One alternative is to multiplex the ports on to a single 16-bit bus. This complicates the design, however, and limits the parallelism

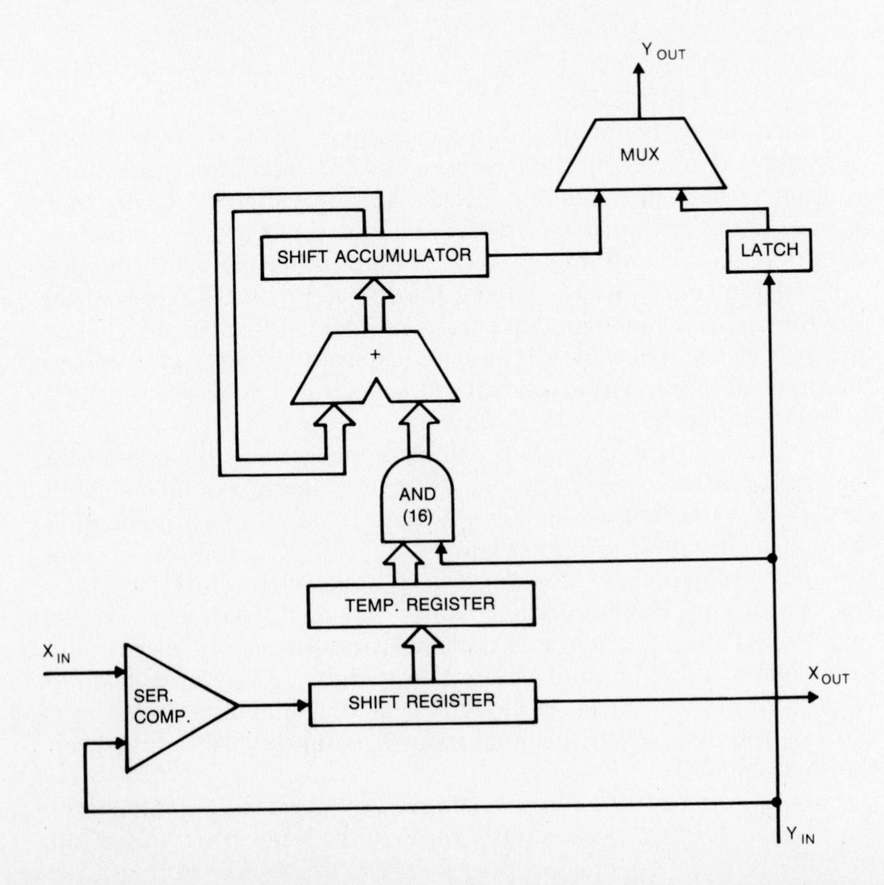

FIGURE 9.4. Special purpose hardware solution

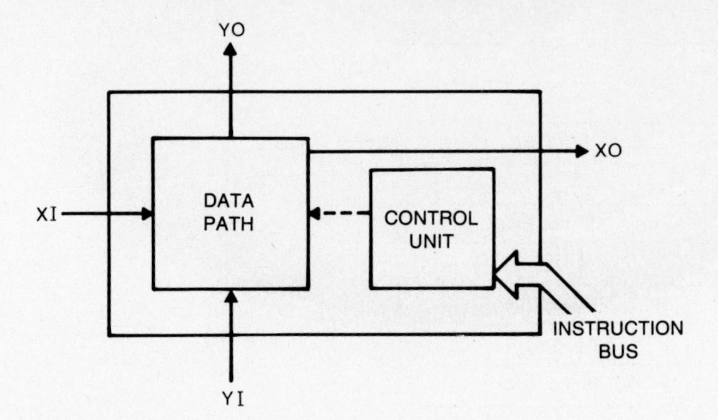

FIGURE 9.5. Programmable processing element

that can be achieved while transferring data. A second alternative (the one that was chosen) is to use serial ports. This drastically reduces the amount of interconnect but limits performance since it takes at least 16 clock cycles to transfer a single datum. Much of this delay can be masked, however, by overlapping data transfer with other iterative operations (e.g., multiply) that must take place within the PE.

One of the first proposals considered was a special purpose PE that would directly implement, in hardware, the required data operations. An example is shown in Fig. 9.4. Serial comparators, adders, and a shift-and-add multiplier operate directly on incoming serial data. No parallel conversions are necessary. An additional advantage of this type of approach is that no external control, other than synchronization, is needed to perform the required operations.

The chief disadvantage of such a special purpose PE is that it is tied to one set of data operations (algorithm). The internal complexity drastically increases if different distance measures or partial sum functions are to be used. Since different algorithms are frequently required in different speech recognition applications, it was decided to develop a programmable PE based on a control and data path, as shown in Fig. 9.5. Note, however, that programmable does not necessarily mean general purpose. The goal of physical implementation requires that even a programmable PE be no more complex than is required to perform the various algorithms associated with the application.

9.2.4.1 Data path

Before designing the data path, it is necessary to consider just what operations need to be performed. Distance measures typically use arithmetic operators such as addition, subtraction, absolute value, multiplication, and, in some cases, logarithmic scaling. The only other operation required (for partial sum calculation) is the ability

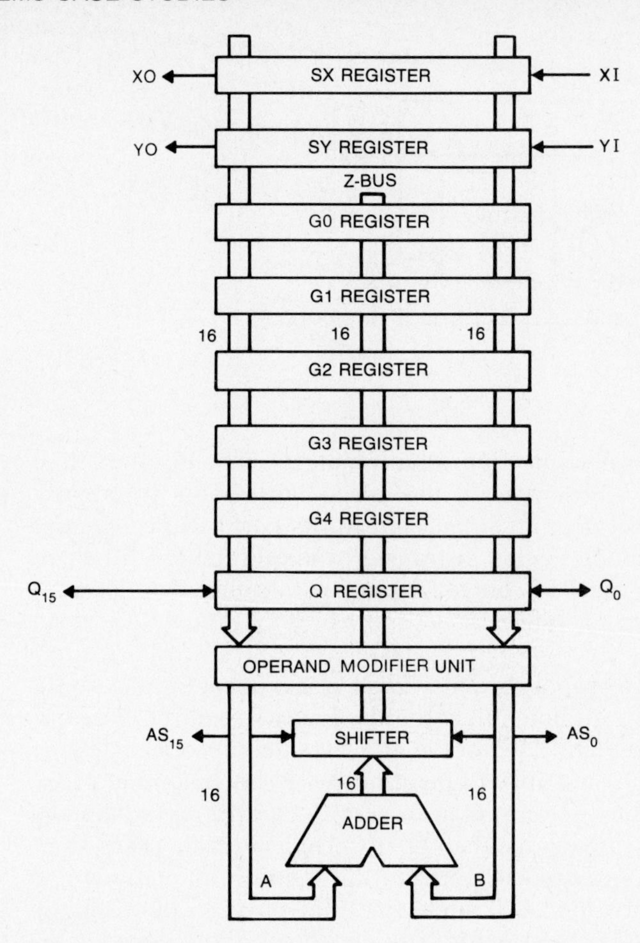

FIGURE 9.6. Data path section

to select the greater (or smaller) of two numbers. The data path must also be able to store a limited number of intermediate results. The structure chosen is shown in Fig. 9.6. It consists of a number of 16-bit registers and a 16-bit arithmetic unit built around three 16-bit busses. A and B serve as operand busses for the arithmetic unit. Z is the result bus that delivers arithmetic results back to the registers.

A number of registers have specific function. SX and SY are shift registers that can be serially loaded from the two input ports XI and YI. Their outputs drive, via an output multiplexer, the two serial output ports XO and YO. This enables the PE to pass input feature data on to adjacent PEs. In addition, the contents of the SX and SY registers can be loaded into the arithmetic unit via the operand busses. G0–G4 are general purpose registers, which are loaded from the result bus and can be read on to the operand busses. They are principally used for storing intermediate results. Q is a special register with both left and right shift capability. Its serial output can be used to transfer results to either of the output ports XO,YO. In addition, it is used as an extension register during iterative multiplication.

The arithmetic unit consists of a 16-bit adder, a 16x1-bit left or right shifter, and an operand modifier unit (OMU). The OMU allows either operand bus to be zeroed and the B bus to be conditionally complemented. This expands the functionality of the adder to include subtraction and register move operations.

The operation of this data path is determined by microcode bits generated in the control section. Before proceeding any further with the design of the data path, let's take a look at the control unit.

9.2.4.2 Control unit

The control unit is responsible for generating the horizontal microcode bits needed to control the operation of the data path. One obvious solution would be to use a conventional microsequencer, complete with program memory, program counter, and conditional branch logic. This approach yields maximum flexibility. It also leads, however, to a complex design that is, in many ways, an overkill, considering the restricted set of algorithms to be implemented. A much simpler solution would be to implement a combinational (i.e., stateless) control unit based on PLA expansion of an externally supplied instruction. The main disadvantage of this approach is that the PE is limited to a fixed simple instruction set.

The approach that was taken is a compromise between these two and is shown in Fig. 9.7. A microinstruction memory is used to decode an external instruction into a horizontal microcode word. Since the memory can be downloaded at runtime, special instructions that take maximum advantage of the parallelism in any one algorithm can be used. The PE functions as a purely combinational processor

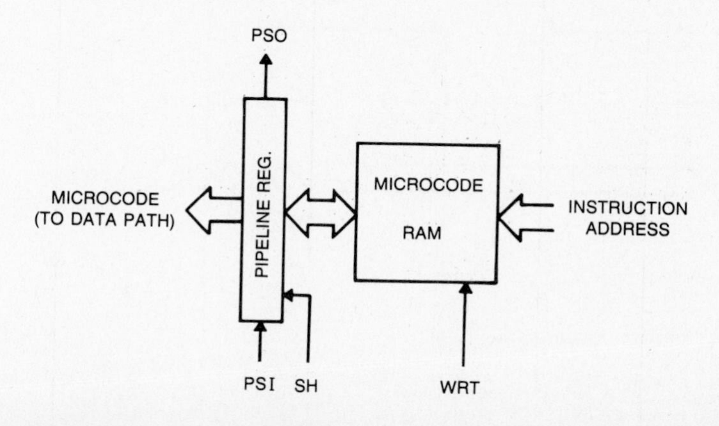

FIGURE 9.7. Microcontrol section

with no decision making capability in its control unit. This avoids unnecessary duplication of control in PEs that lie on the same diagonal and always execute the same instruction (Fig. 9.3).

The pipeline register allows one microinstruction to be executed while another is being fetched from the memory. It also functions as a shift register. This shift capability serves two purposes. Firstly, it allows serial loading of input data to the memory. Secondly, it allows separate testing of the control and data sections of the PE by allowing the user to either read the contents of the microinstruction memory or to test the operation of the data path under direct microcontrol.

9.2.4.3 Timing

A 2-phase synchronous clock scheme was employed, as shown in Fig. 9.8. During CP1, master/slave registers (data path and pipeline) transfer data from master to slave. Output data (XO,YO) changes during CP1. Also, memory address circuitry is precharged during this clock phase. During CP2, data is read into the register masters. Input data (XI,YI) must be valid during CP2. Memory access occurs during CP2 and produces data that is read into the pipeline master.

Arithmetic operations occur from the time new control and data become available on the busses (CP1) to the time when register masters close (end of CP2). Memory address (control instruction) lines must remain valid from the end of CP1 to the end of CP2, since this represents the entire memory access cycle.

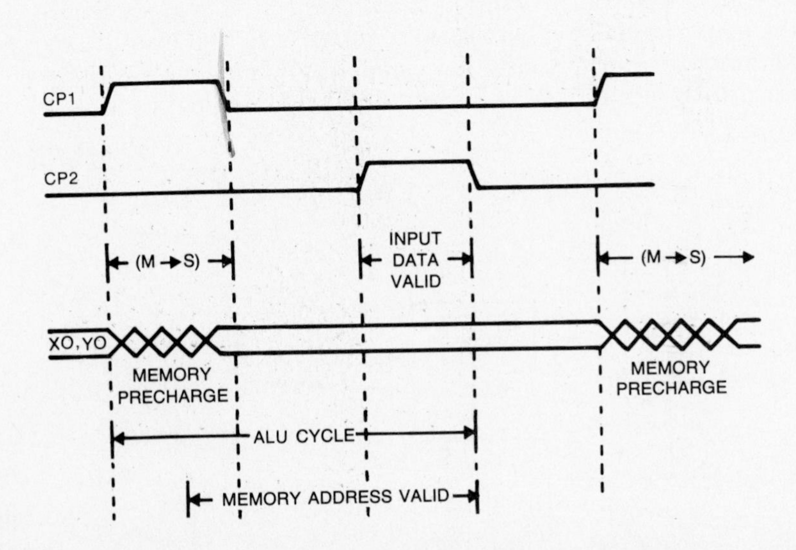

FIGURE 9.8. Pseudo 2-phase timing waveforms

9.2.5 Detailed functional specification

We can now proceed to generate a detailed functional specification of the PE. The decision to design the PE as a combinational processor means that primitive operators which are normally based on decision making algorithms, must be implemented directly as combinational operators. The special operators needed are multiplication, minimum value, absolute value, and logarithm. We will next take a look at how each of these can be implemented directly within the data path.

The "multiply" instruction is the simplest. It gates the B operand bus to zero whenever the least significant bit of the Q register is zero. This feature is used in conjunction with the adder/shifter in the arithmetic unit to generate a 16-cycle shift and add sequence.

A sign flip-flop stores the sign of the result generated during the previous cycle. The "minimum" instruction gates either the A or B operand bus to zero, depending on the state of this flip-flop. The minimum (or maximum) of two operands can be obtained by first subtracting one from the other and then executing this minimum instruction.

The absolute value of an operand v is simply found by evaluating $\max(v, -v)$.

As mentioned previously, some DTW applications require a logarithmically scaled distance measure. Fortunately, absolute accuracy is not important. Rather, what is required is a "log-like" function that is monotonic and sensitive to operand differences of the order of 5–10 percent. A simple log approximation that meets these criteria is as follows. First determine the integer part of the logarithm $L_i(x)$ by searching for the most significant '1' in the binary word. Then approximate the fractional part of the logarithm $L_f(x)$ as the remaining bits once the most significant '1' has been removed:

$$L_f(x) = \frac{x - 2^{L_i(x)}}{2^{L_i(x)}}. \tag{9.2}$$

For example, the log approximation of the 8-bit binary word 00101101 would be

$$L_i(x) = 5$$
$$L_f(x) = 0.01101.$$

This algorithm can be implemented by left-shifting a binary register (in this case, the Q register) into a flip-flop until a '1' is detected. Once this occurs, further shifting is inhibited. The Q register thus retains the fractional part. The state of the flip-flop is also used to set carry-in to the adder. By conditionally adding one to a register

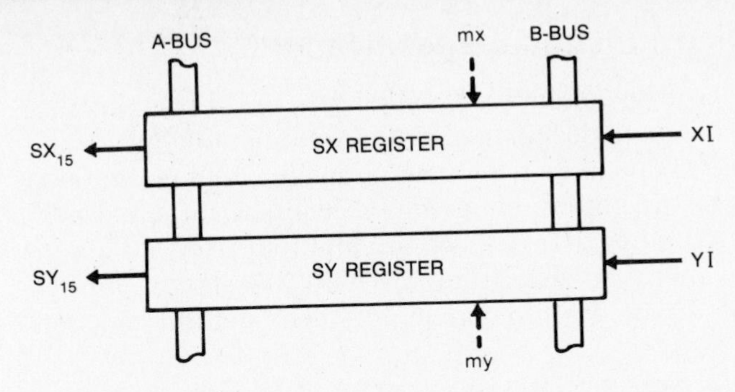

FIGURE 9.9. Shift registers SX and SY

that was initially zero, the position of the most significant '1' (the integer part) can be recorded.

9.2.5.1 Data path

The operation of the data path is controlled by 31 bits of microcode. Six bits control the loading of the operand busses. *sa*0–*sa*2 select the register that will source the A bus. *sb*0–*sb*2 select the register that will source the B bus. The register bus select addresses are given in Table 9.1.

TABLE 9.1. Register bus select addresses

REGISTER	BUS ADDRESS
G0	0
G1	1
G2	2
G3	3
G4	4
SX	5
SY	6
Q	7

9.2.5.1.1 Registers

The shift registers SX and SY are shown in Fig. 9.9. Data is shifted in from the serial input ports XI,YI. Data is shifted out from the MSBs of the registers (SX_{15},SY_{15}). This data passes to the output multiplexers. Operation of these registers is controlled by microcode bits *mx*,*my*. When *mx* = 1, SX shifts left one place. When *mx* = 0, the contents of SX are held unchanged. SY is similarly controlled by *my*.

General purpose registers G0–G4 (Fig. 9.10) are controlled by microcode bits *mg*0–*mg*4. When *mgi* = 1, Gi is loaded with result data from the Z bus. When *mgi* = 0, the contents of the register are held unchanged.

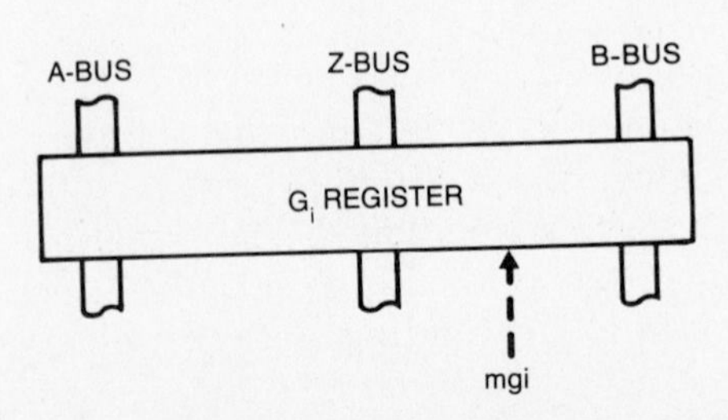

FIGURE 9.10. General purpose registers G0–G4

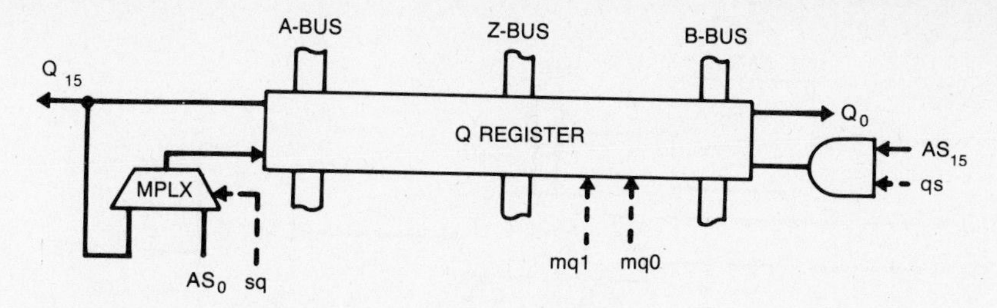

FIGURE 9.11. Q register

The Q register, shown in Fig. 9.11, has four modes of operation according to microcode bits $mq0, mq1$. These modes are summarized in Table 9.2. During right shift operations, the MSB of Q is determined by microcode bit sq. When $sq = 1$, the MSB of Q is loaded with the LSB of the arithmetic shifter (AS_0). This provides a double precision shift capability.

When $sq = 0$, the MSB of Q is loaded with itself; that is, an arithmetic (sign preserved) shift is produced. During left shift, the LSB of Q is determined by microcode bit qs. When $qs = 0$, zero is shifted into Q. When $qs = 1$, the MSB of the arithmetic shifter (AS_{15}) is loaded into the LSB of Q. This gives a double precision shift left capability.

The operation of the Q register is further modified by the log flip-flop shown in Fig. 9.12. Under normal operation, microcode bit $lgr = 0$. This causes the flip-flop to be permanently reset. If $lgr = 1$, the flip-flop is loaded, once each cycle with the MSB of the Q register until a '1' is detected. The flip-flop then remains set and prevents the Q register from performing any further shift left operations.

TABLE 9.2. Operation of Q register

mq1	*mq0*	FUNCTION
0	0	hold
0	1	load from Z bus
1	0	shift left
1	1	shift right

9.2.5.1.2 Arithmetic unit

The 16-bit adder is shown in detail in Fig. 9.13. Each cycle, the MSB of the sum result is stored in the sign flip-flop. Carry-out from the most significant bit is similarly stored in the carry flip-flop. A special sign-extended carry-out (SXC) is also generated. This is used in signed multiply operations.

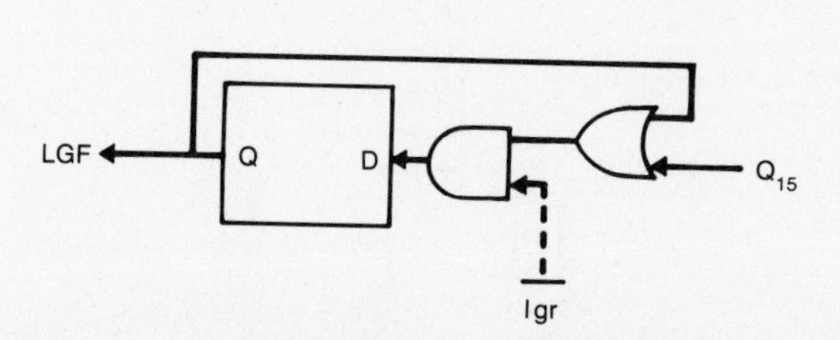

FIGURE 9.12. Logarithm flip-flop

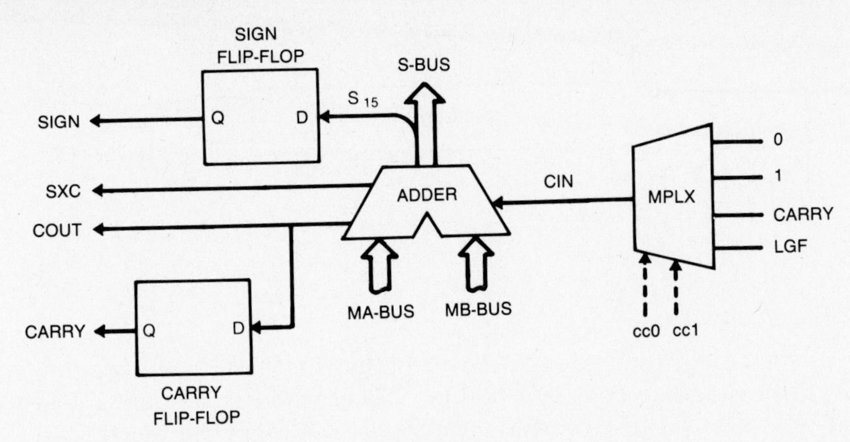

FIGURE 9.13. Adder

TABLE 9.3. Carry-in control

*cc*1	*cc*0	CIN
0	*0*	0
0	*1*	CARRY
1	*0*	1
1	*1*	LGF

Carry-in is determined by the two microcode bits *cc*0,*cc*1, as summarized in Table 9.3. By setting carry-in to be the output of the carry flip-flop, we get multiple precision arithmetic. Carry-in is set to the output of the log flip-flop when generating the integer part of a log approximation.

The OMU, shown in Fig. 9.14, modifies operand data prior to addition. The A bus is modified by control bits *za* and *min*. The B bus is subject to microcode bits *zb*, *min*, *mul*, and *op*. Table 9.4 summarizes the operation of these control bits and describes the arithmetic operations that can be realized.

Basically, the *op* bit is used to generate subtraction, the *min* bit gates one of the busses to zero according to the state of the sign flip-flop, and the *mul* bit gates the B bus according to the LSB of the Q register.

The shifter, shown in Fig. 9.15, takes data from the output of the adder, optionally shifts this data left or right one place, and then loads the result onto the Z bus. Operation is determined by microcode bits *sl*,*sr* and is summarized in Table 9.5. During left-

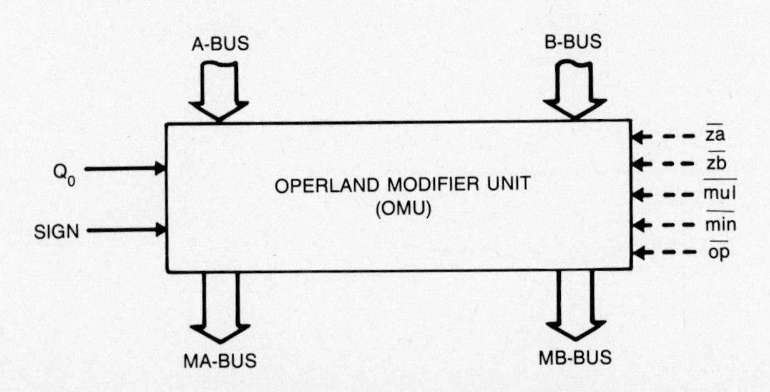

FIGURE 9.14. Operand modifier unit

TABLE 9.4. Operation of operand modifier unit

INSTRUCTION	MICROCODE					DATA		OUTPUTS		
	$\overline{op}$	$\overline{za}$	$\overline{zb}$	$\overline{min}$	$\overline{mul}$	SIGN	Q_0	MA_i	MB_i	S (adder)
add	1	1	1	1	1	X	X	A_i	B_i	A + B + CIN
sub	0	1	1	1	1	X	X	A_i	$\overline{B_i}$	A − B − 1 + CIN
movb	1	0	1	1	1	X	X	0	B_i	B + CIN
negb	0	0	1	1	1	X	X	0	$\overline{B_i}$	− B − 1 + CIN
mova	1	1	0	1	1	X	X	A_i	0	A + CIN
deca	0	1	0	1	1	X	X	A_i	1	A − 1 + CIN
clr	1	0	0	1	1	X	X	0	0	CIN
ones	0	0	0	1	1	X	X	0	1	− 1 + CIN
mim	1	1	1	0	1	0	X	A_i	0	A + CIN
						1	X	0	B_i	B + CIN
mply	1	1	1	1	0	X	0	A_i	0	A + CIN
						X	1	A_i	B_i	A + B + CIN
smyc	0	1	1	1	0	X	0	A_i	1	A − 1 + CIN
						X	1	A_i	$\overline{B_i}$	A − B − 1 + CIN

shift operations, a zero is loaded into the LSB unless microcode bit $sq = 1$. In this case, the LSB of the Z bus is loaded with the MSB of the Q register.

During right-shift operations, the MSB of the Z bus can be selected from one of four sources, as shown in Table 9.6. Carry-out is used during unsigned multiply operations. SXC is used to generate a signed multiply instruction.

TABLE 9.5. Shift control

sl	*sr*	FUNCTION
0	0	no shift
0	1	shift right
1	0	shift left
1	1	illegal

9.2.5.1.3 Output multiplexers

The output multiplexers select serial data sources for the two serial output ports XO and YO, as shown in Fig. 9.16. Output data can come from the shift registers SX,SY or Q, or from the special flip-

TABLE 9.6. Right shift input select

$\overline{mul}$	*qs*	Z_{15}
1	0	S_{15}
1	1	Q_0
0	0	SXC
0	1	COUT

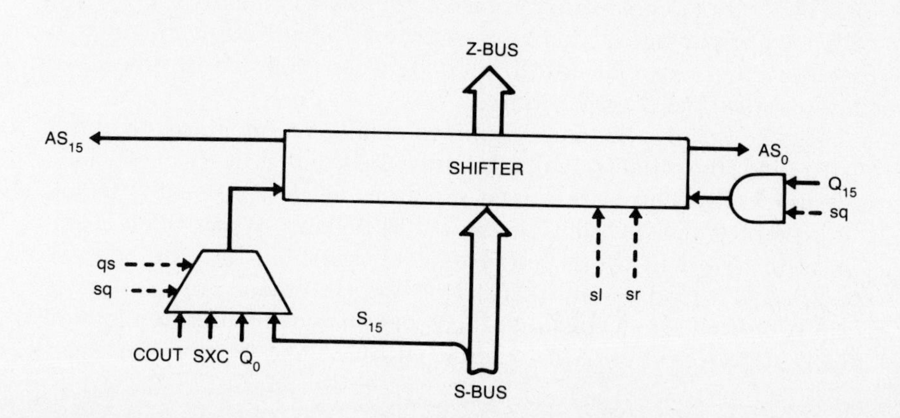

FIGURE 9.15. Arithmetic shifter

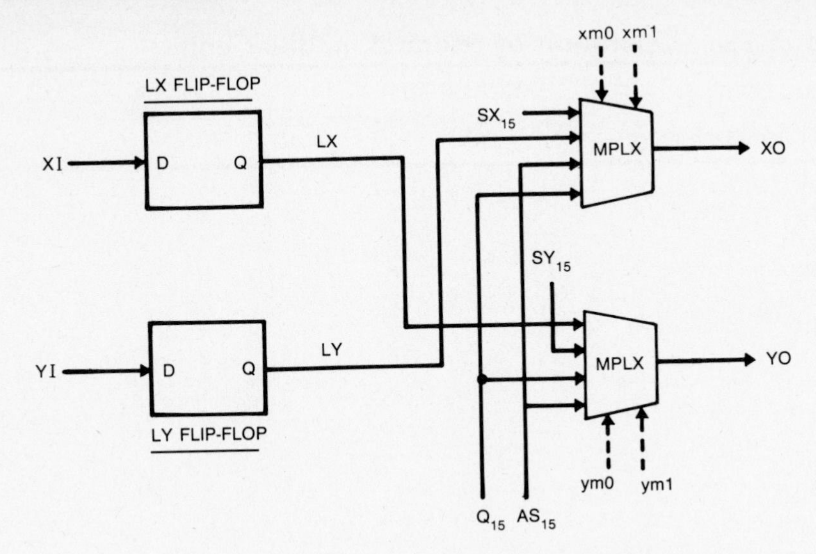

FIGURE 9.16 Output multiplexers

flops LX and LY. These allow serial data to be directly routed from one adjacent processor to another, giving the illusion of a diagonal connection between PEs.

9.2.5.2 Control unit

The control unit generates the 31 microcode bits that feed the data path. The number of microinstruction words that must be stored depends on the number of unique instructions required to perform distance and partial sum calculation. After reviewing a number of existing algorithms, it was felt that a capacity of 32 different codes would be ample. This sets the size of the microinstruction memory at 32 x 31. For the sake of regularity, this was modified to 32 x 32 bits. This size memory array is easily implemented using static RAM cells. An advantage of using a static microinstruction memory is that the clocks can be interrupted between instruction definition (downloading) and execution.

Fig. 9.17 shows the control unit in detail. A 5-bit address (control instruction) bus selects one 32-bit microinstruction every cycle. This is loaded into the master/slave pipeline register. When SH is set, the pipeline register functions as a shift register. When WRT is set, the 32-bit word in the pipeline register is loaded into the address specified. To load a new instruction into the memory, the data is serially shifted (32 cycles) into the pipeline register and then loaded into the appropriate RAM location with a single write cycle.

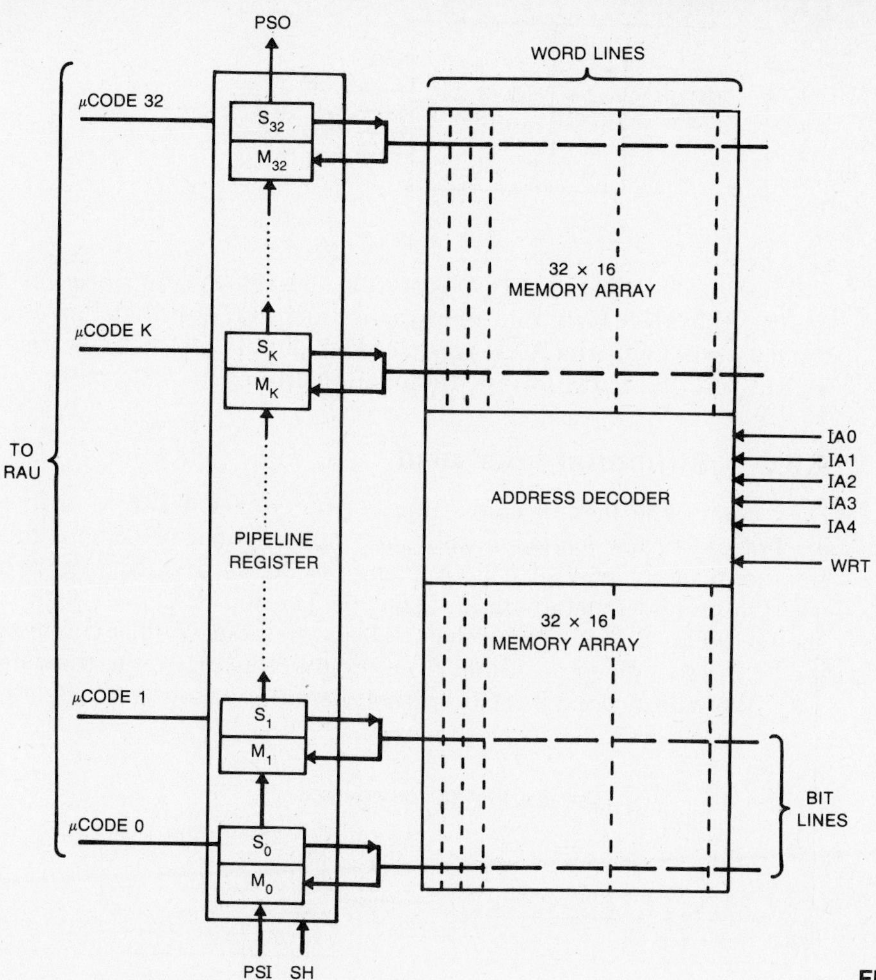

FIGURE 9.17. Control unit

9.2.5.3 Testing

Once the functional specification was complete, a functional simulator was written to verify the completeness of the structure. At the same time, a test sequence that would ultimately be used to test real chips was written. This test sequence was tested on the simulator. In this way we were at least able to verify that the chip we intended to build satisfied our test sequence.

The test sequence first verified the operation of the pipeline register by loading serial data into the PSI port and checking for its appearance on the output PSO, 32 cycles later. The pipeline register

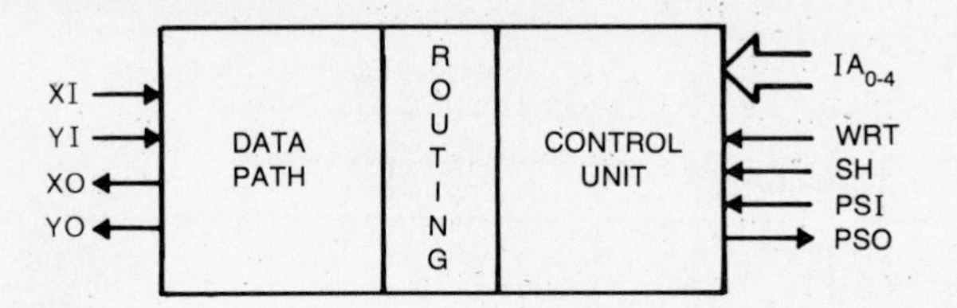

FIGURE 9.18. Floor plan of processing element

was then used to test the RAM by storing data into specific locations, reading that data back into the pipeline register, and then shifting the data out. Once the RAM was checked, simple instructions were loaded into the RAM to test the functionality of the data path.

9.2.6 Structural floor plan

Now comes the task of converting the functional description into a set of physically realizable cells and showing their placement and interconnect by way of a floor plan. The PE is divided into two main modules — data path and control. The top level floor plan is shown in Fig. 9.18. A routing channel between these modules connects the 31 microcode control lines from the pipeline register to the data path. We will next examine the structure of these modules in detail.

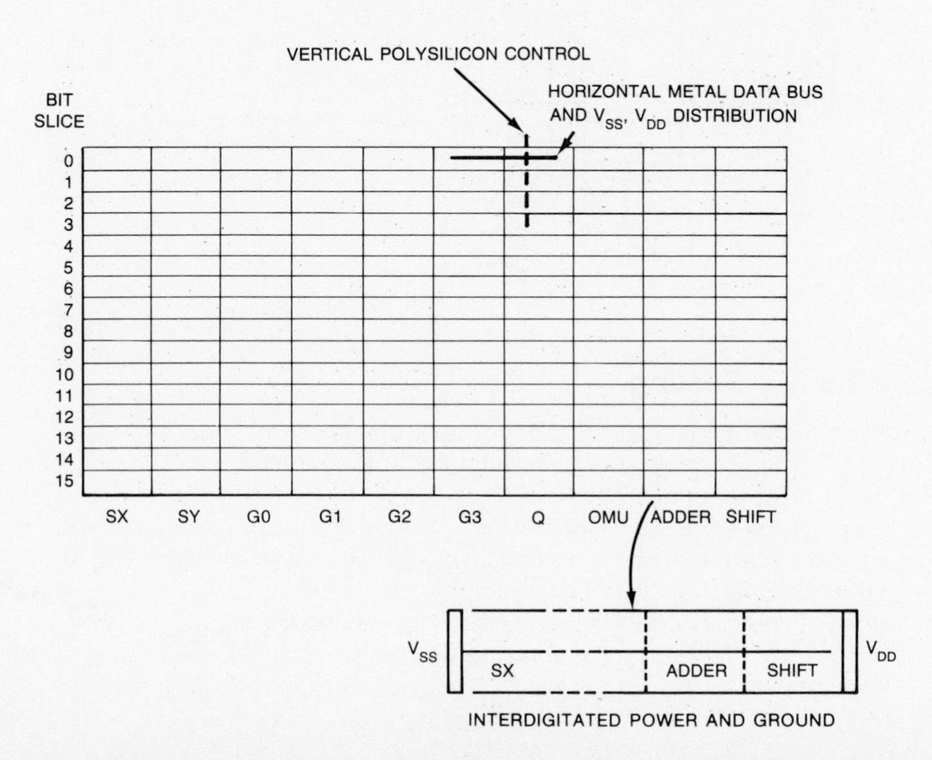

FIGURE 9.19. Data and control flow—data path

9.2.6.1 Data path

Bus oriented data paths lend themselves to highly structured floor plans in which data and control flow right through the leaf cells (rather than between them). Fig. 9.19 shows the basic structure. The module is divided horizontally into vertical strips, each strip representing one functional element (e.g., the G3 register). The module is also divided vertically into 16 horizontal strips, each one representing one entire bit slice through all functional elements in the module. Data is transmitted horizontally along metal bus lines, which are distributed on a bit slice basis. Control is fed vertically on polysilicon lines. Control and data intersect in those cells whose function it is to apply that control to that particular piece of data.

Power is distributed horizontally via interleaved fingers, as shown in Fig. 9.19. Clocks are required by almost all leaf cells and could conceivably be distributed on either layer. We chose to use horizontal metal clock distribution to minimize clock skew delays.

The simple floor plan shown in Fig. 9.19 needs to be augmented with some extra control circuitry to decode the various microcode bits and buffer these decoded signals onto polysilicon control lines. Fig. 9.20 shows how this is done using three horizontal control channels. The central channel is reserved for those control signals whose speed directly limits the cycle time of the PE. An example is the operand bus address decoders. A new PE cycle begins when CP1 causes a new microcode word to appear on the output of the pipeline register. The first operation that must take place is for these address decoders to select the appropriate register sources for the A and B busses. Any delay in this decoding process will delay calculation of the result. By placing this circuitry in the central control channel, we reduce the RC delay of the polysilicon control lines by halving their length.

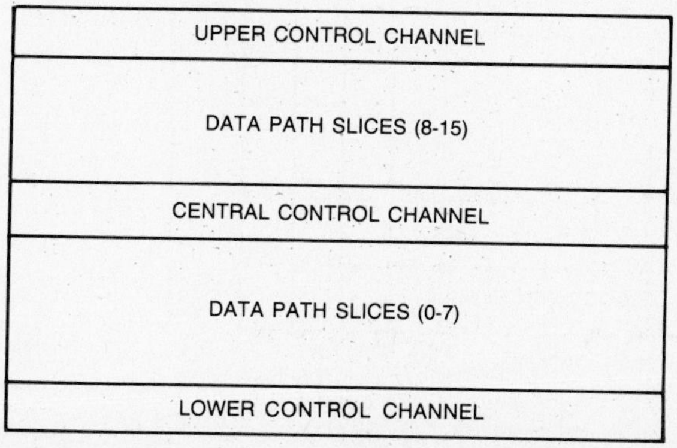

UPPER CONTROL CHANNEL
DATA PATH SLICES (8-15)
CENTRAL CONTROL CHANNEL
DATA PATH SLICES (0-7)
LOWER CONTROL CHANNEL

FIGURE 9.20. Data path and control block arrangement

Noncritical control is placed in the upper or lower channel. An example is the various register load lines. These do not have to be correctly set up until CP2 loads the result from the Z bus into selected registers. Plenty of time is therefore available to decode these signals. These upper and lower channels also contain the various status flip-flops (e.g., SIGN). One set of time critical signals that could not be accommodated in the central control channel were the control lines for the OMU. The solution was to duplicate this circuitry in both the upper and lower channels so that each had only to drive one-half the total polysilicon distance.

The final floor plan is shown in Fig. 9.21. Horizontal channels route microcode bits from the pipeline register to the appropriate control decode cells. These channels also pass serial data and some of the internally generated status signals between registers. Vertical strips on either end of the array complete the power and ground nets.

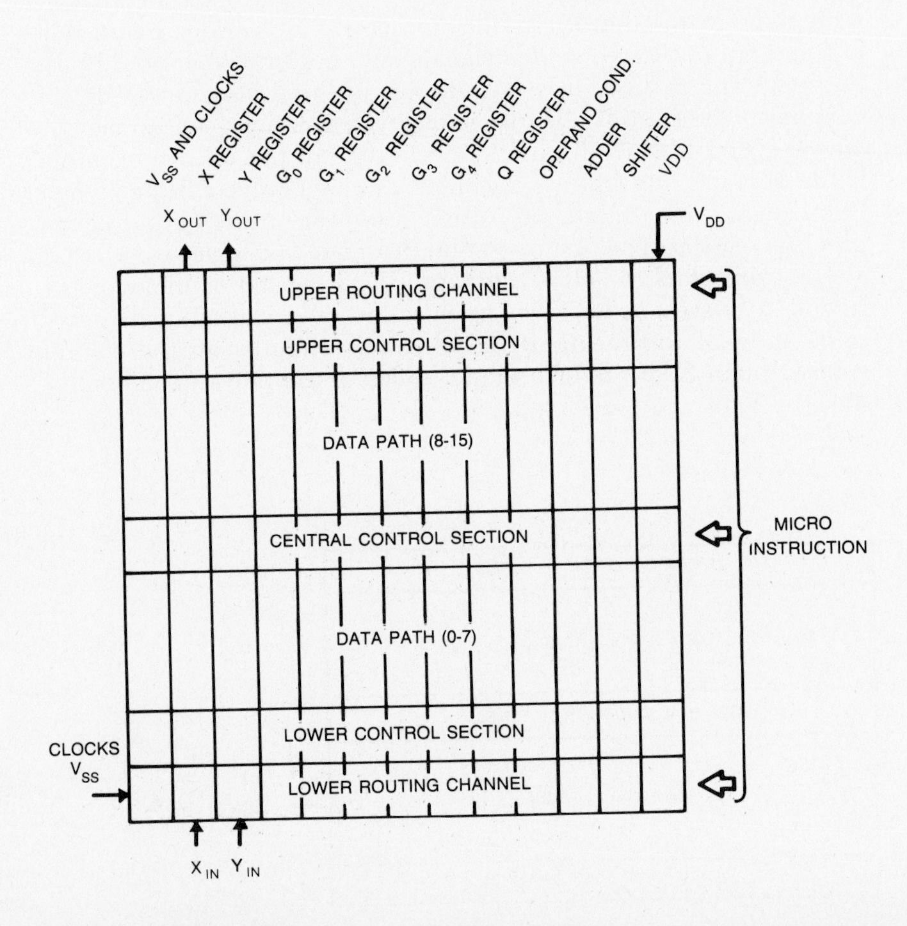

FIGURE 9.21. Detailed floor plan of data path

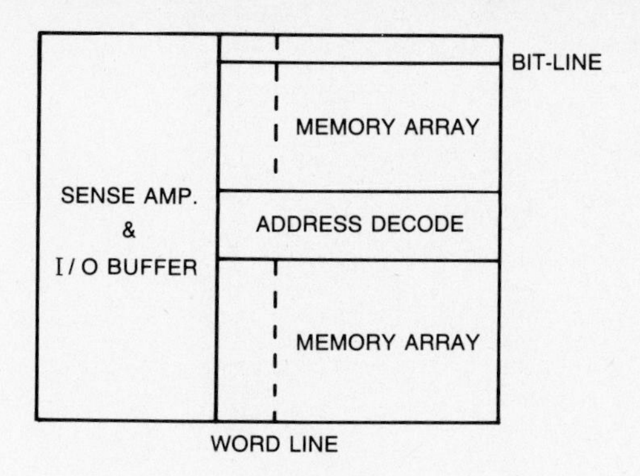

FIGURE 9.22. Data and control flow—microcontrol memory

9.2.6.2 Control unit

The control unit is built around a fairly conventional memory structure shown in Fig. 9.22. Memory cell word (control) lines run vertically in polysilicon. Power and bit (data) lines run horizontally in metal. This structure forms the basis for the floor plan shown in Fig. 9.23.

A central address decoder channel is used to generate the time critical word lines. Note that the pipeline register has been pitch-matched directly on to the left side of the memory array. The pipeline register block also contains sense amplifiers for the memory bit lines and memory write circuitry. The cell labeled CONTROL decodes the external signals WRT (write) and SH (shift) and controls the operation of the pipeline block via a number of vertical polysilicon control lines.

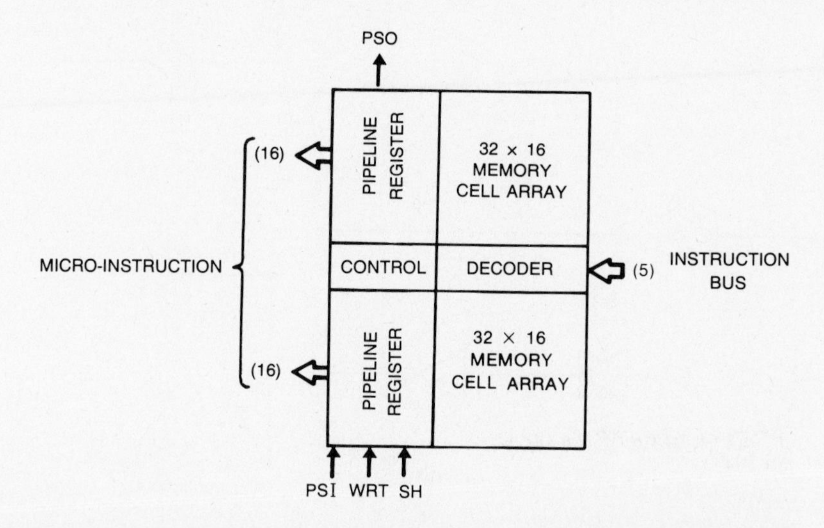

FIGURE 9.23. Detailed floor plan of control unit

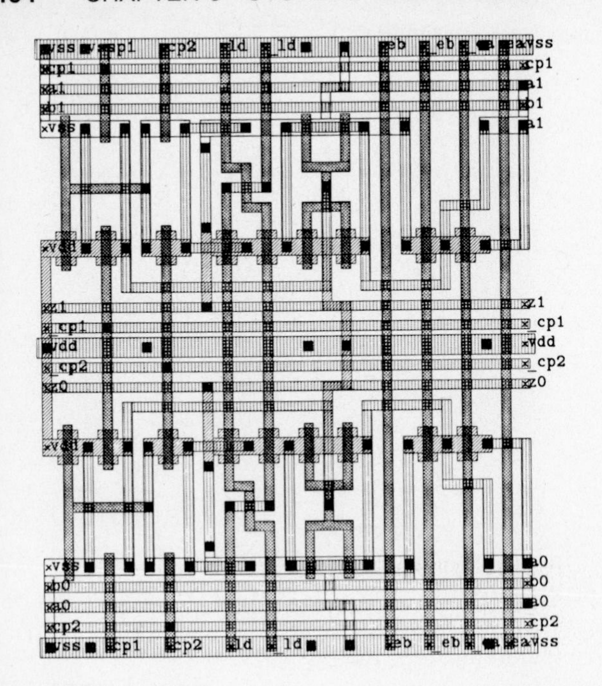

(a) SYMBOLIC LAYOUT

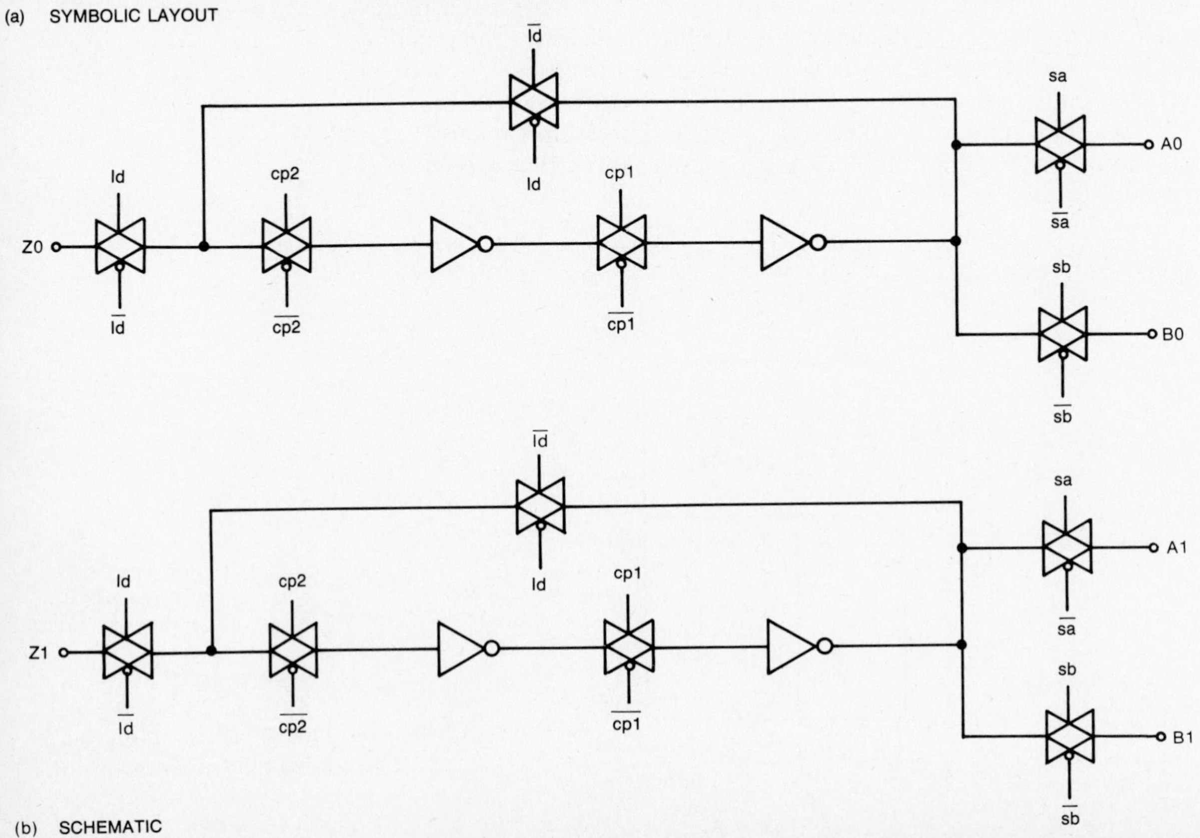

(b) SCHEMATIC

9.2.7 Physical design

Once a structural floor plan has been defined, we can proceed with the symbolic layout of the individual leaf cells in each module. The layout of a given leaf cell must satisfy the boundary connection requirements of its neighbors. An example is shown in Fig. 9.24a. This cell called GR2 is a 2-bit slice of one of the general purpose registers in the data path module. Fig. 9.24b shows a transistor level schematic of this same cell.

Note how the power lines (V_{SS} and V_{DD}) are used as a vertical butting plane between cells. One bit of each of the three data busses A, B, and Z passes through each bit slice. The four clock lines ($cp1$, $cp2$, and their complements) run alongside the power rails. All the control lines run vertically through the cell. These are shared by other slices above and below this cell. The *ld* signal is used to load data from the Z bus into the master register. The *sa* and *sb* lines (ea, eb on symbolic) load slave register data onto the A and B busses respectively.

Fig. 9.25a shows the symbolic layout of one of the leaf cells in the control section. This is a 2 x 2 bit section from the static memory array. The corresponding transistor schematic is shown in Fig. 9.25b. Once again we use the power rails to butt cells together. The two polysilicon words enable lines ($wd0$, $wd1$) to run vertically through the cell. The four bit lines ($b0$, $b1$, and their complements) run horizontally in metal.

The operation of the leaf cells was verified using a timing simulator. Various combinations of leaf cells were put together to represent time critical paths in the PE. For example, the functionality of the microcode RAM and its speed were measured by simulating a test module consisting of one 4 x 4 RAM cell, two bits of the pipeline register, and a two-word slice from the address decoder. Extra capacitance was added to word and bit lines to simulate the effect of the entire RAM array. A few simple timing simulations of the entire data path and control modules were used to verify their correct assembly.

Once all the leaf cells have been physically designed and tested in this manner, the modules can be symbolically assembled and converted into physical mask descriptions. The modules are then routed to each other and to an I/O pad frame to yield a final mask representation.

◀ **FIGURE 9.24. Leaf cell GR2**
a) Symbolic layout
b) Schematic

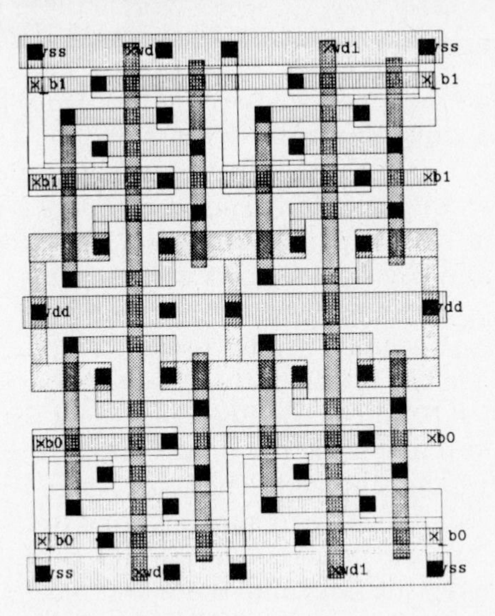

(a) SYMBOLIC LAYOUT

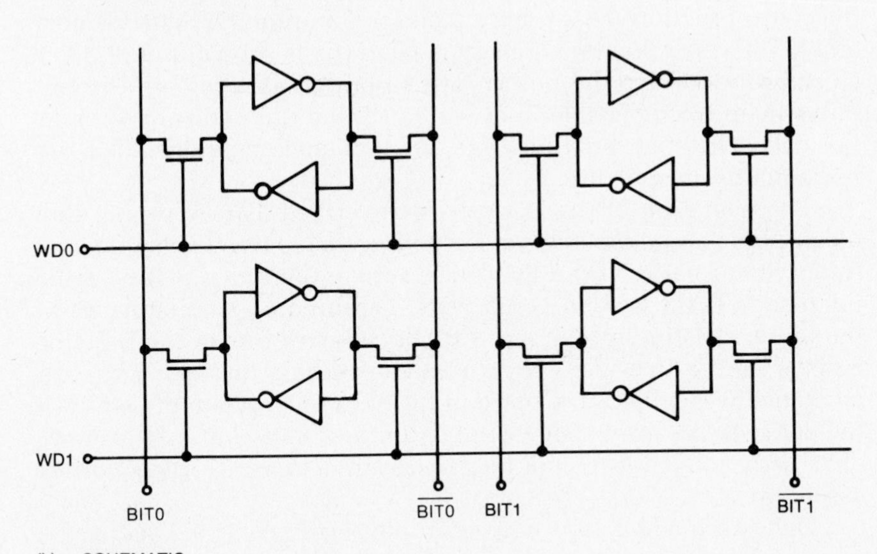

(b) SCHEMATIC

FIGURE 9.25. Leaf cell RAM4 a) Symbolic layout b) Schematic

9.2.8 Fabrication

One of the original design objectives was to put a small array of PEs onto a single chip. For prototype implementation, however, we decided to implement just one PE per chip. In addition, we also implemented two test chips. The first contained only the control module with a number of word and bit lines brought out to pads

FIGURE 9.26. Photomicrograph of fabricated DTW chip

for testing. The second contained only the data path with the Z bus brought out to pads. These test chips were to be used to debug the PE in the event that there were structural errors in the design (there were none).

The PE was fabricated on a 3.5 μ CMOS process. A photomicrograph of the chip is shown in Fig. 9.26. The chips were tested at the wafer level using the test program that was originally developed on the functional simulator. Speed tests indicated a worst-case cycle time of under 200 ns.

9.3 Real time video moment generator chip

R. L. ANDERSSON*

9.3.1 Introduction

This case example illustrates that even high speed video processing can be handled with moderate technology by using appropriate architectures. The design epitomizes many structured design principles, including the use of regularity on a chip basis (regular use of cells) and at the system level (multiple use of the same chip). The application is in a system that takes a TV image and analyzes

* AT&T Bell Laboratories, Holmdel, N.J.

the image to allow a robot to pick objects seen by a TV camera. The example also serves to show the thought processes that are evident in the partitioning of a representative chip.

A TV image contains an enormous amount of information. A 256 * 240 black and white image at 60 Hz, digitized to 8-bits resolution, contains 30 megabits per second. Many computers are hard pressed to generate or accept a continuous 9600 bit per second data stream. Modern robot systems need to be able to process and respond to video signals in real time. The main problem is one of data reduction: how to reduce the string of bits from the TV camera to a smaller number that may be processed by a general purpose processor without losing relevant information.

Image processing techniques have been studied for some time. One of the simplest processing methods relies on the idea of moments, found in physics and probability theory as well as image processing. Moments are defined as follows:

$$M^{m,n} = \sum_{i,j} a_{i,j}\, i^m\, j^n, \tag{9.3}$$

where a is the gray scale intensity and i and j range from 0 to 255. The sum $m + n$ is defined as the order of the moment. The moments of order 0 through 2 allow one to determine the center of gravity, angle to major axis, and standard deviation along major and minor axes of an object in the image. This is illustrated in Fig. 9.27. The object is essentially approximated by an ellipse. This information, although not sufficient in general, is enough to give one a good start on picking up an object viewed from overhead. The moments are sufficient to guide further image analysis and have a number of interesting properties. They are a linear operator with respect to the image, so that the moment of a sum is the sum of the moments. Additionally, given the moments in a given coordinate system, the

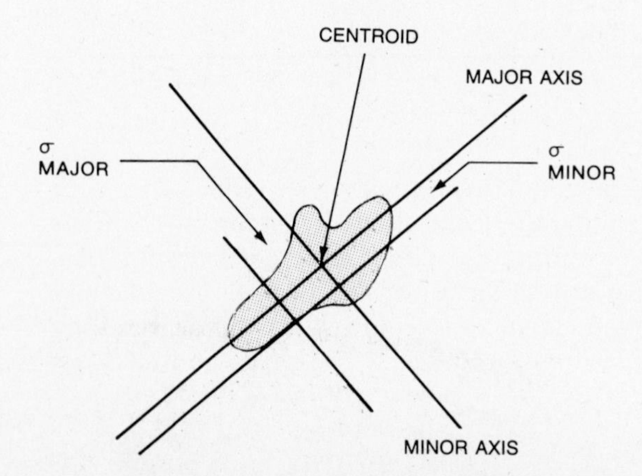

FIGURE 9.27. Moments of an object

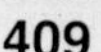

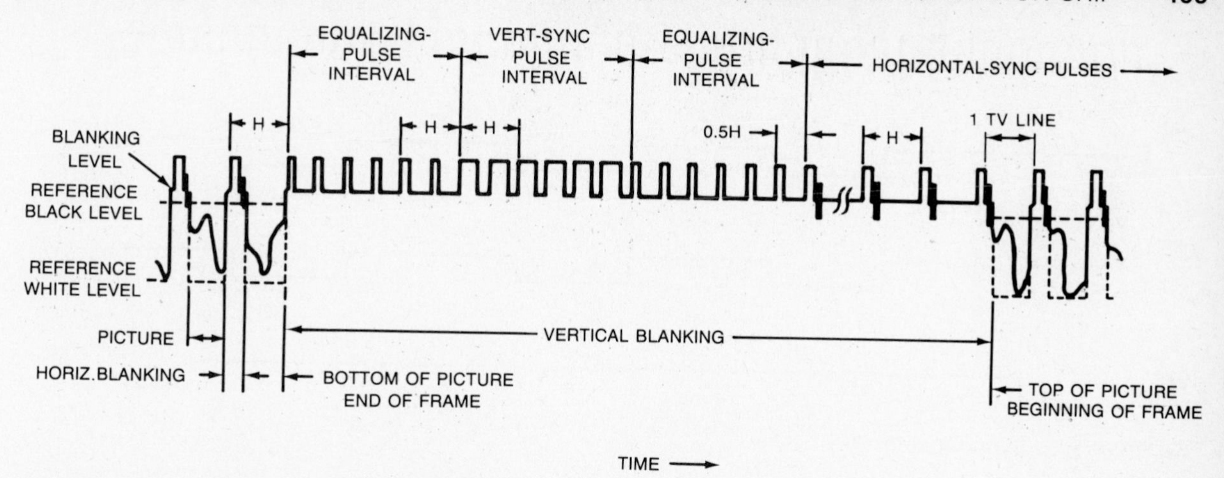

FIGURE 9.28. Analog TV signal waveform

moments in a different coordinate system may be found by linear operations on the original moments.

9.3.2 Review: video formatting

Let us briefly review the format of data coming from a TV camera. A TV camera takes 60 frames a second with a theoretically infinite horizontal resolution, and 262 lines per frame vertically. We ignore interlacing, since for robotics purposes the doubling of the frame rate is more useful than a half line of resolution. Each frame consists of a vertical retrace interval plus about 240 worthwhile lines of video data. Each line is composed of a sync signal, a blanking period, plus actual video. The vertical sync is a longer version of the horizontal signal. The sync, blanking, and video signals are encoded in an analog voltage multiplexing scheme, as shown in Fig. 9.28. In further discussion, we will simply posit the existence of an analog to digital conversion and sync extraction box (Fig. 9.29). The box outputs a

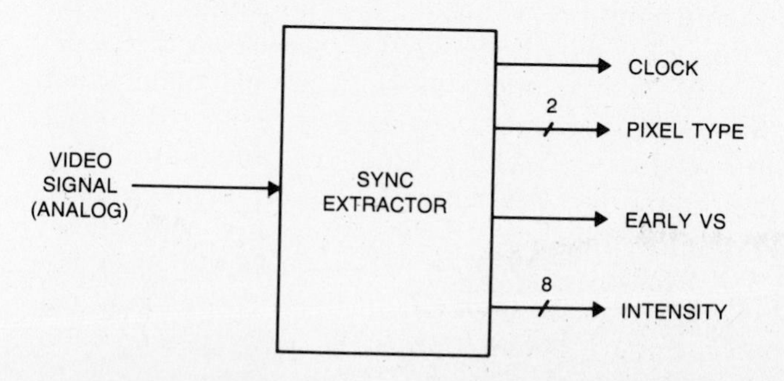

FIGURE 9.29. Sync extractor

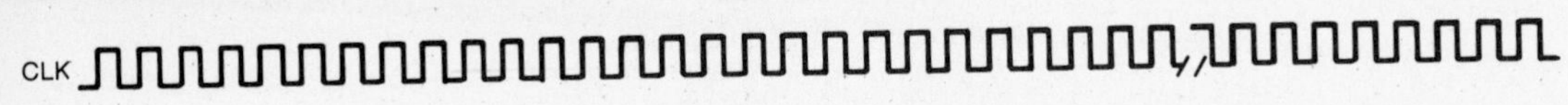

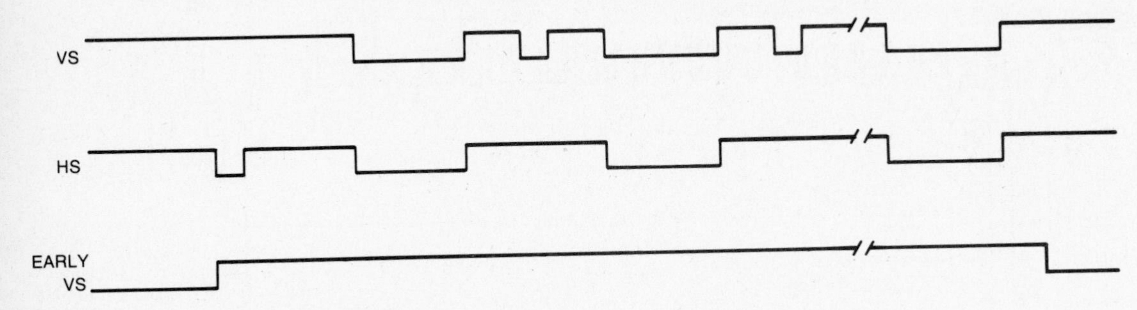

FIGURE 9.30. Sync waveforms

clock, a 2-bit pixel-type code, an early vertical sync signal, and eight bits of intensity data. The clock rate is generally selected to give either a square pixel size (a measuring stick is the same size in pixels independent of direction), or to maximize the horizontal field of view while keeping 256 pixels per line, not including sync or blanking. In the latter case, pixels are significantly wider than they are high; this introduces some complications in later processing. The pixel-type signals are such that there is only one clock with vertical or horizontal sync when the corresponding function occurs. This relationship is shown in Fig. 9.30. This is necessary to avoid double processing of the horizontal or vertical sync. The coding for the pixel type is shown in Table 9.7. Note that the video scan starts at the top and proceeds to the bottom, left to right across each line. This is counter to our standard graphs; it is easy to create errors in processing programs because of this.

9.3.3 Chip architecture

Assuming now that moments are something we would like to compute, how can we set about doing it? Consider first the simple TTL implementation in Fig. 9.31, which shows the calculation of the XY moment ($m = 1$, $n = 1$). The other moments are calculated analogously, with some simplifications. We need an $8 * 8$ multiplier, an $8 * 16$ multiplier, and a 32-bit accumulator, plus appropriate readout and support circuitry. For square pixels, all this equipment must run at 160 ns per pixel. Without getting any really fancy equipment, we will probably have to pipeline the circuit, so a few pipeline registers need to be added. All of these components are pretty big and power hungry, so we'll probably be lucky to get more

TABLE 9.7. Pixel types

11	blanked—ignore this pixel
10	vertical sync
01	horizontal sync
00	active pixel

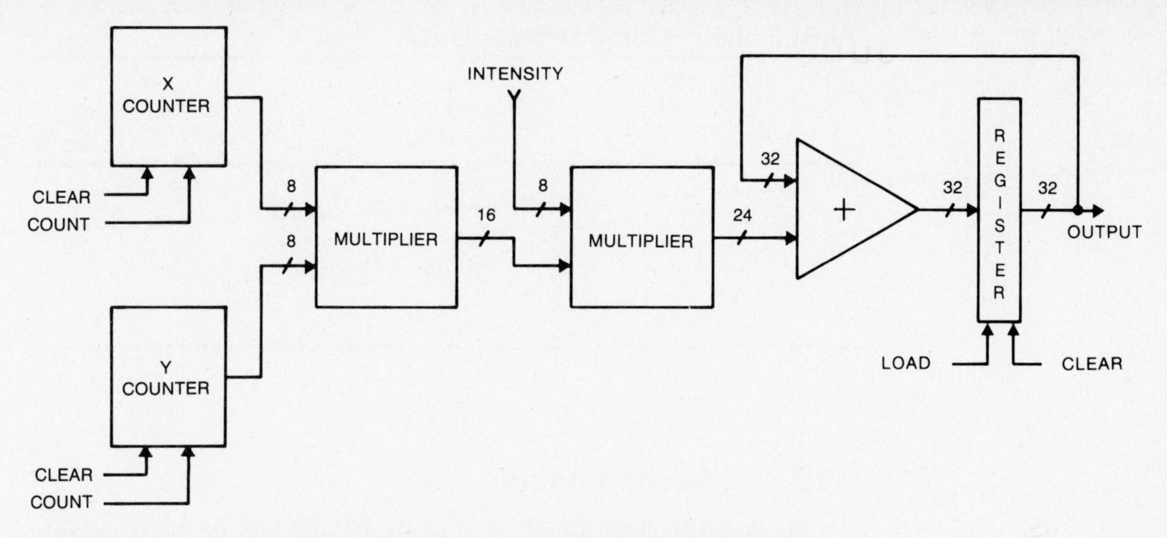

FIGURE 9.31. X-Y moment calculation—conventional architecture

than two or three individual moment calculators on a board, meaning two or three boards in the system. Naturally, we would like to integrate some of these things onto a chip or two. The multipliers are the most obvious impediment to integration since they take an area proportional to the product of the bit sizes. How can we use the structure of the problem to our advantage?

9.3.3.1 Recursion relations

The scan order of the TV camera is fixed; we always generate the same sequence of numbers representing

$$\overline{p}_{i,j}^{m,n} = i^m j^n, \tag{9.4}$$

where for any individual moment, m and n are fixed. We can recursively generate the $\overline{p}$ vector using a different rule for each moment:

$$\overline{p}_{i+1,j}^{2,0} = \overline{p}_{i,j}^{2,0} + 2i + 1 \tag{9.5}$$

$$\overline{p}_{i+1,j}^{1,1} = \overline{p}_{i,j}^{1,1} + j \tag{9.6}$$

$$\overline{p}_{i,j+1}^{0,2} = \overline{p}_{i,j}^{0,2} + 2j + 1. \tag{9.7}$$

There are other cases for horizontal syncs and vertical syncs. Table 9.8 summarizes the next term of the series as a function of the moment being computed and the type of pixel. Note that we do nothing when a "blank" pixel is received.

By relying on the fixed scan order, we can eliminate the need for the 8 * 8 multiplier, at the expense of a counter, shifter, and accumulator, as shown in Fig. 9.32.

TABLE 9.8. Moment calculation

P^{ij}	NEXT FS x=0 y=0	NEXT HS x=0 y->y+1	NEXT PC x->x +1
$P^{0,0}$	1	1	1
$P^{1,0}$	0	0	$P_{1,0} + 1$
$P^{2,0}$	0	0	$P_{2,0} + 2x + 1$
$P^{0,1}$	0	$P_{0,1} + 1$	$P_{0,1}$
$P^{1,1}$	0	0	$P_{1,1} + y$
$P^{0,2}$	0	$P_{0,2} + 2y + 1$	$P_{0,2}$

9.3.3.2 Decomposition

We must still get rid of the 8 * 16 multiplier. To do this, we rely on the linearity property of moments. The incoming gray scale intensity may be viewed as the following sum:

$$\overline{a} = 2^7\overline{a}_7 + 2^6\overline{a}_6 + \cdots + \overline{a}_0, \quad \textbf{(9.8)}$$

where a_n represents the nth bit of the a value. It is then readily apparent that the moment computation can be distributed to

$$M^{m,n} = 2^7(\overline{a}_7, \overline{p}^{m,n}) + 2^6(\overline{a}_6, \overline{p}^{m,n}) + \cdots + (\overline{a}_0, \overline{p}^{m,n}). \quad \textbf{(9.9)}$$

This equation may be rewritten as

$$M^{m,n} = 2^7F_7 + 2^6F_6 + \cdots + F_0 \quad \textbf{(9.10)}$$

and

$$F_k = (\overline{a}_k, \overline{p}^{m,n}). \quad \textbf{(9.11)}$$

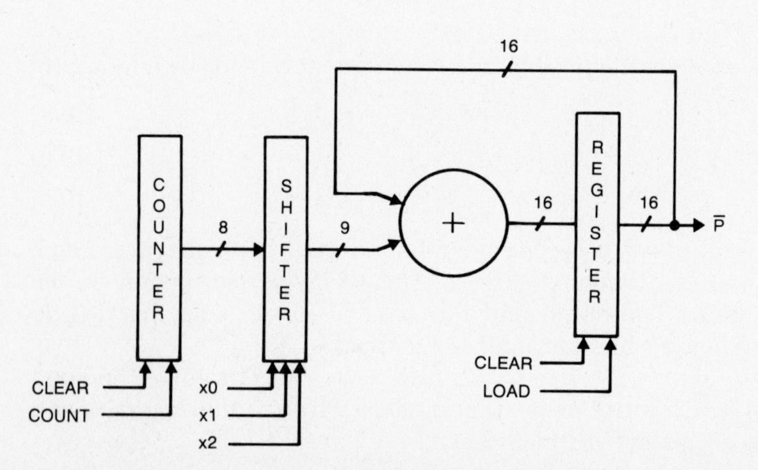

FIGURE 9.32. $\bar{p}$ generator block diagram

The evaluation of Eq. (9.10) — a polynomial in 2 — has to be done only once per frame, and is very quick using Horner's Method, so we can relegate it to the host processor. The moment generation system must then simply compute the primitive terms F_k in Eq. (9.11), which is easy since there are now no multiplications — just a parallel AND, a shift, and a few additions, including the generation of the $\overline{p}$ vector.

9.3.3.3 Carry save addition

The carry propagation delay of a combinatorial ripple carry adder implemented in the CMOS technology used is approximately 5–10 ns/bit. We must perform a 32-bit accumulation in a 160 ns pixel time. To overcome this problem, one approach is to use a carry lookahead adder. The design of the adder would be a major consumer of design time. Instead, we rely on a variant of a scheme used in the days of vacuum tube multipliers (and more recently, too). The technique used here, which is shown in Fig. 9.33, is to insert a flip-flop in the carry chain every eight bits in the accumulators. Now the limiting clock path is only eight bits, or 80 ns, long, allowing a significant margin. As long as several "idle" clock periods are allowed before the result is examined, the adder works just fine. The carry flip-flops introduce convenient gaps in the structure of the chip for control wiring. The only place where the strategy breaks down is in the $\overline{p}$ vector generator, whose result must be fed each clock cycle into the actual moment accumulation cells. We patch this up by putting another adder after the actual accumulator to add in the saved carry, as illustrated in Fig. 9.34.

9.3.3.4 System aspects

Early chip sizing estimates indicated that we could produce a chip capable of computing a single moment of an incoming video stream in real time. A system is therefore comprised of six chips, each computing one of the six zero through second order moments (i.e., 1, x, y, xy, x^2, y^2). It is not generally possible to simply pass the digitized video signal into any kind of moment generator. Background features must be ignored, and one must separate out multiple objects that may be present in the image, to compute their moments individually. We utilize a preprocessor consisting of an intensity map and a location map, each of which is a fast dual ported RAM. By appropriately specifying the data in these maps, the processor may specify what intensities are important (for example, light objects against a dark background) and what region of the image the moments should be calculated over.

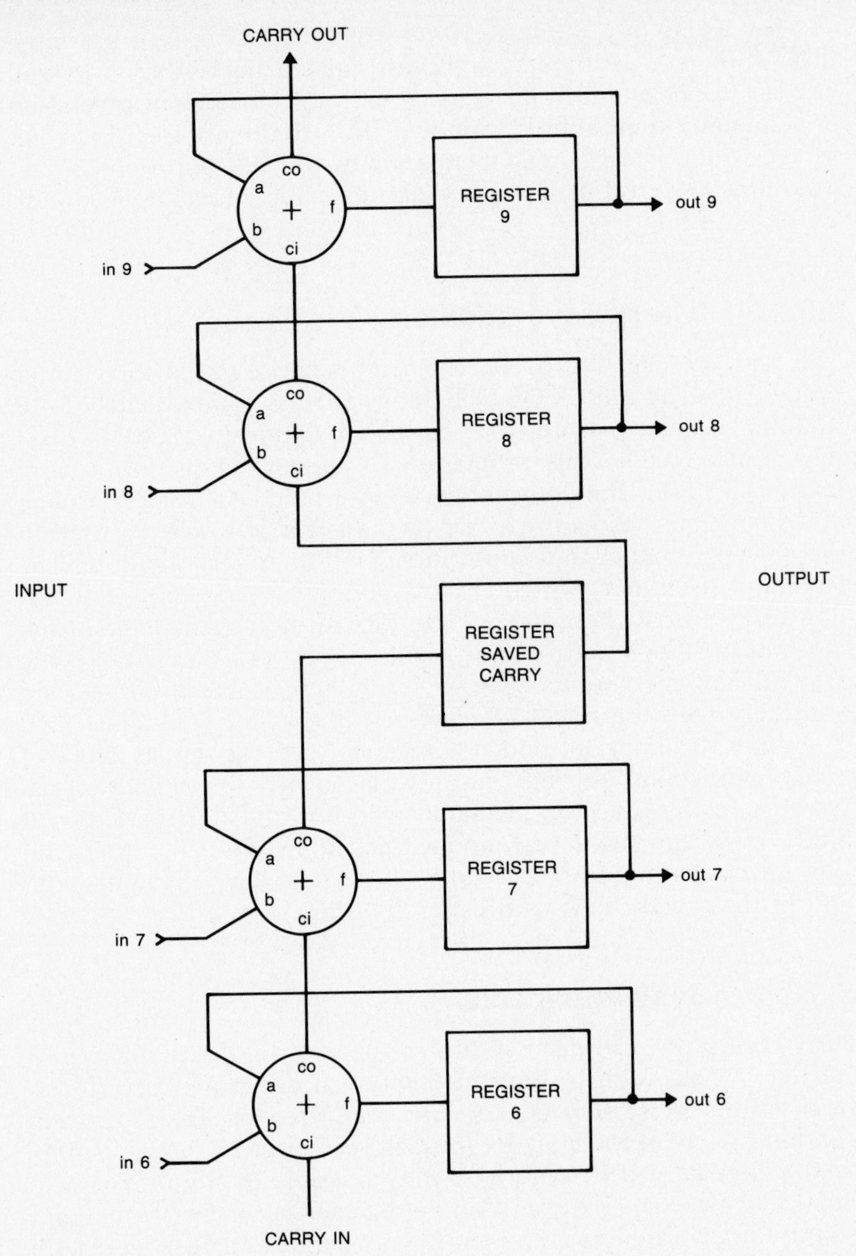

FIGURE 9.33. Carry save adder

The preprocessor, six moment generator chips, and a MultiBus™ (Intel) interface can be placed on a single board (Fig. 9.35), a far cry from the many boards that otherwise would be required.

The structure of the preprocessor and the chip implementation algorithm combine to yield a very useful property. By appropriately

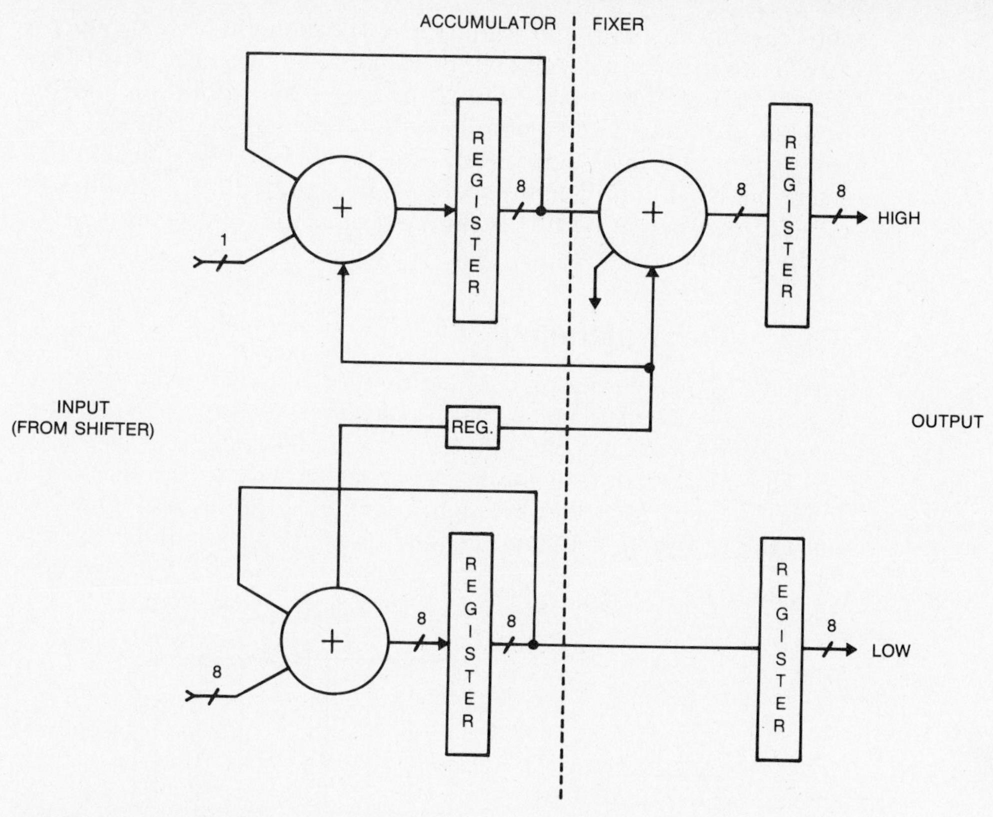

FIGURE 9.34. $\bar{p}$ fixer

FIGURE 9.35. Finished MG board—glue and VLSI

constructing the maps and recombining the moment terms, moments can be calculated on several regions simultaneously. The only trade-off is between the number of bits in gray scale resolution and the number of moment sets computed simultaneously. This means we can track multiple objects in real time which is of advantage in real scene analysis. Equivalent capability in standard TTL would take racks of equipment — this is the kind of advantage we are looking for in VLSI.

9.3.4 Floor planning

We have to pick a floor plan that allows for an easy implementation. Fig. 9.36 shows just such a plan, which was constructed after brief consideration of the computation to be performed.

The construction of the $\overline{p}$ vector begins with a counter on the lower left and proceeds through a shifter, accumulator, and fixer until the complete $\overline{p}$ vector bus (16 bits wide) passes under the Vss

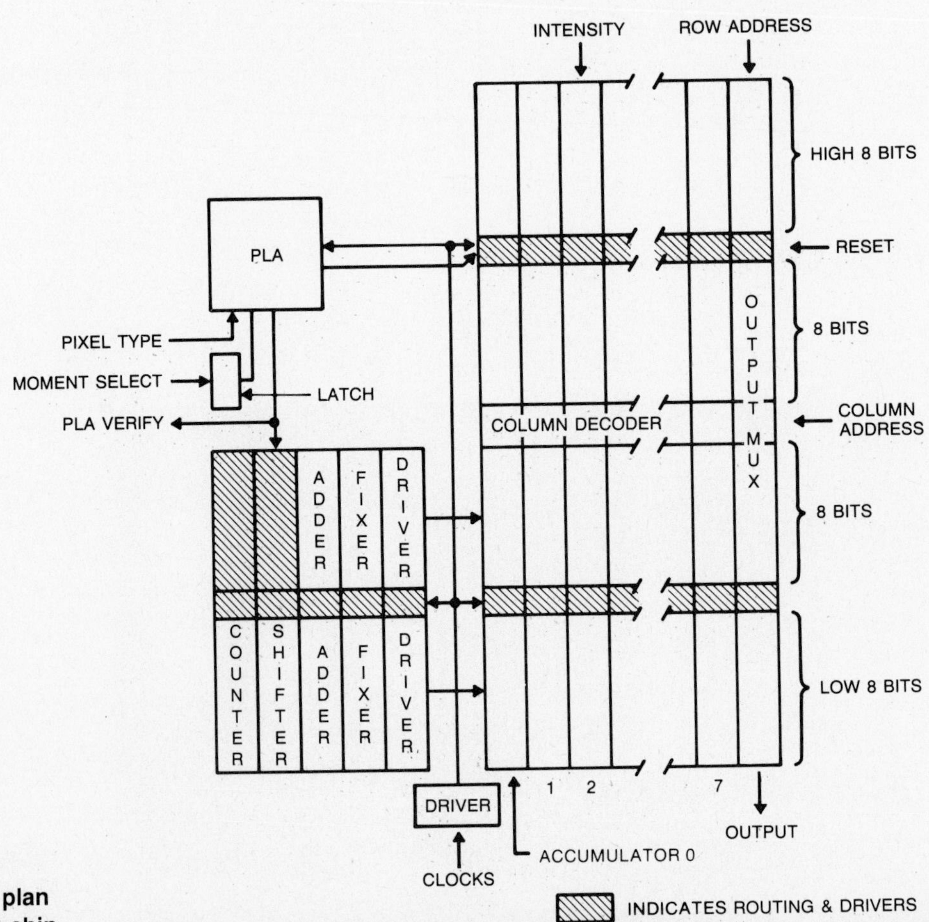

FIGURE 9.36. Floor plan of moment generator chip

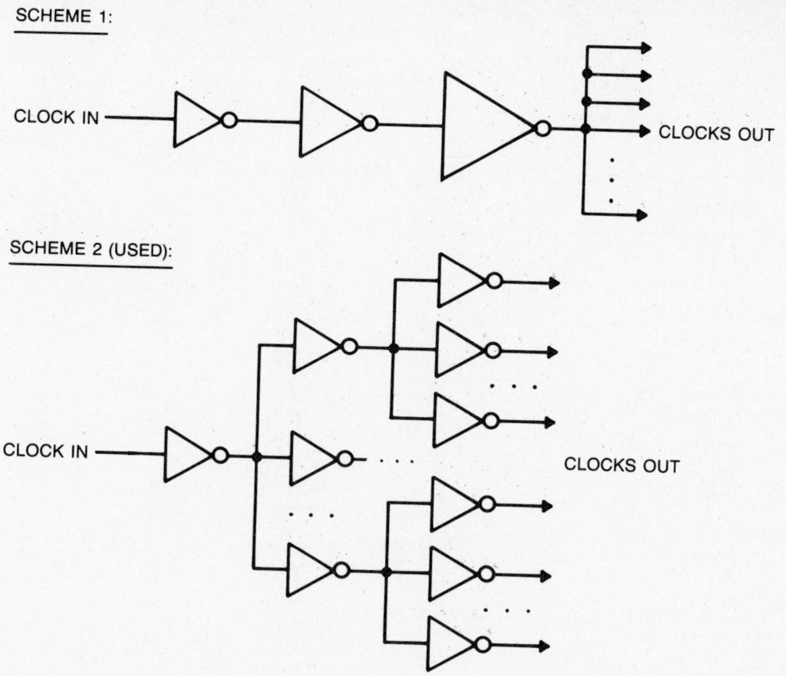

SCHEME 1:

CLOCK IN

CLOCKS OUT

SCHEME 2 (USED):

CLOCK IN

CLOCKS OUT

FIGURE 9.37. Clock distribution

rail. The $\overline{p}$ vector generator is pipelined, so each stage contains a register (but not the shifter). The right two-thirds of the chip is composed of eight accumulators, differing only in an address decoder for readout. Readout is accomplished by a 32-bit parallel tri-state bus that terminates at an output multiplexer, which multiplexes one of four bytes out to the pads. Intensity values flow down from the top of the chip through a series of shift registers to equalize the pipeline delay. The shift registers are necessary to ensure that intensities and control signals arrive at the same locations on the chip at the same time. A programmable logic array (PLA) is included in the chip to generate control signals as a function of the moment to be computed and the pixel type. Control signals from the PLA are routed to the part of the chip needing them while other control signals are brought in directly from the pads. The routing is accomplished through the sections between groups of eight adder bits and along the edges of the chip. Power and ground are distributed via a folded interdigitated scheme, with the ground bar in the center and power around the edge. The central ground bar constitutes another routing area. Clock distribution is handled by a multilevel scheme. Unlike other schemes, which use continually larger drivers until the same net drives all clock inputs on the chip, this scheme uses identical drivers to equalize the load so as to minimize clock skew (Fig. 9.37). The advantage of the scheme is that the relatively small drivers may be distributed within the rest of the circuit.

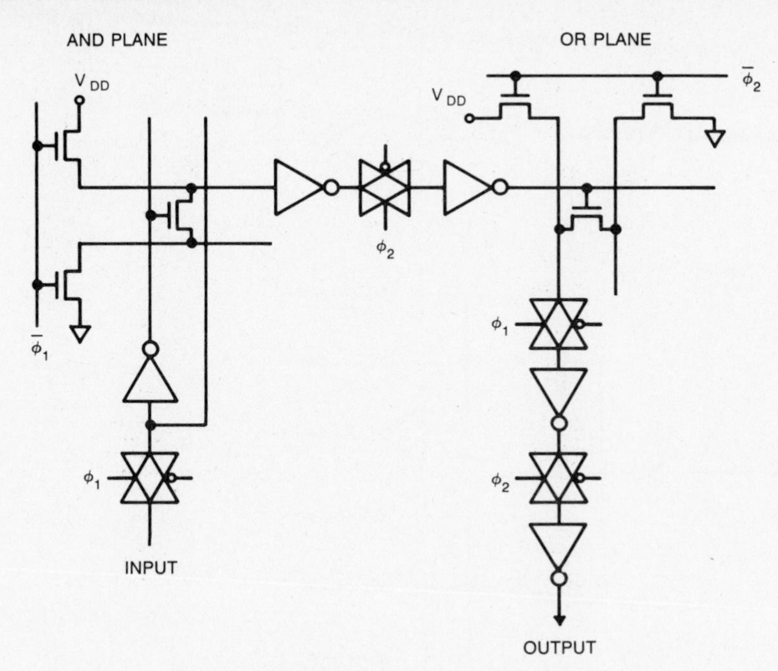

FIGURE 9.38. PLA structure

9.3.5 Component cell examples

9.3.5.1 Programmable logic array

A programmable logic array (PLA) was chosen as the implementation technique for the generation of control signals. CMOS logic complicates the design of PLAs, since a static design requires two complementary transistor structures, which is impractical for large arrays. Instead, a clocked structure shown in Fig. 9.38 was used that requires only a single n-channel transistor array. The timing is shown in Fig. 9.39. The PLA has a two-clock pipeline delay. With simplifications

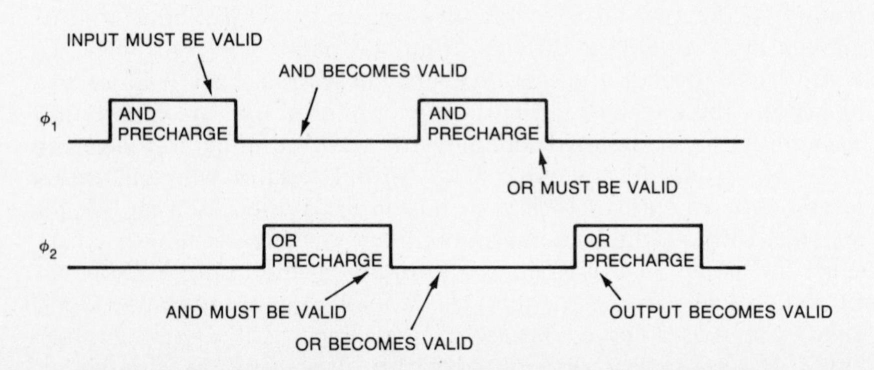

FIGURE 9.39. PLA timing

to the output structure, the delay can be reduced so that outputs may be fed back to inputs to form a finite state machine.

The PLA was specified in a high level LISP-based language (CQ) designed for PLA construction. Five inputs, twelve minterms, and eight outputs were necessary. Although PLAs of this size can clearly be designed by hand, one would like to have complete confidence in the result. Although simulation of large parts of the chip can verify the design, it's better to start with something that is known to be correct. For larger PLAs, a powerful specification language like CQ is essential. The language translator outputs a numeric table specification of the PLA, so different front (or back) ends could be utilized.

9.3.5.2 Accumulator cell

One of the most important cell designs in the moment generator chip is the low order accumulator cell. A layout of this cell is shown in Fig. 9.40. (Actually two bits are shown.)

An AND gate operates on the incoming $\overline{p}$ bit and the a_i intensity bit. This product is then fed into an adder, the other input for which comes from the output of a 2-phase register. Embedded in the register is a second AND gate used to reset the registers. The placement of the gate removes it from the critical timing path of the circuit. Rather than have an inverter after the second transmission gate, the gate inputs on the adder are used directly as a storage node. However, the output bus transmission gate will deplete the storage node, so an inverter was included for this purpose. This cell is used for the lower 16 bits of the F_k accumulators. The upper 16 bits of the accumulators have no $\overline{p}$ inputs, so a simpler cell containing an incrementer rather than an adder is used. The space freed up is used for a pipeline equalizing shift register and an AND gate used to suppress the intensity during blank pixels.

9.3.5.3 Output bussing

The design of the output bus along the right-hand edge of the chip is shown in Fig. 9.41. The basic problem is to connect the two outputs of a driver to the appropriate bits of a bus. Rather than take the brute force approach of designing four different cells (of which several copies would be needed), the design shown requires only a single basic cell. The price being paid is the number of contacts along the wire. If the readout path had to be heavily optimized for speed, this approach would not be applicable. Wherever possible, techniques such as this should be used to reduce the number of cells and encourage modularity. Another method of dealing with this case would be to generate the cell procedurally.

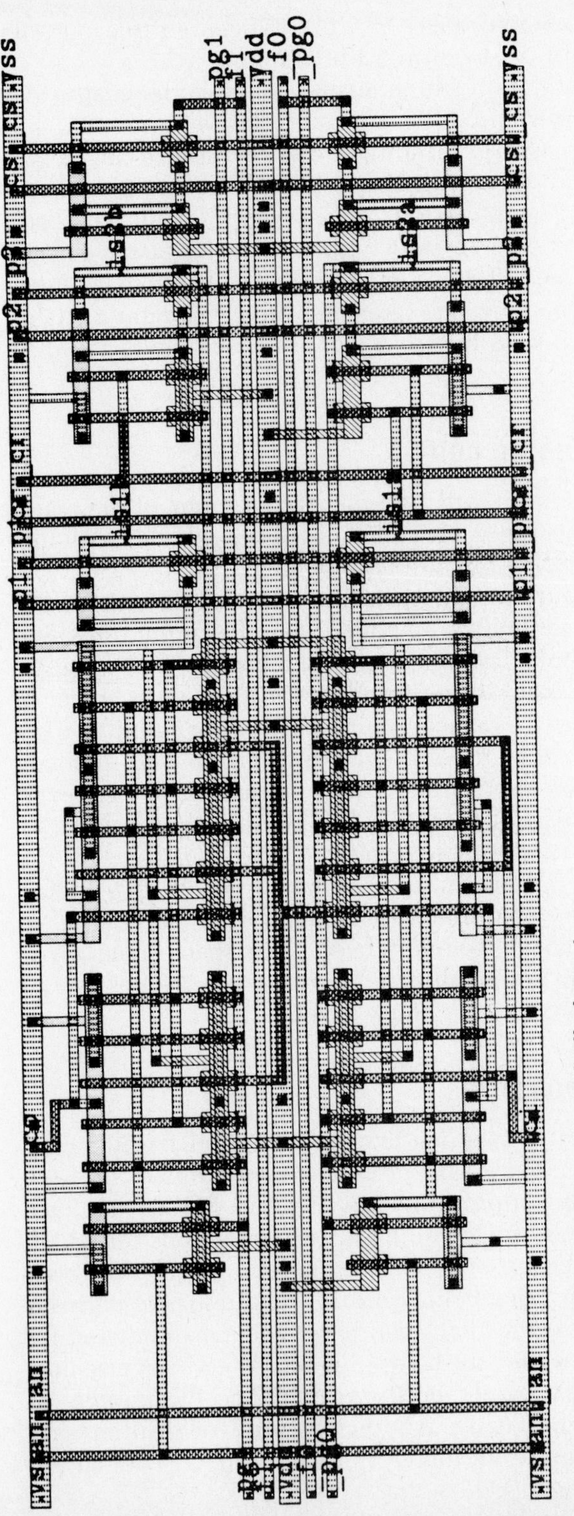

FIGURE 9.40. Accumulator cell symbolic layout

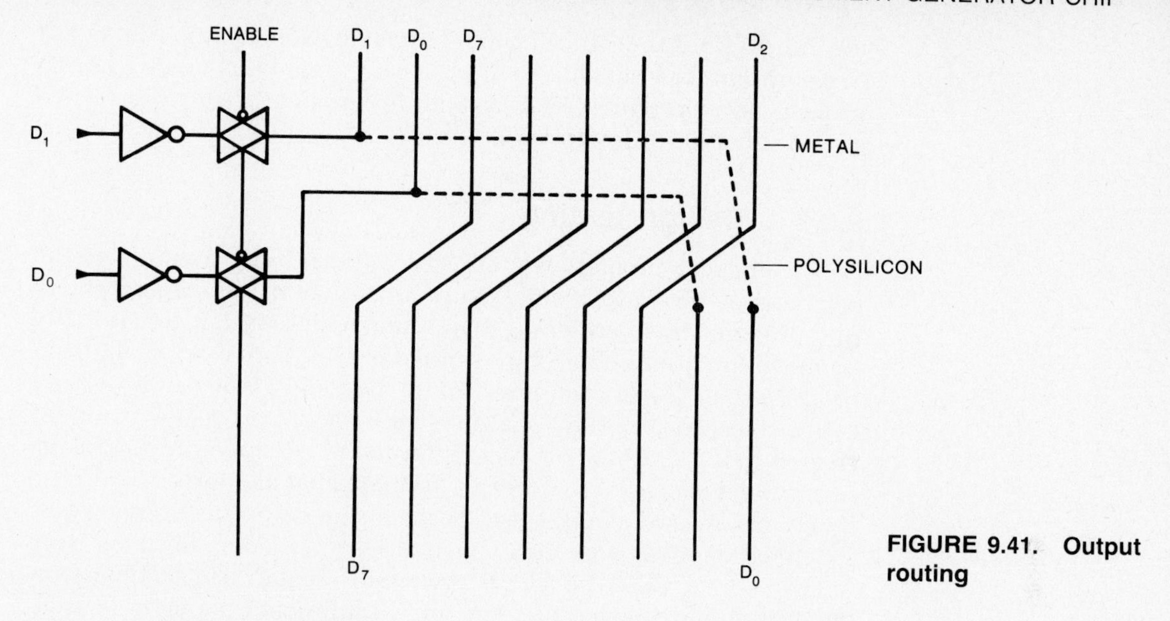

FIGURE 9.41. Output routing

9.3.6 Chip assembly

The moment generator was assembled using a multilevel tiling scheme. Leaf cells were tiled to form major cells; for example, the lower sixteen bits of an F_k accumulator. The intermediate level cells were smashed (converted to primitive components) and then compacted as a unit to minimize cell boundary problems.

Chip assembly was embodied in a special purpose UNIX™ (AT&T) shell script of about 160 lines. It was very worthwhile to have a script, since the assembly was done quite a few times.

9.3.7 Optimization

At a late stage in the design of the chip, the actual board was designed. Based on the design, a number of refinements in the chip were made. Originally, a tri-state external bus was contemplated, but the pad generator was lacking the requisite pads and the ability to route the pad to enable signals. A fair number of external multiplexer chips were required. A standard pad was modified for "open-collector" operation, requiring only an external pull-up resistor. The modified pad has served admirably. A latch was added on the input of the PLA to store the moment number. This eliminated the need for external latches, saving six chips. Two-phase registers would require that clocks be routed to the register cells. A simple unclocked two-gate SR-type latch implemented this function nicely. Inverting the load input to the register would have saved a chip of external inverters. The bottom line from this discussion is that the external circuitry should be definitively designed (not just sketched) before the chip is finalized, since even relatively trivial changes may result in significant external chip savings. (Bear in mind this was a research project, not a product. Full system design should always precede any product chip design.)

9.3.8 Design testing

A "C" language program was constructed that exactly simulated the behavior of the chip (before chip construction was begun, even). The program was verified against a much simpler but functionally identical program. The chip simulator is sufficiently accurate to produce the state of each node on the boundary (with a pin) of any cell in the chip at the end of each clock cycle. A test vector generator created half a million test vectors for the chip, based on processing several real pictures. A software system generated tests based on a time window and a cell subset of the entire chip. Creating new test specifications was very easy. Several tests of major substems were performed: the PLA, the whole $\overline{p}$ generator, an entire F_k accumulator, the output multiplexer, etc., for several different time periods each. The tests were performed concurrently on several VAXs at once, taking 5–10 hours of CPU time apiece. Design rule checks were also performed over the entire chip, at several hours for each run. This level of testing requires a serious expenditure of effort, but seems a relatively small penalty for getting it right the first time.

9.3.9 Physical chip testing

Sooner or later (probably the latter), you will be confronted with a large number of chips and will have to decide which of them, if any, work. The following discussion relates to testing in a non-production environment and represents what might be done in a university environment. There are two stages to the testing problem: first, deciding which of the wafers work, and then later, which of the packaged devices still work. We have found that a logic analyzer with a small (<256 word) pattern memory served acceptably to identify packageable devices for prototyping in this chip. Note that in production, a full test set would be required to test at the wafer probe stage. The (software) test vector generator and chip simulator can generate test vectors for this purpose. The packaged devices must then be tested before they are placed in service. Rather than creating a separate test system for chip testing, the moment generator board has a test mode built into it that allows the computer to simulate the video input. Note that this approach is essentially functional testing. It is not the type of testing that is required to qualify manufactured parts. See Chapter 6.

Chip testing was comprised of three stages, each implemented as a separate computer program. The first stage was a slow functional test, using a 473,000 element test vector. Test vectors were applied at an approximately 500 Hz rate. Blank pixels were inserted to bring the clock rate to about 2000 Hz. Chips operated at 500 Hz at wafer probe; commercial dynamic RAMs are refreshed at this rate. Minimum clock rate is determined by the storage node leakage rate. The verify data was supplied by the simulator. Originally, chips were to be batch tested six-at-a-time, but the odds were one in five that a device would exhibit a certain problem that made all of the devices appear bad. Testing took 15 minutes for each good device, but most devices failed within a second (untested wafers were packaged for expediency). Testing was stopped when a device appeared to be good and repeated to conclusion later. The next test was a high speed test using fake low resolution pictures created with the location map. Test data was also supplied by the simulator. This test allowed speed testing of the devices. The final test was a real time chip comparison — six working chips were compared with each other, using a majority voting scheme, while processing a live incoming video picture.

9.3.10 Conclusions

The final layout of the chip implemented in 2.5 μ twin-well CMOS is shown in Fig. 9.42. The die size is 7 mm * 7 mm for compatibility

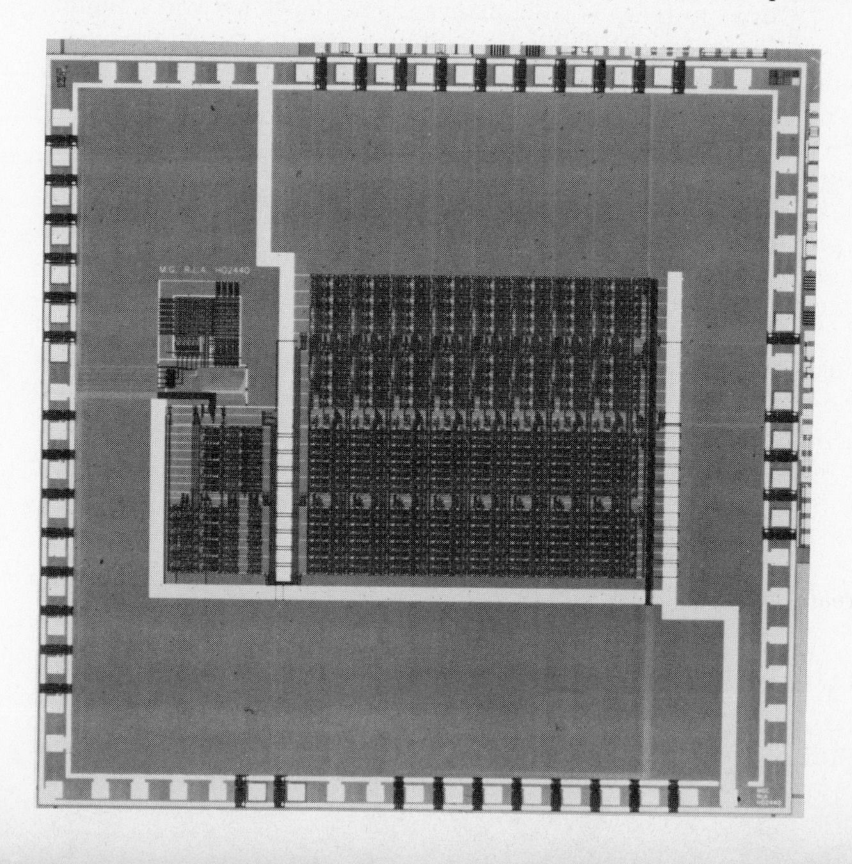

FIGURE 9.42. Microphotograph of fabricated MG chip

with the multi-project shuttle run internally at AT&T-Bell Laboratories. The active area is only about 4.7 mm * 2.9 mm. There are 10354 transistors. The devices that were functionally correct all ran at the maximum test speed of 115 nS.

9.4 Self-routing switching network

S. C. KNAUER,*
J. H. O'NEILL, A. HUANG†

9.4.1 Introduction

This section demonstrates a VLSI architecture for wide-band switching that employs regularity at the cell, chip, and system level to yield an architecture that surpasses conventional architectures. It illustrates some of the architectural innovations that may be required when one converts a design from TTL to VLSI.

Sorting, switching, and data routing (as in parallel computers) are examples of functions performed in parallel by large networks of small identical elements. To efficiently package these elements and their interconnections on custom chips, the topology of each network must be well understood. One topology common to many routing networks is the perfect shuffle [Ston71][WuFe81], which can be used to build sorting, banyan, and Benes networks, among others. The perfect shuffle is also used to interconnect small processors performing fast transform calculations. The case history in this section deals with the problems encountered in designing and testing a set of chips that is used to construct a switching network with the perfect shuffle topology. The switch consists of a sorting network based on Batcher's bitonic sorting algorithm [Batc68], followed by a modified version of Lawrie's Omega network [Lawr75]. A set of three custom chips, two for the Batcher network and one for the Omega, will build a switching network of arbitrary size.

The switch is used to route m 5 Mbit/s serial data streams in a packet format; m is assumed to be a power of 2. Fig. 9.43 illustrates the basic switch design. Each packet is divided into header and data sections; the header contains the address to which the packet is routed, the address bits being ordered most significant bit (MSB) to least (LSB). Preceding the MSB in the header is a single "activity" bit, which is "0" if the packet contains information and "1" if it is "inactive" — occupying an unused time slot. In this design the packets are of equal length and aligned in the same time slots. The sorting network arranges the m packets in each time frame, in ascending order, spatially along its output lines. In typical switch operation, some source inputs may be unused and some destinations

* AT&T Bell Laboratories, Murray Hill, N.J.
† AT&T Bell Laboratories, Holmdel, N.J.

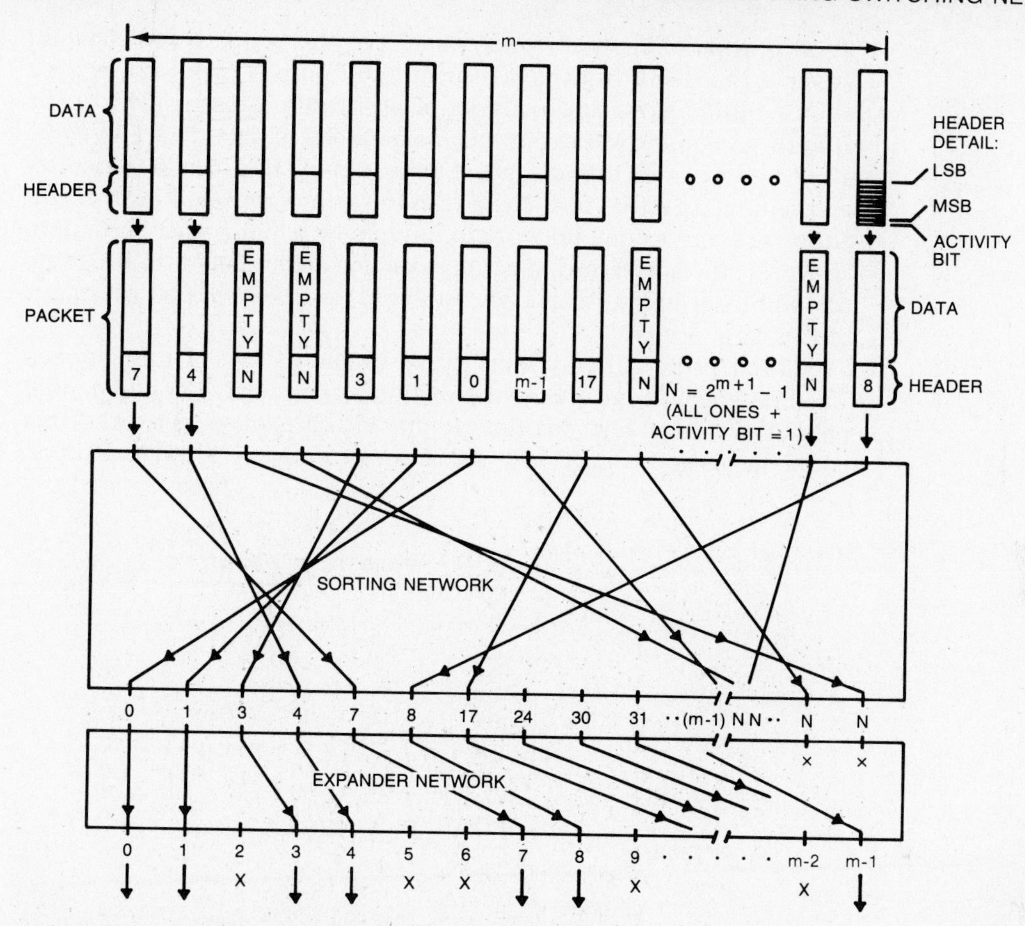

FIGURE 9.43. Basic switch design and packet format

uncalled. By including the activity bits of the inactive packets from unused source inputs as the most significant bit of the sorter's header, these inputs are effectively given header numbers larger than the largest output address. This ensures that they all end up at the high side of the sorting network's output. At the sorter output the packets are ordered relative to one another, and this is not necessarily the consecutive numbered order of the physical connections to their destinations. For example, in Fig. 9.43, no source is sending a packet to destination 2; as a result, all packets destined for higher addresses are offset to the left at the sorter's output. The expander network corrects this by routing the sorted packets past unused outputs.

Both networks are self-routing, which means that the switch nodes set themselves by reading the headers. They are also non-blocking, which means that no matter how many active packets the networks handle, the switch settings required by one packet for connection from input to output won't interfere with those required by any others. The networks also have the property of constant latency; the delay from any input to any output is the same. This allows the construction of links of arbitrarily large bandwidth by simply ganging several 5 Mbit/s channels in parallel. There are other networks in this design, a concentrator to precede the sorter and remove inactive packets, a trap to remove repeated destination addresses, and a copy network to implement a "broadcast" function. However, the sorter and expander designs are sufficient to illustrate the VLSI design problems encountered, so they will be the focus of this case study.

Three principal problems arose in the course of this project. The first was finding a way of partitioning the sorting and expander networks into a few smaller, identical, chip-sized networks that could be used to build any sized sorter and any sized expander.

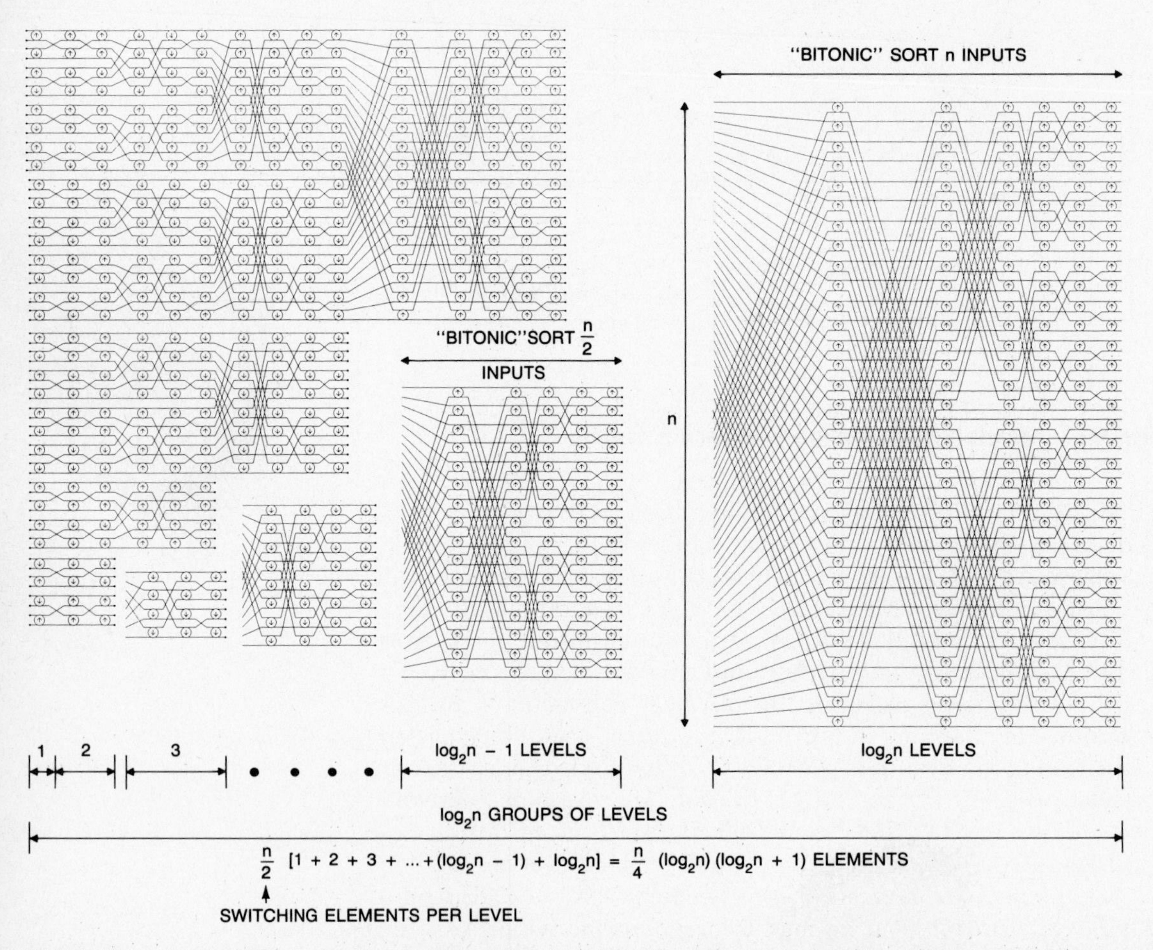

FIGURE 9.44. Batcher's sorting network broken into its component parts—bitonic sorting networks

The second was finding an efficient layout for each custom design in the chip set, paying special attention to the cost in floor plan area of the routing networks required to connect the active elements. The third was developing efficient test procedures for each chip to, first, isolate each piece of the design so that it could be tested and made to work independently, and then (once the design was verified), second, to test the chip's entire functionality as thoroughly and quickly as possible.

9.4.2 Partitioning the sorting and expander networks

Fig. 9.44 illustrates a 64-input Batcher sorting network; data flow is from left to right. Fig. 9.45 gives the detail of the basic 2-input, 2-output sorting element. The function of the sorting element is to

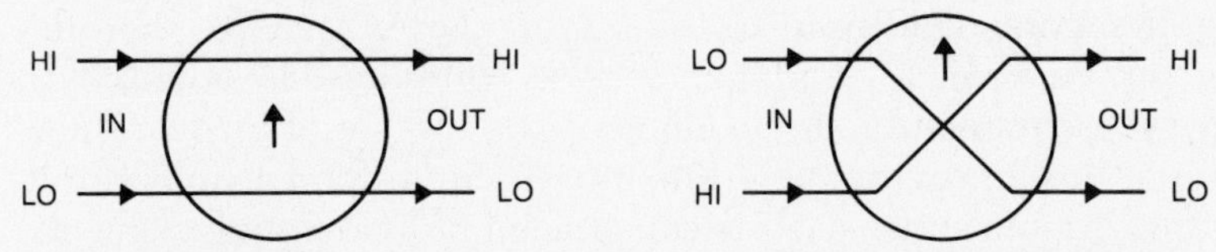

SWITCHING ELEMENT PASSES OR INTERCHANGES LINES TO SORT HIGH NUMBER TO UPPER OUTPUT.

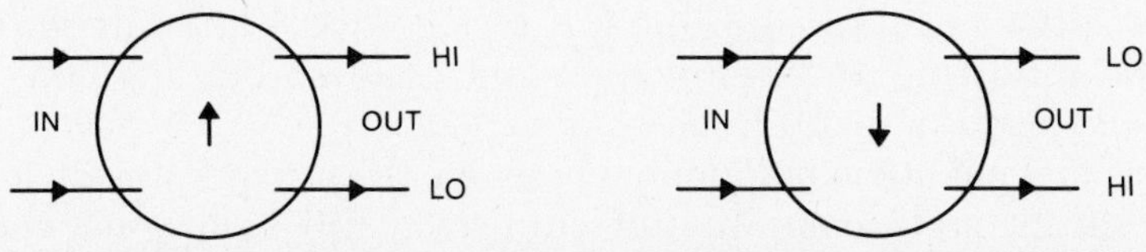

ARROW ON SWITCHING ELEMENT USED AS A CONVENTION TO INDICATE WHETHER THE HIGH NUMBER IS ROUTED TO THE TOP OR BOTTOM LINE.

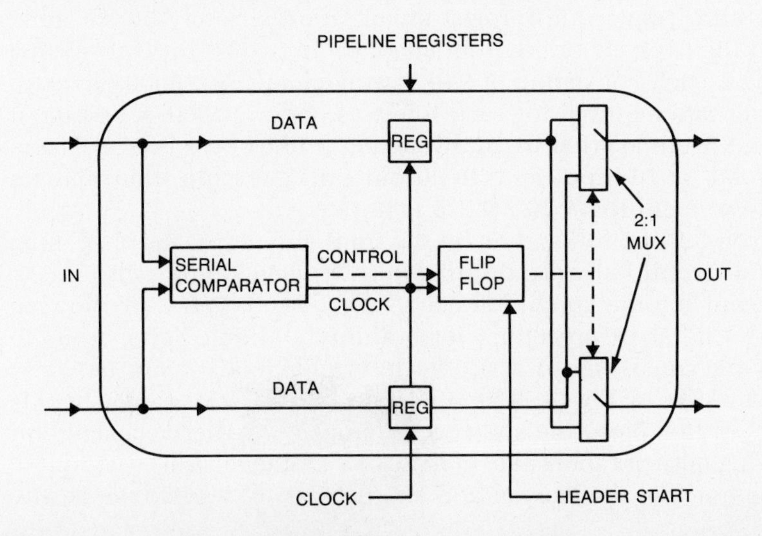

FIGURE 9.45. Block diagram of a 2-input, 2-output sorting element

compare headers and either pass or exchange the two inputs so that the larger one is output at the side indicated by the arrowhead on the symbol of the element. The hardware consists of a serial comparator, a "pass"-"exchange" switch, and registers to pipeline the data. The switch is set to "pass" at the beginning of each packet and either freezes in the pass mode or switches into the exchange mode when it first finds unequal bits (within the header) in the two addresses being compared. If the headers are equal, the switch freezes in the pass mode. This setting is held for the remainder of the packet. The location of the header is supplied by an external control signal that is high for the duration of the header.

The partitioning of the network in Fig. 9.44 is approached by paying attention to its derivation, as given in Batcher's paper [Batc68]. The full sorting network is built of a succession of smaller "bitonic" sorters. These bitonic sorters take two sorted lists of $m/2$ lines each and merge them into one sorted m line list with $\log_2 m$ levels of sorting elements. The input to the bitonic sorter ascends, then descends (or vice versa), and hence is a "bitonic," as opposed to monotonic, sequence. A single element is a full sorter for two lines, and two of these provide the 4-line bitonic input for a 4-line bitonic sorter; this bitonic sorter itself is constructed of four sorting elements. This forms a 6-element full sorter; two of these are shown isolated at the lower left of Fig. 9.44. These two feed a 12-element, 8-line bitonic sorter to make an 8-line full sorter. Two 8-line full sorters feed the following 16-line bitonic sorter, and so on, allowing an arbitrarily large network to be constructed.

As a first cut at the partitioning problem, the network was divided into a "front end" consisting of chip-sized full sorting networks (the largest that would fit) and a "back end" consisting of a series of larger and larger "bitonic" sorting networks. The space available on the chip (done multiproject style), approximately 36 sq. mm., limited the capacity of the front end chip to two 32-line full sorting networks, each consisting of 240 sorting elements (18000 devices). The data rate required for each line was under 5 Mbit/s, so pinout problems were averted by multiplexing 4 lines over each I/O pin, for a total of 16-input and 16-output pins per chip; the pins are capable of handling a 20 Mbit/s data rate.

Given 32-line full sorters on the front end chips, the only thing required to build any sized full sorter was developing a chip set to implement bitonic sorters of size 2^n for $n = 6, 7, 8, \ldots$ Following Batcher's iterative procedure for building a bitonic sorter leads to a network consisting of a shuffle followed by a banyan network; the right side of Fig. 9.44 is a 64-line bitonic sort shown in this format. At this point the ability to reference or re-derive a small bit of routing network lore came in handy (we followed the latter route). It turns out that the banyan and shuffle networks are topologically

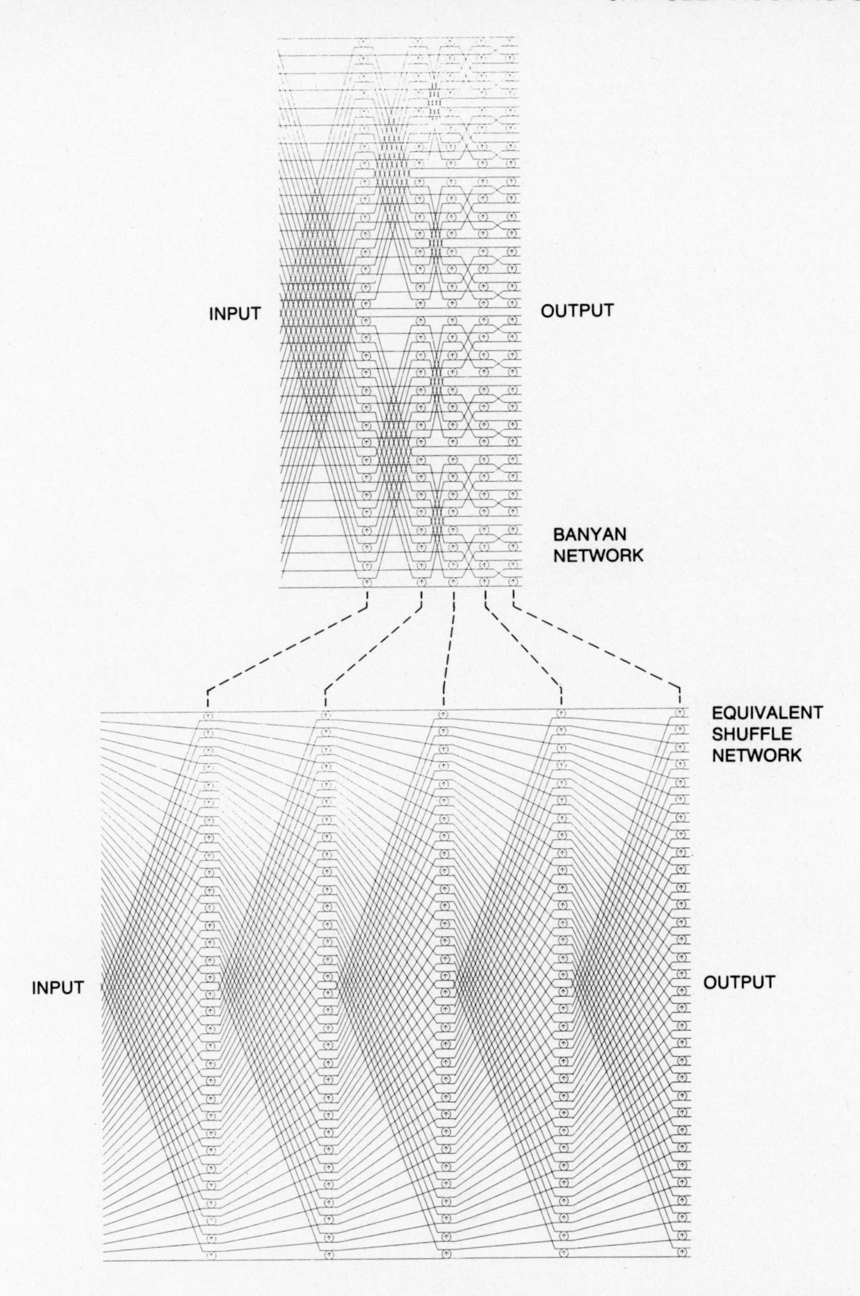

FIGURE 9.46. Equivalence of the perfect shuffle and banyan networks

equivalent — just two ways of drawing the same graph. This is illustrated in Fig. 9.46. This means that an m line bitonic sorting network consists of a shuffle network m inputs wide and $\log_2 m$ levels deep. The problem became finding a way of building shuffle networks of widths 64, 128, 256, etc., and depths of 6, 7, 8, etc., with one chip type.

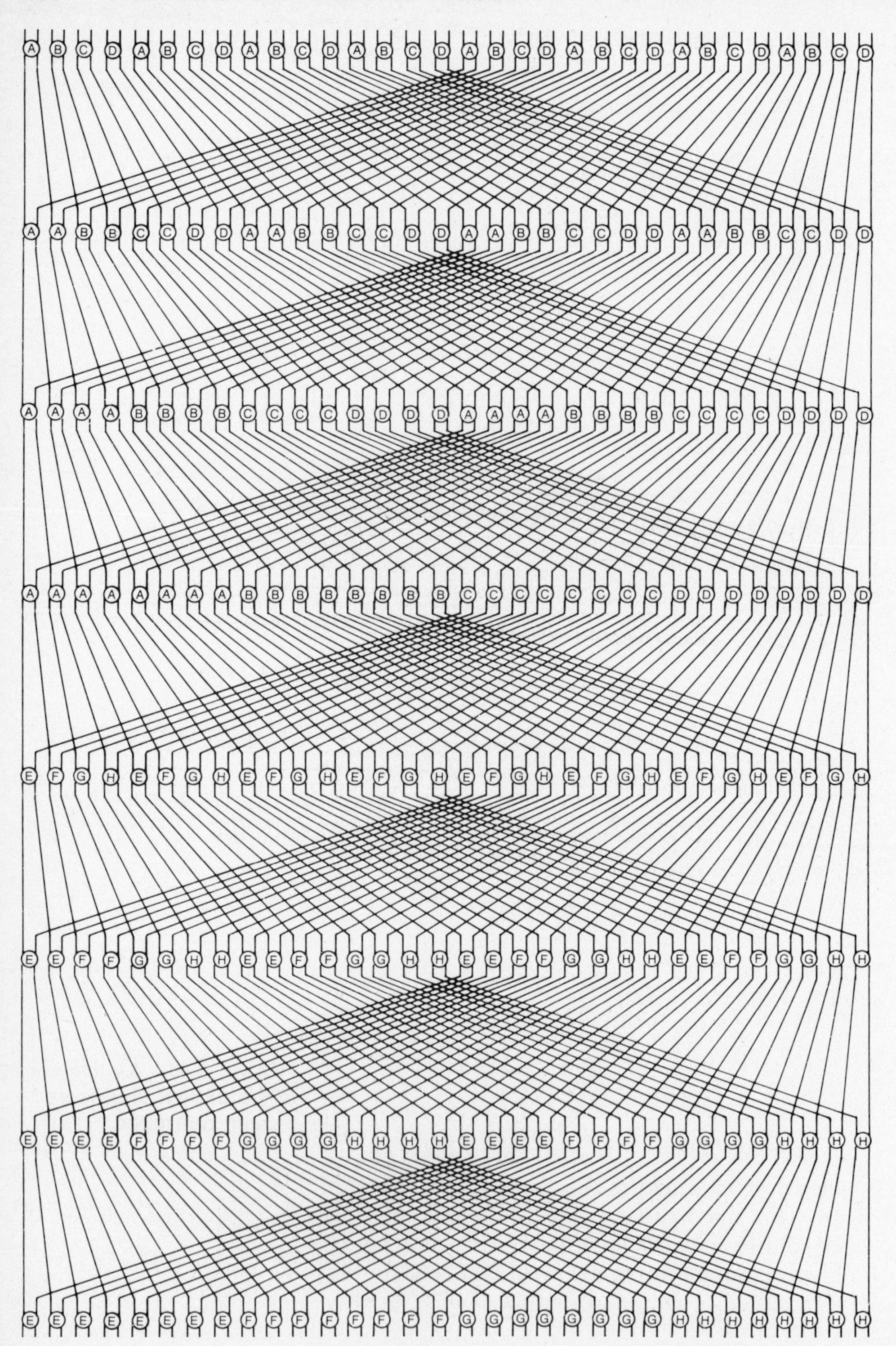
A B C D A B C D A B C D A B C D A B C D A B C D A B C D A B C D
A A B B C C D D A A B B C C D D A A B B C C D D A A B B C C D D
A A A A B B B B C C C C D D D D A A A A B B B B C C C C D D D D
A A A A A A A A B B B B B B B B C C C C C C C C D D D D D D D D
E F G H E F G H E F G H E F G H E F G H E F G H E F G H E F G H
E E F F G G H H E E F F G G H H E E F F G G H H E E F F G G H H
E E E E F F F F G G G G H H H H E E E E F F F F G G G G H H H H
E E E E E E E E F F F F F F F F G G G G G G G G H H H H H H H H

FIGURE 9.48. Network of Fig. 9.47 with elements grouped into subsets

Re-deriving another bit of routing lore (with colored pencils and copies of shuffle graphs), it can be shown that a large shuffle network m lines wide and $p < \log_2 m$ levels deep can be partitioned into $m/2^p$ shuffles, each of width 2^p. Fig. 9.47 illustrates the partitioning of a 64-wide 4-level network into $64/16 = 4$ smaller shuffles, and Fig. 9.48 shows the smaller shuffles on separate chips interconnected to make up the larger network. Using this procedure, it is possible to start with a chip containing a 64-wide shuffle to build arbitrarily large shuffle networks of width 2^n for $n \geqslant 6$. As in the case of the front end chip, the 64 inputs and outputs are multiplexed 4 lines per I/O pin.

◀ **FIGURE 9.47. A 64-input, 8-level shuffle network labeled to reflect a partitioning into 8 subsets labeled A, B, C, D (first 4 rows) and E, F, G, H (last 4 rows)**

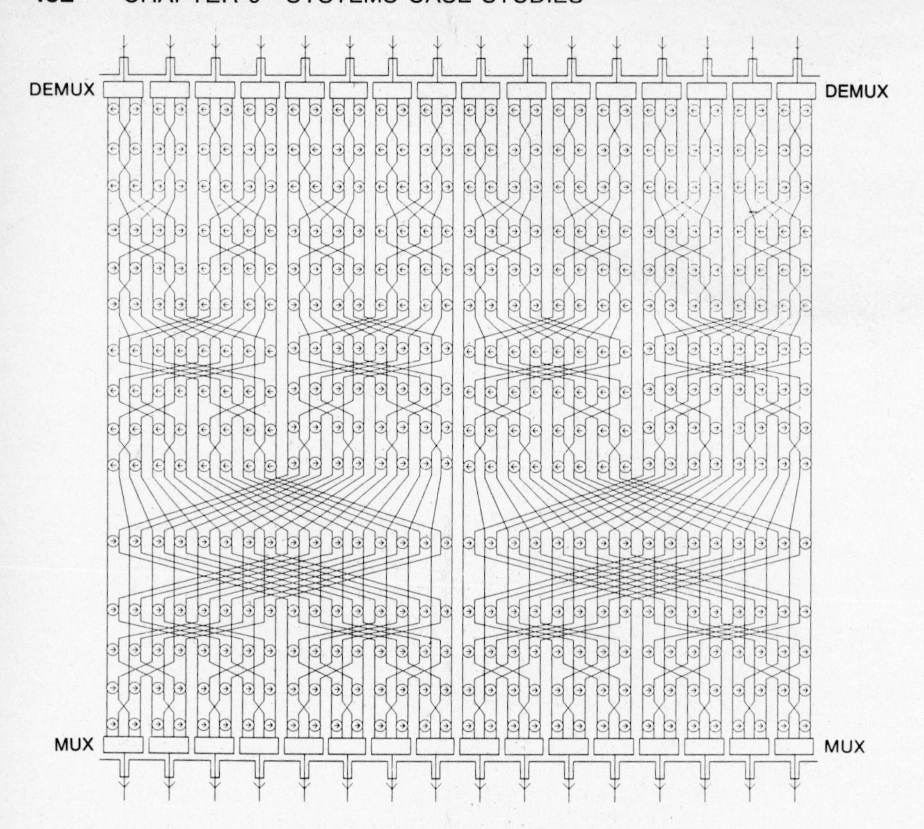

FIGURE 9.49. Front end multiplexed Batcher sorting chip

Fig. 9.49 and Fig. 9.50 illustrate the front end and back end chips in schematic form. The back end Batcher sorting chip contains 128 sorting elements and about 10000 devices. Fig. 9.51 illustrates the interconnection of the two types of Batcher chips to build a 256-line Batcher sorting network. The ordering (ascending or descending) of the output of both 32-line sorters on each front end chip will be the same, as it depends on the direction in which the output multiplexers scan the sort element outputs in the bottom level, and these multiplexers have a common control. Since the sorting algorithm merges ascending and descending sequences, different front end chips on row 0 would feed the back end chips on row 1 (not shown). The back end chip is used to build shuffles of varying depth, therefore it must be able to vary its depth. Thus the last one, two, or three levels of shuffle followed by sort element are designed to be bypassed by entering a 2-bit binary representation of the number of levels to bypass. In Fig. 9.51, the back end chips in row 2 have their last two levels bypassed and those in row 4 have one level bypassed. Rows 1 and 2 comprise four 64-wide, 6-

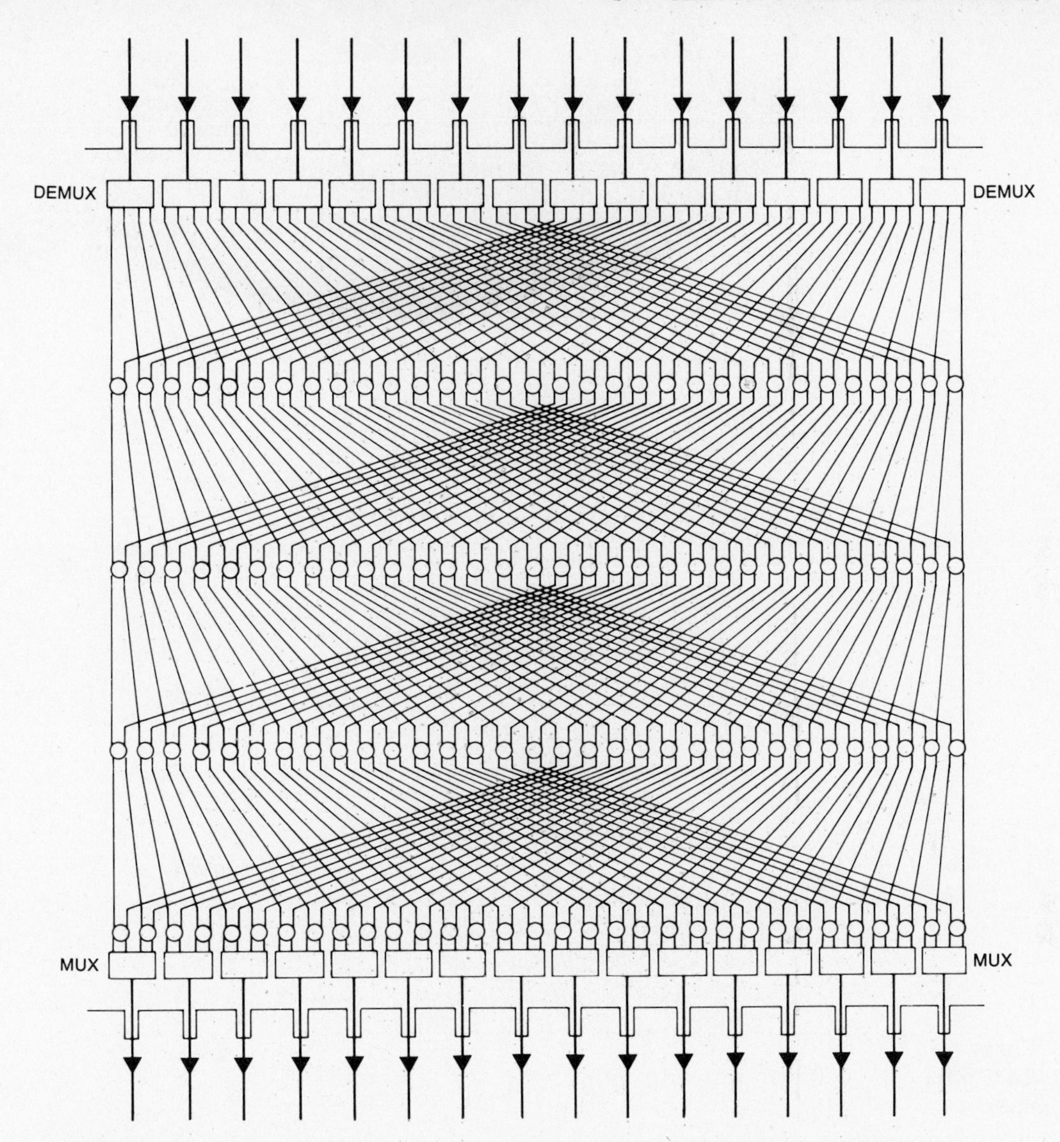

FIGURE 9.50. Back end multiplexed Batcher sorting chip (perfect shuffle)

level bitonic sorters that merge the output of the dual 32-line sorters in each of the front end chips. Rows 3 and 4 comprise two 128-wide, 7-level bitonic sorters that merge the outputs of the four 64-wide sorters directly above them. Finally, rows 5 and 6 comprise a single 256-wide, 8-level bitonic sorter.

The multiplexing of inputs on the chips not only reduces the pin count of the chips by a factor of four, but also reduces the number of wires required to interconnect the chips by the same

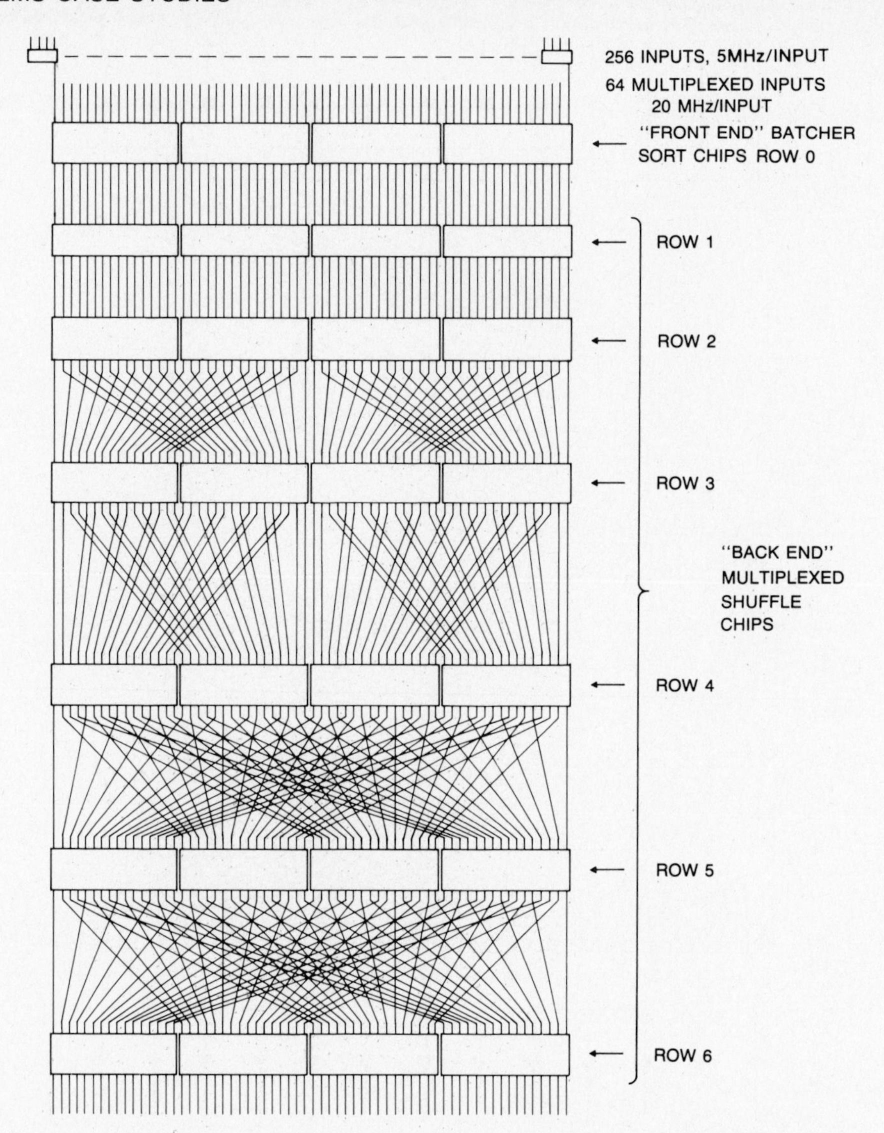

FIGURE 9.51. Chips required to sort 256 5 MHz channels (28 chips)

factor. Thus the interconnects between rows in Fig. 9.51 are reduced from 256 wires to 64. A factor of four in interconnect complexity has been pushed *inside* the chip, along with all the interconnects between the levels inside the chips. It may be noted that a full depth m line shuffle network is $\log_2 m$ levels deep, and the back end chip has only four levels with room for several more. The fact that there are only 16 physical pins limits the number of levels to $\log_2 16 = 4$. It can be shown that using more levels per chip would time multiplex on a single output pin data that is required by two or more different chips in the next row down. This is a rather obscure point, but it represents a class of subtle pitfalls to be avoided.

Making physically small networks act like large ones is treated generally in [FiFi82].

The expander network has exactly the same shuffle-banyan topology as the bitonic sorting network; the only change is in the algorithm used in the switching elements. Fig. 9.52 illustrates the operation of the modified Omega network with the banyan form of the graph. The first header bit, the most significant bit (MSB), switches

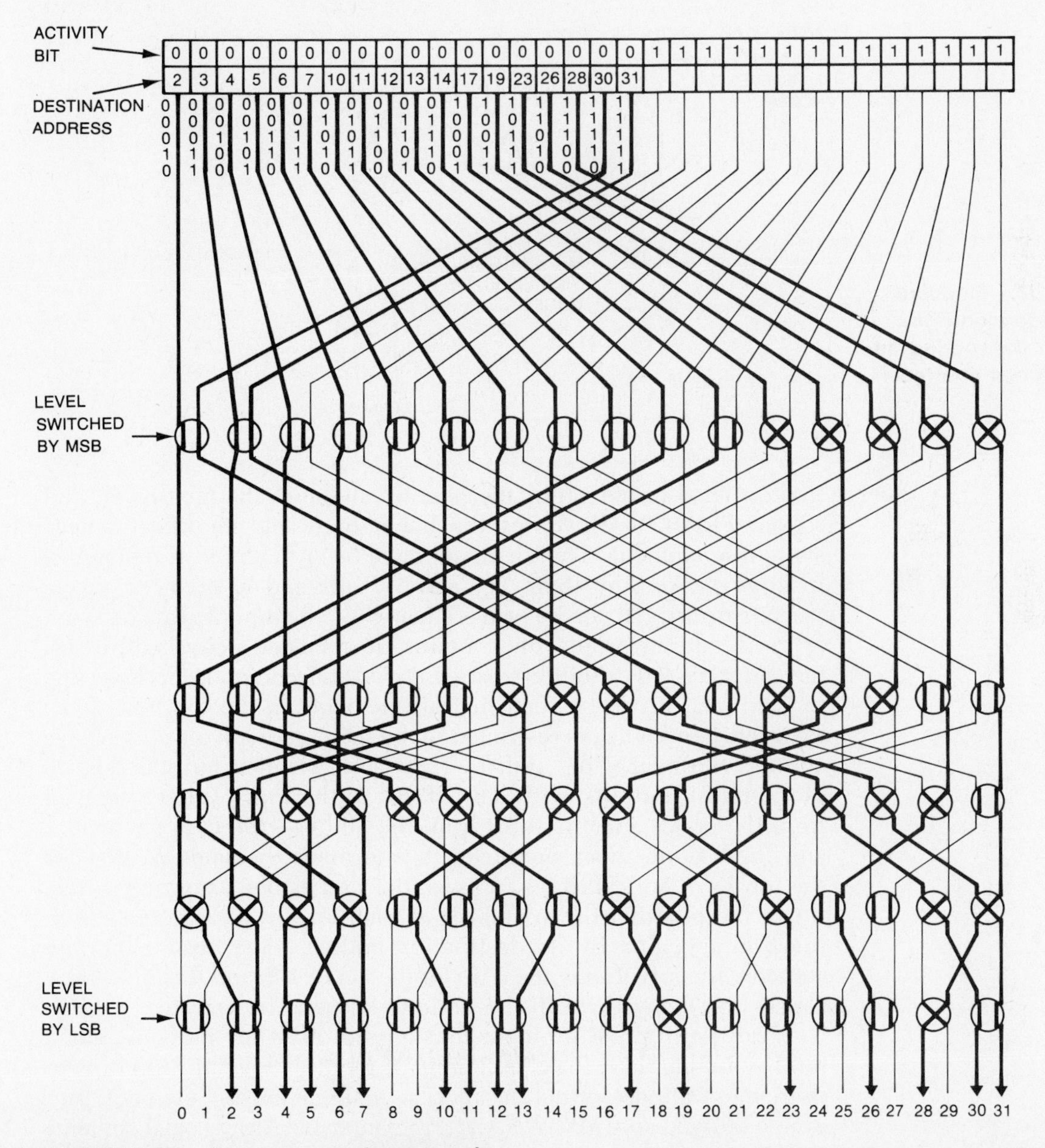

FIGURE 9.52. Omega expander network

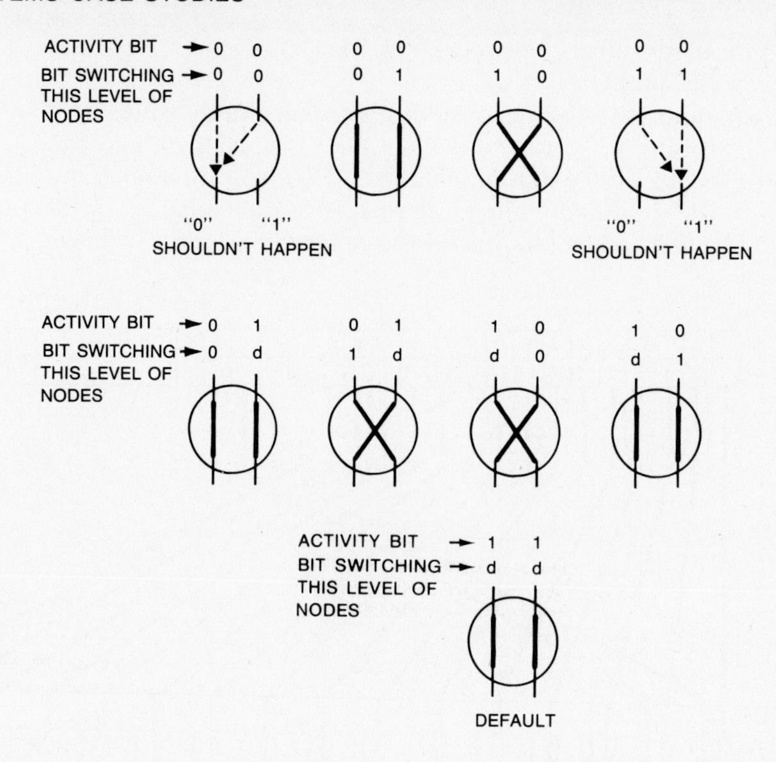

FIGURE 9.53. Modified Omega node algorithm; each bit in the routing address switches one level of nodes

the top level of elements, the next bit switches the next level, and so on. Fig. 9.53 illustrates the behavior of the modified Omega switching element. The elements have been modified to recognize inactive packets by their "1" activity bits and to give the active packets priority. It can be proved that when the input to this network is in ascending order and contains no repeated active addresses, the cases labeled "shouldn't happen" will indeed not occur. Referring to Fig. 9.52, it can be seen that all elements in the top level have their left-hand outputs restricted to the set of network outputs whose address MSBs are "0," and the right-hand element outputs lead to the network outputs with "1" MSBs. If the top level is removed from the graph and the MSB from the address, the case for level 2 and the second most significant bit becomes the same as that for the top level and MSB on an expander half as big. Continuing this, it can be seen that the switching at each level routes the packet to the line specified by its destination address. The header bits are shifted circularly to line them up for the next level, and this procedure causes the delay through the node to be two clock cycles.

Partitioning is a problem that doesn't exist for most networks built with standard logic functions. It requires several SSI chips to implement a small function such as a single sorting element, and even a design using MMI™ PALs (Programmable Array Logic) contains only two sorting elements per chip. In such cases, all routing is obviously external to the chips. In contrast, when hundreds of sorting elements fit on a chip, much of the routing between them does too. In fact, putting a large portion of the routing problem on the chip is a savings comparable to putting hundreds of elements on the chip.

At the same time the VLSI designs were in progress, a dual 32-line sorter was constructed on a 15″ × 16″ wire-wrap board holding 270 chips (240 MMI PALs plus miscellaneous registers and inverters). The cost of board and chips is over 3,800 dollars and it draws 35 A at 5 V. The PAL implementation does run faster at 15 Mbit/s per line compared to the equivalent front end Batcher chip's speed of 5 Mbit/s per line; this is because four lines share a 20 Mbit/s pin in the chip design. However, the application requires only 2–3 MHz/line, as each user's bandwidth is built of several parallel lines to give it a modular structure. The front end Batcher chip (equivalent to the wire-wrap board) draws only 20 ma and costs about 200 dollars per working chip as part of a low volume multiproject shuttle; this could drop by a factor of 10 for large production runs. It should be noted that this switching scheme was not regarded as very practical prior to the VLSI design, due to its hardware complexity.

Algorithms have been changed as a result of attempting to "move" TTL designs into VLSI. Self-routing PM2I networks [SiMc81] were used originally in a TTL PAL implementation of the expander. However, the PM2I network partitions into the form of a shuffle with additional lateral interconnects that double for each level of elements included on a custom chip. This was downright unaesthetic and required an additional 8 pins (using 4:1 multiplexing on some). This inspired a continuing search for new algorithms, resulting in the discovery that the Omega network was nonblocking for sorted, nonrepeated inputs if it was modified to handle inactive packets. The moral of the story is that a TTL design may need to be reworked all the way back to the algorithm selection stage before it is "moved" to VLSI.

9.4.3 Chip layout

The key building block for the parallel sorting network is the 2-line sorting element shown in Fig. 9.45. Recall that the sorting element serially compares the input lines during a "header" period, sets a pass/cross switch when the first difference is encountered, and remains set until a reset pulse reinitializes the element. This behavior is summarized in the state transition diagram of the sorting element shown in Fig. 9.54.

Since data constantly flows through the switch, simple 8-transistor 2-phase clocked CMOS shift registers were chosen to pipeline the

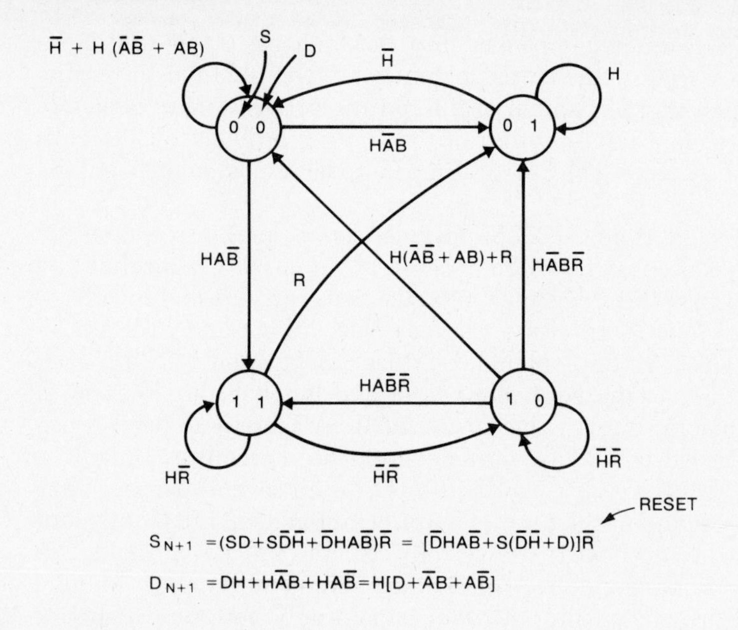

S	D	H	A	B	S_{N+1}	D_{N+1}
0	0	1	0	0	1	—
		1	0	1		
0	0				—	1
		1	1	0		
0	1	X	X	X	0	—
0	1	1	X	X	—	1
		1	1	0		
1	0	0	X	X	1	—
1	0	1	0	1	—	1
		1	1	0		
1	1	X	X	X	1	—
1	1	1	X	X	—	1

S — STATE OF SWITCH 0 = PASS, 1 = CROSS
D — DIFFERENCE HAS BEEN FOUND (1), CAUSES S TO BE HELD SAME
A — INPUT LINE
B — INPUT LINE
H — HEADER PULSE (1 WHEN HEADER BITS)
R — RESET (BYPASS), CAUSES SWITCH TO REMAIN IN PASS POSITION

FIGURE 9.54. Switch element state diagram

data. The choice of a memory element to hold the switch position depends on the frequency at which the packets arrive. Since the length of time a packet takes to pass through the switch element (which is the length of time a switch setting must be held) can be arbitrarily long, a static memory element is required. The use of a continually refreshing dynamic shift register meets this requirement, keeps the device count and cell area to a minimum, and keeps the timing compatible with the data pipeline.

The serial comparator may be implemented in several ways. An initial design was a feedback logic circuit (a complex R-S latch), which had race conditions and timing uncertainty. The idea of a synchronous state machine developed into an alternative implementation consisting of a pipeline register and two bits of state. The result is a finite-state machine with four inputs and four outputs

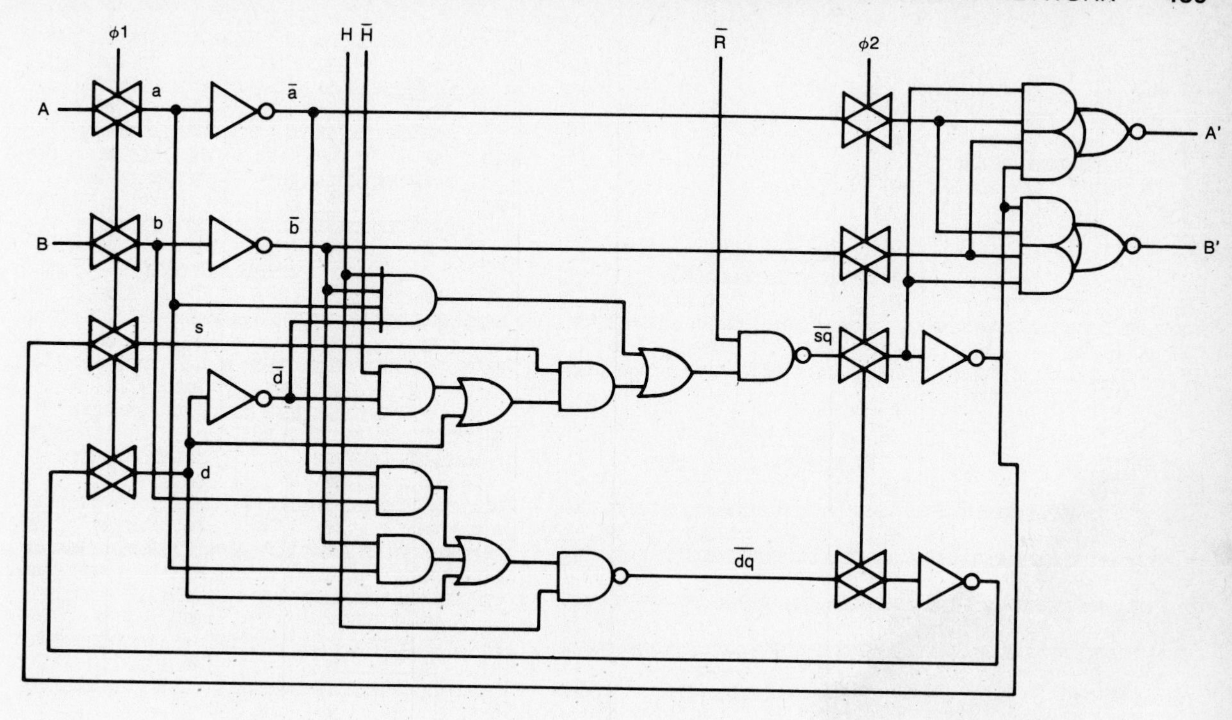

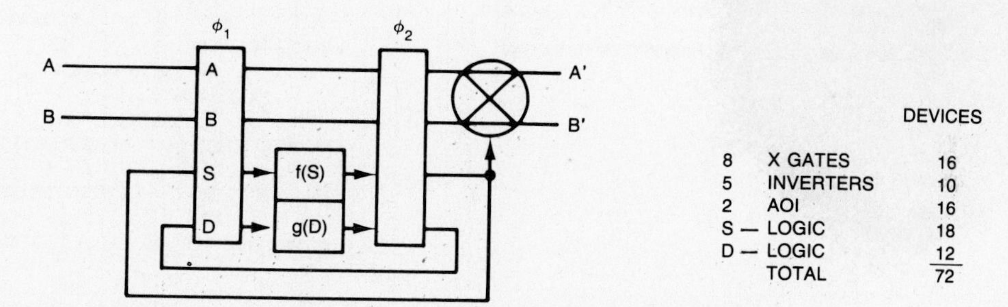

		DEVICES
8	X GATES	16
5	INVERTERS	10
2	AOI	16
S —	LOGIC	18
D —	LOGIC	12
	TOTAL	72

FIGURE 9.55. Switch element schematic for CMOS VLSI implementation

(four shift registers, two to feedback the state). One state bit indicates the switch position (S). The other indicates whether a difference has been detected since the last reset pulse (D). The logic equations required to update the state information are shown in Fig. 9.54. The schematic of the final implementation in CMOS (containing 72 devices) is shown in Fig. 9.55. The two complex gate functions used are a good example of what can be realized in a customized cell.

Another important consideration concerns the density of the interconnections between levels. The minimum area required to layout an m line perfect shuffle routing networks grows asymptotically

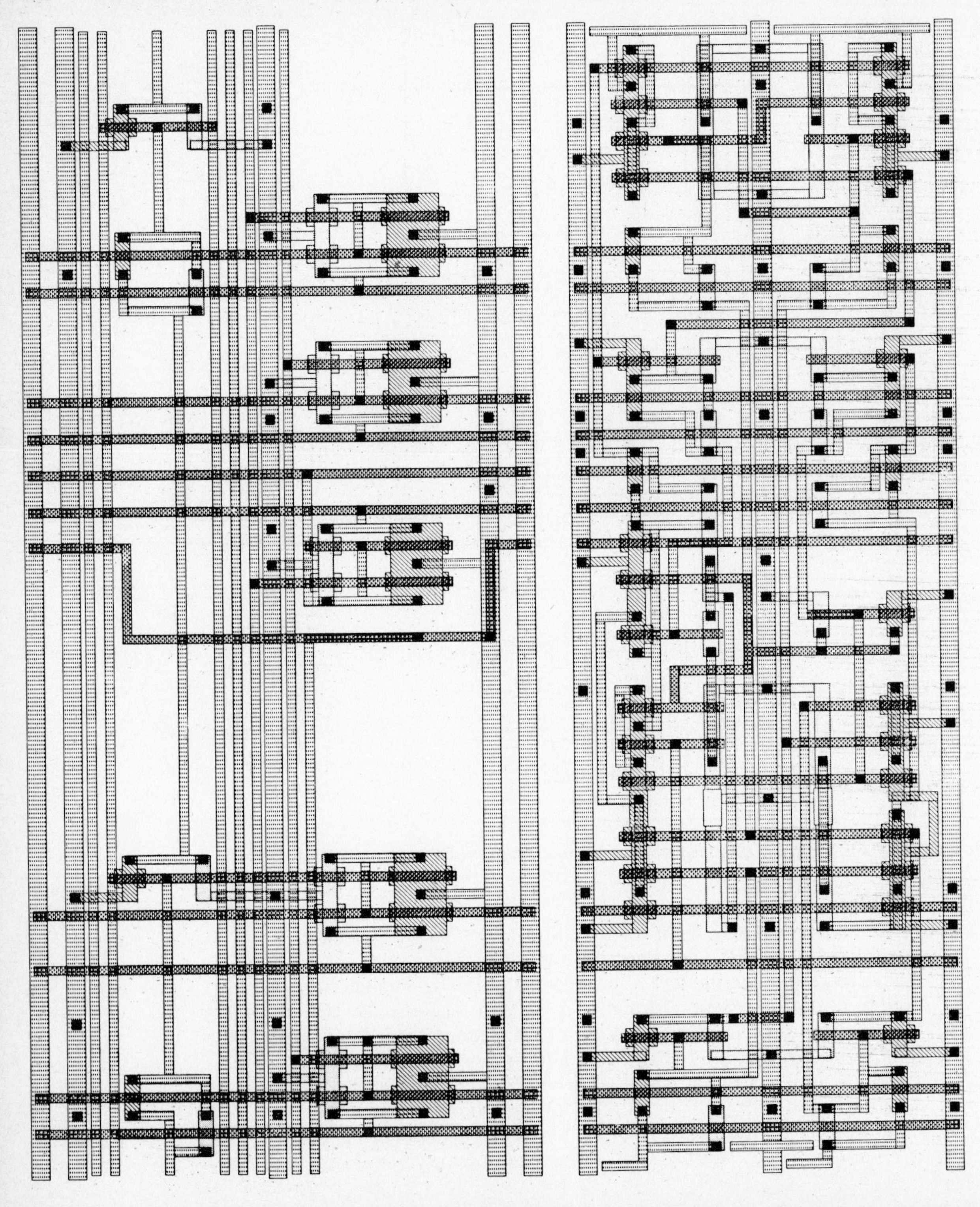

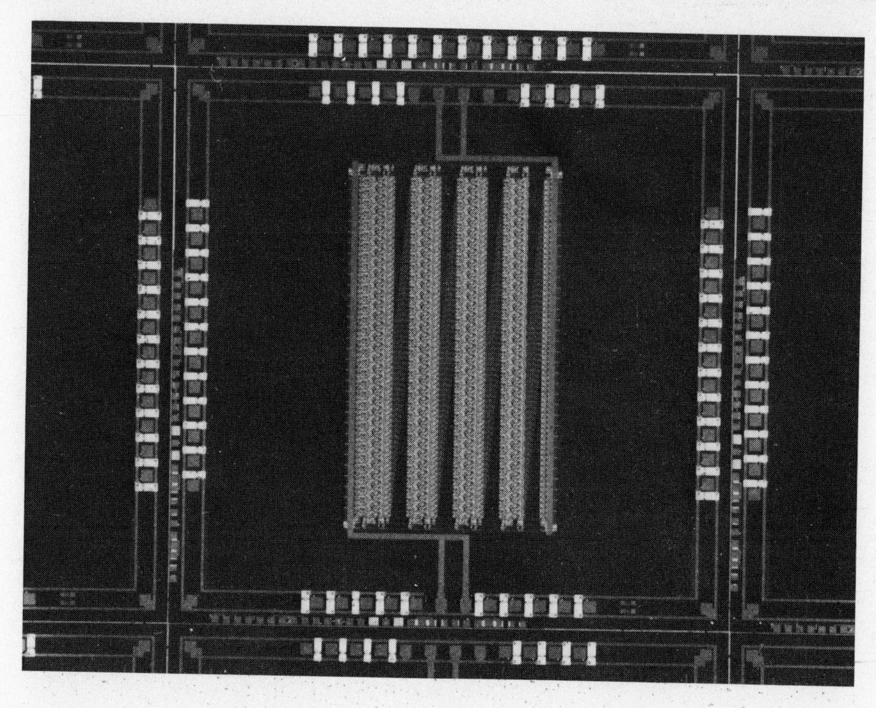

FIGURE 9.57. Back end Batcher chip microphotograph

as $m^2/(\log_2 m)^2$, and for the simple regular layout used here the growth is m^2. It is thus desirable to minimize the total area of the routing, and this is accomplished by keeping the distance between lines as small as possible. The cell height of the circuitry associated with each line must be at least as high as a CMOS inverter with a couple of bus wires through it. With this as the goal for the pitch of signal lines, the sorting element was laid out with an aspect ratio of about 2.5:1. The layout is shown in Fig. 9.56 along with the buffer cell that drives the control lines in the sorter. The height of the element was made small enough to allow 32 elements to be stacked within 6 mm. Without routing, almost 18 elements could be placed horizontally in the same distance.

The front end Batcher chip, consisting of two 32-channel sorters, requires 15 levels of 32 elements with associated control circuitry and routing. It fits in a 6 mm square area. With independent 4:1 multiplexing on the inputs and outputs, 44 pins are required.

The photograph of the back end Batcher chip shown in Fig. 9.57 shows the 4 levels of 32 sorting elements and the 64-line perfect shuffle routing networks at the beginning and between each level. The routing networks cover almost half of the active area. To allow

◀ **FIGURE 9.56. Symbolic layout for switch element and switch element driver**

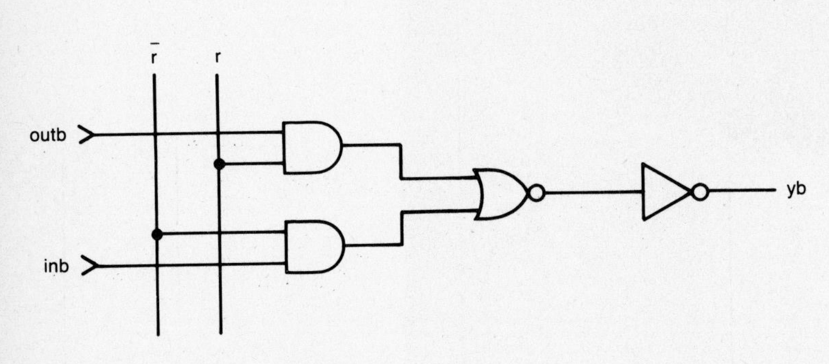

(a) BYPASS CIRCUIT SCHEMATIC

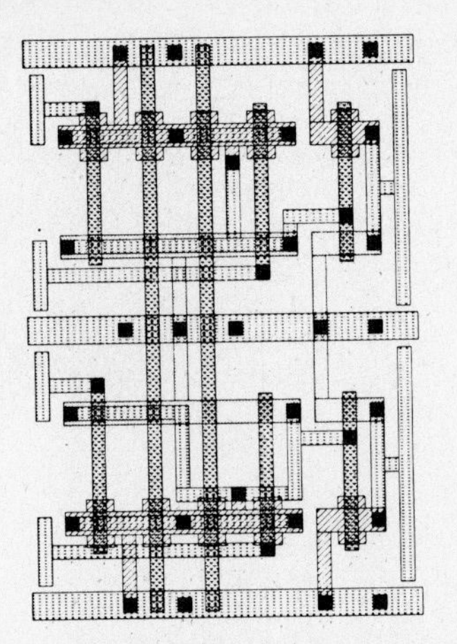

(b) BYPASS CIRCUIT SYMBOLIC LAYOUT

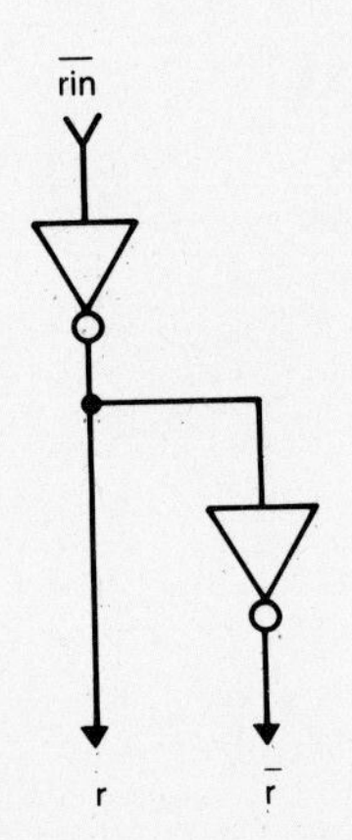

(c) BYPASS DRIVER SCHEMATIC

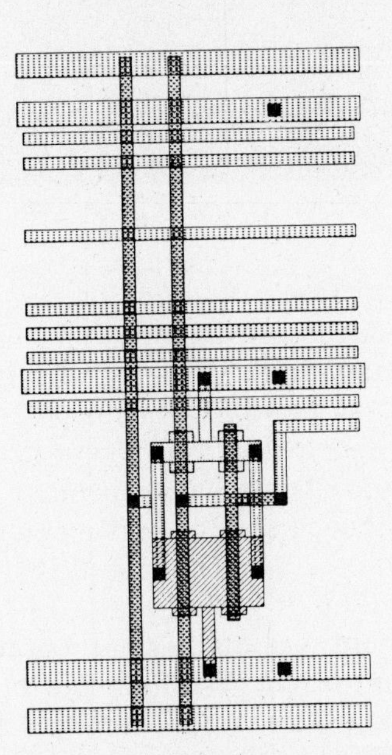

(d) BYPASS DRIVER SYMBOLIC LAYOUT

FIGURE 9.58. Level bypass and driver circuits and symbolic layouts

the chip to be configured as one, two, three, or four levels, additional circuitry was added to each sorting element in the last three levels. This bypass circuitry and associated symbolic layouts are shown in Fig. 9.58. At each level it allows the preceding routing network to be bypassed and locks all 32 switches in the pass state.

Additional circuitry includes multiplexers, clock drivers, header stagger, and level bypass. Fig. 9.59 gives the schematic and layout

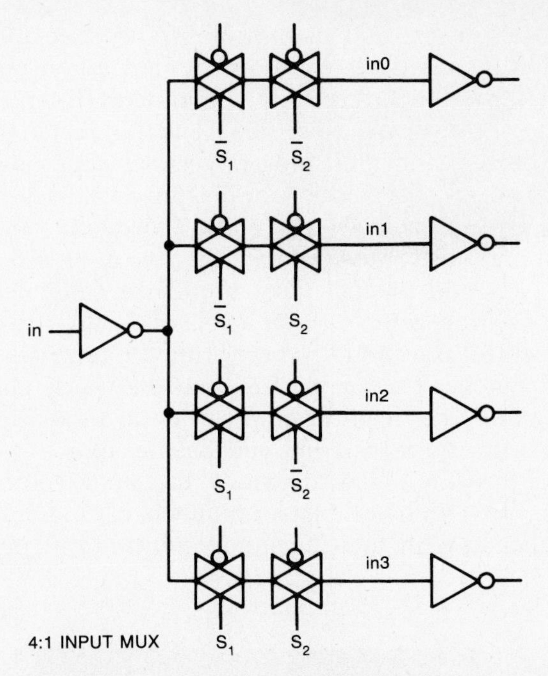

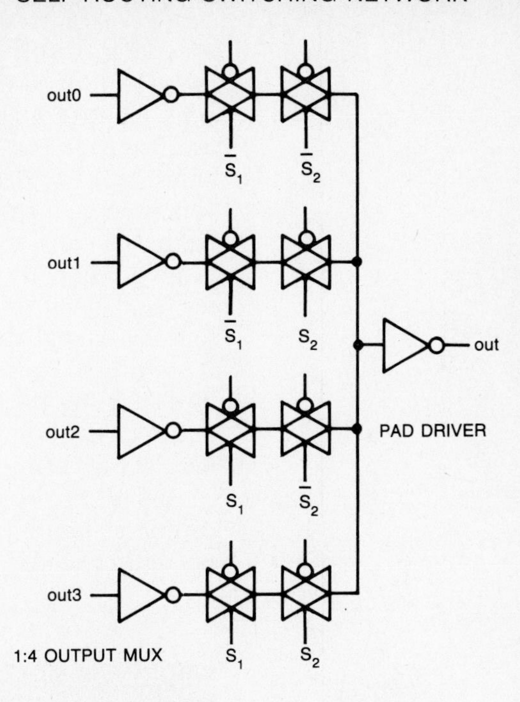

(a) INPUT AND OUTPUT MULTIPLEXER SCHEMATICS

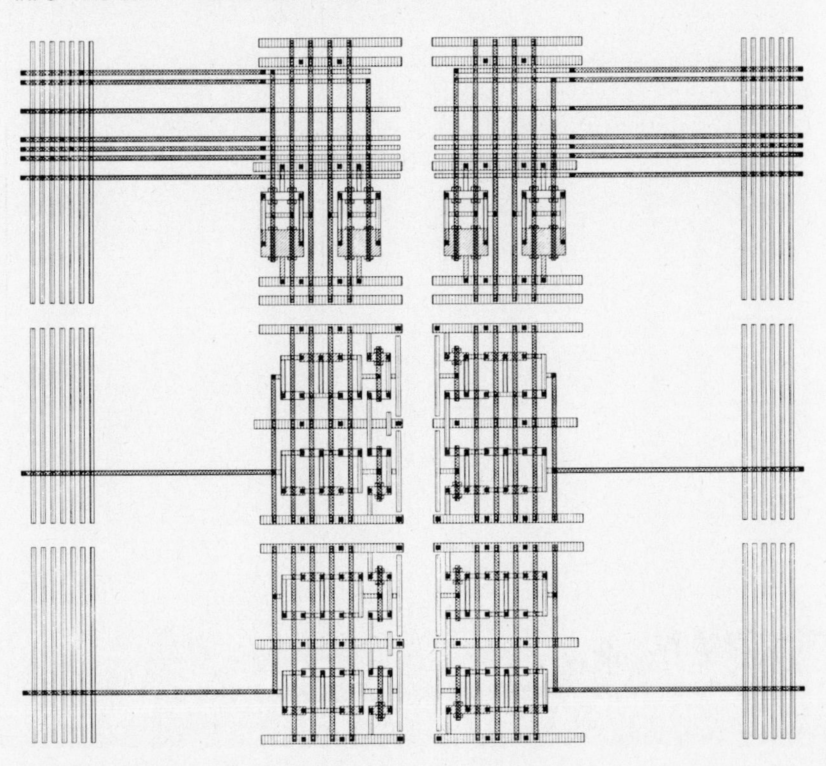

(b) INPUT AND OUTPUT MULTIPLEXER SYMBOLIC LAYOUTS

FIGURE 9.59. Input and output multiplexers and associated drivers (circuit and symbolic layouts)

diagrams of the 4:1 multiplexers and demultiplexers used to put the 64-input and 64-output channels on 16-input and 16-output pins. Control requirements are two pins for the input demultiplexer and two for the output multiplexer; this direct and independent control of the demultiplexers and multiplexers was designed as a feature to facilitate testing. A 2-phase clock (two pins) was distributed to driver cells at each end of each level of sorting elements. The same driver cells buffer and invert the header signal (two more pins) for each level. In the back end Batcher chip, the driver cells also buffer a signal to bypass from zero to three levels of sorting (two more pins). These 10 control pins and 2 power bring the pin count to 44 for the back end Batcher (42 for the front end Batcher). The entire chip fits in a 48-pin DIP or leadless chip carrier package. Fig. 9.60 shows a microphotograph of the front end Batcher chip.

As VLSI progresses towards higher densities, this architecture can persist and even improve to handle larger numbers of channels per chip. Faster pin speeds with more channels multiplexed per

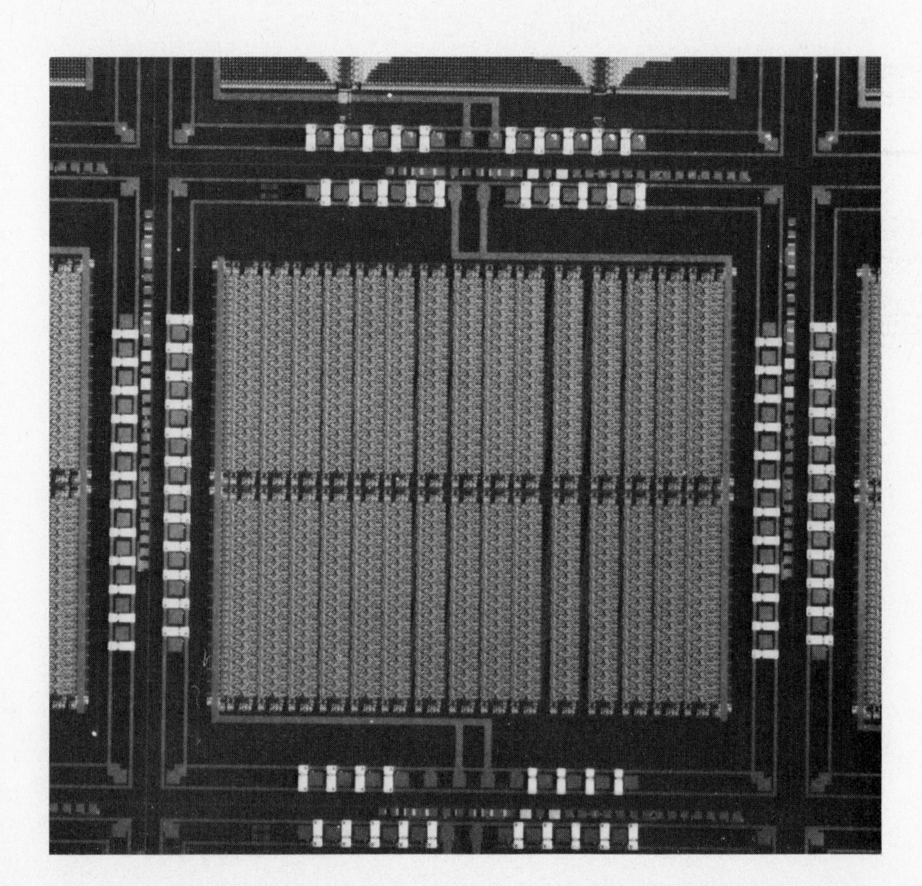

FIGURE 9.60. Front end Batcher chip microphotograph

pin and chip carriers with more pins will handle the chip I/O. The problem becomes the area required by the larger on-chip shuffle interconnect in the back end Batcher chip and in the latter stages of the front end chip; this area grows as the square of the number of lines in the shuffle for a simple regular layout. By multiplexing the lines in the same manner as the I/O pins are multiplexed, the m line shuffles are reduced to m/r lines, where r is the multiplexing factor, and the area requirement is reduced by r^2. The cost is the addition of demultiplexers and multiplexers at every level. In the case of 64-wide networks ($m = 64$) multiplexed 4:1 ($r = 4$), the costs of extra multiplexers and wider shuffles are roughly equal, and the use of the latter is preferred for simplicity. In contrast, the former would be quite attractive for a chip with 128 line internal shuffles and 8:1 multiplexing.

9.4.4 Chip circuit simulation

The switching element was extensively simulated for proper operation at 10, 20, and 40 MHz and with various clock timings (nonoverlapping and normal). Proper function of the entire network was modeled by a C program. These steps brought confidence that the cells would function properly when placed together.

9.4.5 Chip testing to isolate functions

The first step in checking newly fabricated chips is isolating the various functions of each chip and verifying that each works correctly. This is usually easier said than done, but it proved relatively easy to do for the back end chip. Tests were developed to check the operation of each input multiplexer, each 2-input sorting element, each bypass circuit, and each output multiplexer. When the timing associated with each of these components was verified, it was possible to test the back end chip as a whole, and likewise the front end chip, since it uses the same components.

Fig. 9.61 illustrates the test used to isolate each sorting element in the first level of elements on the back end Batcher chip. The last three levels are bypassed. Header pattern "A" was fed to the left-hand 8-input pins and pattern "B" to the right-hand input pins. The patterns contain headers designed to require the sorting element to pass, then exchange, then pass, etc. The direct availability of the input demultiplexer and output multiplexer controls proved to be a key factor in isolating components within the chip. With the demultiplexers switched to the left, the eight sorting elements shown

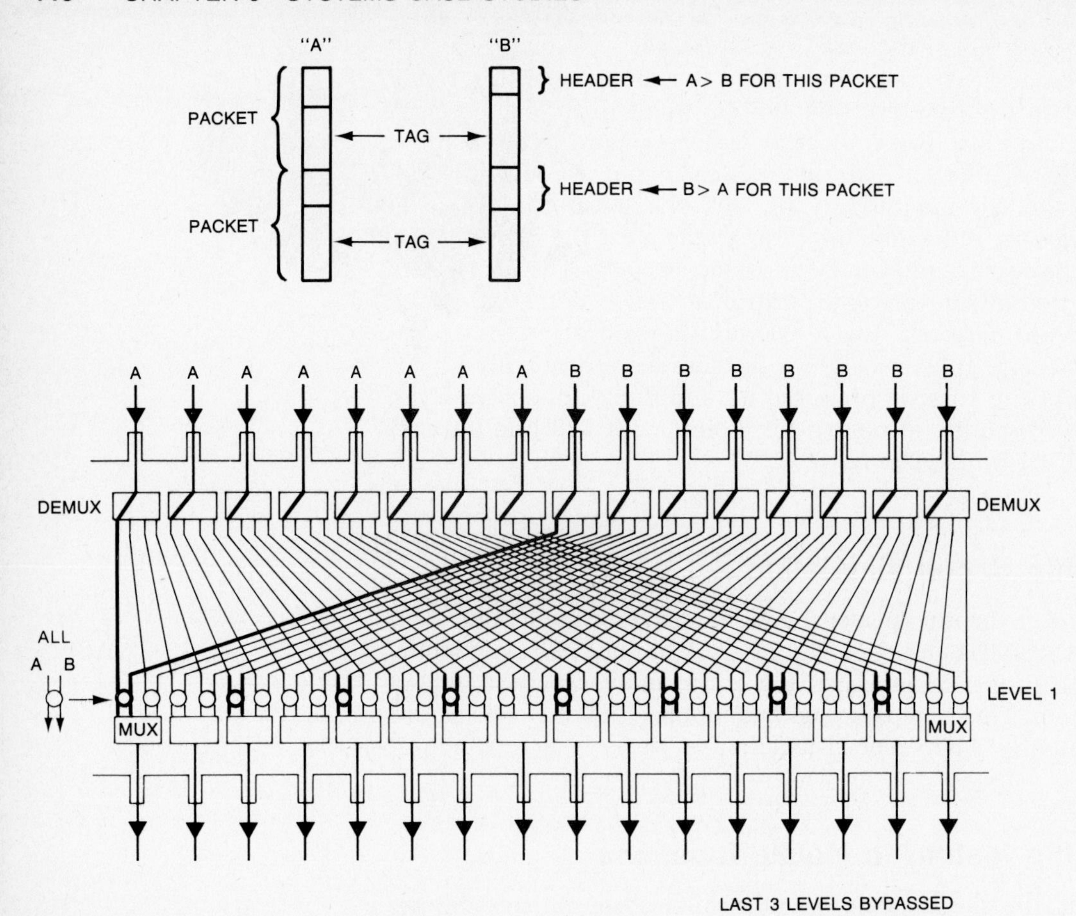

FIGURE 9.61. Test signals applied to check operation of sorting elements in level 1

with dark lines were supplied with "A" "B" inputs. The output multiplexers were stepped slowly to read each output. The outputs of elements with inputs not fed by the demultiplexer were also checked to verify that no changing signals were present. This was to ensure that the demultiplexers weren't shorted to multiple outputs. Once this test was complete, the demultiplexers were switched one position right and the test repeated to check the next set of eight sorting elements; two more repetitions of the test finished the test for the first level.

Figs. 9.62, 9.63, and 9.64 show how the input test pattern was varied to give each level tested "A" "B" inputs and each level above the tested level either "A" "A" or "B" "B" inputs. The testing progressed from level 1 to 4, so the nodes above a level being tested

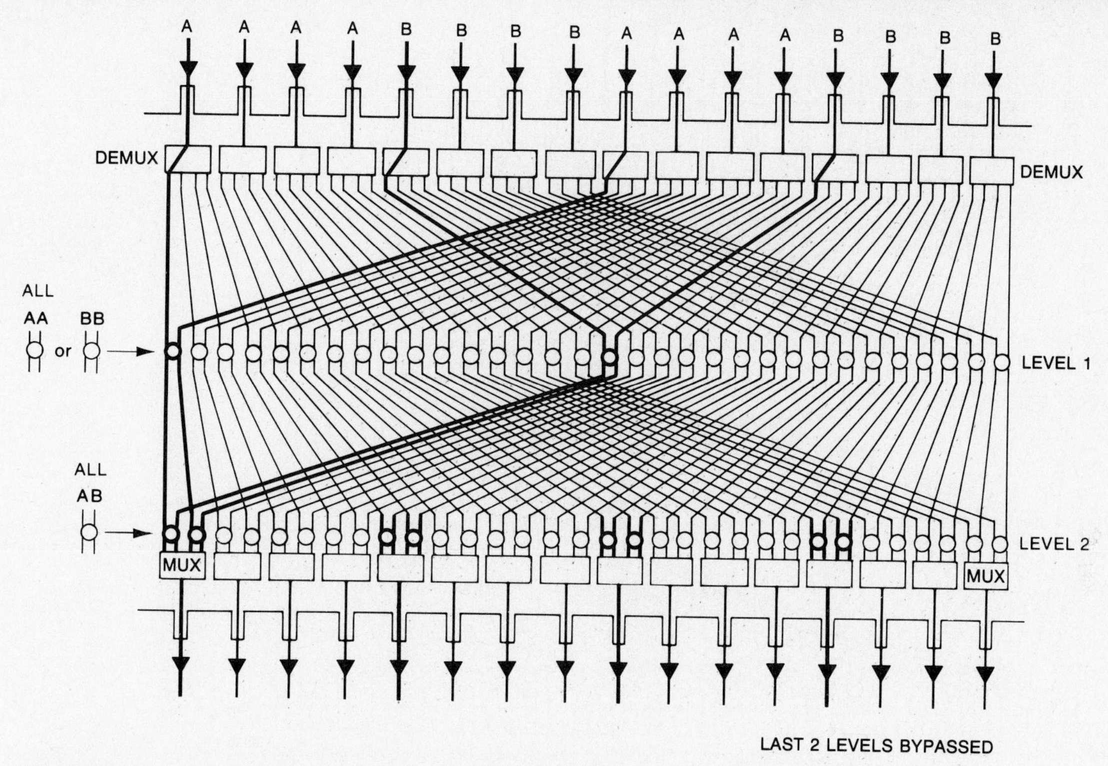

FIGURE 9.62. Test signals applied to check operation of sorting elements in level 2

had passed the test for their level. Even if switching elements in the levels preceding the level being tested have failed, the fact that they are fed equal inputs ensures that they will have equal outputs for the most common failure modes (pass/exchange switch frozen, inputs shorted, or outputs shorted). This allows testing of the elements on the lower level to proceed in most cases. The results of this test confirmed the operation of the switching element as simulated. The only bad switch elements were on chips that were later discarded for processing flaws (yield).

9.4.6 Summary

This design illustrates some important points in VLSI design. To achieve a highly regular layout, the architecture had to be worked from the algorithm level, not from previous implementations. The resulting set of chips provide a powerful set of components for building advanced wide-band switches.

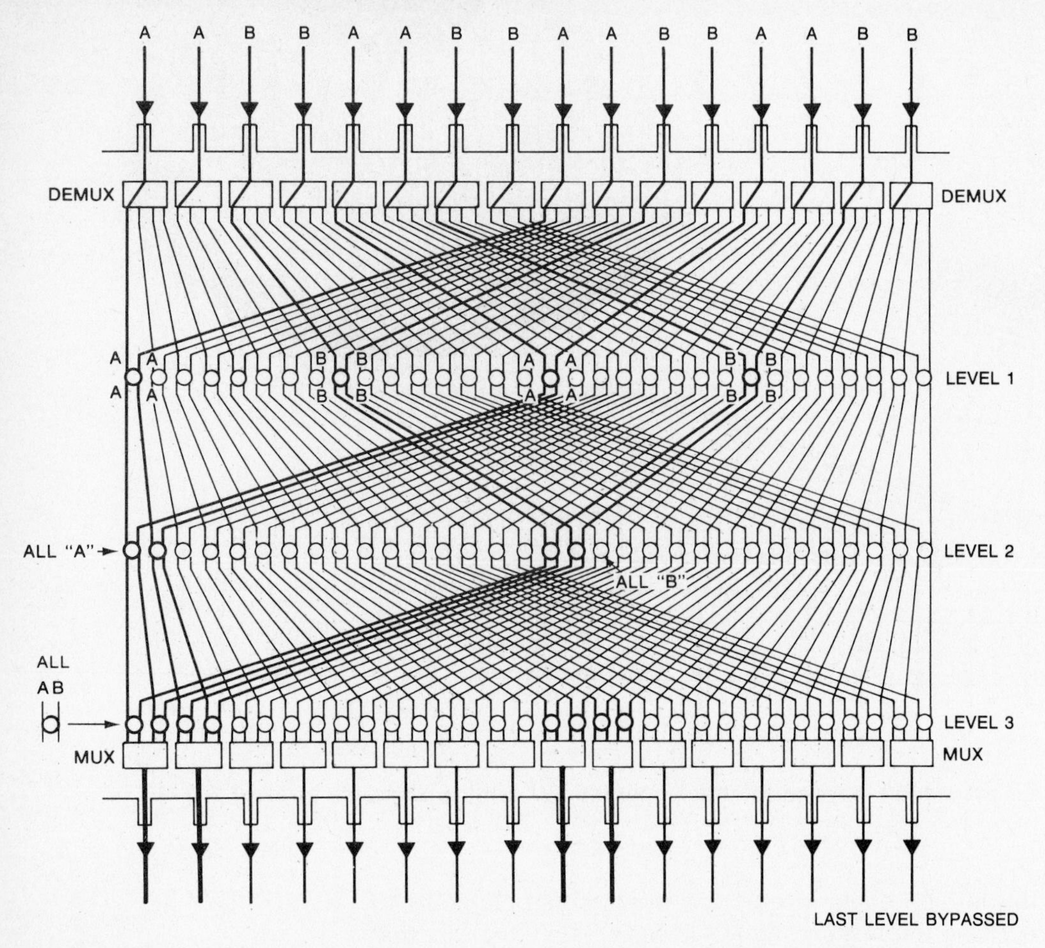

FIGURE 9.63. Test signals applied to check operation of sorting elements in level 3

9.5 Pixel-planes graphic engine

J. W. POULTON, H. F. FUCHS,*
A. PAETH†

9.5.1 Introduction

This section describes a design of a CMOS memory for a VLSI-based machine that can support real-time graphic interaction with three-dimensional shaded and colored images. Called *Pixel-planes*, this design promises to provide the realism of million-dollar "digital scene generators," such as those used in flight simulation, but at a

* Department of Computer Science, University of North Carolina at Chapel Hill.
† University of Waterloo, Ontario; formerly at Xerox Palo Alto Research Center.

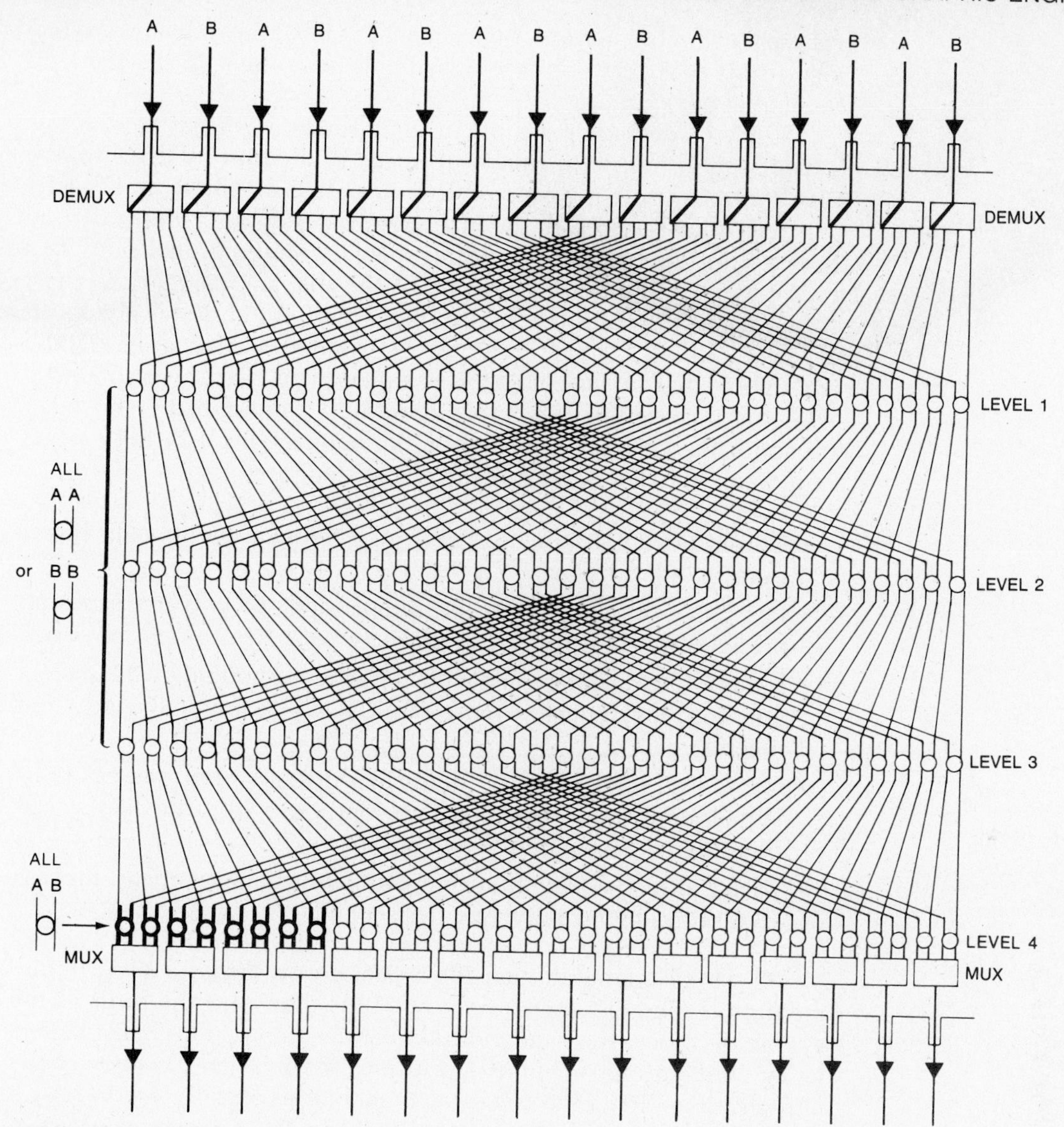

FIGURE 9.64. Test signals applied to check operation of sorting elements in level 4

price comparable to affordable systems that at present can generate only "wire-frame" line drawings. A graphic system based on this design should be useful in a variety of applications, including computer-aided design, medical imaging, simulators for flight and navigational training, tactical displays, and even video games. The design addresses the general problem of image generation using the most common scene description (convex planar polygons) as input and the most common output format (standard video). Pixel-planes graphic algorithms make no assumptions about the content of the scene to be displayed.

Much of current research in experimental graphic systems is aimed at improving the speed of image generation by dividing the display into small regions, each of which is handled by separate concurrent processors. In the Pixel-planes design, this idea is carried to the limiting case by providing special computational hardware for every pixel (picture element) in the display, so that the most time-consuming calculations can be carried out simultaneously for 2^{18} to 2^{20} pixels. Calculations are distributed so that only a small amount of circuitry is needed for each pixel. This circuitry is combined with the memory required for a graphic "frame buffer" and takes up about the same silicon area as the memory itself in the current implementation. Key features of the system are:

- Graphic calculations are performed polygon-by-polygon rather than in scanline order.
- Polygon processing time is independent of polygon size and orientation and grows only linearly with the number of edges.
- Lines and two-dimensional objects can be handled by the system and can be processed more rapidly than smooth-shaded solid objects.
- Composed of an array of identical chips, a system can easily be expanded for increased screen resolution; a working display can be built with any number of chips.
- Based on conservative speed estimates, the system can process 15,000 to 30,000 polygons per second.

The Pixel-planes design fits especially well into a VLSI implementation. The enhanced memory chips that form the heart of the system overcome the bandwidth limitations imposed on frame buffers using conventional memories. Further, the internal circuitry of the chip is, like that of conventional memories, highly regular with minimum area required for interconnection wiring. A complete system consists of an array of *identical chips* connected together by a very simple bus structure, so that the complete system is extremely regular at board and backplane levels.

The fundamental ideas for graphic algorithms and for the design of the Pixel-planes computational circuits have been previously published [FuPo81][FPPB82]. Two iterations of the design have been carried out in nMOS technology. Second-generation chips, whose layout was designed by Alan Paeth, formerly at Xerox PARC, have been used to build a small working prototype display. The prototype has successfully run the basic Pixel-planes graphic algorithms outlined below. Third-generation chips are currently being used in a somewhat larger prototype system, which will be used to develop algorithms and hardware for a full-scale display. Layout for these chips was carried out by Alan Paeth and by Scott Hennes (University of North Carolina, Department of Computer Science). The CMOS design described here represents a second iteration of the third-generation design and is intended to form the basis of a full-scale working display.

Before describing the design and operation of the system, we briefly outline the general problem of generating realistic images in a graphic system.

9.5.2 Raster-scan graphic fundamentals

One of the many ways of rendering a raster-scan image from a polygonal data base [NeSp79] is outlined in Fig. 9.65. The data base describes a scene containing one or more objects by a set of convex

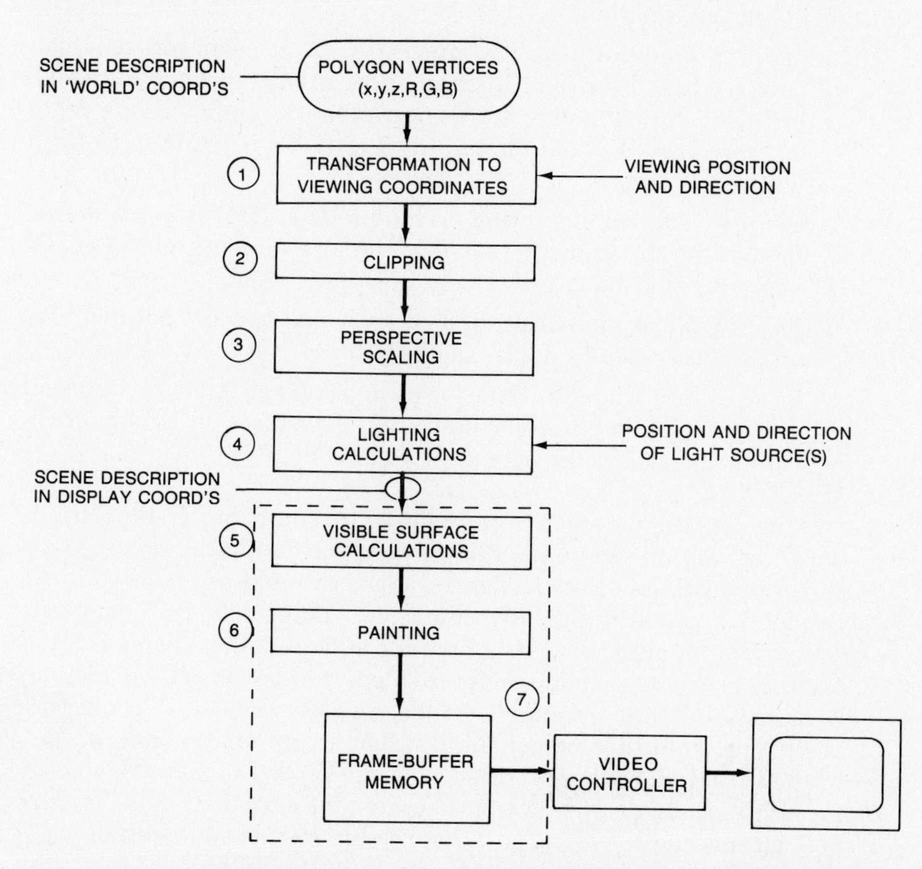

FIGURE 9.65. Steps required in raster-scan rendering of solid objects

planar polygons that approximate the surfaces of the objects. Polygons are processed by the graphic system one at a time and in any order. Each polygon is described by a sequence of vertices whose x, y, z coordinates are expressed in a 'world' coordinate system. A triple *R, G, B* is associated with each polygon, specifying its intrinsic color; if smooth shading [NeSp79][Gour71] is required, this triple must be specified for each *vertex* of a polygon. Polygon processing consists of these steps:

1. *Viewing transformations.* Vertex coordinates are transformed according to the position and direction of the viewer; this information is typically supplied to the system by means of graphic input devices such as joysticks, trackballs, and the like.
2. *Clipping.* Polygons, now described in *viewing coordinates,* are clipped to remove portions outside the field of view.
3. *Perspective scaling.* Polygons are scaled for an appearance of perspective; objects that are farther away from the viewer are displayed smaller.
4. *Lighting calculations.* The precise color at each vertex is calculated, based on the direction and distance to light source(s), intrinsic vertex color, surface reflectivity, and perhaps other factors. The calculations modify the *R, G, B* triple associated with each vertex.
5. *Visible surface calculations.* This process identifies pixels that lie within the current polygon *and* that are not obscured by polygons that have previously been processed.
6. *Painting.* The visible portion of each polygon is "painted" for an appearance of smooth shading.
7. *Storage and display.* Data for each pixel is stored in a special memory (frame buffer), which can be accessed by a video controller to refresh the screen of a color display.

For processing steps 1 through 4, the number of calculations depends only on the number of polygon vertices in the scene. Many currently available line-drawing systems can perform real-time processing on graphic data involving only polygon vertex data; any such system could serve as a host for a Pixel-planes display. Steps 5 and 6, however, require that for each polygon processed, calculations be carried out for *all* pixels in the display. Clearly this places a vastly greater burden on a graphic system's computational hardware. In the Pixel-planes design, these steps are carried out simultaneously for all pixels by computational circuitry at each pixel. This circuitry is combined with the memory required in step 7, so that the system comprises both a graphic processor and a frame buffer.

9.5.3 Pixel-planes system overview

The Pixel-planes system itself consists of three parts:

1 A *preprocessor*, which accepts polygon descriptions in display coordinates from the host, and which outputs data and instructions for "rendering" each polygon to the engine.

2 An *engine*, which performs steps 5 through 7.

3 A *video controller*, which scans the pixel memory elements in the engine, converts the digital data representing a scene into video format, and drives a color display.

9.5.3.1 Pixel-planes engine

We first describe a conceptual (but impractical) model for the Pixel-planes engine. This model, shown in Fig. 9.66, consists of an array of identical processing elements (PEs), one for each pixel in the display. For simplicity, imagine that these PEs are arranged so that the pixels are in scan-line order. The PEs operate bit-serially on a common input data stream, and on a given machine microcycle, all PEs perform the same instruction and access memory bits having identical addresses. As shown in the inset of Fig. 9.66, each PE consists of four functional blocks:

1 A serial-parallel multiplier, which takes in data from each of three bit streams A, B, and C, and outputs the function $F(x, y) = Ax + By + C$, where x and y are the coordinates of the pixel in the display. These coordinates are the coefficients applied in parallel to the multiplier circuits and can, for the moment, be imagined as being hard-wired into the circuits of each PE as a multiplying constant. The three inputs A, B, C, coefficients of planar equations, are the only data needed by the engine. These data streams are common to all PEs.

2 A 1-bit ALU, which operates on data from the multiplier and on data fetched from memory. The ALU can output its results to memory and to an Enable register. (The instruction repertoire of this ALU will be outlined below.)

3 An *Enable* circuit, consisting of a 1-bit Enable register and control circuits. When Enable is *true*, the ALU can access and modify the contents of the PEs' memory; when *false*, memory cannot be altered.

4 An array of memory bits used to store the data associated with a pixel.

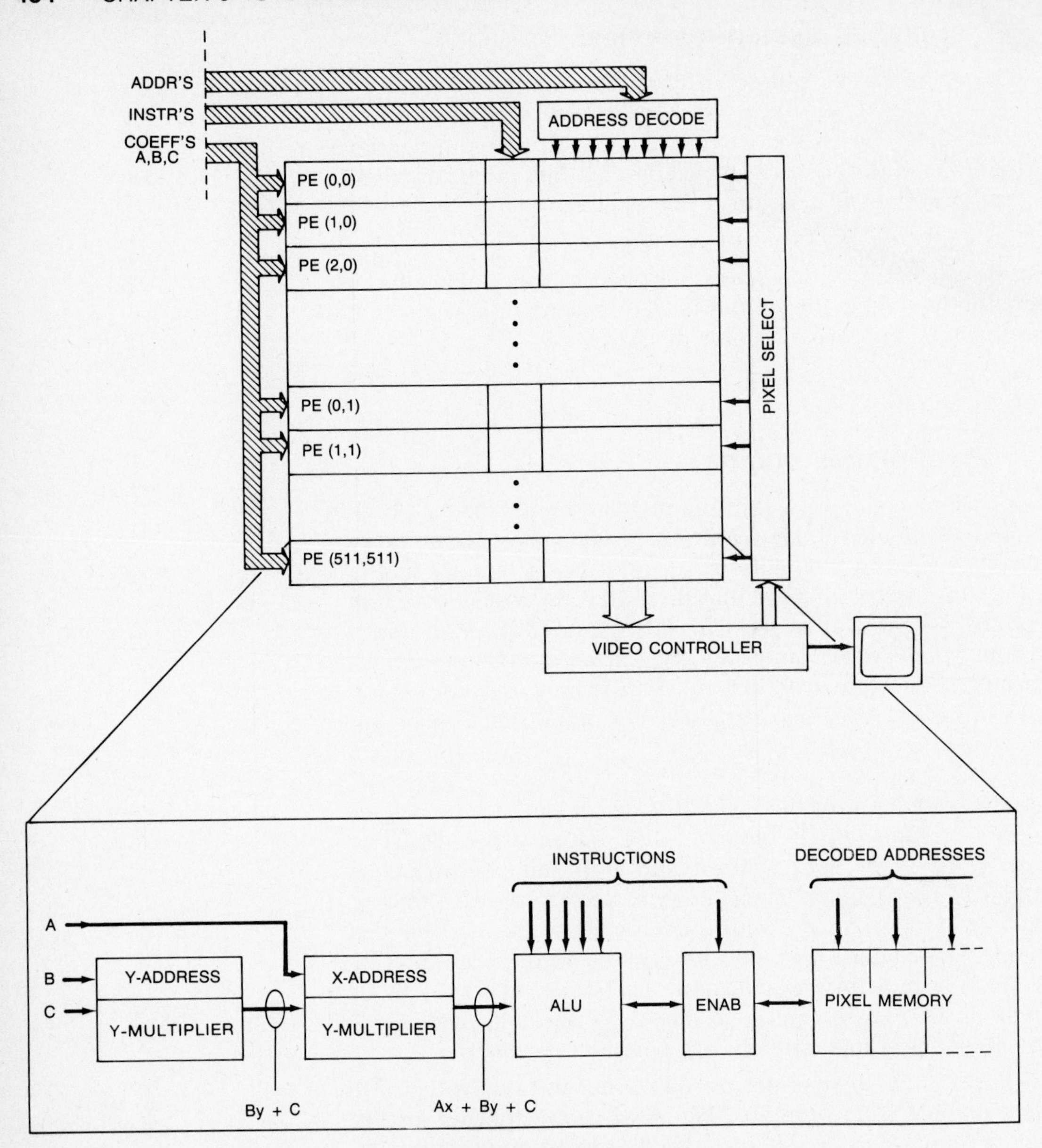

FIGURE 9.66. A conceptual model of the Pixel-planes engine

A separate (and orthogonal) addressing mechanism in the engine allows the video controller to scan out data from each pixel for refreshing a color display, without interrupting image calculations.

The fundamental operation of the engine is calculating, simultaneously in every PE, the planar function $F(x, y) = Ax + By + C$. Rather than building special hardware to carry out particular graphic algorithms, these algorithms are instead recast into the form of solutions to planar equations, as will be shown below.

The system described by this model has an important simplifying feature. Since no connections are needed between PEs, the system can easily be partitioned into multiple chips. Apart from the hard-wired multiplier constants, which represent the PEs' address (pixel's coordinates), all chips in a system would be identical. The actual implementation, to be described in the next section, preserves this useful property and removes the need for hard-wired constants.

9.5.3.2 Algorithms

A basic set of graphic operations will be described:

- Edge definition for polygons (determining which pixels are inside a polygon).
- Hidden-surface elimination (determining which pixels within a polygon are not hidden by previously processed objects).
- Polygon painting (determination of red, green, and blue intensity at each visible pixel within a polygon).

The method to be described for hidden-surface elimination makes use of the common "depth-buffer" or "z-buffer" algorithm [NeSp79]. This algorithm requires storage at each pixel for maintenance of the z-coordinate of the last polygon processed at that pixel. Memory bits within each PE can be allocated in any convenient way; for this discussion, imagine that they have been divided into four buffers: ZBuff, for depth-buffer storage, and RedBuff, GreenBuff, and BlueBuff, for storage of red, green, and blue color-intensity values. (Often it is desirable to establish two sets of buffers for color-intensities. One can be used for image-generation, while the other provides unchanging data for screen refresh. The two buffers are swapped invisibly during retrace, when an image has been completed.) The basic operations required to render an image are listed below:

New image. When a new image is to be constructed, the Enable at each PE is set *true* by a global instruction, z-buffers are preset to infinite depth (all 1's), and the color-intensity buffers are cleared.

Enable ← true ZBuff ← MaxZ (MaxZ = all 1's in ZBuff)

RedBuff ← 0 BlueBuff ← 0 GreenBuff ← 0

Note that constants such as *MaxZ* and 0 can be transmitted to all PEs by setting the *A* and *B* coefficients to 0, and setting *C* = *constant*.

New polygon. When a new polygon is to be painted, Enables at all PEs are preset *true*.

$$\text{Enable} \leftarrow \text{true}$$

Edge definition. Each edge of a polygon is encoded in the equation for a line, $F = Ax + By + C = 0$. PEs at pixels whose (x, y) lies on one side of the line calculate a value of F, which is greater than 0, while those on the other side of the line calculate a negative F. The operation

$$\text{Enable} \leftarrow \text{Enable } \textit{and} \ (F < 0)$$

(AND the sign-bit of F with Enable) therefore leaves enabled *only* those pixels in the half-plane on one side of the line. As each successive edge is transmitted to the system, pixels outside the polygon are eliminated (disabled) by half-planes. When all edges have been processed, only those pixels that are inside the polygon remain enabled.

Z-buffer calculation. The z-buffer algorithm encodes the z-coordinate of each pixel in the current polygon by means of the planar equation $z = F = Ax + By + C$. The operation

$$\text{Enable} \leftarrow \text{Enable } \textit{and} \ (F < \text{ZBuff})$$

leaves enabled only those PEs whose pixels are visible. If the value of F calculated at a PE is greater than the value stored in its z-buffer, then a previously processed polygon containing that pixel was closer to the viewer than the current polygon, and the pixel is hidden. Next, the z-buffer at still-enabled pixels must be updated with the new value of z; $z = F$ is recalculated and

$$\textit{if} \ (\text{Enable}) \ \text{ZBuff} \leftarrow F$$

Pixel painting. Finally, the pixels that are visible in the current polygon must be painted. Pixel-planes can provide Gouraud-like [NeSp79] smooth shading by representing the color-intensity at each pixel location x, y as a planar function $Red = F$, and similarly for *Blue* and *Green*. The operation

$$\textit{if} \ (\text{Enable}) \ \text{RedBuff} \leftarrow F$$

puts the proper Red-intensity value into visible pixels of the current polygon. Similar operations are required for Blue and Green.

These basic algorithms are sufficient to process realistic colored, shaded images; other algorithms include:

Lines. In the edge-definition algorithm, the numerical value of F calculated at each pixel is not needed and only the sign of F is used in processing. The value of F for an edge can of course be retained and can be interpreted as the distance of the pixel from the edge. This information can be used to enable all pixels that lie along a "stripe" within a given distance from a line. A stripe of any desired width can be painted, so that processing two successive stripes defines a parallelogram. In this way line segments and rectangles can be defined with only two sets of coefficients A, B, C.

Circles. A clever algorithm due to F. P. Brooks allows circles to be processed by the system with only a single set of coefficients. The equation of a circle

$$(x - x_1)^2 + (y - y_1)^2 = r^2$$

can be factored into

$$(x^2 + y^2) - 2x_1x - 2y_1y + x_1^2 + y_1^2 - r^2 = 0. \qquad \textbf{(9.12)}$$

Apart from the leading quadratic term, this expression has a form which can be expressed as $F = Ax + By + C$. Circles of arbitrary center-position and radius can be processed with a single set of coefficients if some memory bits at each pixel are set aside for storage of this quadratic term. Upon system initialization, the distinct value of the quadratic is transmitted to each pixel (this can be done fairly rapidly using a variation of the line algorithm). Subsequently, operations of the form

$$\text{Enable} \leftarrow \text{Enable } \mathit{and}\ ((\text{F} + \text{QuadBuff}) < 0)$$

leave enabled only pixels within the current circle; the operation requires that each PE's ALU add to the incoming F the local value of the quadratic term (stored in QuadBuff).

Other graphic algorithms are being developed for the machine. For example, a single PE can be used to process and store data for two or more pixels, trading processing speed and color-intensity resolution for screen resolution. Since multiple pixels can be processed at a given PE, it is possible to maintain some information about an image at higher resolution than the display itself. This "subpixel" data can be used to display anti-aliased lines and fog and transparent-surface effects.

The conceptual scheme for the Pixel-planes engine, described in the previous section, is not practical because of the large amount of circuitry needed for the multipliers. The upper part of Fig. 9.67 shows a design for the serial-parallel multiplier in the conceptual PE. Two sections are shown; the first responsible for storing the

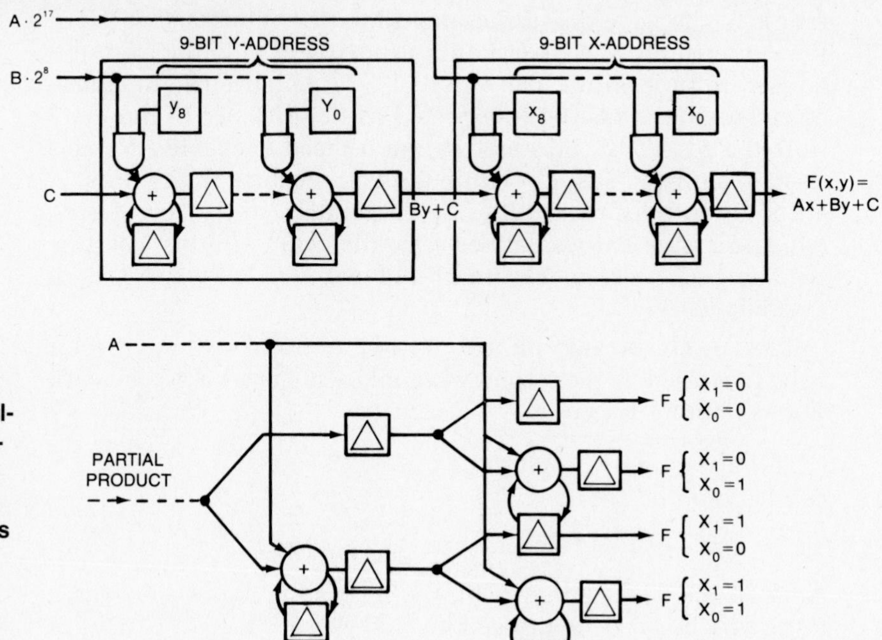

FIGURE 9.67. Serial-parallel multipliers used to calculate F(x,y) for a given PE. (Circles are 1-bit full adders with carry, squares are 1-bit latches.) Lower part of figure shows modification to last two stages to form binary-tree multiplier.

x-address of the pixel and calculating $Ax + C$, and the second storing the y-address and calculating $By + (Ax + C)$. In a full-scale 512 x 512 display, for example, each such multiplier would require nine stages each for the x and y parts of function; nearly five million stages would be needed for a complete system.

The solution to this severe difficulty is due to Fuchs [FuPo81]. Examining the multipliers of neighboring pixels in the engine (see Fig. 9.66), note that all constant bits $x_8 \cdots x_0$ and $y_8 \cdots y_1$ are *identical*. Only the bit y_0 differs between these two PEs. If the y_0 bit is a '1', the coefficient stream B is added to the partial product arriving at the last stage; if it is a '0', the partial product is simply shifted out of the stage with one unit-delay. A simple modification to the last multiplier stage allows *both* neighboring PEs to share a *single* multiplier, so that the work of both multipliers for the neighboring PEs is done by a single structure. If the idea is extended to all neighboring pairs of PEs in the engine, the number of multiplier stages is halved.

Each pair of PEs that share a multiplier now have a neighboring pair of PEs whose shared multiplier differs in only the y_1 stage. These neighbors can share a single multiplier by iterating the modification to the multiplier, thereby reducing the amount of circuitry again by half. The scheme is shown for the last two stages of the multiplier in the lower part of Fig. 9.67, and clearly can be further extended. Ultimately all PEs in the engine could be attached to the output leaves of a "binary-tree multiplier," an extremely sparse structure that both calculates F for all PEs and distributes the individual values of F to each PE. With this scheme, the number of multiplier stages has been reduced from $N \log(N)$ to $(N - 1)$, where N is the number of pixels. Note that it is no longer necessary to hard-wire pixel coordinates into each PE, as in the conceptual model. Each PE receives the appropriate value of F simply by virtue of its position on the tree.

In a multichip implementation of an engine of useful size, where all chips are to be identical, it is inconvenient to provide a complete binary-tree multiplier covering all PEs. Instead, the tree on each chip is sized to cover only those PEs on the chip, and the rest of the multiplication task is carried out by providing a single conventional serial-parallel multiplier on each chip.* On system initialization, a constant that is, effectively, the address of the *chip* (the x, y coordinates of its pixel array on the display) must be loaded into each chip's serial-parallel multiplier. A serial shift path threading all chips in the system provides a simple and efficient means of loading addresses into the multipliers. Since address loading is only done at system initialization, the same shift path can be used as a scan-out mechanism for video data when the system is running.

An important simplification in chip layout results from the ability to arrange the elements of the binary-tree multiplier in a single column. This "flat tree" arrangement allows efficient pitch-matching of the various elements in a PE. The flat-tree wiring, and the interconnections and layout of PEs is shown in Fig. 9.68.

To produce an image on a display, the video data stored at each PE must be read out in the proper sequence. Two important features characterize the difference between the scan-out process in Pixel-planes and that in conventional frame buffers:

1 *Scan addressing.* In a frame buffer, memory elements are always addressed in the same *scan-line* order. This property is used to advantage in Pixel-planes by replacing the usual address bus/address decoders required in conventional memories by a system of serial-shift tokens. A serial-shift token chain threads all chips in the system and carries a "global" token that determines which chip is enabled for output to a common video data bus. A second "local" token chain on each chip selects which pixel's video data is to be output from that chip. The local token replaces

* The serial-parallel multiplier above the tree-multiplier on each chip could be replaced by providing a *single* uncommitted tree node, with chip pads for I/O. This scheme is not used because 1) output pad delay would reduce overall speed, 2) interconnections at board level would become more complex, and 3) such a system could not readily be reconfigured for changes in system size and display resolution.

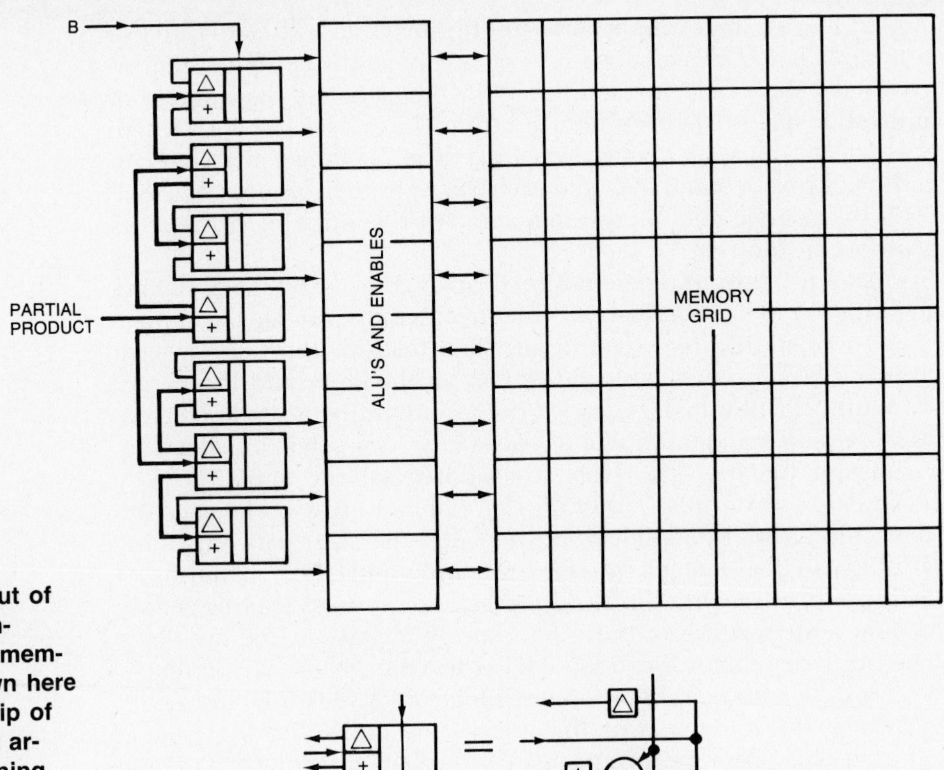

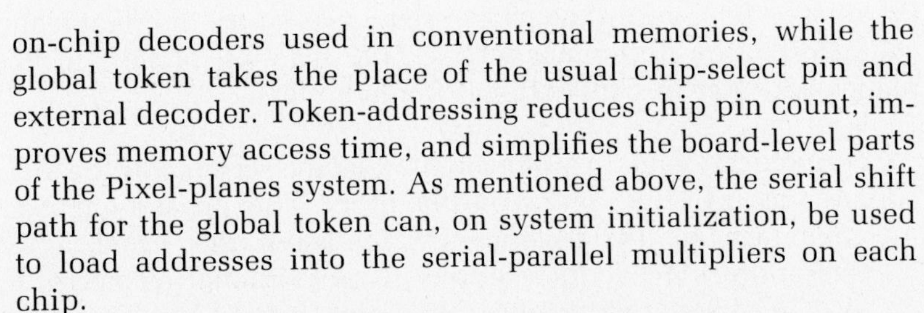

FIGURE 9.68. Layout of flat tree allows pitch-matching to grid of memory cells. Tree shown here covers a vertical strip of pixels, a convenient arrangement for scanning video data from multiple chips.

on-chip decoders used in conventional memories, while the global token takes the place of the usual chip-select pin and external decoder. Token-addressing reduces chip pin count, improves memory access time, and simplifies the board-level parts of the Pixel-planes system. As mentioned above, the serial shift path for the global token can, on system initialization, be used to load addresses into the serial-parallel multipliers on each chip.

2 *Scan data.* For image generation, each PE must have read and write access to its row of memory cells one bit at a time. Simultaneously (but asynchronously), all of the data from the pixel (row) selected by the local token must be available in parallel to the display refresh controller. The current design uses a "shadow register" into which the data from the selected pixel is copied. This copy is made one bit at a time and requires that every memory bit be "visited" at least once per scan line. The shadow register is double-buffered to overcome the latency in the bit-wise copy. The double buffering also serves to isolate

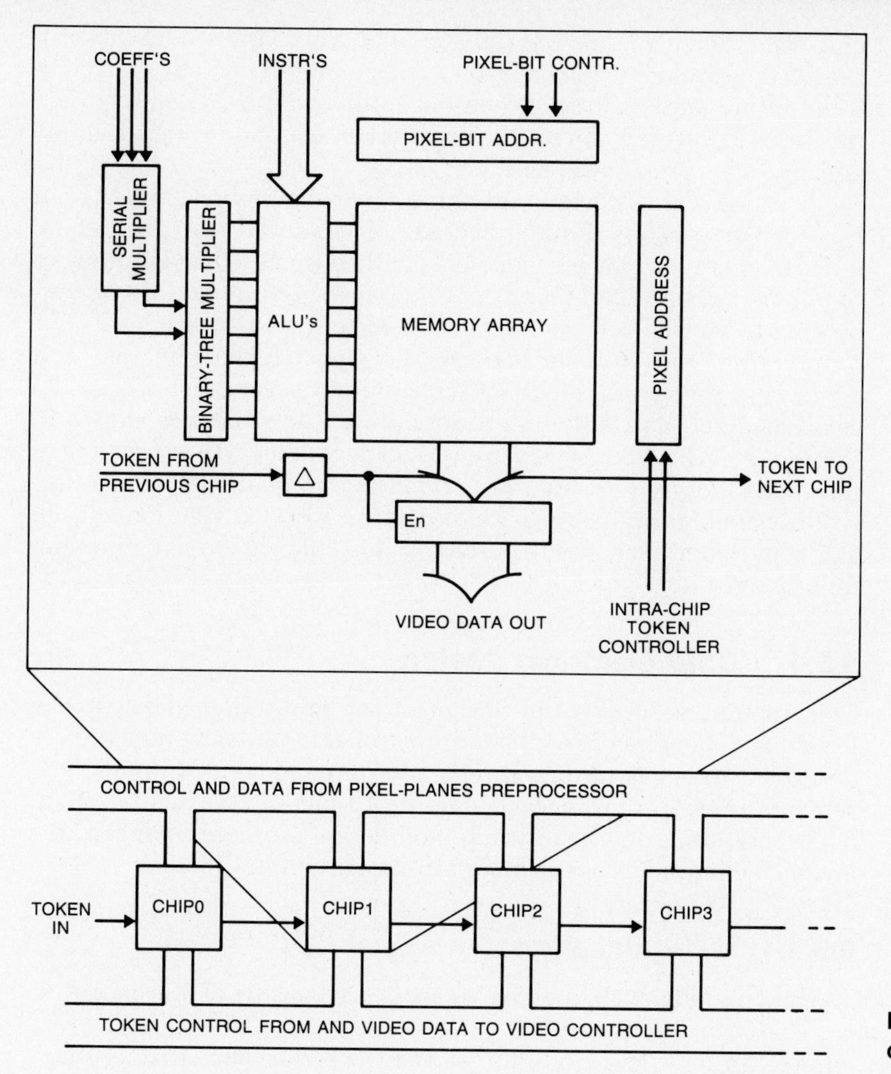

FIGURE 9.69. Overall view of system organization

the timing of the scan-out process from image-generation; the two processes are controlled by independent clocks.

The method used to scan data from an array of chips is shown schematically in Fig. 9.69. The pixels within a given chip lie on a vertical strip in the image, so that each chip is accessed only once per horizontal scan line. The bandwidth needed at the chip output port is thereby minimized. In operation, the video controller initializes the local token to point to the lowest PE and puts a token into the global token register of the first chip in a chain, enabling it for output. During the first scan-clock cycle, data is output from this chip onto the video data bus (all other chips are disabled for output). During the second cycle, the token moves to the second chip, enabling it for output, and so forth. When the entire chain has been scanned, the video controller moves the local token up one PE in all chips, and the scan process repeats.

In a full scale system, scan-out is somewhat more complex than this for two reasons. Firstly, because of pin-count restrictions, the number of output pins for video data will be smaller than the number of bits in a pixel. Pixel memory is therefore partitioned into several byte-wide words, one of which is selected by a byte address for output from the chip. In general, the global token remains on a given chip for several clock cycles to output several bytes from the selected pixel. Secondly, the bandwidth of the video bus shown in the figure is limited by existing MOS technology to a fraction of the speed required for high-performance graphic displays. In a full-scale system, a number of such chains are scanned simultaneously, and data from their local busses is multiplexed onto a common high-speed video bus.

9.5.4 Chip electrical design

This section describes the design of the chips that make up the Pixel-planes engine. First, major functional modules of the chip are described in terms of data inputs and outputs, control inputs, and a textual description of each module's function. Next, the circuit and operational details for each module are presented. Finally, the layout techniques used in the chip design are outlined.

9.5.4.1 Organization

As shown in Fig. 9.70, the Pixel-planes chip consists of four modules:

1 *Multiplier*. This module receives the serial data streams *ADat*, *BDat*, and *CDat* as inputs and produces 64 Tree outputs, one for each PE on the chip. As described above, these outputs are the leaves of a binary-tree multiplier, and they represent 64 distinct values of the expression $F = Ax + By + C$, (x, y) being the coordinates of a pixel. The multiplier module also receives the XLSB control signal. When asserted, this signal initiates a sequential clear of the carry registers in the multiplier structure, allowing fully overlapped operations on successive sets of coefficients. A serial data path enters and exits this module (XYIn, XYOut); its operation is controlled by the signal XYShft, which causes data to be shifted through or saved in an internal shifter register. This shift register stores coefficients which represent the (x, y) address of the chip's pixel array in the image.

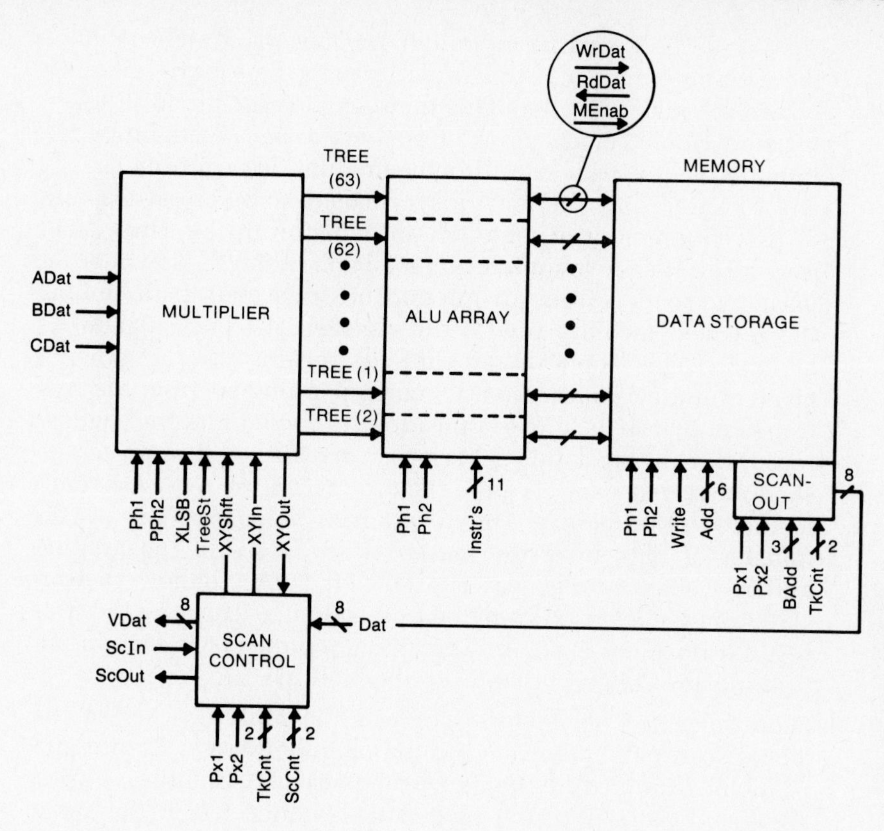

FIGURE 9.70. Block diagram of the enhanced memory chip used to construct the engine. Data enters and leaves each module on left and right of its block, control enters at the bottom.

2 *ALU array*. This module contains an array of 64 1-bit ALUs. Each of the 64 submodules receives data from one of the outputs of the multiplier, and each also has read/write access to a row of memory bits via RdDat and WrDat. An output MEnab is passed to the memory module from each ALU; this output determines whether data from the ALU can be written to memory. All of the ALUs in the array receive a common set of instruction bits, which determine the operation to be performed on the current microcycle. The ALU instruction set is described below.

3 *Memory system*. The memory module consists of two submodules, the larger of which stores data for each pixel. The data store is organized as 64 rows (one for each PE) by 64 columns (each PE is allocated 64 bits). Control for data storage consists of a 6-bit address, Add, which determines which column is to be accessed during the current microcycle, and the control signal Write. Data from an ALU is written to the currently selected bit of its memory row if Write and MEnab are both asserted. If Write is not asserted, data is read from the selected memory bit, and is available for input to each ALU.

The second, smaller module in the memory system contains the scan-out circuitry which provides video data for the system's screen refresh controller. This module contains the local token register, which selects a row of memory to be copied, bit by bit, into the shadow register within the module. The controls TkCnt0 and TkCnt1 allow the local token register to be preset to point to the bottom row of memory, shifted upward by a row, and saved. During a given microcycle, one bit of the shadow register master is loaded from an internal bit-serial data path linking the selected memory row to the register. The particular bit to be loaded is selected by the Add control lines, so that when a given column is addressed, its corresponding bit from the row (selected by the local token position) is copied into the shadow register. This mechanism performs the bit-by-bit copy to the shadow register and is pipelined so as not to reduce overall memory performance. The local-token shift function causes transfer of data from the master to the slave of the double-buffered shadow register. The output of the shadow register slave can be accessed in bytewise fashion, selected by the four BAdd (byte address) control lines. Data from the register is output to the Scan Control module on the eight Dat lines.

4 *Scan control.* This module controls the multiple uses of the global scan path, connected to the module through ScanIn and ScanOut. Module function is selected by two ScCnt lines. During system initialization, the scan path is connected to the serial shift path through the multiplier in order to load the chip's address in the image. The multiplier shift paths in all chips are thus linked into a single very long multiplier configuration register. The scan control module contains a one bit global token register. When the system is running, presence of a token in this register enables the module to pass data from the memory module through to the chip's output pins. When the token is not on-chip, the output pins are disabled. Global token shifting is enabled by one of the four states encoded in the TkCnt control lines. The module also contains an alive register. When cleared, this register disables the chip, shorting ScIn to ScOut, and thus effectively removing the chip from the system. This mechanism is intended to provide a method for building very large, fault-tolerant chips. The circuitry of Fig. 9.70 would become one of many 'super-modules' on such a large chip; inoperative super-modules could be cut out using the alive mechanism, leaving good super-modules active.

9.5.4.2 System timing

Timing on the chip is governed by two sets of 2-phase nonoverlapping clocks. The Ph clock controls image generation, timing the multiplier,

ALU, and data storage modules. The Px clock controls the video data scan-out process, timing the scan-out portion of the memory module and the scan control module. Both control and data inputs to the chip are required to be stable throughout Ph1 (Px1) and may change during Ph2 (Px2). This convention requires that control and data for a given microcycle be settled on the chip pins during the previous microcycle.

It is often necessary to stop the multiplier, preserving its current state, while clocking data through the ALU and memory. A signal TreeSt qualifies the Ph2 clock in the multiplier producing PPh2 = Ph2.TreeSt.

The only data outputs from the chip are the eight VDat pins. These outputs are connected to an 8-bit video data bus linking all chips in a system-level module called a "logical board." Electrically, this bus is a dynamic wired-NOR that is precharged on Px1 and evaluated on Px2. All of the chips connected to the bus participate in the precharge phase, while only the selected chip (the one currently enabled by the global token) can pull down the bus wires. The data on the logical board's video bus is valid at the end of Px2 and can be latched on that edge.

All modules (and submodules) on the chip obey the following timing convention: Data inputs and outputs are stable during Ph1(Px1) and may change during Ph2(Px2). Inputs, therefore, are latched on Ph1(Px1) and outputs are latched on Ph2(Px2). Control inputs to modules, as discussed above, are stable on Ph1(Px1) because of the timing convention for all chip input pins. Some control inputs to modules, however, are qualified Ph2(Px2) clocks; in these cases, the chip control inputs used to generate these signals must be latched on Ph1(Px1).

9.5.4.3 Circuit and operational details

9.5.4.3.1 Multiplier

The multiplier module consists of an x-multiplier, which forms the expression $Ax + C$, and a y-multiplier, which calculates $By + (Ax + C)$. Both multipliers have ten stages (x and y can be represented as 10-bit unsigned integers), allowing expansion of the engine to $1024 * 1024$ pixels. The 10-stage x-multiplier and the first four stages of the y-multiplier are conventional serial-parallel multipliers whose parallel coefficients represent the 10-bit x-address of the chip's pixel array and the most-significant four bits of the y-address. These coefficients are loaded into coefficient registers in the multipliers at system initialization. The last six stages of the y-multiplier form a 6-level binary tree multiplier, as described earlier. Each stage of the multiplier contains an LSB register, which allows multiplier-stage carries to be cleared sequentially in such a way that multiplier operations can be fully overlapped.

Arithmetic in the multiplier is two's complement, with coefficients x and y unsigned 10-bit integers and with ADat, BDat, and CDat signed numbers. The x-multiplier takes in an N-bit-long ADat (including sign), a (10 + N)-bit CDat, and produces a (10 + N + 1)-bit result. The y-multiplier accepts an N-bit BDat and the output of the x-multiplier producing 64 (10 + N + 2)-bit results on the Tree outputs.

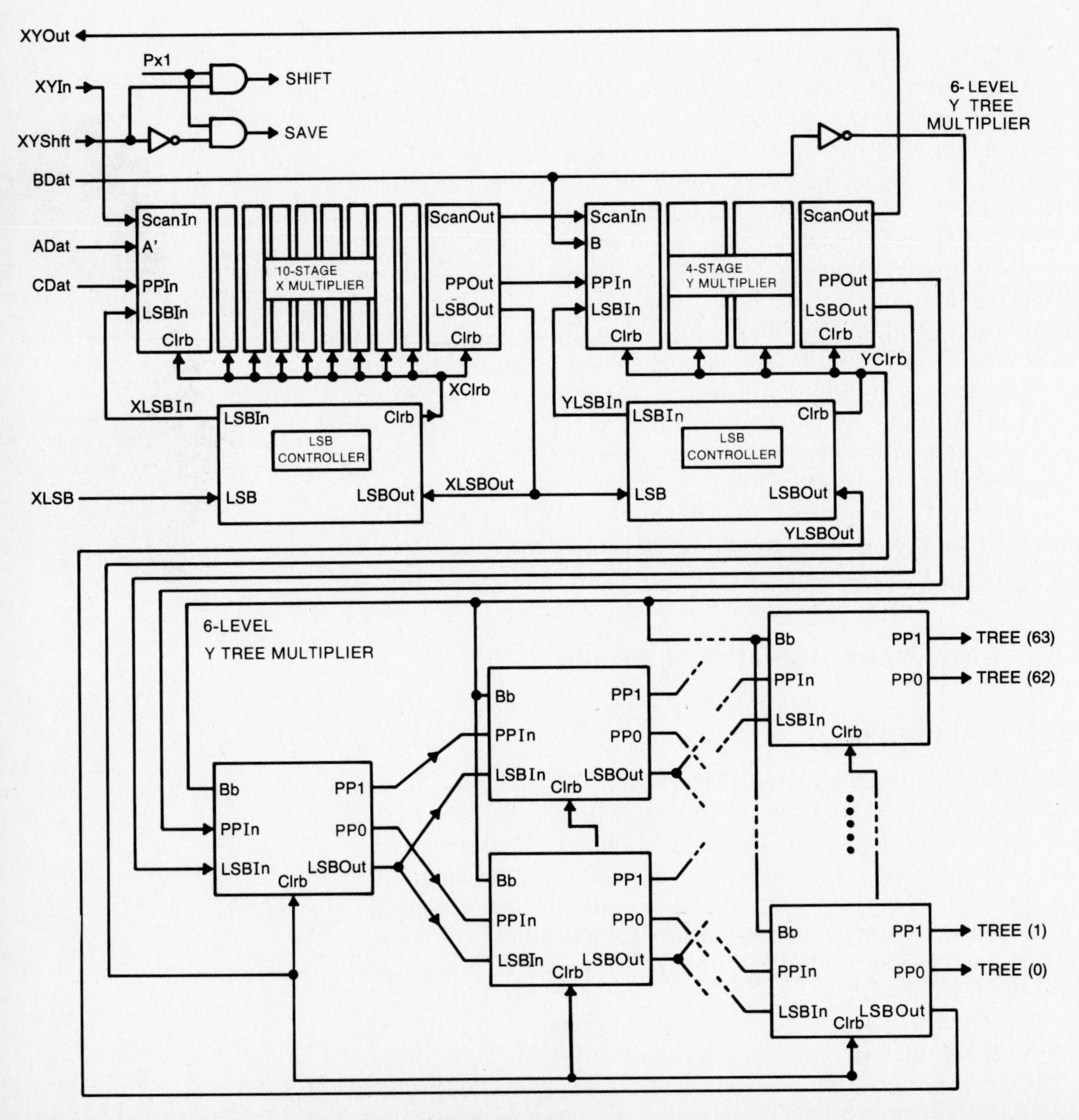

FIGURE 9.71. Block diagram of multiplier module

Fig. 9.71 shows a block diagram of the multiplier module. The 10-bit x-multiplier and the 4-stage y-multiplier are built from identical multiplier stages, whose schematic is shown in Fig. 9.72a. The 6-level y-multiplier tree is built using 63 identical tree stages, shown

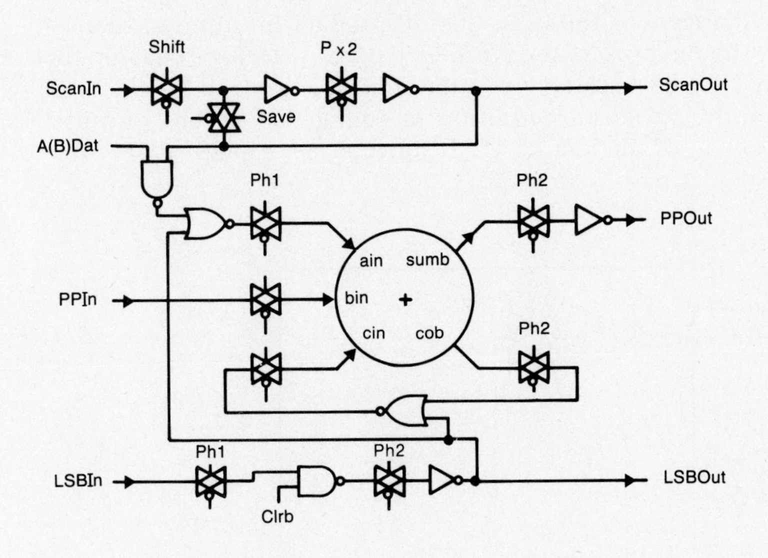

FIGURE 9.72a. Schematic of one of the 14 identical stages of the 10-stage x-multiplier and 4-stage y-multiplier

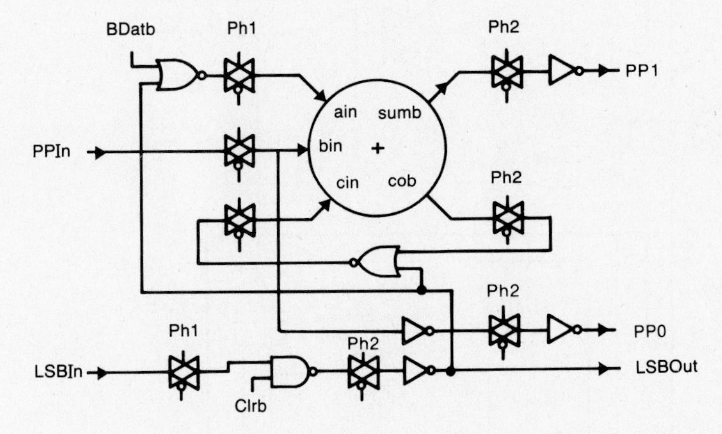

FIGURE 9.72b. One of the 63 identical stages of the 6-level y-multiplier binary tree

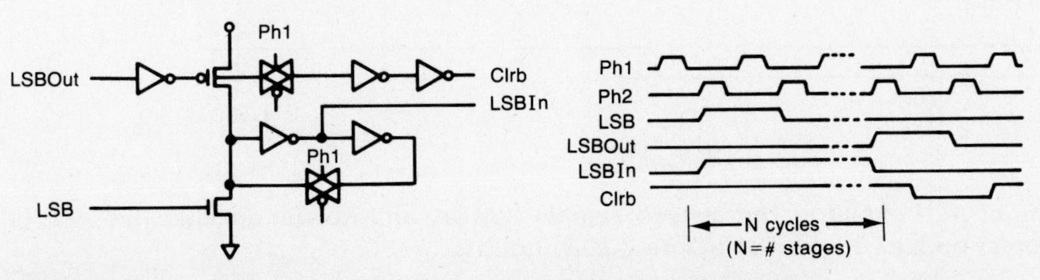

FIGURE 9.72c. Schematic and timing for LSB control controller

in Fig. 9.72b. LSB control for the x- and y-multipliers is performed by an RS controller, which is set by XLSB (YLSB) and cleared by XLSBOut (YLSBOut). The control signal XLSB, normally low, is asserted for one microcycle, one cycle ahead of the arrival of the LSB of CDat at the multiplier. The XLSB controller then inserts a stream of 1's into the LSB shift register of the x-multiplier until CDat's LSB arrives at the PPIn (partial product in) input of the last stage. On the next cycle, the XLSBOut signal goes high causing the LSB controller to assert Clrb; this control clears all of the LSB register, allowing normal addition to commence to form the first partial product of the result. XLSBOut thus stays high for only one

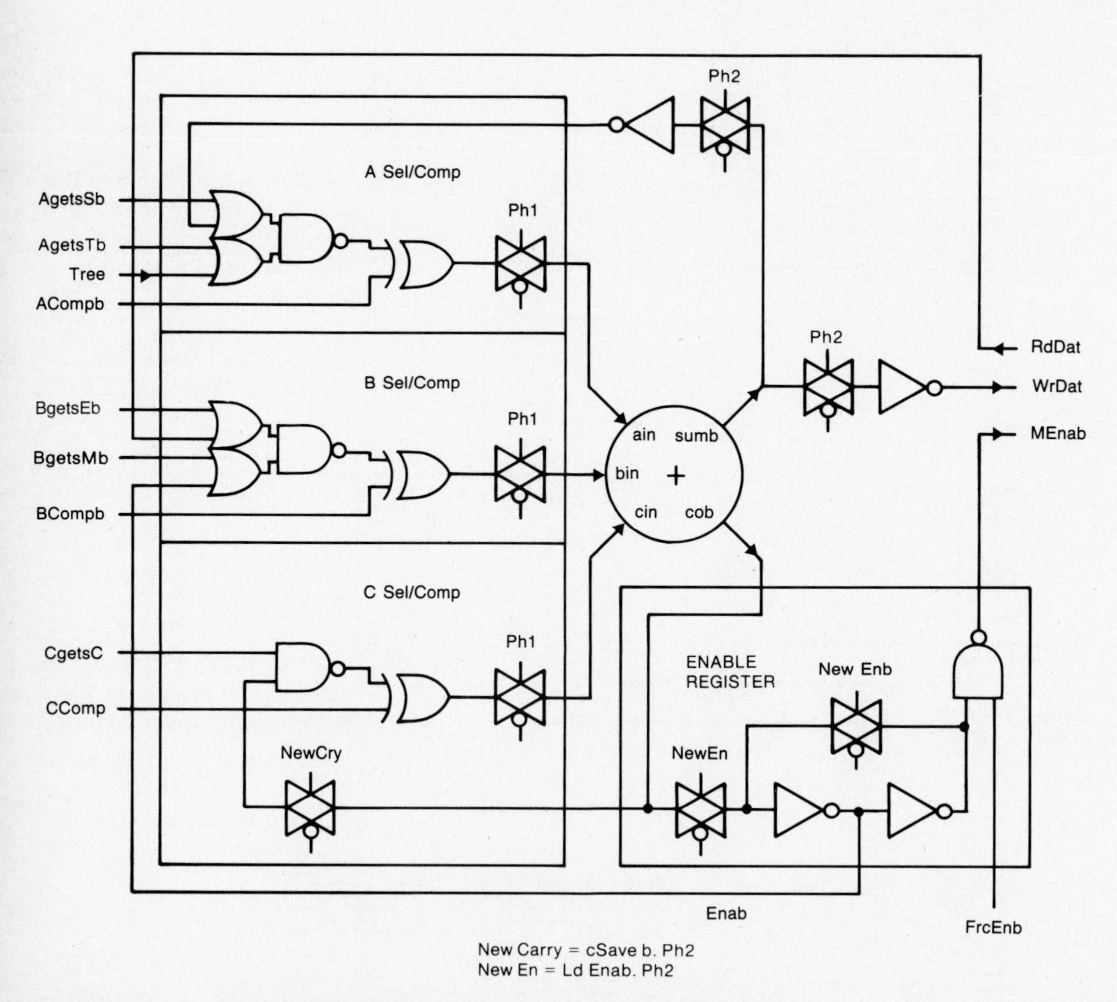

FIGURE 9.73. Details of ALU circuitry. The derived signals NewCry and NewEn are qualified clocks; CSave and LdEnab must be latched on Ph1 before qualifying Ph2.

cycle and serves as the YLSB strobe for the y-multiplier, whose LSB control structure is identical. The LSB controller and its timing are shown in Fig. 9.72c.

9.5.4.3.2 ALU

The ALU consists of an adder and an Enable register. The adder is equipped with circuitry at each of its three inputs to select the data source for the input, to complement the data, and to override the data with a constant. The selector/complementer at the carry input differs from the other three in having only a single data source (the carry output). The contents of the (dynamic) carry register can be preserved on any microcycle or loaded from the carry output. The Enable register is a quasi-static storage element whose contents can be modified, by loading from the carry output of the adder, or saved. A detailed schematic of the ALU circuitry is shown in Fig. 9.73.

Functions of the three data selectors/complementers, the carry register, and the Enable register are controlled by the 11 instruction inputs to the chip. The instruction word is organized as four fields, three of which contain the control inputs for the selectors/complementers for the ain, bin, and cin inputs to the adder; the fourth field contains the controls for the Enable register. The functions of the four instruction fields are outlined in Table 9.10 on page 470.

ALU function blocks and their associated instruction fields are organized to provide the following types of instructions (where B is a Boolean function):

- Enable ← B(Enable, Tree). This type of instruction allows the contents of the Enable register to be modified according to data arriving at an ALU from its binary-tree multiplier connection.
- Enable ← B(Enable, RdDat). Enable can be modified locally according to data stored in memory, and can be stored and retrieved.
- WrDat ← B(Tree, RdDat). Memory store can be modified according to data incoming from multiplier and data retrieved from store.
- WrDat ← B(RdDat, RdDat). Memory-to-memory operations. These operations involve fetching first operand and passing through

TABLE 9.9. ALU 11-bit instruction word

A FIELD			B FIELD			C FIELD			ENABLE FIELD	
AgetsT	AgetsS	AComp	BgetsM	BgetsE	BComp	CgetsC	CComp	CSave	LdEnab	FrcEn

TABLE 9.10. ALU operations

A FIELD				B FIELD			
AgetsT	AgetsS	AComp	Function	BgetsM	BgetsE	BComp	Function
0	0	0	ain-1	0	0	0	bin-1
0	0	1	ain-0	0	0	1	bin-0
0	1	0	ain-Sum	0	1	0	bin-Enable
0	1	1	ain-Sumbar	0	1	1	bin-Enablebar
1	0	0	ain-Tree	1	0	0	bin-RdDat
1	0	1	ain-Treebar	1	0	1	bin-RdDatbar
1	1	0	ain-Sum•Tree	1	1	0	bin-Enable•RdDat
1	1	1	ain-(Sum•Tree)-bar	1	1	1	bin-(Enable•RdDat)-bar

C FIELD				ENABLE FIELD		
CgetsC	CComp	CSave	Function	LdEnab	FrcEn	Function
0	0	0	cin-1, cry-cout	0	0	Enable saved
0	0	1	cin-1, cry saved	0	1	Enable saved, force Mem write
0	1	0	cin-0, cry-cout	1	0	Enable-cout
0	1	1	cin-0, cry saved	1	1	Enable-cout, force Mem write
1	0	0	cin-cry, cry-cout			
1	0	1	cin-cry, cry saved			
1	1	0	cin-crybar, cry-cout			
1	1	1	cin-crybar, cry saved			

Note: The 'cry' referred to in the C-Field function table is the (dynamic) register formed by the data input to the NAND in the carry selector/complementer; if CSave is asserted, NewCry is not asserted, and the last data transferred to this storage node is preserved.

sum output of ALU to A control. Second operand is fetched into B control.*

The adder in the ALU, together with the three adder control blocks, provide all of the needed Boolean functions of two variables in addition to sums and differences with carry.

9.5.4.4 Memory

The memory module consists of two main parts: a data storage module and a data scan-out module.

9.5.4.4.1 Data store

The data store is based on a 4-transistor dynamic RAM cell. The organization of the RAM array and circuit details are shown in Fig. 9.74.

---

* A more straight-forward implementation of memory-to-memory operations is possible; RdDAT could simply be input at both A and B control blocks through separate pass gates. This method was not used because of restrictions in the number of wiring channels through an ALU; see the next section on layout for details.

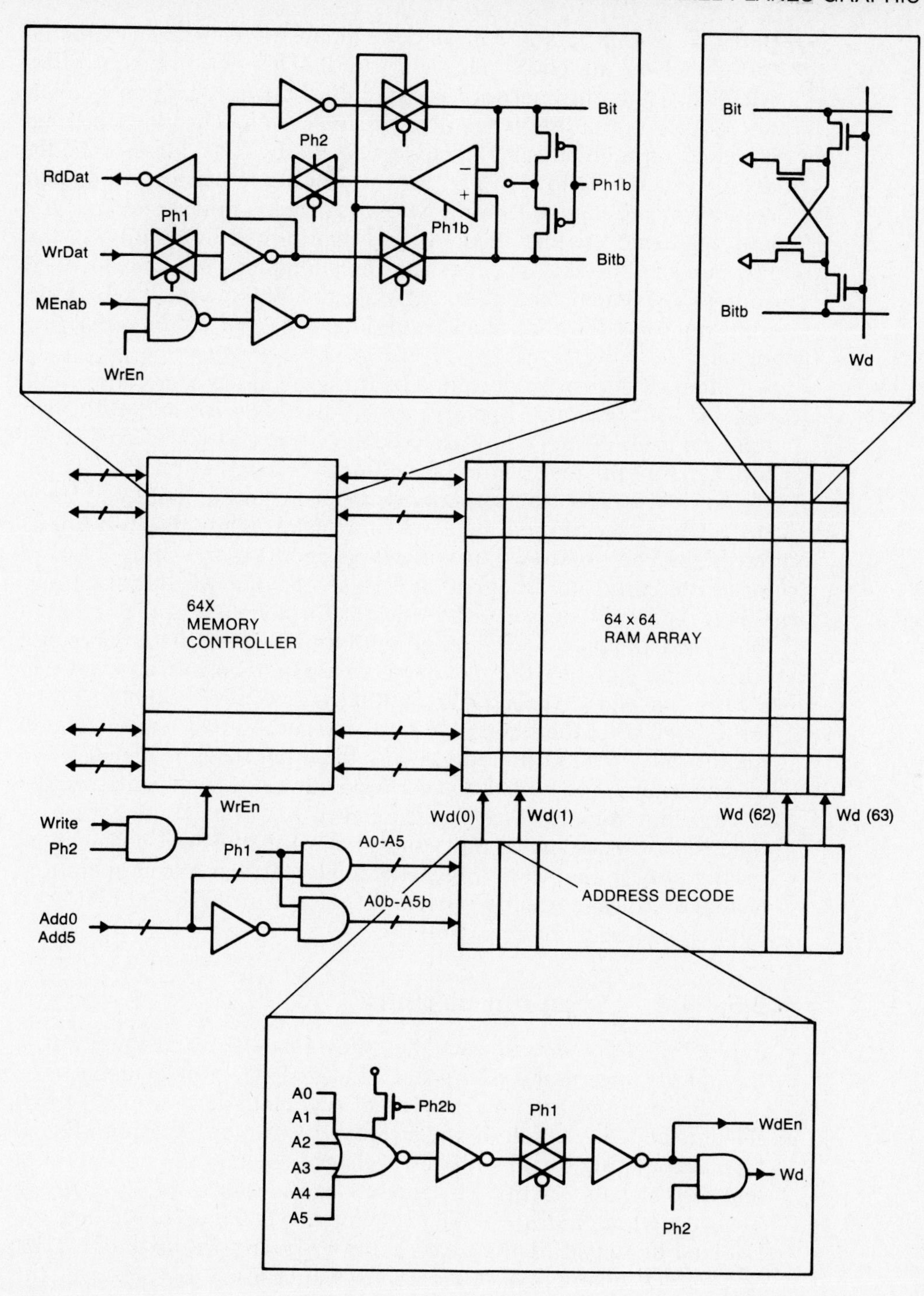

FIGURE 9.74. Organization and schematic details of data storage portion of memory module

Each row of the RAM array corresponds to the storage elements for one PE in the chip, and each row has its own row controller. All PEs in the chip access corresponding memory bits on a given microcycle, these bits lying in a single column. The column to be accessed on a given microcycle is selected by the address decoder detailed at the bottom of Fig. 9.74. Decoding is performed at each column by a dynamic 6-input NOR; this gate is precharged on Ph2, and evaluated on Ph1. Evaluation is performed by qualifying the Ph1 clock by the six Add lines and their complements. The decoded address is latched on Ph1 to generate the WdEn signal, which, in turn, qualifies the Ph2 clock to produce the Wd column-enabling signal.

Memory access is controlled by the array of memory controllers, one for each PE, and operates as follows: The Bit and Bitb data lines for each memory row are precharged on Ph1. On a read cycle, the Wd line for one of the 64 columns is asserted on Ph2. If the stored data is a '0', the uppermost transistor in the RAM cell pulls the Bit line low through its access transistor, while the Bitb line is left high. The difference in voltage between Bit and Bitb is sensed by a differential amplifier and a '1' is latched onto the RdDat output. When reading a store '1', the roles of Bit and Bitb are reversed, a '0' is sensed, and a '1' is latched onto RdDat. The differential sense amplifier is gated by the inverse clock Ph1b to conserve power. On a write, the signal WrEn (=Write.Ph2) is asserted. If MEnab is high for a given row, the data on WrDat, latched on Ph1, causes one of the Bit lines to be pulled low, while the other remains high. As Wd for one column is asserted, the corresponding RAM cell is overwritten to store the value of WrDat. In memory rows for which MEnab is low, the two pass-gates that control the Bit lines are not enabled during Ph2, so that these rows effectively perform a read operation. Note that RdDat correctly reflects the data at the accessed RAM cell following each memory operation.

9.5.4.4.2 Scan-out circuitry

Fig. 9.75 details the circuitry that allows data to be scanned out of the data storage module to refresh the system's video screen.

The scan-out module consists of two main parts: a pixel (row) selector and the 'shadow' register. Row selection is accomplished by manipulating the local token register, which produces the set of one-shot signals Tok(n). The register can be preset to point to the lowest row, shifted up by one row, or saved. These operations are selected by decoding the token-control signals TkCnt0 and TkCnt1 to produce qualified Px1 clocks for the register. Decoding of the token-control signals is shown in Table 9.11.

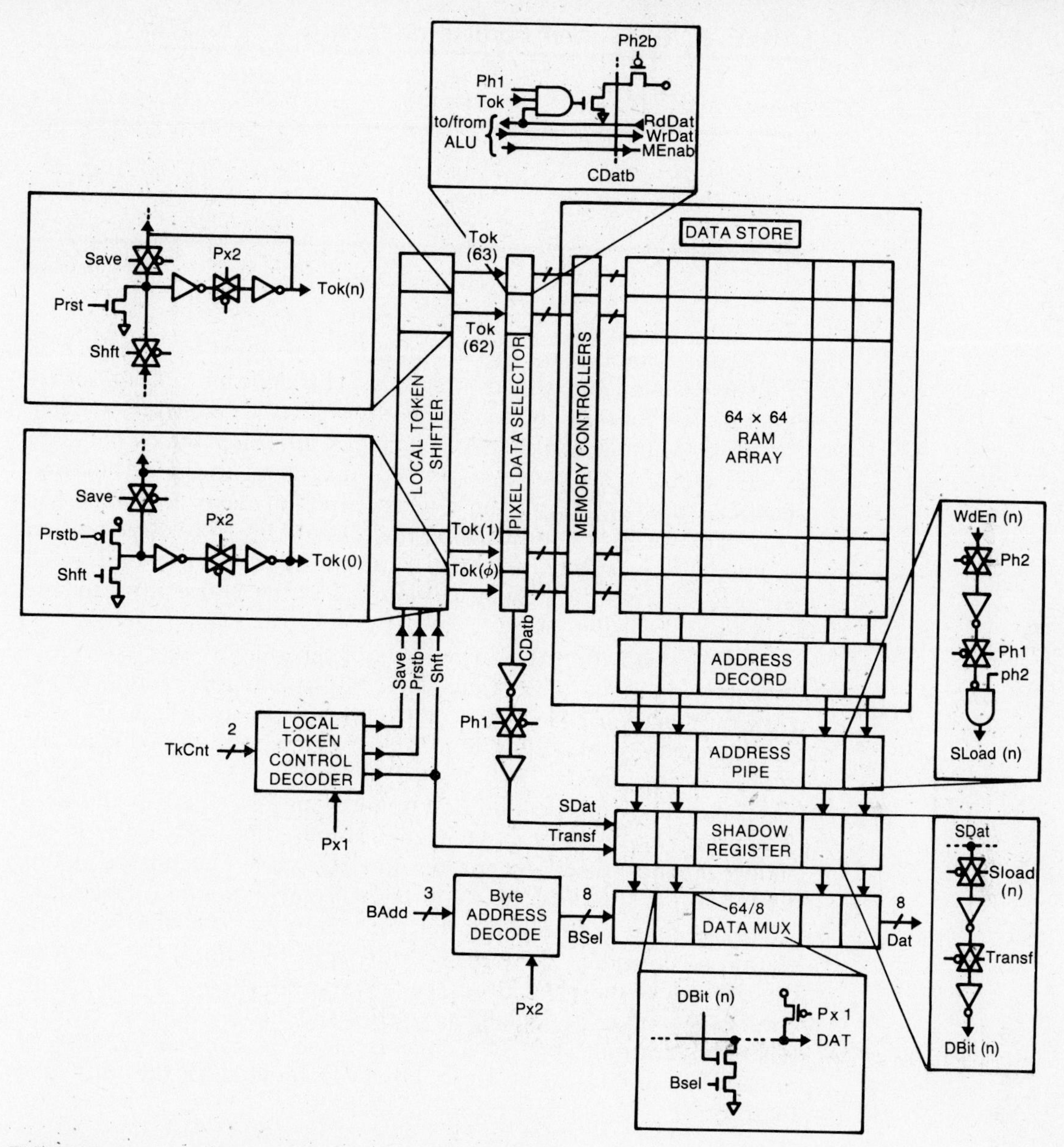

FIGURE 9.75. Circuit details of scan-out module

The selector itself consists of a 64-input dynamic NOR, which is precharged on Ph2 and evaluated on Ph1. Evaluation is performed by the row that has a high Tok. On that row RdDat, Tok, and Ph1 are ANDed; if all three are high, then the output of the dynamic NOR, CDatb, is pulled low.

TABLE 9.11. Token control decoding

TkCnt1	TkCnt0	LOCAL TOKEN	GLOBAL TOKEN	SHADOW REGISTER
0	0	Save	Save	x
0	1	Shift	Save	Transfer
1	0	Preset	Save	x
1	1	Save	Shift	x

CDatb is latched on Ph1 to produce the signal SDat, which is input as data to the shadow register. This register consists of 64 dynamic latches, which are enabled by a set of 64 signals SLoad(n). The SLoad signals, qualified MPh2 clocks, are generated by delaying by one clock cycle the outputs WdEn(n) from the RAM address decoder, thus providing the correct pipeline delay to match the delay between SDat and RdDat on the selected row. The slave portion of the shadow register is loaded from the master when the Transf signal is asserted; this signal is identical to the Shft control for the local token register.

Data is read from the 64 outputs of the shadow register onto 8 internal video data wires Dat0-Dat7. Multiplexing is accomplished using a dynamic AND-OR-INVERT, which is precharged on Px1 and evaluated by Px2-qualified BSel lines, which are decoded from the three BAdd signals.

The data scan-out module contains the only part of the chip design that connects the two asynchronous processes in the Pixel-planes engine, image generation and scan-out. The timing of this module must therefore be examined carefully to avoid difficulties arising from synchronization failure in the data transfer from one regime to the other. Specifically, this type of failure will occur in the data selector when the local token is shifted or preset (asynchronously with the selector circuit) and in the shadow register when the trailing edge of Transf is close to the leading edge of SLoad. These failures will not, however, be fatal for the following reasons:

1. Transfers of data from master to slave occur on Px1, while the token outputs change on Px2. The slave therefore receives a faithful copy of the master, except when the value being written to the master is changing when the transfer takes place.
2. The image-generation algorithms are designed to 'visit' each bit in each pixel more often than once per scan line, so that between successive assertions of Transf sufficient time has passed to copy all bits from the data store to the shadow register. The incorrect

transfer, which occurs subsequent to moving the local token, will therefore be guaranteed to have been overwritten.

3 The remaining possibility for failure occurs when the image-generation algorithm is in the act of changing one of the video data (color) bits at exactly the time when that bit is to be read from the shadow register. This problem is inherent in any frame buffer system and can be avoided by double-buffering the video information in the RAM array. In this scheme, the (color) bits that are being accessed on scan-out can only change during retrace, times when the shadow register is not being read.

9.5.4.5 Scan control module

The scan control module handles the chip's connection to the serial shift path used by the system to 1) initialize the address registers in each chip's multiplier, and 2) pass the global token from chip to chip for scan-out selection. Fig. 9.76 shows the circuitry for this module. The upper part of the module schematic is a multiplexer that selects one of four paths connecting the scan input ScIn to the scan output ScOut. This selection is controlled by the two ScCnt signals and by the state of the Alive register in the module as shown in Table 9.12.

Fig. 9.77 details the generation of internal control signals ScScnt and TkCnt. Referring to Fig. 9.76 the signal OutEn is asserted on Px1 when Alive and GlTok are both true. Data, latched from the internal Dat bus on Px2, is then put onto the chip's VDat bus, which may connect one or more super-modules (Fig. 9.74) to the chip's output pins. The VDat bus is, like the Dat bus, a dynamic wired-NOR, precharged on Px2 and evaluated on Px1. Circuitry for the output pads on the chip is shown in Fig. 9.78. The video data bus external to the chip is electrically similar to the data busses within the chip. It also uses a dynamic wired-NOR and is precharged on Px1 and evaluated on Px2. This external bus design somewhat simplifies bus control by eliminating the need for an enabling control signal required for a tri-state implementation and can provide somewhat better speed performance at the expense of increased switching energy.

9.5.5 Chip organization and layout

The overall organization of the Pixel-planes memory chip is shown in Fig. 9.79. The layout contains three main sections: the multiplier, the ALU, and the memory system. Each of these sections is built as two nearly identical half-planes, with a central control section.

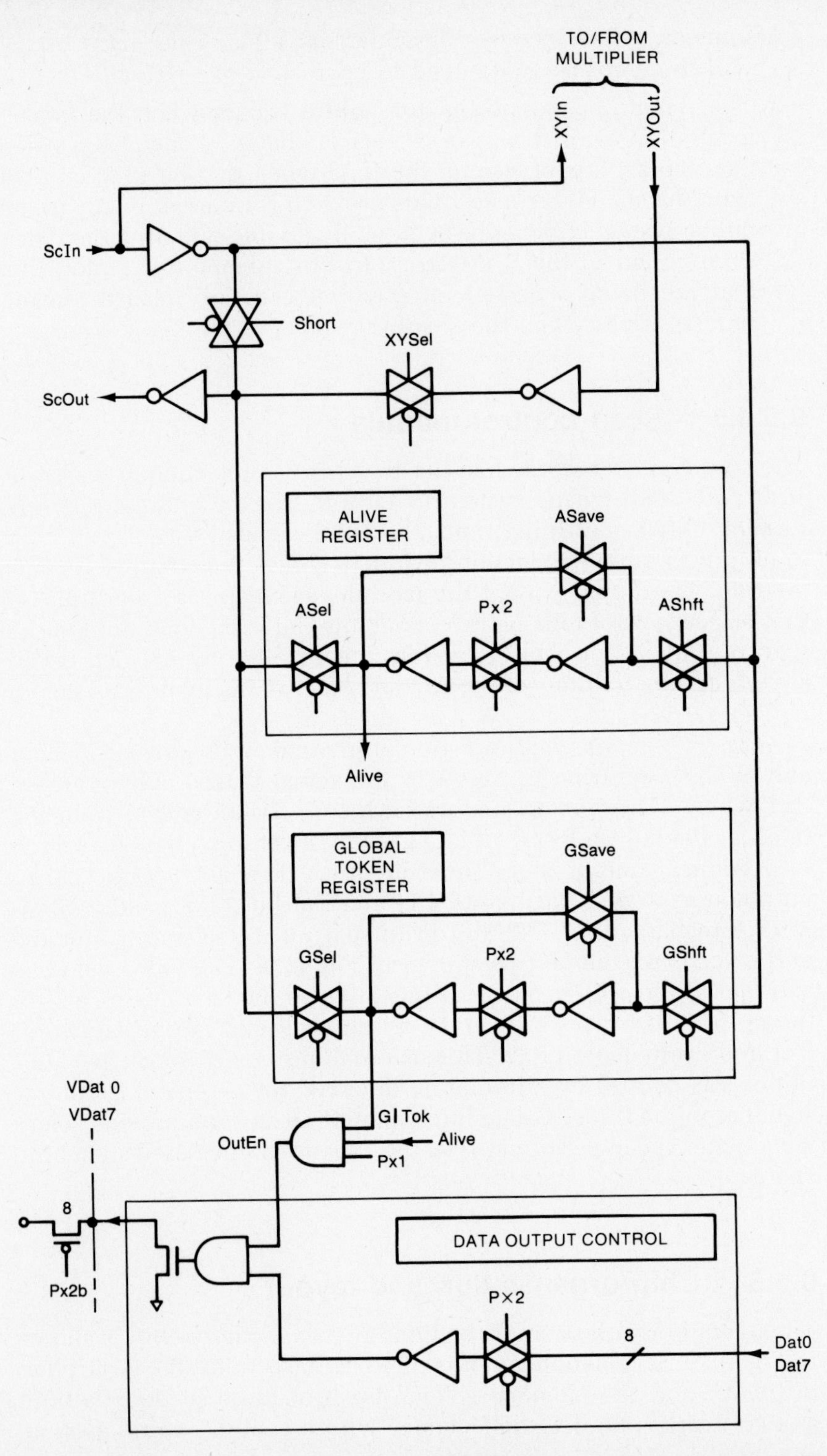

FIGURE 9.76. Circuitry for the scan control module

TABLE 9.12. Scan control multiplexer modes

ScCnt1	ScCnt0	ALIVE	PATH
0	0	x	Connect ScIn to ScOut (test)
0	1	x	Through Alive register
1	0	0	Connect ScIn to ScOut
1	0	1	Through XY-address registers
1	1	0	Connect ScIn to ScOut
1	1	1	Through global token register

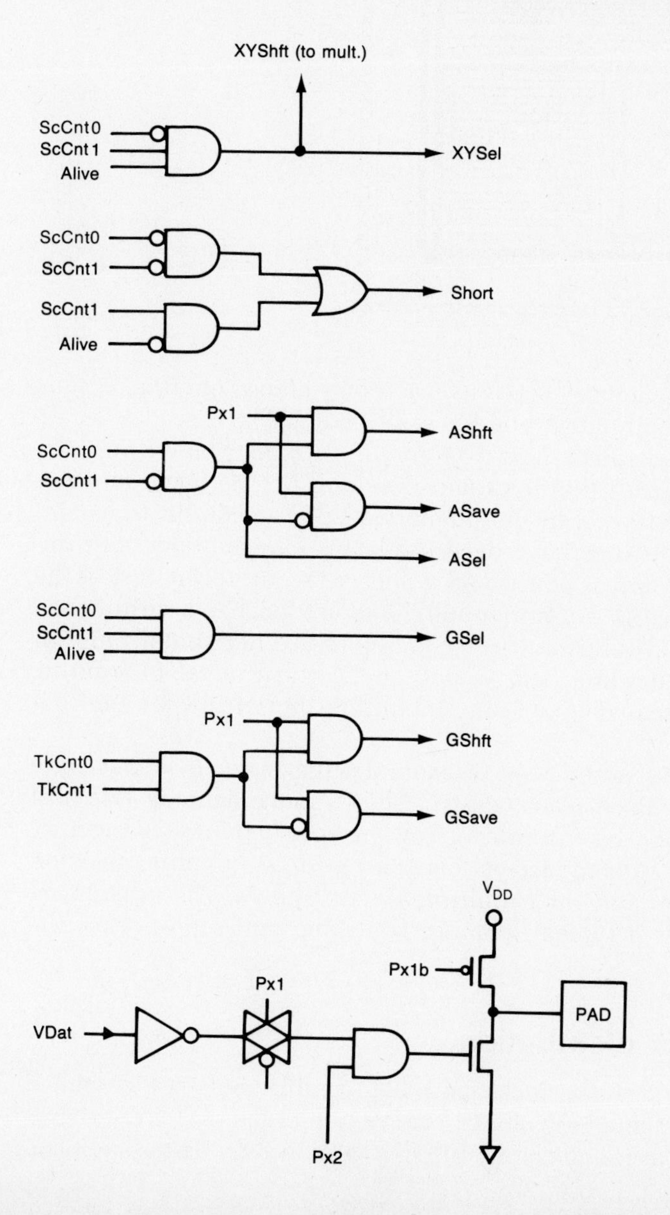

FIGURE 9.77. Control signal decoding in scan control module

FIGURE 9.78. Video data output pad schematic

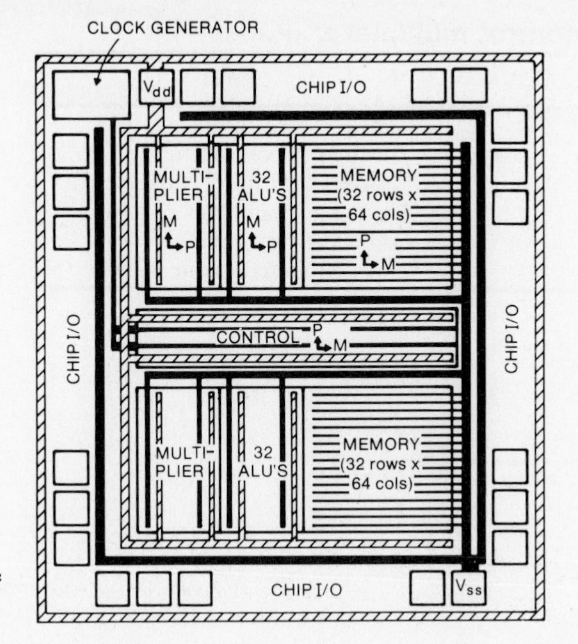

FIGURE 9.79. Overview of chip layout and routing

In the ALU and memory sections, each half-plane contains 32 horizontal strips of circuitry, one for each pixel.

Fig. 9.79 also shows how V_{DD} and V_{SS} are distributed over the layout and indicates the dominant direction for the metal and polysilicon wiring layers. In the memory section, metal bit lines and polysilicon word lines have been used; this choice offers both improved memory access speed and a denser layout. In the rest of the circuitry, data paths run horizontally and are localized within each module. The ALU and multiplier therefore use horizontal poly for intra- and intermodule data wiring and vertical metal for routing control and global data signals, including the wiring for the 'flat tree'.

Fig. 9.80 shows, by way of example, the leaf-cell layout style employed in the Pixel-planes design. Shown in the figure is a vertical slice through one bit column of the memory module. A memory cell and its associated driver, decoder, address pipeline register, shadow register, and data multiplexer are shown. Also displayed in the figure are a typical memory controller and a portion of the ALU.

9.5.6 Clock distribution

The two sets of 2-phase clocks (and their complements) are generated from off-chip single-phase clocks. Nonoverlap of the two clock phases is assured by a technique similar to that described by Seitz [MeCo80].

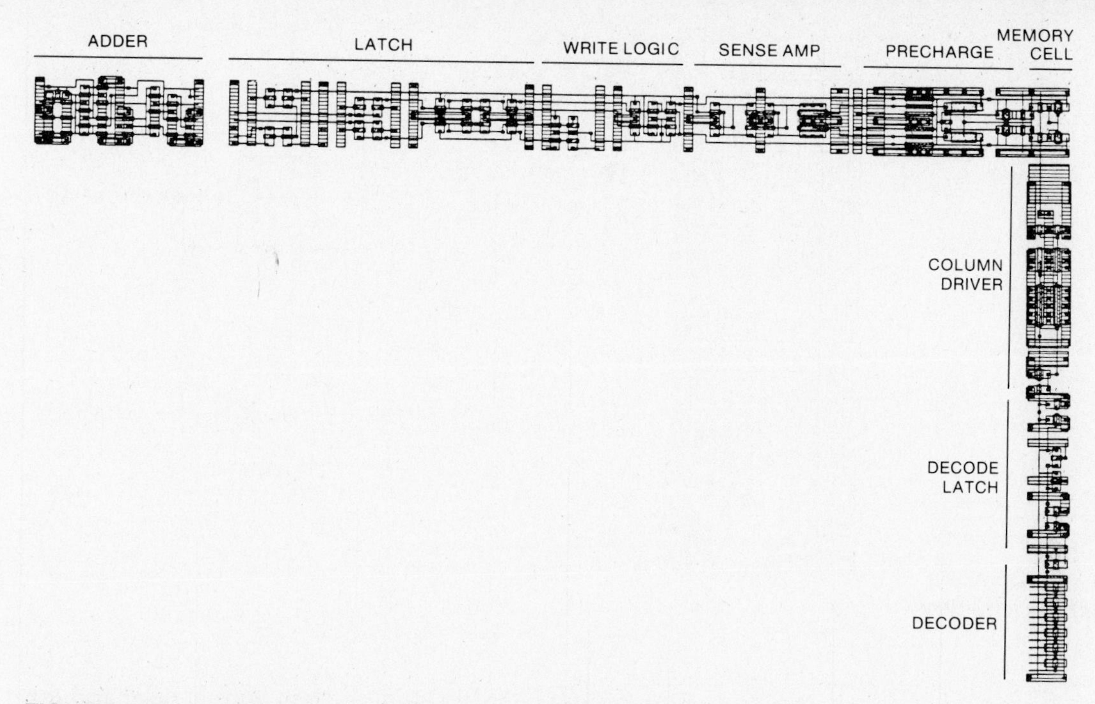

FIGURE 9.80. Layout of a vertical slice through the memory module, and of portion of ALU pitch-matched to a memory cell

Large clock buffers located in the pad ring are sized to drive all of the clock loads within the chip. Their outputs are distributed in metal, interrupted only at power busses where they are jumpered with short, wide polysilicon wires.

The Pixel-planes design has a large number of instances of clocks and qualified clocks, which drive an array of pass-gates. In order for the machine to be fast and free from timing hazards, timing skew among these clocks and qualified clocks must be minimized. A conventional approach to generating qualified clocks uses combinational logic to combine a control signal and a clock, followed by amplifiers that effectively regenerate each qualified clock locally. To minimize the skew between clocks and qualified clocks, unqualified clock signals must be regenerated in a similar way to match the delay through the clock qualifiers, thus leading to a pyramid-like clock distribution/regeneration scheme. If such a design is to be free from timing hazards, great care must be taken to equalize the delay through each clock regenerator and each clock qualifier.

Timing control in the Pixel-planes chip is greatly simplified by using clock qualifiers of the type shown in Fig. 9.81. In this circuit, complementary pass transistors are inserted between one of the clocks and a qualified output. The gates of these transistors are

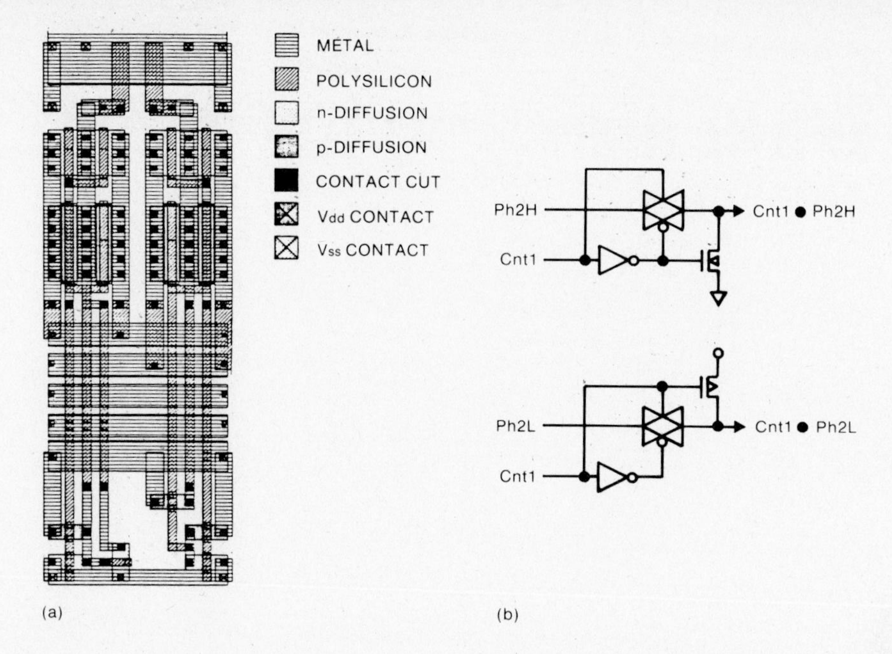

FIGURE 9.81. Qualified clock generator circuitry

driven by the qualifying control signal (or its complement) and are latched on the opposite phase. These qualifiers therefore insert very little delay between the clock input and the qualified output, and deskewing separate qualified clocks can be accomplished simply by appropriate sizing of the pass transistors. In this scheme, control of the skew between a clock and its complement has been centralized in the chip's global clock driver; the problem therefore need only be solved once, rather than at every clock regenerator.

9.6 Hierarchical layout and design of a single chip 32-bit CPU

R. H. KRAMBECK*

9.6.1 Introduction

This final section describes the design methodology and some of the techniques used to design a 32-bit microprocessor. It represents the largest case example in this text, so rather than concentrate on the architecture (which would fill a book alone), the design methodology is highlighted. While the previous examples were completed

* AT&T Bell Laboratories, Murray Hill, N.J.

by small teams or individuals, this design is a large team design. This results in some changes in the design style to ensure good communication between all groups involved with the project.

The first microprocessor designed at Bell Labs, the WE 8000* microprocessor had 10,000 transistors and ROM sites. The most recent is the WE 32000 microprocessor, which has 146,000 transistors and ROM sites. An earlier version of this chip was described previously by B. T. Murphy et al. [METM81]. This design used a control section comprised predominantly of standard cells connected by an automatic router. As a result, it was possible to produce a layout and working models quickly, but area and performance were not optimized. As the first silicon of this version was being produced, work was proceeding on a hierarchical design aimed at implementing the same instruction set with more nearly optimum area and performance.

The driving force for this new design comes from potential system users. They need the best possible performance and the shortest possible design times in order to achieve timely introduction of competitive products. As a result, though the chip complexity has increased more than an order of magnitude over the complexity of state of the art chips of a few years ago, the time available for laying out and debugging the chip has actually decreased. In addition, maximum performance requires optimum use of area and use of the very latest set of design rules at the time of product introduction.

These needs lead to two serious problems. First, we need a way of designing and laying out the chip that requires less time than that used for design of a chip one order of magnitude simpler, without wasting area. Second, we need to begin the layout work before the design rules that will ultimately be used on the first models are available.

The problem of getting the extremely complex design and layout done quickly is solved by use of a top-down hierarchical approach. The job is divided up in such a way that numerous parts may be designed in parallel with a small core of people assigned the job of making sure it will all fit together in the end. This, of course, is the way complex jobs have always been handled, both in software and in hardware. In addition, to insure that the logic design is consistent with optimum utilization of chip area, the logic design, timing, and layout work are done in parallel. This is in contrast to a silicon compiler approach, where layout work begins only after logic design is finished, and timing is scarcely considered at all. With this parallel approach, the interval from system requirements to working chip is minimized.

---

* WE is a trademark of AT&T-Technologies.

The problem of how to begin layout work before design rules are fixed has been handled in the past by the use of the latest set of design rules followed by a uniform shrink of the layout when a finer set of design rules became available. An example is the Mead-Conway [MeCo80] design rules, which chose a rigid set of relationships among the various design rules. The difficulty with this approach is that pursuit of very fine design rules leads to qualitative changes to processing, which in turn yields a nonproportional shrink of various features.

The approach adopted for this chip involves the use of a class of layout techniques developed at AT&T Bell Laboratories called technology updatable layout techniques. These are LTX [METM81], gate matrix [LoLa80], technology updatable polycell [LeCJ81] and automatically generated programmed logic arrays (PLA) [LaSh82]. All share certain basic features in common: The layout engineer does not work with actual geometries such as rectangles and paths. Instead, his or her layout consists of a description that shows only topologies and interconnections. The detailed geometries needed for layout are then created by a computer program that uses as inputs the description created by the layout engineer and a design rule file based on the current technology. In this way, when design rules change a new layout can be made quickly, even if design rules do not shrink uniformly.

The highly structured design in 2.5 μm design rules yielded a typical clock rate of 8 MHz at 70°C, as opposed to the 2 MHz clock rate of the 3.5 μm design rule prototype. In addition, the improved cycle architecture made more extensive use of pipelining, reducing the required cycle count by approximately 50 percent. The resulting single chip, 32-bit CMOS microprocessor has a performance eight times that of the prototype models while retaining a low power dissipation (0.7 Watt at 8 MHz).

The first processor system designed around the WE 32000 microprocessor, running at 6.2 MHz, will have a performance of at least 75 percent that of a high end 32-bit midicomputer.

9.6.2 Design methodology

While the main thrust of this chapter is towards a description of the design and layout methodology, it is useful to introduce a brief description of the chip architecture. For a more complete description, see [BGKL82]. The WE 32000 CPU has a very rich instruction set oriented towards support of a high level programming language and operating systems. The hardware that implements this instruction set is both structured and modular. This insures compatibility with the hierarchical design approach.

The CPU supports an orthogonal instruction set that allows the full range of opcodes to be used with the full range of operand descriptors. This decoupling of opcode and addressing mode simplifies code generation. Most instructions can operate on all three main data types (word, half word, and byte) and a few special instructions can handle three other data types (bit fields, blocks, and C strings). There are fifteen addressing modes including several variations each of the generic addressing modes, immediate, register, absolute, and displacement. Arithmetic (including multiply and divide) and logic instructions have both dyadic and triadic forms.

Several features have been included to facilitate the design of an operating system. Among these are the provision for four levels of execution privilege, the existence of a four-level exception structure, explicit process switching instructions, and an interrupt mechanism that can be used to cause a process switch.

The block diagram that implements these capabilities is shown in Fig. 9.82. There is a full 32-bit internal architecture. The control is organized hierarchically with a MAIN control coordinating the operation of separate FETCH and EXECUTE controls. These latter two controls are designed to operate in parallel to maximize throughput. To permit the maximum possible parallel operation,

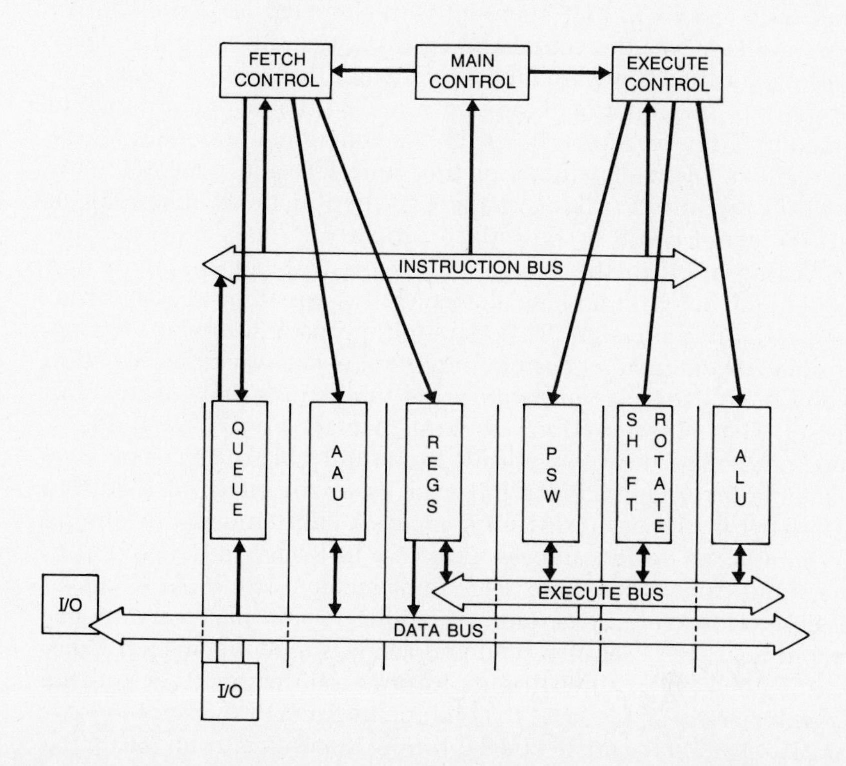

FIGURE 9.82. Block diagram of internal architecture of WE-32000 CPU

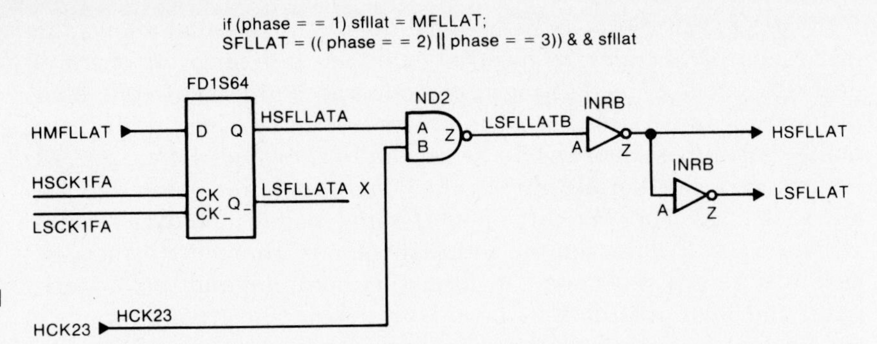

FIGURE 9.83. C language description with associated logic schematic

two 32-bit buses are provided, one for use of the execute section and one for general use.

This block diagram of the architecture suggests a layout structure and hierarchy and can be used as the starting point for the hierarchical design and layout of the chip. At the highest design level, the microprocessor was functionally partitioned into superblocks and inter-superblock protocols were defined. These superblocks were in turn partitioned into blocks. A functional simulator, written in the C language, described the operations of the device on a per clock phase basis. An example of a piece of C code is shown in Fig. 9.83, along with the logic that eventually was created to implement it. This simulator was the vehicle for insuring that all functional specifications for the chip were met. Throughout the design process, the hierarchy of the functional simulator and of the layout were made to match. Thus for every block in the functional simulator, there was a block in the layout with the same I/O and contents. This sometimes required making changes to the functional hierarchy so that the layout could be optimized properly.

At this point in the design process, with block functions and I/O defined, but with no logic designed, it is possible to begin layout work and timing design. This parallel approach to layout, timing, and logic design made possible important exchanges of information among them. Next, layout strategy for the various parts of the chip was set. The control section was implemented primarily with PLAs. These can be laid out automatically from truth tables using technology updatable techniques. PLAs have the great advantage of requiring no manual layout effort or timing analysis effort and can be altered very late in the design process with very little cost in time or effort. As a result, control functions were implemented with PLAs wherever possible. However, in certain situations it was necessary to use random logic. For example, random logic was used when a particular control function required inputs from two different PLAs and the PLAs were already too large to combine into one PLA. Random logic was also needed to satisfy special timing requirements on particular control signals. PLA outputs remain in their state for a full-state machine cycle, but control signals often are active for only a fraction of the machine cycle. For the part of the control requiring random logic the gate matrix [LoLa80] layout style was adopted to allow flexibility in transistor size design while still obtaining high packing density and an updatable layout. This has been covered in more detail in Chapter 7. The cost was a substantial manual effort both in drawing the initial layout and in doing the timing analysis. The data path requires a 32-bit bus throughout its length and two such busses for the execute section. This can be handled with equally spaced metal busses running horizontally and polysilicide control lines running vertically. The I/O uses a standard cell approach [LeCJ81] because of the need to meet standard requirements on current drive, voltage levels, latch-up protection, and immunity to static discharge.

Connecting together all of these parts can be done with an automatic wiring program (in our case, LTX2). This can be done even before any parts are actually laid out because all that is needed are the I/O list and terminal placement for each piece and an estimate for the size of each piece.

An example of an LTX description of a piece of circuitry is shown in Fig. 9.84a for a part of the Arithmetic Logic Unit (ALU). The size of this piece was estimated from the number of transistors, multiplied by a density factor for each transistor, based on past experience for this type of circuit. The positioning of the I/O is tentative at this point and can be changed to facilitate interconnections during the assembly process. Descriptions like this are written for all blocks. An automatic wiring program (LTX2) is used to assemble groups of blocks into larger entities called superblocks such as the complete ALU. The physical position of each block is given to LTX2 and any mismatch between adjacent blocks is reported as an error. Also, LTX2 compares the description of the block to the description used by the functional simulator. Any discrepancy between the layout description and the functional description of the block is flagged at this point. This step checks that where blocks butt together, signals with identical names also butt together, and that all I/Os exist that should exist. The program also verifies that power supplies and clocks connect properly from one adjacent block to another. After verifying that all corrections internal to the newly created superblock are correct, the program generates a description of all I/O locations for the superblock. The description for the ALU is shown in Fig. 9.84b. As can be seen, this has the same format as the blocks that were used to construct it. This makes it possible to assemble superblocks into still larger structures.

Where blocks or superblocks have edges that do not butt to another block or superblock, LTX2 creates a wiring channel. Fig.

```
RAM (8308020932)   :
  BLOB N11 0, 0    22.5, 0    22.5, 62    0, 62   ;
* T# T      X          Y  D   EE   LE   WIDTH   CAP   RES  MASK  NAME/CLMP   LOG   ;
   1 I    11.25       62  U    1    0       2  0.05     0   N40  RS          ;
   2 O     22.5     31.5  R    2    0       2  0.03     0   N70  QB          ;
   3 O     22.5      6.5  R    3    0       2  0.03     0   N70  Q           ;
   4 O        0      6.5  L    3    0       2  0.03     0   N70  Q           ;
   5 O        0     31.5  L    2    0       2  0.03     0   N70  QB          ;
   6 I    11.25        0  D    1    0       2  0.05     0   N40  RS          ;
   7 F     22.5     55.5  R    7    0       2  0.01     0   N70  MET6        ;
   8 F        0     55.5  L    7    0       2  0.01     0   N70  MET6        ;
  -1 C     22.5   60.625  R   -1    0    2.75  0.13     0   N70  VDD;
  -2 C    11.25       62  U   -1    0    22.5  0.13     0   N70  VDD;
  -3 C        0   60.625  L   -1    0    2.75  0.13     0   N70  VDD;
  -4 C        0    1.375  L   -4    0    2.75  0.14     0   N70  VSS;
  -5 C    11.25        0  D   -4    0    22.5  0.14     0   N70  VSS;
  -6 C     22.5    1.375  R   -4    0    2.75  0.14     0   N70  VSS;
* END OF MODULE TYPE LIBRARY FOR LTX2;
```

FIGURE 9.84a. LTX source code description for a part of ALU

```
RAM5 (8407250901)   :
  BLOB N11 0, 1794.25    22.5, 1794.25    22.5, 2107.5    0, 2107.5   ;
* T# T      X          Y  D   EE   LE   WIDTH   CAP   RES  MASK  NAME/CLMP   LOG   ;
   1 I    11.25  1794.25  D    1    0       2  0.25     0   N40  RS          ;
   2 O     22.5  1849.75  R    2    0       2  0.03     0   N70  Q           ;
   3 O     22.5  1824.75  R    3    0       2  0.03     0   N70  QB          ;
   4 O     22.5  1862.75  R    4    0       2  0.03     0   N70  Q           ;
   5 O     22.5  1887.75  R    5    0       2  0.03     0   N70  QB          ;
   6 O     22.5  1973.75  R    6    0       2  0.03     0   N70  Q           ;
   7 O     22.5  1948.75  R    7    0       2  0.03     0   N70  QB          ;
   8 O     22.5  1986.75  R    8    0       2  0.03     0   N70  Q           ;
   9 O     22.5  2011.75  R    9    0       2  0.03     0   N70  QB          ;
  10 O     22.5  2097.75  R   10    0       2  0.03     0   N70  Q           ;
  11 O     22.5  2072.75  R   11    0       2  0.03     0   N70  QB          ;
  12 O        0  1849.75  L    2    0       2  0.03     0   N70  Q           ;
  13 O        0  1824.75  L    3    0       2  0.03     0   N70  QB          ;
  14 F     22.5  1800.75  R   14    0       2  0.01     0   N70  MET6        ;
  15 F        0  1800.75  L   14    0       2  0.01     0   N70  MET6        ;
  16 O        0  1862.75  L    4    0       2  0.03     0   N70  Q           ;
  17 O        0  1887.75  L    5    0       2  0.03     0   N70  QB          ;
  18 F     22.5  1911.75  R   18    0       2  0.01     0   N70  MET6        ;
  19 F        0  1911.75  L   18    0       2  0.01     0   N70  MET6        ;
  20 O        0  1973.75  L    6    0       2  0.03     0   N70  Q           ;
  21 O        0  1948.75  L    7    0       2  0.03     0   N70  QB          ;
  22 F     22.5  1924.75  R   22    0       2  0.01     0   N70  MET6        ;
  23 F        0  1924.75  L   22    0       2  0.01     0   N70  MET6        ;
  24 O        0  1986.75  L    8    0       2  0.03     0   N70  Q           ;
  25 O        0  2011.75  L    9    0       2  0.03     0   N70  QB          ;
  26 F     22.5  2035.75  R   26    0       2  0.01     0   N70  MET6        ;
  27 F        0  2035.75  L   26    0       2  0.01     0   N70  MET6        ;
  28 O        0  2097.75  L   10    0       2  0.03     0   N70  Q           ;
  29 O        0  2072.75  L   11    0       2  0.03     0   N70  QB          ;
  30 I    11.25   2107.5  U    1    0       2  0.25     0   N40  RS          ;
  31 F     22.5  2048.75  R   31    0       2  0.01     0   N70  MET6        ;
  32 F        0  2048.75  L   31    0       2  0.01     0   N70  MET6        ;
  -1 C     22.5 1795.625  R   -1    0    2.75  0.13     0   N70  VDD;
  -2 C        0 1854.875  L   -2    0    2.75  0.28     0   N70  VSS;
  -3 C     22.5 1916.875  R   -3    0    2.75  0.26     0   N70  VDD;
  -4 C        0 1978.875  L   -4    0    2.75  0.28     0   N70  VSS;
  -5 C     22.5 2040.875  R   -5    0    2.75  0.26     0   N70  VDD;
  -6 C     22.5  2104.25  R   -6    0     5.5  0.14     0   N70  VSS;
  -7 C    11.25  1794.25  D   -1    0    22.5  0.13     0   N70  VDD;
  -8 C        0 1795.625  L   -1    0    2.75  0.13     0   N70  VDD;
  -9 C     22.5 1854.875  R   -2    0    2.75  0.28     0   N70  VSS;
 -10 C        0 1916.875  L   -3    0    2.75  0.26     0   N70  VDD;
 -11 C        0 1857.625  L   -2    0    2.75  0.28     0   N70  VSS;
 -12 C     22.5 1857.625  R   -2    0    2.75  0.28     0   N70  VSS;
 -13 C     22.5 1919.625  R   -3    0    2.75  0.26     0   N70  VDD;
 -14 C        0 1919.625  L   -3    0    2.75  0.26     0   N70  VDD;
 -15 C     22.5 1978.875  R   -4    0    2.75  0.28     0   N70  VSS;
 -16 C        0 2040.875  L   -5    0    2.75  0.26     0   N70  VDD;
 -17 C        0 1981.625  L   -4    0    2.75  0.28     0   N70  VSS;
 -18 C     22.5 1981.625  R   -4    0    2.75  0.28     0   N70  VSS;
 -19 C     22.5 2043.625  R   -5    0    2.75  0.26     0   N70  VDD;
 -20 C        0 2043.625  L   -5    0    2.75  0.26     0   N70  VDD;
 -21 C        0  2104.25  L   -6    0     5.5  0.14     0   N70  VSS;
* END OF MODULE TYPE LIBRARY FOR LTX2;
```

FIGURE 9.84b. LTX source code description for all of ALU

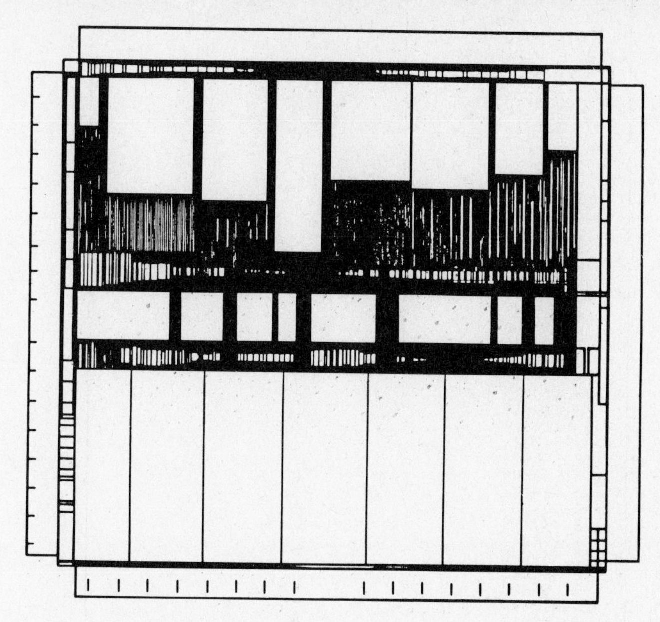

FIGURE 9.85. Initial inter-block layout of CPU

9.85 shows the end result of the first pass at assembling the chip. These are examples of both butted connections and wiring channel connections. An important output of LTX2 at this point is a node-by-node list of the capacitance of the wiring that it has laid out. Such a list is shown in Fig. 9.86. This is very useful for simulating chip timing and for pointing out potential timing problems very early in the design process.

Information from this layout points out problem areas in the design, such as wasted space, excessively long wires, and nodes that are so heavily loaded that timing problems are likely. Now functions can be moved from one block to another to optimize the design. This was done for several iterations until a layout was obtained that used the chip area efficiently and was acceptable for timing design. This version of the layout is shown in Fig. 9.87. It has 30 blocks in addition to the blocks in the I/O frame. The improvement in area utilization as a result of this interactive process is 50 percent as measured by packing density.

As the interblock layout was being optimized, the clock distribution and power supply distribution strategic were also being determined using a technique described by M. Shoji [Shoj82]. It is important to do this part of the design early so that when the design of individual blocks begins, the position of all clock and power supply connections is known.

The clocks must be distributed in such a way that the clock skew from one part of the chip to another is minimized. This requires

```
1    *OPTION: ROUTCAP: CHIP.M32A ;
2    CHIP.M32A.TIO@.LCK23 = 9.242669
3    CHIP.M32A.LMCK3 = 6.517823
4    CHIP.M32A.HSCK1 = 6.217084
5    CHIP.M32A.HFCL34 = 5.027742
6    CHIP.M32A.LSCK1 = 4.555871
7    CHIP.M32A.HMCL23 = 4.519038
8    CHIP.M32A.HMCK3 = 4.076818
9    CHIP.M32A.LMSRST = 3.531932
10   CHIP.M32A.LMSHDT = 3.394909
11   CHIP.M32A.HMSHDT = 3.263511
12   CHIP.M32A.HFCK4 = 3.261212
13   CHIP.M32A.LMALA1 = 3.241445
14   CHIP.M32A.HMAK1L = 3.197358
15   CHIP.M32A.HMCK23 = 2.965212
16   CHIP.M32A.HMFLT2 = 2.928152
17   CHIP.M32A.HMPLD0 = 2.881644
18   CHIP.M32A.HMDTP1 = 2.831347
19   CHIP.M32A.HMDSDQ = 2.805450
20   CHIP.M32A.HMDTP0 = 2.772406
21   CHIP.M32A.HMADML = 2.707605
22   VDD = 2.685592
23   CHIP.M32A.HMCMD5 = 2.673723
24   CHIP.M32A.HMCMD1 = 2.619017
25   CHIP.M32A.HMSSR = 2.592901
26   CHIP.M32A.HMC32 = 2.559732
27   CHIP.M32A.LMTB01 = 2.550672
28   CHIP.M32A.HMFCD0 = 2.536238
29   CHIP.M32A.HMCMD4 = 2.504722
30   CHIP.M32A.LMTC01 = 2.501904
31   CHIP.M32A.HMCMD3 = 2.469775
32   CHIP.M32A.HMLDB2 = 2.452347
33   CHIP.M32A.HMINAD = 2.421526
34   CHIP.M32A.HMLETR = 2.409682
35   CHIP.M32A.LMALON = 2.396606
36   CHIP.M32A.HMCPD = 2.387244
37   CHIP.M32A.HMFCD1 = 2.378612
38   CHIP.M32A.HMCMD2 = 2.338278
39   CHIP.M32A.HMLATQ = 2.319883
40   CHIP.M32A.HMRCB = 2.315299
41   CHIP.M32A.HMCMD0 = 2.307493
42   CHIP.M32A.HMFCD4 = 2.290522
43   CHIP.M32A.HMSD = 2.208425
44   CHIP.M32A.HMFCD3 = 2.195796
45   CHIP.M32A.HMILB6 = 2.175265
46   CHIP.M32A.HMILB7 = 2.173387
47   CHIP.M32A.LMLA01 = 2.173270
48   CHIP.M32A.HMBSWR = 2.168253
49   CHIP.M32A.HMTA01 = 2.154186
50   CHIP.M32A.HMFCD5 = 2.150144
51   CHIP.M32A.LSHAD1 = 2.137809
52   CHIP.M32A.HMILB5 = 2.127848
53   CHIP.M32A.HMILB4 = 2.126204
54   CHIP.M32A.HMFCD2 = 2.125657
55   CHIP.M32A.HMILB3 = 2.124949
56   CHIP.M32A.HMILB2 = 2.116951
```

FIGURE 9.86. Node-by-node list of the capacitance of wiring done by LTX2

that the series resistance in the various paths the clock signals follow must be equal. In addition, the driver size must be tuned to the capacitive load on each clock. One difficulty in achieving this is because there are many clocks used on the chip. The large number of clocks on the chip results from the specialized needs of the various circuits used to implement it. For example, the PLAs form a vital part of the control section and greatly reduce manual layout and timing efforts; however, they require several clocks. The OR plane of the PLA must be precharged during phase 1 of the 4-phase timing and is discharged for the remaining three phases. The AND plane of the PLA must be precharged until the address is valid, so its precharge occurs during phase 2, with discharge extending through phases 3, 4, and 1. The latch at the output of the PLA locks in the AND plane after it is valid at the beginning of phase 4, so a third clock is needed for the latches. Similarly, domino CMOS [KrLL80] (Chapter 5) circuits require precharge and discharge at carefully chosen parts of the cycle, which are not the same for all functions

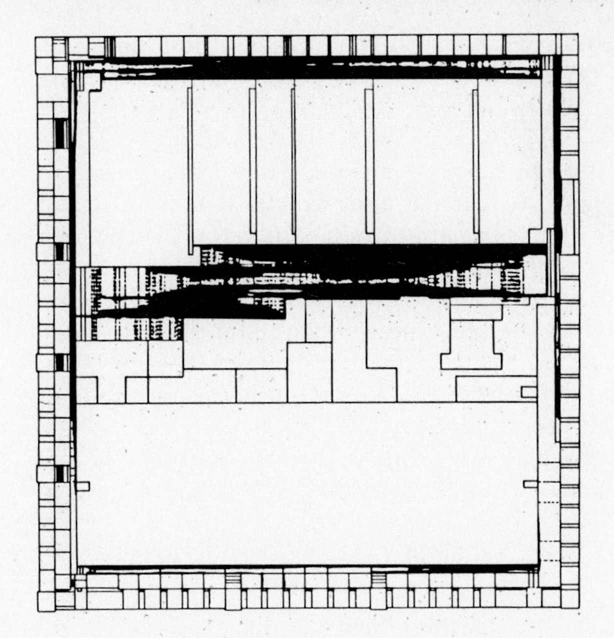

FIGURE 9.87. Optimized interblock layout of CPU

that use domino circuits. Several other clocks are needed to satisfy the needs of the I/O circuits. In order to insure optimum interface with memories and other peripheral chips, the switching of outputs and inputs must be set to minimize time required to access these chips, even if it means creating additional chip clocks. The flexibility in the creation of clock signals is quite important in maximizing chip performance. The time that elapses from when a signal is valid to when it is gated on to its next use is always less than one-quarter of a machine cycle and usually can be made a small fraction of that. This results in efficient use of the circuits' capability. All together there are fourteen different clock signals needed by various latching, precharging, and I/O circuits.

Routing fourteen different clocks with minimum skew would be very cumbersome and consume considerable wiring channel area. Instead, a distributed clock scheme was used. Only two clocks were distributed to the entire chip. A diagram of these is shown in Fig. 9.88. One is high for the fourth and first phases of the cycle. The

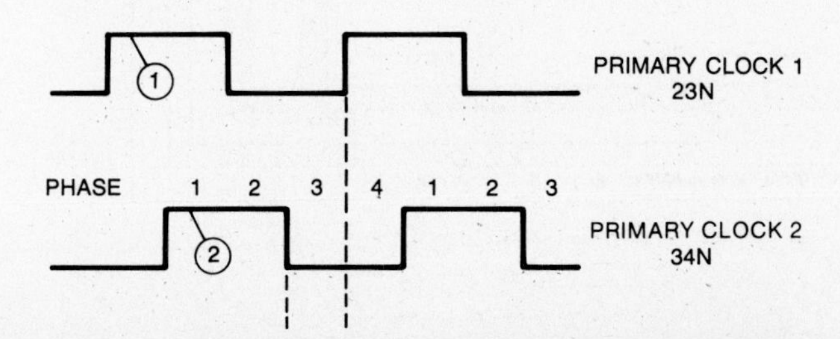

FIGURE 9.88. Diagram of the two primary clocks

other is high for the first two phases of the cycle. Using these two signals a clock can be generated locally, which is high for any combination of cycle phases.

The distribution of these two clocks is shown in Fig. 9.89. The two primary clocks start in the upper right. They are laid out so that every part of the chip is reached after exactly two crossings of the power supply. The clocks use aluminum conductors everywhere except where they cross the power supplies. For these crossings, tantalum silicide is used. Since the resistance of tantalum silicide is far greater than that of the aluminum, equalizing the resistance of the silicide crossunders assures equal resistance in each clock path. To minimize the delay in these clocks, buffers were provided immediately after each crossunder.

The power supplies also were laid out very carefully at this early stage of the design. Fig. 9.89 shows how V_{DD} and V_{SS} were

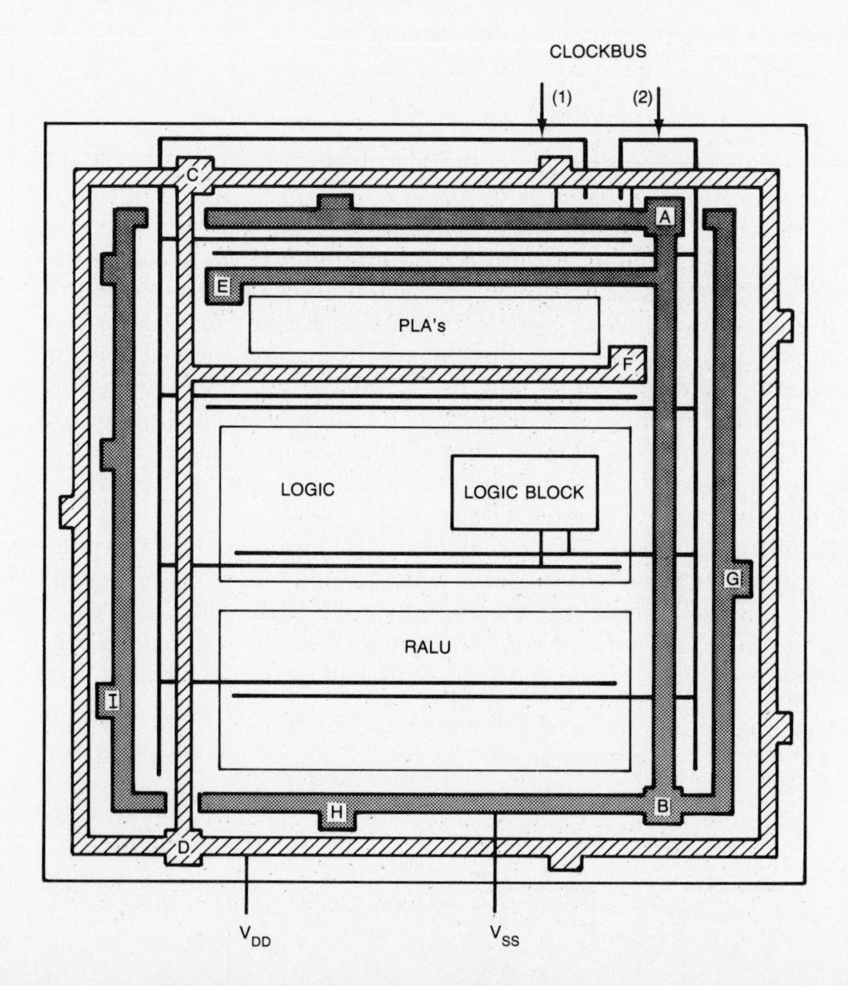

FIGURE 9.89. Clock and power supply distribution schemes that eliminate measurable skew

routed. The critical objective in arranging the power supplies is to insure that power bus voltage noise spikes are small enough so that they have no significant effect on chip operation. This is difficult to achieve because of the numerous sources of current spikes both on and off the chip. For example, the PLAs were originally designed to precharge during the first quarter of each cycle. Simulations showed a voltage spike of over two volts as a result of the precharge. Several steps were taken to reduce this. First, a power pad was added at the ends of the PLA power supply busses. These are shown as points E and F in Fig. 9.89. This reduced the spikes to about one volt, which was still excessive. Finally, the beginning of the precharge of the two parts of the PLA (the decoder (OR plane) and the ROM (AND plane)) were separated in time. The decoder was precharged in the first quarter cycle, while the beginning of the ROM precharge was moved to the second quarter cycle. This made the voltage noise spikes acceptably small. Fig. 9.90 shows the noise spikes on both V_{SS} and V_{DD} for all three stages of the design. Similar design work was carried out to suppress noise spikes on other parts of the chip. In the frame of the chip, I/O switching is the principal source of noise. This noise was suppressed through the use of numerous power supply pads. Also, the output drives were designed to have only enough drive capability to meet external requirements. Excess drive capability tends to make noise problems worse and must be avoided. In the final power supply scheme, there were nine V_{SS} pads and eight V_{DD} pads. The resulting noise is illustrated in Fig. 9.91. All power lines have acceptable simulated noise levels for both internal precharge and external I/O switching. Reaching this point required three months from the initial statement of system requirements.

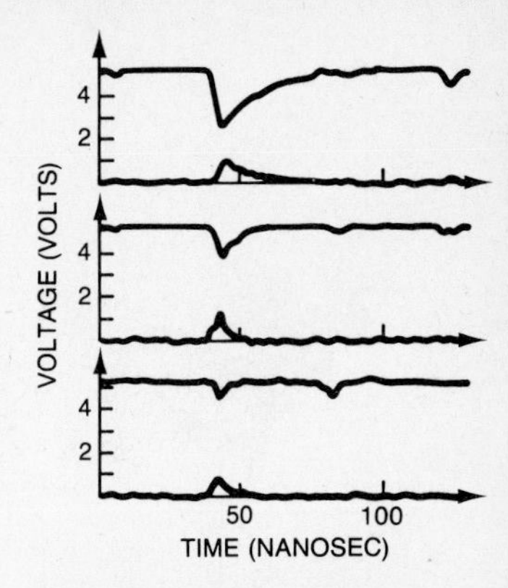

FIGURE 9.90. Noise on the power supplies for PLAs at three stages of the design

From this point in the design process, there are two more stages leading to the final layout. First is initial logic design, layout, and timing analysis. Second is optimization. In the first stage, logic for each block is designed in parallel. Layout and timing analysis on each block also begin. Where necessary, special circuit techniques such as domino CMOS are used to maximize performance. An example of the use of domino CMOS was in the Arithmetic Logic Unit (ALU).

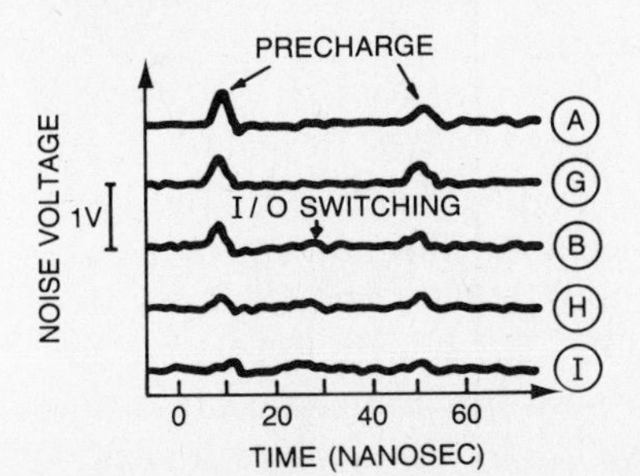

FIGURE 9.91. Noise voltage simulations at various points on the chip in the final design

The ALU is well suited to use of domino circuits because the highest possible speed is essential for good CPU performance, and because the timing of input and output signals to the ALU is very well defined and controlled. One limitation of this circuit technique is that all of the gates are non-inverting. This may seem serious, since an XOR cannot be made and an XOR is essential to do arithmetic. However, circuits that need the XOR function can be arranged so the XOR is the last gate in the chain. Then all of the rest of the circuit can be made with domino logic. The output of the last domino gate can then be used to drive the necessary XOR gate. The timing analysis is aided by the availability of the list of interblock wiring capacitances obtained from the LTX layout done previously. During the early part of this timing analysis, some repartitioning of the blocks occurred to ease timing problems. This becomes necessary when a critical path that is allocated only a fraction of a cycle is found to use gates in several different parts of these chips. Use of the wiring channel more than once for a critical path can lead to excessive delays. When a signal is transferred from Block A to Block B, combined with a second signal, transferred to Block C, and then further combined, the delay will not be minimized. To reduce the delay, all gates could be moved to the same block, thereby eliminating two crossings of the wiring channel.

The logic design is done block by block and is verified using the C level simulator as a software test bed. The layout is done using exactly the same hierarchy as the logic design and is begun even before the logic verification is complete. The layout is also verified block by block. Test vectors for verifying each block layout are generated from the block logic description. Then the layout is converted into a transistor level circuit diagram automatically. Next, using MOTIS [ABKN81], the correspondence between the layout and the logic for each block can be verified. At the same time, design rule errors in the layout can be checked for and eliminated, and a computer readable description of transistor sizes and parasitics is automatically produced. This is used to analyze timing with MOTIS. The initial analysis showed approximately 400 timing paths with worst-case speeds between 3.0 MHz and 7.2 MHz at T_A = 85°C and V_{DD} = 4.75 volts. In addition, there were some instructions that did not work properly in the simulator.

This set the stage for the optimization process. All 400 timing paths were examined and simulated with ADVICE [Nage80]. Transistor-size adjustments were made, as well as delay reducing logic changes. An example of the care with which these adjustments are made is shown in Fig. 9.92 (from [Shoj82]). This figure shows two alternative layouts of a domino CMOS gate whose speed we are trying to maximize. Layout I contains uniform transistor sizes and is the arrangement usually found in MOS circuits. However, where

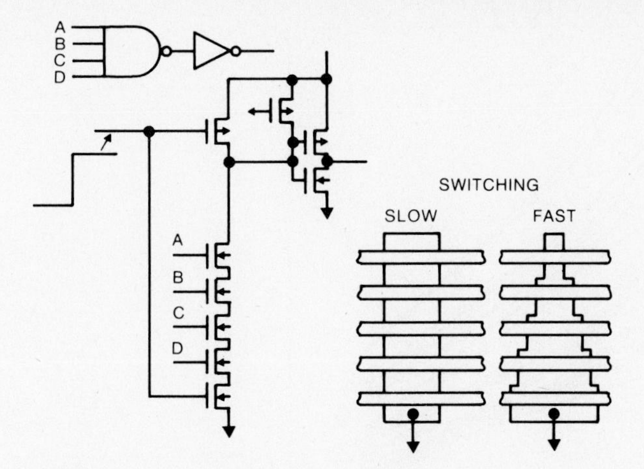

FIGURE 9.92. Domino CMOS circuit and speed up method

maximum speed is needed, simulations and measurements show that Layout II, with graded transistors, can actually be faster, even if the largest transistor in Layout II is equal to the size of the transistors in Layout I. This happens because the parasitic capacitances within the four transistors in series result in a larger capacitive load, and therefore a larger current flow in the transistor nearest to ground. Meanwhile, the transistor nearest the output has a relatively light load. It is therefore advantageous to make the size of the transistor nearest the output smaller than the transistor nearest the ground and to grade those in between. Use of this technique can reduce a gate delay by as much as 30 percent, depending on the type of gate and the loading conditions. At the end of the timing optimization process, all 400 paths had been speeded up to at least 7.2 MHz for worst-case processing, and $T_A = 85°C$, $V_{DD} = 4.75V$. Fig. 9.93 shows the distribution predicted for various paths that had been simulated using worst case, best case, and medium process corners.

The local clocks were also optimized at this time. The delay associated with the most heavily loaded clock was determined by simulation to be approximately 15 nsec. Therefore, our objective was to make all local clocks, even the most lightly loaded, have a delay as close to 15 nsec as possible. Fig. 9.94 [Shoj82] gives an example of how this was accomplished. For a light load, two extra stages of delay are inserted to slow the clock down, and transistor sizes are chosen near to minimum geometry. For a heavy load, large transistors are used and the number of gate delays is minimized. In this way, both circuits have close to a 15 nsec delay. The actual simulated delays are shown in Fig. 9.95 [Shoj82]. No clock was more than 3.5 nsec from the desired 15 nsec delay. At the same time, logic changes were made to correct all functional shortcomings.

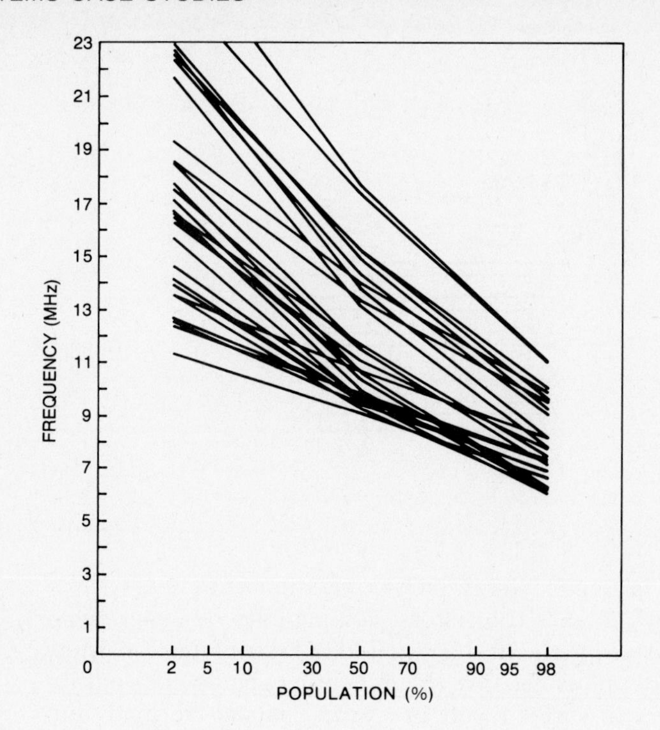

FIGURE 9.93. Predicted yield vs. frequency, determined from critical path studies

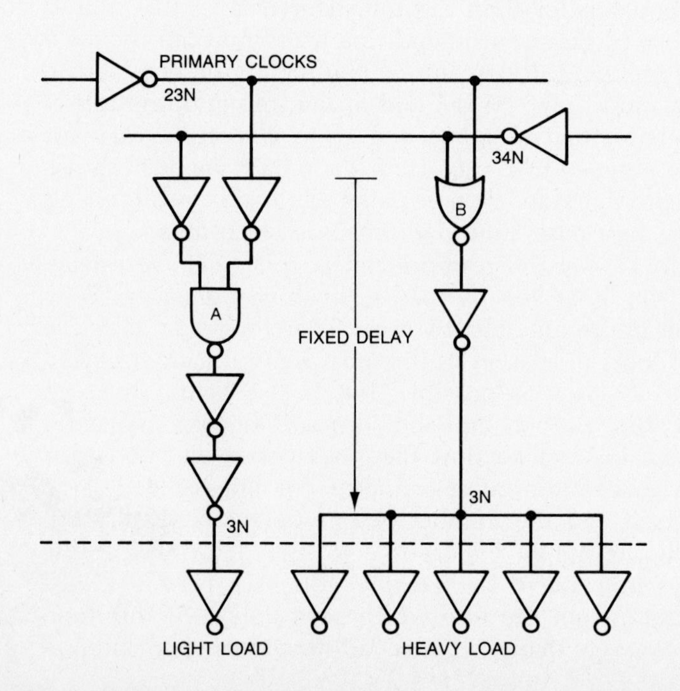

FIGURE 9.94. Local clock decoding scheme

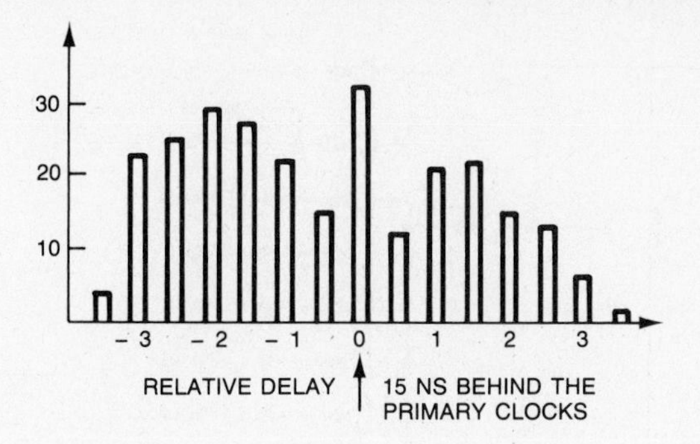

FIGURE 9.95. Simulated local clock delay for all clock edges on the chip

The timing and functional corrections led to 225 change requests. These were implemented by changes to block layouts. Again a highly parallel approach was used with many blocks being corrected simultaneously. To avoid changes to the interblock wiring, these changes were implemented wherever possible by making corrections entirely within the block affected. The gate matrix layout technique was very well suited to this kind of approach and substantial changes were made to blocks without any adjustments to the block size, shape, or I/O locations. An example taken from [LoLa80] (Fig. 9.96) shows a logic block before and after the implementation of changes. Fig. 9.97 shows the corresponding layouts.

The use of the hierarchical methodology permitted us to converge quickly on an optimum design. This happened because the overall chip structure and interblock wiring were known and virtually unchanging during the optimization of the blocks. A top-down implementation, bottom-up implementation approach also permitted accurate predictions and close control of chip area. Fig. 9.98 shows area estimates vs. time during the twelve month interval described above. Note that area was very stable once the number of transistors was determined. In addition, the area changed only 2 percent during the final assembly process. The actual die size is 99 mm^2 in 2.5 μm CMOS rules. A photograph of the chip is shown in Fig. 9.99.

9.6.3 Technology updatability and layout verification

The use of technology updatable layout techniques is also vital to the design of complex state-of-the-art chips. The basic feature of a technology updatable layout technique is that all layout work is done with a symbolic description that contains no design rule information. The symbolic description specifies only topology. A separate design rule file is then used with the symbolic description to

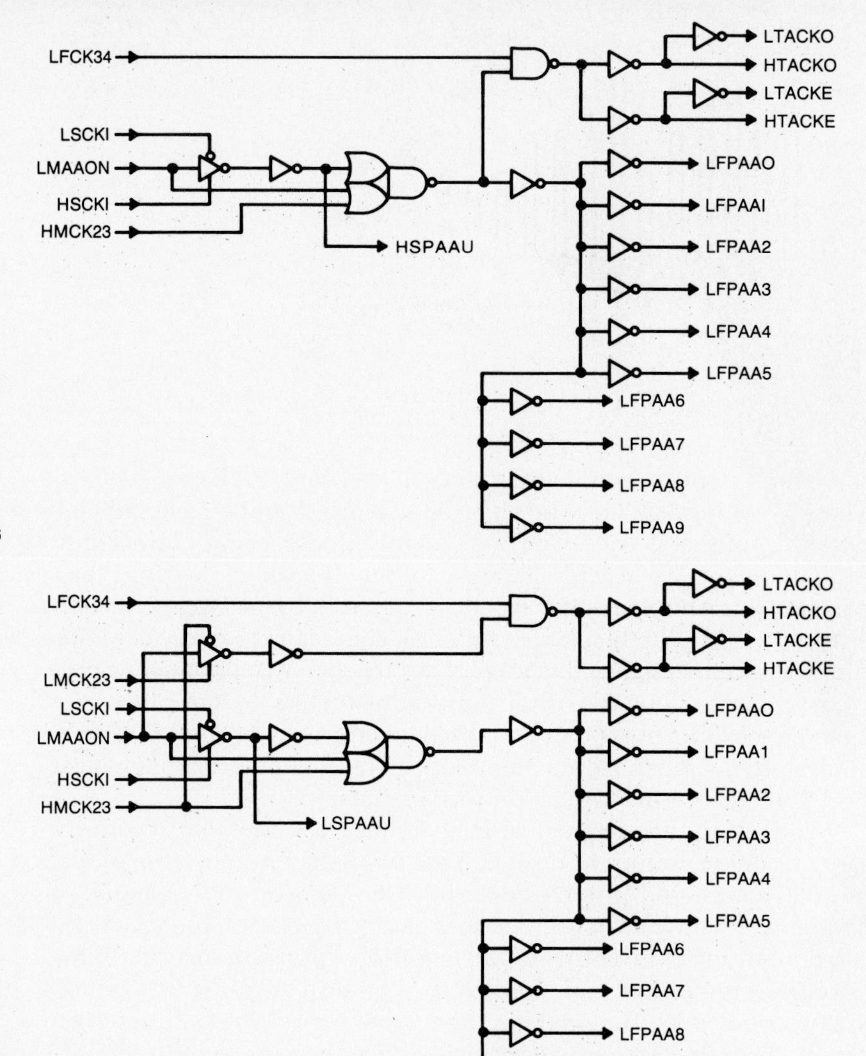

FIGURE 9.96a. Logic diagram of block before the implementation of changes

FIGURE 9.96b. Logic diagram of block after the implementation of changes

produce mask geometries. The advantage of this is that when design rules change, it is only necessary to change the design rule files to produce new mask geometries. All of the symbolic description can be reused in the new design rules without change.

Several technology updatable layout techniques have been developed at Bell Laboratories and different parts of the layout used different techniques. The I/O used technology updatable polycells [LeCJ81]. These were chosen because I/O must be highly standardized to insure that there will be no problems with latch-up or static

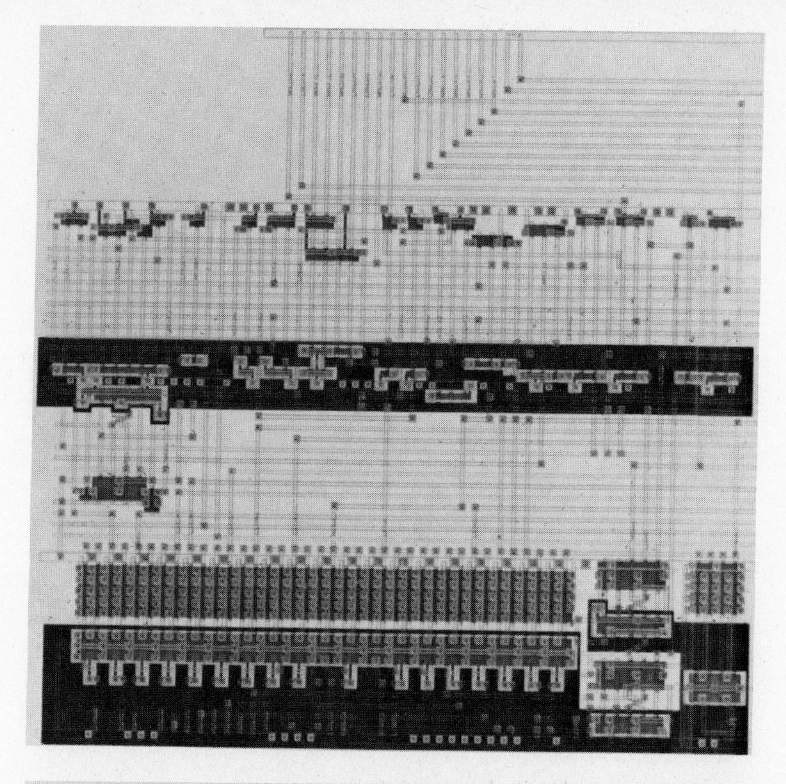

FIGURE 9.97a. Layout of the circuit shown in Fig. 9.96a

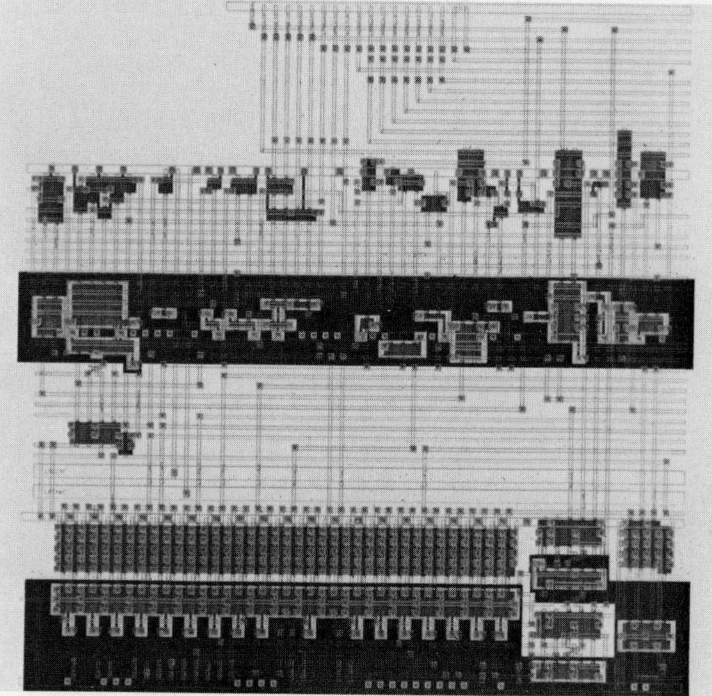

FIGURE 9.97b. Layout of the circuit shown in Fig. 9.96b

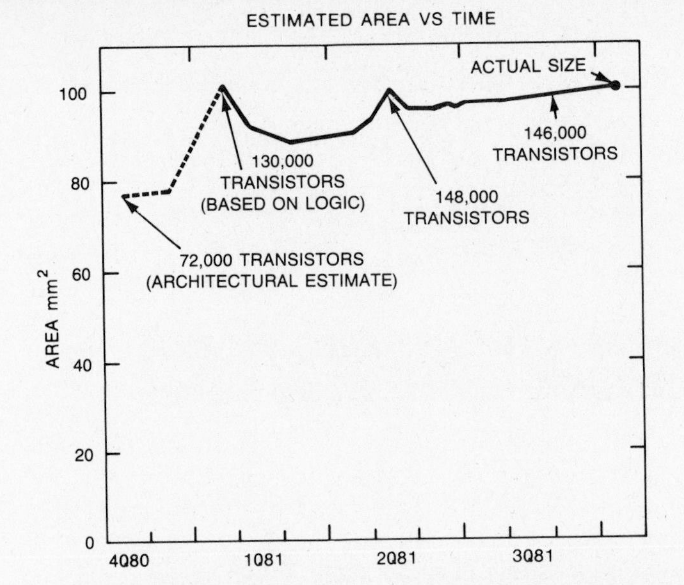

FIGURE 9.98. Area estimate vs. time during design of chip

FIGURE 9.99. A microphotograph of the WE-32000 chip. The die size in 2.5μm rules is 99mm^2.

discharge and to accommodate the wire bond. The PLAs are laid out by a program [LaSh82] that uses a truth table and a list of inputs and outputs as source code. The interblock wires, including clocks and power supplies, are laid out by LTX. Here again the symbolic description contains no design rule information. Only the required connections are used in the description.

For the remaining parts of the chip, gate matrix was used for layout. This technique is technology updatable like the other three layout styles. It has the further advantage of being quite flexible in terms of circuits that can be implemented and transistor sizes that can be used. This made it very compatible with the optimization techniques described earlier. An example of how gate matrix achieves technology updatability is shown in Figs. 9.100 through 9.102. A circuit diagram and symbolic layout for a small section of the arithmetic logic unit are shown in Figs. 9.100a and 9.100b.

Conversion of the symbolic layout in Fig. 9.100b into actual mask geometries requires use of a technology file based on current design rules. This is obtained by defining each symbol for the particular design rules to be used. Fig. 9.101 shows the actual geometries called for by the symbol "N" in both 3.5 μm and 2.5 μm design rules. A photographic shrink would not be effective since the metal covering the contact window is actually larger in the 2.5 μm technology. Also, the spacing between contact window and polysilicon gate is unchanged. However, because of the use of an updatable layout technique the same symbolic source file shown in Fig. 9.100b can be used to produce layouts in both technologies. These are shown in Fig. 9.102. Similarly the layout could be produced in future smaller technologies without changing the source description. This insures that the chip can continue to use state-of-the-art processing in the future.

These four layout techniques are also quite compatible with automatic verification procedures as shown in Fig. 9.103. In this figure, a solid line indicates an automatic step, while a broken line is manual. The same source description that defines PLA contents and interblock wiring connections for the layout is used to define these things for the functional simulator. This insures that for these parts of the chip, the circuit that is laid out is identical to the circuit verified by the functional simulator.

For the I/O cells and gate matrix generation, there is a manual step involved in creating a symbolic character description for layout purposes. However, the mask geometries are verified automatically by use of a verification program that determines the circuitry represented by the layout geometries and runs test vectors on the circuit block to verify that it is the same as the circuit being used in the functional simulator. In this way, all geometries that are put on the

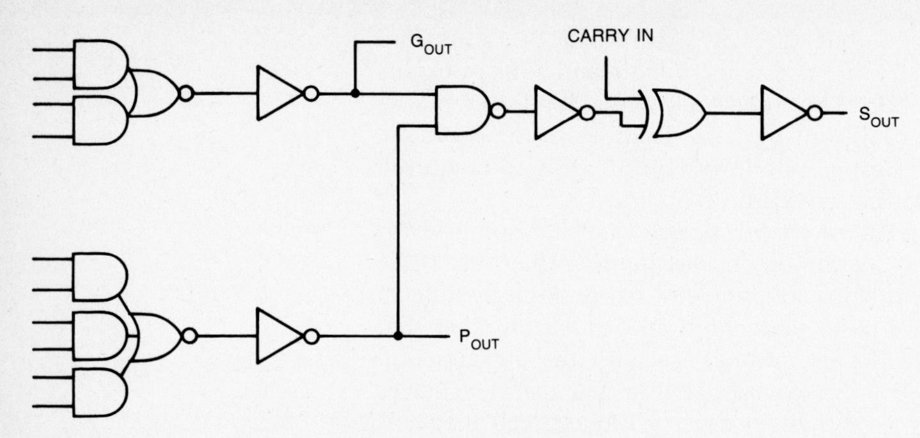

FIGURE 9.100a. Circuit diagram for portion of ALU

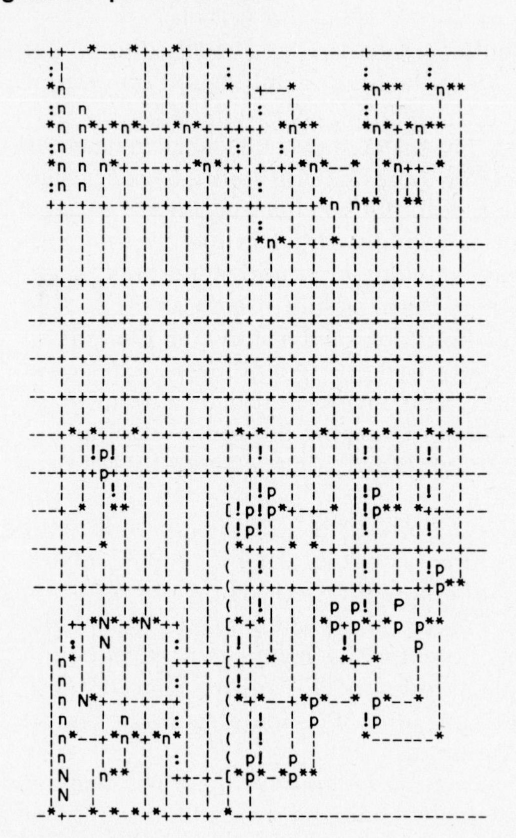

FIGURE 9.100b. Gate matrix code for portion of ALU

chip are verified to give the circuit called for by the functional simulator. Thus the only possibility of errors would be from shortcomings in the verification tools or methodology. This means that while there will be some errors the first time this scenario is followed, as each gap in the verification process is found, it can be eliminated

DEFINITION OF CLUMP "N"
FOR TWO SETS OF DESIGN RULES

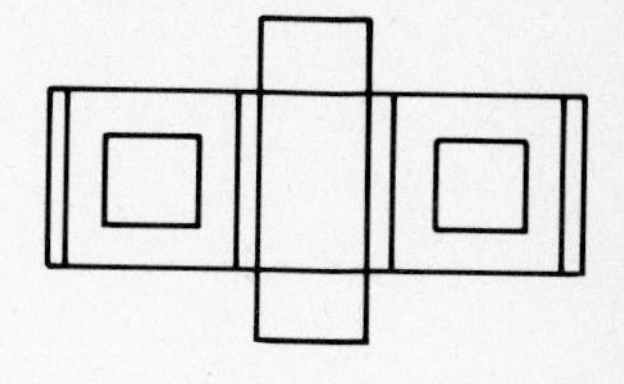

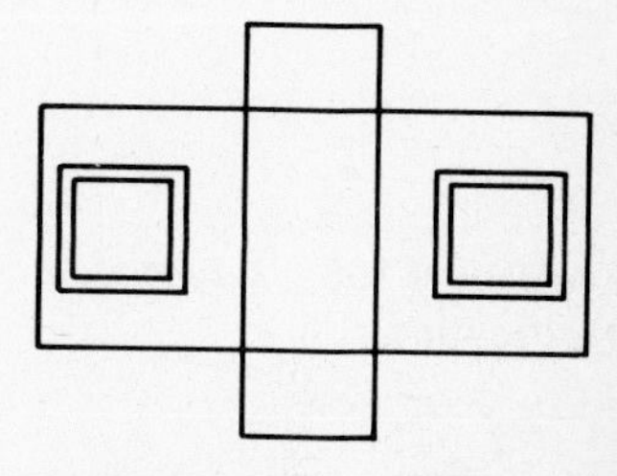

FIGURE 9.101. Mask geometries for symbol "N" in 3.5μm and 2.5μm design rules

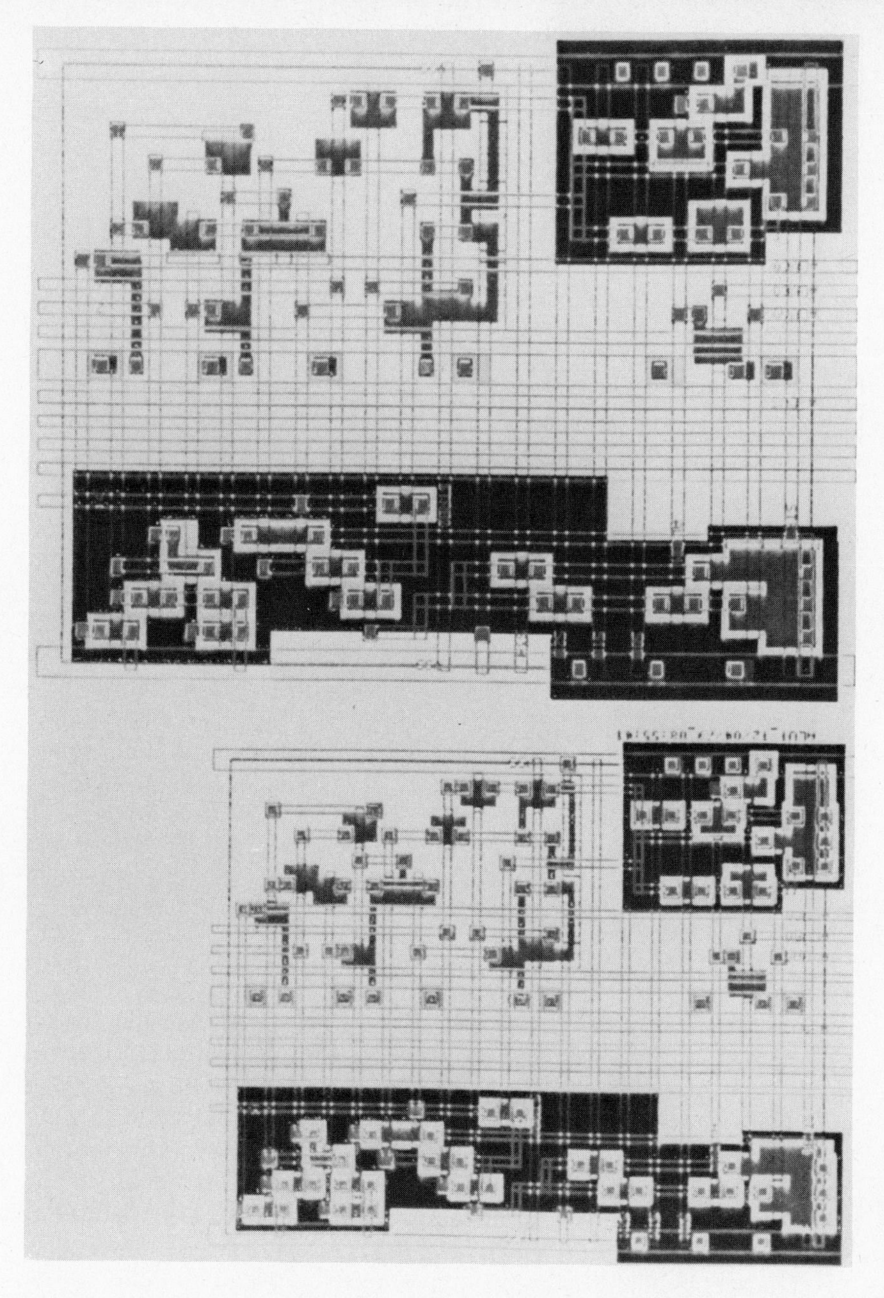

FIGURE 9.102. Mask geometries for circuit of Fig. 9.100b in 3.5 μm and 2.5 μm design rules

for all future chips. This is in sharp contrast to manual verification techniques, where the error rate can be expected to remain unchanged from chip to chip.

The timing also undergoes a final verification at this time. The block verification program also produces code readable by MOTIS for use in verifying timing. The LTX program produces interblock

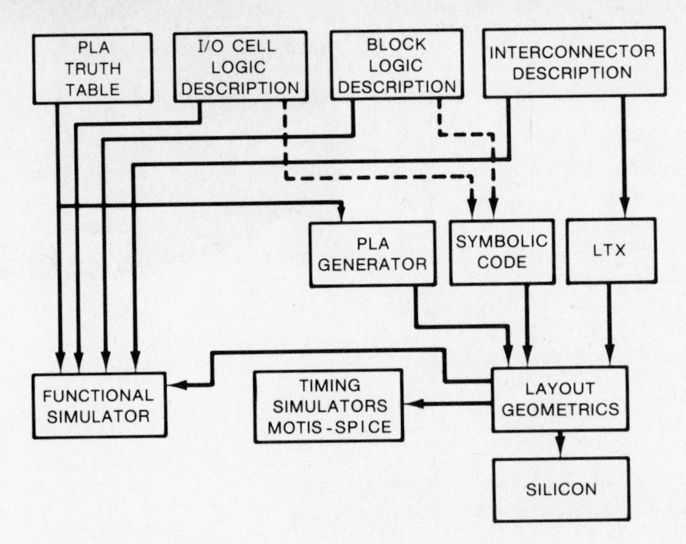

FIGURE 9.103. Layout verification

capacitances, which are also used by MOTIS. The PLA speed is verified separately by an automatically produced ADVICE [Nage80] run. In addition to these automatic timing verifications, numerous ADVICE runs are coded manually to check on critical paths pointed out by the MOTIS run. As a result, when the design is committed to silicon, a minimum of problems with layout errors and timing are expected.

9.6.4 Results

Considering the complexity of the chip, the verification procedure was quite effective at eliminating layout errors. There were three types of errors found, all involving problems not checked for by the verification. These all involved the unwanted intersection of mask levels near the interface between two blocks. The verification algorithms have since been improved to find these errors as well. Nevertheless, the error rate was sufficiently low that working models were running code two months after the initial mask date, so the total time from initial system requirements to a working computer was 14 months.

The accuracy of the timing simulations was even more impressive. Initial models with typical processing ran at 7.2 MHz for an ambient temperature of 85°C and a V_{DD} of 4.75 volts as expected. No tweaking of any timing path was necessary to achieve the predicted speed. Moreover, internal clock skew and power supply bounce were both within expectation. This shows the value of incorporating the timing design into the logic and layout process right from the beginning.

The use of CMOS technology kept the power dissipation down to 630 mW at 7.2 MHz.

The computer in which the chip is used was operated at 6.2 MHz. Its performance at this frequency was tested with seven benchmark programs and the result was a throughput equal to 75 percent of that of a high end midicomputer.

9.6.5 Conclusion

A hierarchical design methodology for complex VLSI chips has been developed and used on a 146,000 transistor 32-bit single chip microprocessor. When used in conjunction with technology updatable layout techniques and CAD tools suited to complete verification of functionality, timing, and layout, it was possible to produce full speed chips and have them running code in a midicomputer 14 months after initial system requirements were formulated.

Despite the success, there is clearly considerable room for improvement in both the CAD tools and in the methodology. The design of complex VLSI will continue to be a fertile field for the development of new design aids. The need for improvement is most obvious in the intensively manual parts of the job. For example, the initial gate matrix layout is drawn manually as a stick diagram with pencil and paper. The interactive symbolic layout methods described in Chapter 7 will be of advantage in this task in future designs. A completely integrated work environment for each designer will also greatly aid communication of design goals and progress. All of the ADVICE code was generated manually at considerable cost, and all transistor size adjustments were arrived at and made by manual means. As each of these bottlenecks is analyzed and suitable algorithms generated to automate each process, the design process can become more efficient while still being used at all times to generate commercial product.

9.7 Conclusion

This book has progressed from dealing with idealized CMOS transistors to a case study of a CMOS microprocessor that uses 146,000 transistors. It is the hope of the authors that the material in the intervening chapters will add to the body of knowledge relating to CMOS design and spur the generation of new VLSI architectures and the advanced tools to create these designs.

APPENDIX: COMPUTING NOISE MARGINS FOR CMOS AND NMOS INVERTERS

K. S. TRIVEDI*
N. S. VASANTHAVADA†

* Department of Computer Science, Duke University.
† Department of Electrical Engineering, Duke University.

Introduction

This appendix compares the CMOS inverter and NMOS inverter noise margins.

CMOS inverter

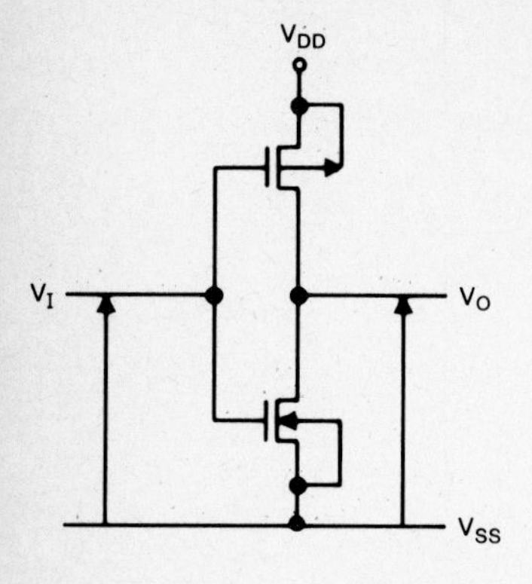

FIGURE A.1. CMOS inverter schematic

Fig. A.1 shows a typical CMOS inverter. The transfer characteristic of the inverter and the definitions of voltage levels V_{IL}, V_{OL}, V_{IH}, and V_{OH} are shown in Fig. A.2.

We define the noise margins as

$$NM_L = V_{IL} - V_{OL}$$

$$NM_H = V_{OH} - V_{IH}.$$

To derive expressions for V_{IL} and V_{IH}, we set the derivative of the output voltage, V_O, with respect to V_I equal to -1 in the appropriate regions of operation (B and D respectively) and solve for V_I.

Let $x = \beta_n/\beta_p$. Then:

$$\frac{d}{dV_I}[(V_I - V_{tp}) + \{(V_I - V_{tp})^2 - 2(V_I - V_{DD}/2 - V_{tp})\,V_{DD} - x(V_I - V_{tn})^2\}^{1/2}] = -1 \text{ at } V_I = V_{IL}.$$

After differentiating, squaring, and collecting terms, we get

$$V_{IL}^2[3 - 2x - x^2] + V_{IL}[(6 + 2x)(-V_{tp} - V_{DD} + xV_{tn})] + [3V_{tp}^2 + 3V_{DD}^2 + V_{tn}^2(-4x - x^2) + 6V_{tp}V_{DD} + 2xV_{tn}V_{tp} + 2xV_{tn}V_{DD}] = 0.$$

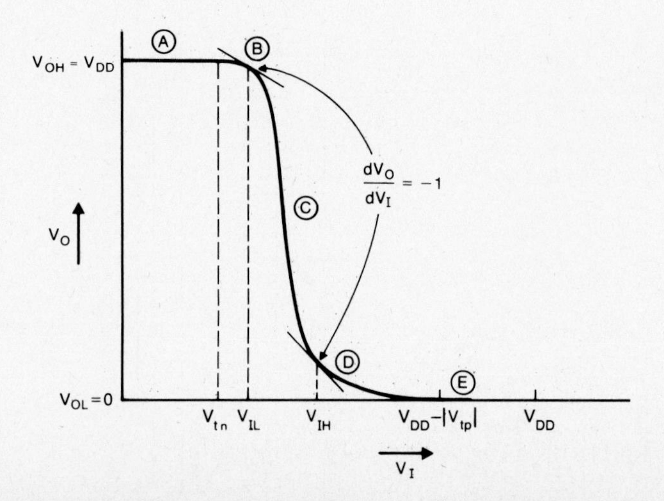

FIGURE A.2. CMOS inverter transfer characteristic

For $x = 1$, upon simplification of the above expression and solving for V_{IL}, we get (noting that V_{tp} is negative)

$$V_{IL} = \frac{3V_{DD} - 3|V_{tp}| + 5V_{tn}}{8}.$$

Note that since V_{OL} is zero, $NM_L = V_{IL}$. Substituting typical values: $V_{DD} =$ 5 Volts, $V_{tn} = |V_{tp}| = 1$V, the quadratic reduces to $V_{IL}^2(3 - 2x - x^2) + V_{IL}(-24 - 2x + 2x^2) + (48 + 4x - x^2) = 0$. Solving this quadratic yields

$$NM_L = V_{IL} = \begin{cases} \dfrac{x - 4 + 6\sqrt{\dfrac{x}{x+3}}}{x - 1}, & \text{if } x > 0,\ x \neq 1 \\[2ex] 17/8, & \text{if } x = 1. \end{cases}$$

[Note that the root of the quadratic equation that is outside the permissible range (0,5) is discarded.]

In a similar fashion, we obtain the quadratic equation for V_{IH} as

$$\begin{aligned} V_{IH}^2[3 - 2/x - 1/x^2] + V_{IH}\Bigg[&-8V_{tn} + \frac{8V_{DD}}{x} + \frac{8V_{tp}}{x} \\ &- 2(1 - 1/x)\left(\frac{V_{DD}}{x} + \frac{V_{tp}}{x} - V_{tn}\right)\Bigg] \\ &+ \Bigg[4V_{tn}^2 - (4/x)(V_{DD}^2 + V_{tp}^2 + 2V_{DD}V_{tp}) \\ &- \left(\frac{V_{DD}}{x} + \frac{V_{tp}}{x} - V_{tn}\right)^2\Bigg] = 0. \end{aligned}$$

For the special case, $x = 1$,

$$V_{IH} = \frac{5V_{DD} - 5|V_{tp}| + 3V_{tn}}{8}$$

and hence

$$NM_H = V_{DD} - V_{IH} = \frac{3V_{DD} + 5|V_{tp}| - 3V_{tn}}{8}, \qquad \text{if } x = 1.$$

For the chosen values of V_{DD}, V_{tn}, and V_{tp}, the above quadratic reduces to

$$V_{IH}^2\left(3 - \frac{2}{x} - \frac{1}{x^2}\right) + V_{IH}\left(-6 + \frac{22}{x} + \frac{8}{x^2}\right) + \left(3 - \frac{56}{x} - \frac{16}{x^2}\right) = 0,$$

which yields

$$V_{IH} = \begin{cases} \dfrac{4/x - 1 - 6\sqrt{\dfrac{1/x}{1/x + 3}}}{1/x - 1}, & \text{if } x > 0,\ x \neq 1 \\[2ex] 23/8, & \text{if } x = 1. \end{cases}$$

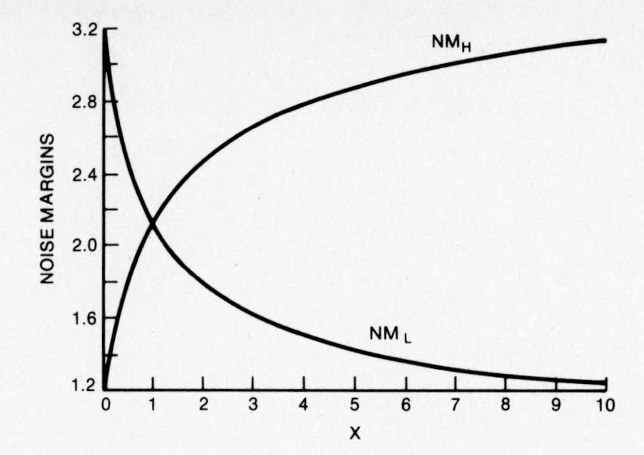

FIGURE A.3. Noise margins vs. x for a CMOS inverter

Hence

$$NM_H = \begin{cases} \dfrac{1/x - 4 + 6\sqrt{\dfrac{1/x}{1/x + 3}}}{1/x - 1}, & \text{if } x > 0,\ x \neq 1 \\ 17/8, & \text{if } x = 1. \end{cases}$$

Fig. A.3 shows the two noise margins NM_L and NM_H as functions of the parameter x. Notice that $1 < NM_L, NM_H < 4$ as x varies in the range (0 ∞).

NMOS inverter with depletion mode pull-up

Figs. A.4 and A.5 show a typical NMOS inverter with depletion load and its transfer characteristic respectively.

Let V_{tpd} and V_{tpu} be the threshold voltages of the pull-down and pull-up transistors, respectively. Also let $x = Z_{pu}/Z_{pd}$.

To compute V_{IL}, we assume that the pull-down is in saturation and the pull-up is in linear region. Then, using first order equations and setting the derivative of V_O with respect to V_I to -1, we get

$$\frac{d}{dV_I}\left[V_{DD} + V_{tpu} + \sqrt{V_{tpu}^2 - x(V_I - V_{tpd})^2}\right] = -1 \text{ at } V_I = V_{IL}.$$

After simplifying, we get

$$V_{IL} = V_{tpd} + \frac{|V_{tpu}|}{\sqrt{x^2 + x}}.$$

For typical values: $V_{tpu} = -4$ V, $V_{tpd} = 1$ V, and $V_{DD} = 5$ V, we have

$$V_{IL} = 1 + \frac{4}{\sqrt{x^2 + x}}.$$

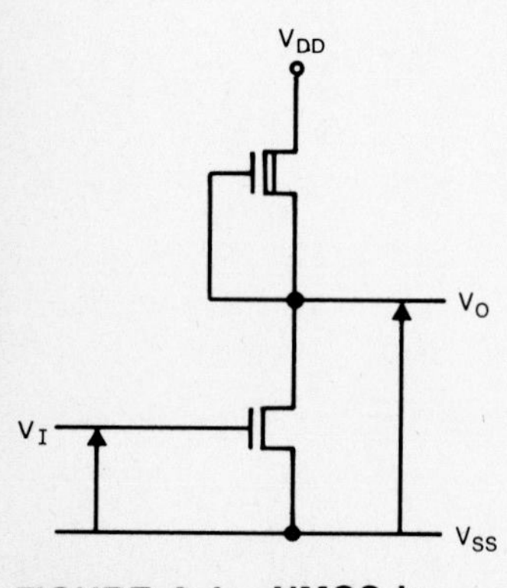

FIGURE A.4. NMOS inverter schematic

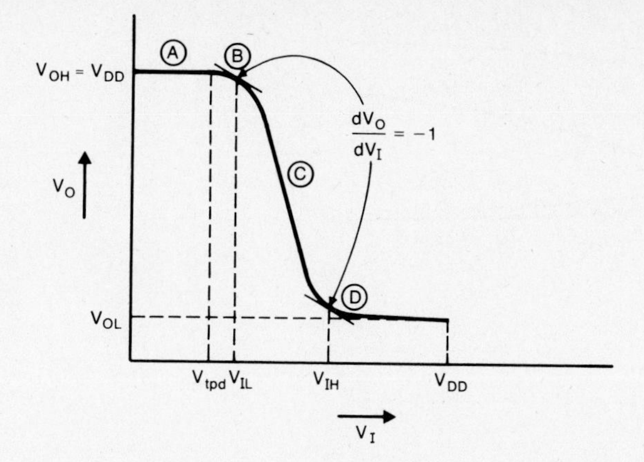

FIGURE A.5. NMOS inverter transfer characteristic

[This value of V_{IL} is slightly greater than the actual value, as body effect is not considered. In actuality, V_{tpu} becomes less negative due to body effect and V_{IL} will be less than the value given by the expression.]

To compute V_{OL}, it is reasonable to assume that the pull-down is in linear region and the pull-up is in saturation. Then, equating the currents in the pull-up and pull-down transistors, and solving for V_{OL}, yields

$$V_{OL} = (V_{DD} - V_{tpd}) - \sqrt{(V_{DD} - V_{tpd})^2 - (1/x)V_{tpu}^2}, \qquad (V_I = V_{DD}).$$

For the chosen values of V_{DD}, V_{tpu}, and V_{tpd}

$$V_{OL} = 4 - 4\sqrt{1 - 1/x}.$$

Hence

$$NM_L = \frac{4}{\sqrt{x^2 + x}} + 4\sqrt{1 - 1/x} - 3.$$

To obtain V_{IH}, we use the same procedure and get

$$V_{IH} = V_{tpd} + 2\frac{|V_{tpu}|}{\sqrt{3x}}.$$

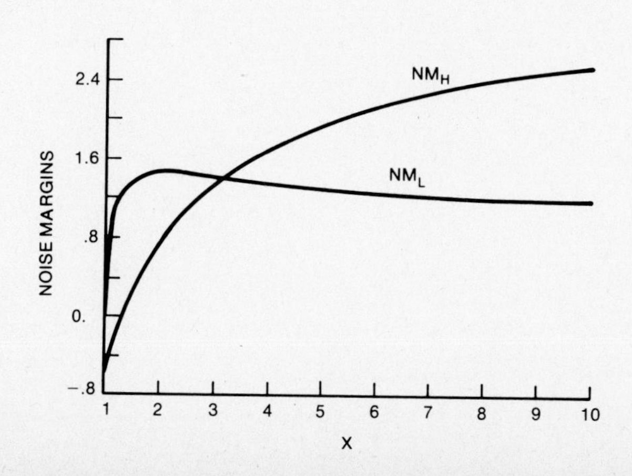

FIGURE A.6. Noise margins vs. x for an NMOS inverter

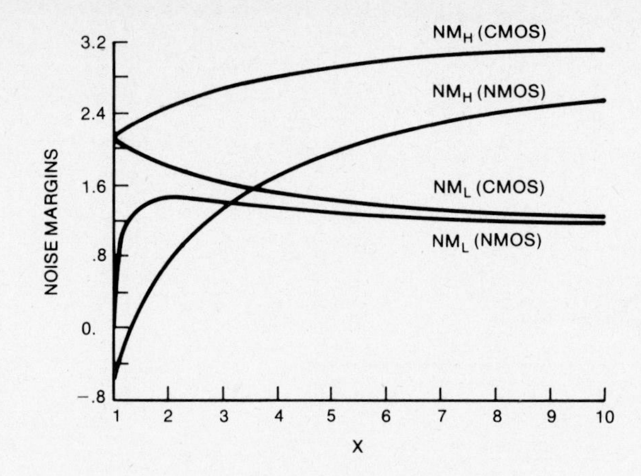

FIGURE A.7. Noise margins vs. x for NMOS and CMOS inverters

For the chosen values of V_{DD}, V_{tpu}, and V_{tpd}

$$NM_H = 4 - 8/\sqrt{3x}.$$

The plot of NM_H and NM_L for the NMOS inverter as functions of the ratio x is shown in Fig. A.6. It is to be noted that the values of NM_H and NM_L for small values of x ($x < 2$) are not very accurate because the assumptions made regarding the regions of operation may not strictly hold for such small values of x. However, no practical NMOS inverter will have $x < 2$ and hence results for these cases do not have great significance. Fig. A.7 shows the noise margins for CMOS and NMOS inverters together on the same plot. This plot indicates the achievable superiority of CMOS for this measure of system performance.

REFERENCES

[ABKN81] V. D. Agrawal, A. K. Bose, P. Kozak, N. N. Nham and E. Pacas-Skewes, "Mixed Mode Simulation in the MOTIS System," *Journal of Digital Systems,* 1981, p. 383.

[AcWe83] B. Ackland and N. Weste, "An Automatic Assembly Tool for Virtual Grid Symbolic Layout," *Proc. VLSI '83,* North Holland Publishing, August 1983, Trondheim, pp. 457–466.

[AOYY83] S. Akiyama, S. Ogawa, M. Yoneda, N. Yoshii and Y. Tervi, "Multilayer CMOS Device Fabrication Laser Recrystallized Silicon Islands," *Technical Digest,* IEDM, December 1983, pp. 352–355.

[Bair77] H. S. Baird, "Fast Algorithms for LSI Artwork Analysis," *Proceedings of the 14th Design Automation Conference,* 1977, pp. 303–311.

[Batc68] K. Batcher, "Sorting Networks and Their Applications," *AFIPS Conf. Proc. 1968 SJCC,* pp. 307–314.

[BaTe80] C. M. Baker and C. J. Terman, "Tools for Verifying Integrated Circuit Designs," *Lambda Magazine (VLSI Design),* 4th quarter 1980, pp. 22–30.

[Bell57] R. E. Bellman, *Dynamic Programming,* Princeton University Press, 1957.

[BGEK83] M. A. Brown, M. J. Gasper, J. W. Eddy and K. D. Kolwicz, "CMOS Cell Arrays — An Alternative to Gate Arrays," *Proceedings of the Custom Integrated Circuit Conference,* May 1983.

[BGKL82] D. E. Blahut, A. K. Goksel, R. H. Krambeck, H-F. S. Law, P. W. Lu, W. F. Miller and H. C. So, "Architecture and Implementation of a 32-Bit Microprocessor with Midicomputer Performance," Spring *COMPCON,* 2/23/82, San Francisco, CA.

[BMSS81] J. Batali, N. Mayle, H. Shrobe, G. Sussman and D. Weise, "The DPL/Daedalus Design Environment," *VLSI 81,* Edinburgh, 1981, pp. 183–192.

[BoWe83] D. Boyer and N. Weste, "Virtual Grid Compaction Using the Most Recent Layers Algorithms," *Proc. IEEE International Conf. on CAD,* September 1983, pp. 92–93.

[BrEw82] R. P. Brent and R. R. Ewin, "Design of an nMOS Parallel Adder," *TR-CS-82-06,* Department of Computer Science, The Australian National University, 1982.

[BrKu82] R. P. Brent and H. T. Kung, "A Regular Layout for Parallel Adders," *IEEE Transactions on Computers,* Vol. C-31, No. 3, March 1982, pp. 260–264.

[BrTF83] H. Brown, C. Tong and G. Foyster, "Palladio: An Exploratory Environment for Circuit Design," *Computer,* Vol. 16, No. 12, December 1983, pp. 41–58.

[Brya81] R. E. Bryant, "MOSSIM: A Switch-Level Simulator for MOS LSI," *Proceedings of the 18th Design Automation Conference,* July 1981, pp. 786–790.

[BuAW84] D. Burr, B. Ackland and N. Weste, "Array Configurations for Dynamic Time Warping," *IEEE Transactions on Acoustics, Speech and Signal Processing,* Vol. ASSP-32, No. 1, February 1984, pp. 119–128.

[Buch80] I. Buchanan, "Modelling and Verification in Structured Integrated Circuit Design," Ph.D. Thesis, University of Edinburgh, Scotland, 1980.

[BuCM83] M. R. Buric, C. Christensen and T. G. Matheson, "PLEX: Automatically Generated Microcomputer Layouts," *Proc. IEEE International Conf. on Computer Design (ICCD '83),* Port Chester, NY, October 1983, pp. 181–184.

[CaMi72] W. N. Carr and J. P. Mize, *MOS/LSI Design and Application,* New York: McGraw-Hill, 1972.

[CCKS77] J. A. Cooper, J. A. Copeland, R. H. Krambeck, D. C. Stanzione and L. C. Thomas, "A CMOS Microprocessor for Telecommunications Applications," *ISSCC Digest of Technical Papers*, p. 138, 1977.

[ChGK75] B. R. Chawla, H. K. Gummel and P. Kozak, "MOTIS — An MOS Timing Simulator," *IEEE Transactions on Circuits and Systems*, Vol. 22, No. 12, December 1975, pp. 901–910.

[Chow83] T. P. Chow, "A Review of Refractory Gates for MOS VLSI," *Technical Digest, International Electron Devices Meeting*, December 1983, pp. 513–517.

[ChVr83] Kuang-Wei Chiang and Zvonko G. Vranesic, "On Fault Detection in CMOS Logic Networks," *Proceedings of the 20th Design Automation Conference*, June 1983, pp. 50–55.

[ClKS80] D. Clary, R. Kirk and S. Sapiro, "SIDS—A Symbolic Interactive Design System," *Proceedings of the 17th Design Automation Conference*, June 1980, pp. 292–295.

[Cobb66] R. S. C. Cobbold, "Temperature Effects on MOS Transistors," *Electronics Letters*, Vol. 2, No. 6, June 1966, pp. 190–192.

[Cobb70] R. S. Cobbold, *Theory and Application of Field Effect Transistors*, Wiley Interscience, 1970.

[Comb81] R. S. Combs, "Scaleable Retrograde P-Well CMOS Technology," *IEDM81*, 1981, pp. 346–348.

[CoSo82] B. W. Colbry and J. Soukup, "Layout Aspects of VLSI Microprocessor Design," *International Symposium on Circuits and Systems*, Rome, Italy, May 1982.

[Denn73] R. H. Dennard et al. in *Semiconductor Silicon Electrochemical Society*, H. R. Huff and R.R. Burgess, Editors, 1973.

[Denn74] R. H. Dennard et al., *IEEE J. Solid-State Circuits SC-9*, 1974.

[DeRB82] P. Denyer, D. Renshaw and N. Bergmann, "A Silicon Compiler for VLSI Signal Processors," *Proceedings of the 1982 European Solid State Circuits Conference*.

[DiSt79] A. G. F. Dingwall and R. G. Stewart, "16K CMOS/SOS Asynchronous Static RAM," *IEEE JSSC*, Vol. SC-14, No. 5, October 1979, pp. 867–872.

[Dunl80] A. Dunlop, "SLIM — The Translation of Symbolic Layouts into Mask Data," *Proceedings of the 17th Design Automation Conference*, June 1980, pp. 595–602.

[EbZa83] C. Ebeling and O. Zajicek, "Validating VLSI Circuit Layout by Wirelist Comparison," *Proceedings of IEEE Int. Conf. on CAD*, September 1983, pp. 172–173.

[EiWi77] E. B. Eichelberger and T. W. Williams, "A Logic Design Structure for LSI Testing," *Proc. 14th Design Automation Conference*, June 1977, pp. 462–468.

[ElCl81] Y. M. El-ziq and R. J. Cloutier, "Functional Level Test Generation for Stuck-Open Faults in CMOS VLSI," *Digest of Papers, IEEE Int. Test Conf.*, October 1981, pp. 536–546.

[Elzi81] Y. M. El-ziq, "Automatic Test Generation for Stuck-Open Faults in CMOS VLSI," *Proceedings of the 18th Design Automation Conference*, June 1981, pp. 347–354.

[EsDu82] D. B. Estreich and R. W. Dutton, "Modeling Latch-Up in CMOS Integrated Circuits and Systems," *IEEE Transactions on CAD*, Vol. CAD-1, No. 4, October 1982, pp. 157–162.

[Estr80] D. B. Estreich, "The Physics and Modeling of Latch-Up in CMOS Integrated Circuits," *Tech. Report No. G-2-1-9*, Integrated Circuits Laboratory, Stanford Electronics Lab., Stanford University, November 1980.

[FGLM84] E. Fujishin, K. Garrett, M. P. Louis, R. F. Motta and M. D. Hartranft, "Optimized ESD Protection Circuits for High Speed

MOS/VLSI," *IEEE Proceedings of the Custom Integrated Circuits Conference*, May 1984, pp. 569–573.

[FiFi82] J. Fishburn and R. Finkel, "Quotient Networks," *IEEE Trans. on Computers*, Vol. C-31, No. 4, April 1982, pp. 288–295.

[FPPB82] H. Fuchs, J. Poulton, A. Paeth and A. Bell, "Developing Pixel-Planes, A Smart Memory-Based Raster Graphic System," *Proceedings of MIT Conference on Advanced Research in VLSI*, January 1982, pp. 137–146.

[Frie73] A. D. Friedman, "Easily Testable Iterative Systems," *IEEE Trans. on Computers*, Vol. C-22, December 1973, pp. 1061–1064.

[FrLi84] V. Friedman and S. Liu, "Dynamic Logic CMOS Circuits," *IEEE JSSC*, Vol. SC-19, No. 2, April 1984, pp. 263–266.

[FuPo81] H. Fuchs and J. Poulton, "Pixel-Planes: A VLSI-Oriented Design for a Raster Graphics Engine," *LAMBDA (VLSI Design)*, Vol. II, No. 3, Third Quarter 1981.

[GaCV80] J. Galiay, Y. Crouzet and M. Vergniault, "Physical versus Logical Fault Models MOS LSI Circuits: Impact on Their Testability," *IEEE Transactions on Computers*, Vol. C-29, No. 6, June 1980, pp. 527–531.

[GEIn83] GE Intersil, "The IGC 20000 Series CMOS Gate Arrays," *Data Sheet*, 1983.

[GiBo83] E. F. Girczyc and A. R. Boothroyd, "A One-Dimensional DC Model for Nonrectangular IGFETs," *IEEE Journal of Solid State Circuits*, Vol. SC-18, No. 6, December 1983, pp. 778–784.

[GiLe80] J. F. Gibbons and K. F. Lee, *Electron Device Letters*, Vol. EDL-1, 1980, p. 117.

[GiNa76] D. Gibson and S. Nance, "SLIC — Symbolic Layout of Integrated Circuits," *Proceedings of the 13th Design Automation Conference*, June 1976, pp. 434–440.

[GoDM83] Nelson F. Goncalves and Hugo J. De Man, "NORA: A Racefree Dynamic CMOS Technique for Pipelined Logic Structures," *IEEE Journal of Solid-State Circuits*, Vol. SC-18, No. 3, June 1983, pp. 261–266.

[Gold79] L. H. Goldstein, "Controllability/Observability Analysis of Digital Circuits," *IEEE Trans. on Circuits and Systems*, Vol. CAS-26, No. 9, September 1979.

[GoTh80] L. H. Goldstein and E. L. Thigpen, "SCOAP: Sandia Controllability/Observability Analysis Program," *Proc. 17th Design Automation Conf.*, June 1980, pp. 190–196.

[Gour71] H. Gouraud, "Computer Display of Curved Surfaces," University of Utah Computer Science Department, UTEC-CCs-71-113, June 1971. Abridged version in *IEEE Trans. on Computers*, C-20, Vol. 6, No. 623, June 1971.

[GrHi83] W. R. Grifton and J. A. Hiltebeitel, "CMOS Four-Way XOR Circuit," *IBM Technical Disclosure Bulletin*, Vol. 25, No. 11B, April 1983, pp. 6066–6067.

[Gris82] T. W. Griswold, "Portable Design Rules for Bulk CMOS," *VLSI Design*, September-October 1982, pp. 62–67.

[GrLN83] M. A. Grimm, K. Lee and A. R. Neureuther, "SIMPL-1 (SIMulated Profiles from the Layout-Version 1)," *Proc. IEDM 1983*, December 1983, pp. 255–258.

[HeHo76] Ernst Hebenstreit and Karlheinrich Horninger, "High-Speed Programmable Logic Arrays in ESFI SOS Technology," *IEEE Journal of Solid-State Circuits*, Vol. SC-11, No. 3, June 1976, pp. 370–374.

[HGDT84] L. G. Heller, W. R. Griffin, J. W. Davis and N. G. Thoma, "Cascode Voltage Switch Logic: A Differential CMOS Logic Family," *IEEE International Solid State Circuits Conference*, February 1984, pp. 16–17.

[HJLV83] D. E. Hackleman, N. L. Johnson, C. S. Lage, J. J. Victor and R. L. Tillman, "CMOSC: Low Power Technology for Personal Computers," *Hewlett-Packard Journal*, Vol. 34, No. 1, January 1983.

[HoDu83] M. Horowitz and R. W. Dutton, "Resistance Extraction from Mask Layout," *IEEE Transactions on CAD*, Vol. CAD-2, No. 3, July 1983, pp. 145–150.

[HoLa83] M. Hofmann and V. Lauther, "HEX: An Instruction Driven Approach to Feature Extraction," *Proceedings of the 20th Design Automation Conference*, June 1983, pp. 331–336.

[Hou83] J. C. L. Hou, "Design of a Fully Associative Cache Memory Controller," Department of Electrical Engineering, MIT, VLSI memo 83-133.

[HsPe79] M. Y. Hsueh and D. O. Pederson, "Computer-Aided Layout of LSI Circuit Building Blocks," *Proceedings of the 1979 International Symposium on Circuits and Systems*, July 1979, pp. 474–477.

[IEDM83] "Device Technology—Isolation and Dielectrics," *Technical Digest, IEDM*, Session 2, December 1983, pp.19–46.

[JaAg83] Sunii K. Jain and Vishwani D. Agrawal, "Test Generation for MOS Circuits Using the D-Algorithm," *Proceedings of the 20th Design Automation Conference*, June 1983, pp. 64–70.

[Joha78] D. Johannsen, "Bristle Blocks: A Silicon Compiler," *Proceedings of the 16th Design Automation Conference*, June 1978, pp. 310–313.

[Joup83] N. P. Jouppi, "Timing Analysis for nMOS VLSI," *Proceedings of the 20th Design Automation Conference*, June 1983, pp. 411–418.

[Kang81] Sung Mo Kang, "A Design of CMOS Polycells for LSI Circuits," *IEEE Transactions on Circuits and Systems*, Vol. CAS-28, No. 8, August 1981, pp. 838–843.

[KeWa83] G. Kedem and H. Watanabe, "Graph Optimization Techniques for IC Layout and Compaction," *Proceedings of the 20th Design Automation Conference*, June 1983, pp. 113–120.

[KiMc83] J. Kim and J. McDermott, "TALIB: An IC Layout Design Assistant," *Proceedings of the AAAI Conference*, Washington, DC, 1983, pp. 197–201.

[KKLL82] S. M. Kang, R. H. Krambeck, H-F. S. Law and A. D. Lopez, "Gate Matrix Layout of Random Control Logic in a 32-Bit CMOS Chip Adaptable to Evolving Logic Design," *Design Automation Conference*, Las Vegas, NV, June 1982.

[KMSW83] W. R. Kraft, V. S. Moore, W. L. Stahl and T. J. Wylie, "PLA Else Clause Implementation," *IBM Technical Disclosure Bulletin*, Vol. 25, No. 12, May 1983, p. 6502.

[KoMZ79] B. Koenemann, J. Mucha and G. Zwiehoff," Built-in Logic Block Observation Techniques," *Digest 1979 Test Conference*, 79CH1509-9C, October 1979, pp. 37–41.

[KoTh83] T. J. Kowalski and D. E. Thomas, "The VLSI Design Automation Assistant: Prototype System," *Proceedings of the 20th Design Automated Conference*, June 1983.

[KoWe84] P. W. Kollaritsch and N. Weste, "TOPOLOGIZER: An Expert System Translator of Transistor Connectivity to Symbolic Layout," *Proceedings ESSCIRC*, 1984.

[KrLi83] M. Krejak and R. Lipp, "Logic Design with CMOS Gate Arrays," *VLSI Design*, October 1983, pp. 86–98.

[KrLL82] R. H. Krambeck, C. M. Lee and H-F. S. Law, "High Speed Compact Circuits with CMOS," *IEEE Journal of Solid State Circuits*, June 1982.

[KSIM83] S. Kawamura, N. Sasaki, T. Iwai, R. Mukai, M. Nakano and M. Takagi, "3-Dimensional SOI/CMOS IC's Fabricated by Beam Recrystallization," *Technical Digest, International Electron Devices Meeting*, December 1983, pp. 364–367.

[Lars78] R. P. Larson, "Versatile Mask Generation Techniques for Customer

Microelectronic Devices," *Proceedings of the 15th Design Automation Conference,* June 1978, pp. 193–198.

[LaSh82] H-F. S. Law and M. Shoji, "PLA Design for the BELLMAC-32A Microprocessor," *Proceedings of the ICCC,* 1982, pp. 161–164.

[Latt82] Lattice Logic Ltd., "Designing with Gate Arrays," *Edition 3.2 LLL,* Edinburgh, Scotland, 1982.

[Lawr75] D. Lawrie, "Access and Alignment of Data in an Array Processor," *IEEE Trans. on Computers,* Vol. C-24, No. 12, December 1975, pp. 1145–1155.

[LeCJ81] C. M. Lee, B. R. Chawla and S. Just, "Automatic Generation and Characterization of CMOS Polycells," *Proceedings of the 18th Design Automation Conference,* June 1981, pp. 220–224.

[LoLa80] A. D. Lopez and H-F. S. Law, "A Dense Gate Matrix Layout Style for MOS LSI," *IEEE Journal of Solid-State Circuits,* Vol. SC-15, No. 4, August 1980, pp. 736–740.

[LuKu82] J. Luhukay and W. J. Kubitz, "A Layout Synthesis System for nMOS Gate-Cells," *Proceedings 19th DAC,* June 1982, pp. 307–314.

[Lyon76] R. F. Lyon, "Two's Complement Pipeline Multiplier," *IEEE Trans. on Communications,* Vol. COM-24, April 1976, pp. 418–425.

[MaSi64] H. M. Manasevit and W. I. Simpson, "Single Crystal Silicon on a Sapphire Substrate," *J. App. Phys.,* Vol. 35, 1964, pp. 1349–1351.

[McCB81] E. J. McCluskey and S. Bozorgui-Nesbat, "Design for Autonomous Test," *IEEE Transactions on Computers,* Vol. C-30, No. 11, November 1981, pp. 866–875.

[MeCo80] C. A. Mead and L. A. Conway, *Introduction to VLSI Systems,* Reading, MA: Addison-Wesley, 1980.

[METM81] B. T. Murphy, R. Edwards, L. C. Thomas and J. J. Molinelli, "A CMOS 32-Bit Single Chip Microprocessor," *ISSCC Digest of Technical Papers,* 1981, pp. 230–231.

[MMSN80] O. Minato, T. Masuhara, T. Sasaki, H. Nakamura, Y. Sakai, T. Yasui and K. Vehihori, "2K x 8 Hi-CMOS Static RAM's," *IEEE Journal of Solid State Circuits,* Vol. SC-15, No. 4, August 1980.

[Most81] R. C. Mosteller, "Rest — A Leaf Cell Design System," *VLSI '81,* 1981, pp. 163–172.

[Mudg80] J. C. Mudge, "A VLSI Chip Assembler," *Design Methodologies for Very Large Scale Integrated Circuits,* NATO Advanced Summer Institute, Belgium, 1980.

[Murp64] B. T. Murphy, "Cost-Size Optima of Monolithic Integrated Circuits," *Proc. IEEE,* Vol. 52, December 1964, pp. 1537–1545.

[Murp83] Bernard T. Murphy, "Microcomputers: Trends, Technologies, and Design Strategies," *IEEE Journal of Solid-State Circuits,* Vol. SC-18, No. 3, June 1983, pp. 236–244.

[Nage75] L. W. Nagel, "SPICE2: A Computer Program to Simulate Semiconductor Circuits," *Memo ERL-M520,* University of California, Berkeley, CA, May 9, 1975.

[Nage80] L. W. Nagel, "ADVICE for Circuit Simulation," *IEEE International Symposium on Circuits and Systems,* Houston, TX, April 1980.

[NeSp79] W. M. Newman and R. F. Sproull, *Principles of Interactive Computer Graphics,* 2nd edition, New York: McGraw-Hill, 1979.

[OhKD79] N. Ohwada, T. Kimura and M. Doken, "LSI's for Digital Signal Processing," *IEEE Journal of Solid-State Circuits,* April 1979, Vol. SC-14, No. 2, pp. 221–239.

[Ohzo80] T. Ohzone et al., "Silicon-Gate n-Well CMOS Process by Full Ion-Implantation Technology," *IEEE Trans. Electron Devices,* Vol. ED-27, September 1980, pp. 1789–1795.

[OYIH80] T. Ohzone, J. Yasui, T. Ishihara and S. Horiuchi, "An 8K*8 Static MOS RAM Fabricated by nMOS/nWell CMOS Technology," *IEEE JSSC*, Vol. SC-15, No. 5, October 1980, pp. 854–861.

[Parr80] L. C. Parrillo et al., "Twin-Tub CMOS — A Technology for VLSI Circuits," *IEEE Int. Electron Devices Meeting*, 1980, pp. 752–755.

[PBHK82] M. Pomper, W. Beifuss, K. Horninger and W. Kaschite, "A 32-Bit Execution Unit in an Advanced nMOS Technology," *IEEE Journal of Solid State Circuits*, Vol. SC-17, No. 3, June 1982, pp. 533–538.

[PeDS77] G. Persky, D. N. Deutsch and D. G. Schweikent, "LTX — A Minicomputer-Based System for Automated LSI Layout," *Journal of Design Automation and Fault Tolerant Computing*, Vol. 1, No. 3, May 1977, pp. 217–255.

[PeLa73] W. M. Penny and L. Lau, *MOS Integrated Circuits: Theory, Fabrication, Design and Systems Applications of MOS LSI*, New York: Van Nostrand, Reinhold, 1973.

[PZSB84] C. Piguet, J. Zahnd, A. Stauffer and M. Bertarionne, "A Metal-Oriented Layout for CMOS Logic," *IEEE JSSC*, Vol. SC-19, No. 3, June 1984, pp. 425–436.

[RBDD83] J. Rosenberg, D. Boyer, J. Dallen, S. Daniel, C. Poirier, J. Poulton, D. Rogers and N. Weste, "A Vertically Integrated VLSI Design Environment," *Proceedings of the 20th Design Automation Conference*, June 1983, pp. 31–38.

[ReIv83] M. C. Revett and P. A. Ivey, "ASTRA — A CAD System to Support a Structured Approach to IC Design," *VLSI '83*, 1983, pp. 413–422.

[Ride79] V. L. Rideout, "One-Device Cells for Dynamic Random-Access Memories: A Tutorial," *IEEE Trans. Electron Devices*, Vol. ED-26, June 1979, pp. 839–852.

[RoBS67] J. P. Roth, W. G. Bouricus and P. R. Schneider, "Programmed Algorithms to Compute Tests to Detect and Distinguish Between Failures in Logic Circuits," *IEEE Trans. on Electronic Computers*, Vol. EC-16, October 1967, pp. 547–580.

[Rose83] J. Rosenberg, "A Vertically Integrated VLSI Design Environment," Ph.D. Thesis, Duke University, 1973.

[RoWe82] J. Rosenberg and N. Weste, "The ABCD Language," *Microelectronic Center of North Carolina Tech Report 82-01*, 1982.

[Rows83] J. Rowson, "Understanding Hierarchical Design," Ph.D. Thesis, California Institute of Technology, 1983.

[Rubi83] S. Rubin, "An Integrated Aid for Top-Down Electrical Design," *VLSI 83*, Norway, August 1983, pp. 63–72.

[RuBr73] A. E. Ruehli and P. A. Brennan, "Accurate Metallization Capacitances for Integrated Circuits and Packages," *IEEE JSSC*, Vol. SC-8, No. 4, August 1973, p. 289.

[RuDi83] A. E. Ruehli and G. S. Ditlow, "Circuit Analysis, Logic Simulation and Design Verification for VLSI," *Proc. IEEE*, Vol. 71, No. 1, January 1983, pp. 34–48.

[Rung81] R. D. Rung, "Determining IC Layout Rules for Cost Minimization," *IEEE Journal of Solid State Circuits*, Vol. SC-16, No. 1, February 1981, pp. 35–43.

[SaAr82] K. Saito and E. Arai, "Experimental Analysis and New Modeling of MOS LSI Yield Associated with the Number of Elements," *IEEE Journal of Solid State Circuits*, Vol. SC-17, No. 1, February 1982, pp. 28–33.

[SaCh71] H. Sakoe and S. Chiba, "A Dynamic Programming Approach to Continuous Speech Recognition," *Proceedings 7th Int. Conf. Acoustics*, Budapest, Hungary, August 1971, pp. 65–68.

[Sah64] C. T. Sah, "Characteristics of the Metal-Oxide-Semiconductor Transistor," *IEEE Trans. ED,* ED-11, July 1964, pp. 324–345.

[Savi80] J. Savir, "Syndrome-Testable Design of Combinational Circuits," *IEEE Trans. on Computers,* Vol. C-29, June 1980, pp. 442–451.

[Savi81] J. Savir, "Syndrome-Testing of 'Syndrome-Untestable' Combinational Circuits," *IEEE Trans. on Computers,* Vol. C-30, August 1981, pp. 606–608.

[ScVe70] J. J. H. Schatorge and C. G. Verkuylen, "Local Oxidation of Silicon and Its Application in Semiconductor-Device Technology," *Philips Res. Rep.,* Vol. 25, No. 2, April 1970, pp. 118–132.

[Seed67] R. B. Seeds, "Yield and Cost Analysis of Bipolar LSI," *Proc. IEEE Int. Electron Devices Meeting,* Paper 1.1, October 1967.

[Shiv83] Sajjan G. Shiva, "Automatic Hardware Synthesis," *Proceedings of the IEEE,* Vol. 71, No. 1, January 1983, pp. 76–87.

[Shoj82] M. Shoji, "Electrical Design of the BELLMAC-32A Microprocessor," *Proc. ICCC,* New York, January 1982, pp. 112–115.

[Shro82] H. E. Shrobe, "The Data Path Generator," *Proceedings of Conference on Advanced Research in VLSI,* January 1982, pp. 175–181.

[SiMc81] H. Siegel and R. H. McMillen, "Using the Augmented Data Manipulator Network in PASM," *Computer,* Vol. 14, February 1981, pp. 25–33.

[SiSC82] J. M. Siskind, J. R. Southand and K. W. Crouch, "Generating Custom High Performance VLSI Designs from Succinct Algorithm Descriptions," *Proceedings of Conference on Advanced Research in VLSI,* January 1982, pp. 28–40.

[Smit83] K. F. Smith, "Design of Regular Arrays Using CMOS in PPL," *Proceedings IEEE Int. Conference on Computer Design,* ICCD '83, November 1983, pp. 158–161.

[SKPS82] R. W. Sherburne, M. G. H. Katevenis, D. A. Patterson and C. H. Sequin, "Datapath Design for RISC," *Proc. Conf. on Advanced Research in VLSI,* MIT, January 1982, pp. 53–62.

[SpNe83] R. L. Spickelmier and A. R. Newton, "Wombat: A New Netlist Comparison Program," *Proc. IEEE Int. Conf. on CAD,* September 1983, pp. 170–171.

[SrHa81] T. Sridhar and J. P. Hayes, "Design of Easily Testable Bit-Sliced Systems," *IEEE Trans. on Computers,* Vol. C-30, No. 11, November 1981, pp. 842–854.

[Ston71] H. Stone, "Parallel Processing with the Perfect Shuffle," *IEEE Trans. on Computing,* Vol. C-20, February 1971, pp. 153–161.

[StSD72] K. U. Stein, A. Sihling and E. Doering, "Storage Array and Sense/Refresh Circuit for Single-Transistor Memory Cells," *IEEE JSSC,* Vol. SC-7, October 1972, pp. 336–340.

[SuOA73] Yasoji Suzuki, Kaichiro Odagawa and Toshio Abe, "Clocked CMOS Calculator Circuitry," *IEEE Journal of Solid-State Circuits,* Vol. SC-8, No. 6, December 1973, pp. 462–469.

[SzVW83] T. G. Szymanski and C. J. Van Wyk, "Space Efficient Algorithms for VLSI Artwork Analysis," *Proceedings of the 20th Design Automation Conference,* June 1983, pp. 734–739.

[Tana84] S. Tanaka, J. Iwamura, J. Ohno, K. Maeguchi, H. Tango and K. Doi, "A Sub-Nanosecond 8K-Gate CMOS/SOS Gate Array," *Proceedings ISSCC,* February 1984, pp. 260–261.

[TIIF84] M. Takechi, K. Ikuzaki, T. Itoh and M. Fujita, "A CMOS 12K Gate-Array with Flexible 10Kb Memory," *Proceedings ISSCC,* February 1984, pp. 258–259.

[Tsai83] M. Y. Tsai, "High Density Parity-Checking Circuits with Pass Transistors," *IBM Technical Disclosure Bulletin*, Vol. 26, No. 3A, August 1983, pp. 959–960.

[UeVC79] T. Uehara and W. M. van Cleemput, "Optimal Layout of CMOS Functional Arrays," *IEEE Trans. on Computers*, Vol. C-30, No. 5, May 1981, pp. 305–311.

[UyKY84] M. Uya, K. Kaneko and J. Yasui, "A CMOS Floating Point Multiplier," *IEEE International Solid State Circuits Conference, Digest of Technical Papers*, February 1984, pp. 90–91.

[VaGr66] L. Vadasz and A. S. Grove, "Temperature of MOS Transistor Characteristics Below Saturation," *IEEE Trans. on Electron Devices*, Vol. ED-13, No. 13, 1966, pp. 190–192.

[Veen84] H. J. M. Veendrick, "Short-Circuit Dissipation of Static CMOS Circuitry and Its Impact on the Design of Buffer Circuits," *IEEE Journal of Solid State Circuits*, Vol. SC-19, No. 4, August 1984, pp. 468–473.

[Wads78] R. L. Wadsack, "Fault Modelling and Logic Simulation of CMOS and MOS Integrated Circuits," *Bell System Technical Journal*, Vol. 57, No. 4, May-June 1978, pp. 1449–1474.

[WeAc81] N. Weste and B. Ackland, "A Pragmatic Approach to Topological Symbolic IC Design," *Proc. VLSI '81*, Edinburgh, Scotland, August 1981, pp. 117–129.

[WeBA83] N. Weste, D. Burr and B. Ackland, "Dynamic Time Warp Pattern Matching Using an Integrated Multiprocessing Array," *IEEE Trans. on Computers*, Vol. C-32, No. 8, August 1983, pp. 731–744.

[Wein67] A. Weinberger, "Large Scale Integration of MOS Complex Logic: A Layout Method," *IEEE JSSC*, Vol. SC-2, December 1967, pp. 182–190.

[Wern84] J. Werner, "Custom IC Design in Europe," *VLSI Design*, January 1984, pp. 28–33.

[West81a] N. Weste, "Virtual Grid Symbolic Layout," *Proc. of 18th Design Automation Conference*, Nashville, TN, June 1981, pp. 225–233.

[West81b] N. Weste, "MULGA — An Interactive Symbolic Layout System for the Design of Integrated Circuits," *Bell System Technical Journal*, Vol. 60, No. 6, July-August 1981, pp. 823–858.

[Whit81] T. Whitney, "A Hierarchical Design Analysis Front End," *VLSI 81*, 1981, pp. 217–225.

[Whit83] Sterling Whitaker, "Pass-Transistor Networks Optimize n-MOS Logic," *Electronics*, September 22, 1983, pp. 144–148.

[Will78] J. Williams, "STICKS — A Graphical Compiler for High Level LSI Design," *Proceedings NCC*, May 1978, pp. 289–295.

[WiPa83] Thomas W. Williams and Kenneth P. Parker, "Design for Testability — A Survey," *Proceedings of the IEEE*, Vol. 71, No. 1, January 1983, pp. 98–112.

[Wing82] O. Wing, "Automated Gate-Matrix Layout," *Proc. IEEE International Symposium on Circuits and Systems*, 1982, pp. 681–685.

[Wing83] Omar Wing, "Interval-Graph-Based Circuit Layout," *Proceedings ICCAD*, 1983, pp. 84–85.

[WuFe81] C. Wu and T. Feng, "The University of the Shuffle-Exchange Network," *IEEE Trans. of Computing*, Vol. C-30, No. 5, May 1981, pp. 324–332.

[YMKP83] T. Yamaguchi, S. Morimoto, G. H. Kawamoto, H. K. Park and G. C. Eden, "High-Speed Latchup-Free p.5um Channel CMOS Using Self-Aligned TiSi2 and Deep-Trench Isolation Technologies," *Technical Digest, IEDM*, December 1983, pp. 522–525.

BIBLIOGRAPHY

Alexander, D. R. et al., "SPICE 2 Modeling Handbook," *Report BDM/A-77-071-TR*, New Mexico: BDM Corporation.

AMI Staff, "MOS Process," *IEEE Trans. on Consumer Electronics*, CE24, 1978, pp. 155–167.

Antognetti, P., Antoniadis, D. A., Dutton, R. W. and Oldham, G. William (Eds.), "Process and Device Simulation Les MOS-VLSI Circuits," NATO ASI Series, Netherlands: Martinus Nijhoff Publishers, 1983.

Appels, J. A., Kovi, E., Daffen, M. M., Schatorge, J. J. H. and Verkuylen, C. G., "Local Oxidation of Silicon and Its Application in Semiconductor-Device Technology," *Philips Res. Rep.*, Vol. 25, No. 2, April 1970, pp. 118–132.

Armstrong, G. A. and Magowan, J. A., "Pinch-off in Insulated-Gate Field Effect Transistors," *Solid State Electron.*, Vol. 14, 1971, pp. 760–764.

Armstrong, G. A. and Magowan, J. A., "The Distribution of Mobile Carriers in the Pinch-off Region of an Insulated-Gate Field-Effect Transistor and Its Influence on Device Breakdown," *Solid State Electronics*, Vol. 14, 1971, pp. 723–733.

Bailey, C. M., "Basic Integrated Circuit Failure Mechanisms," *Large Scale Integrated Circuit Technology: State of the Art and Prospects*, Hague: Martinus Nijhoff Publishers, 1982.

Barbe, D. F. (Ed.), *Very Large Scale Integration: VLSI Fundamentals and Applications*, Berlin, New York: Springer-Verlag, 1982.

Bean, K. E., "Anisotropic Etching of Silicon," *IEEE Trans. Electron Devices*, Vol. ED-25, October 1978, pp. 1185–1193.

Burns, J. R., "Switching Response of Complementary Symmetry MOS Transistor Logic Circuits," *RCA Review*, December 1964, pp. 627–661.

Carr, W. N. and Mize, J. P., *MOS/LSI Design and Application*, New York: McGraw-Hill, 1972.

Cobbold, R. S. C., *Theory and Applications of Field Effect Transistors*, New York: Wiley, 1970.

Dennehy, W. J. et al., "Transient Radiation Response of Complementary Symmetry MOS Integrated Circuits," *IEEE Trans. Nuclear Science*, NS-16, December 1969, p. 114.

DeTroye, N. C., "Digital Integrated Circuits with Low Dissipation," *Philips Tech. Rev.*, 1975, pp. 910–911.

El-Mansy, Y. A. and Boothroyd, A. R., "A New Approach to the Theory and Modeling of Insulated-Gate Field-Effect Transistors," *IEEE Transactions on Electron Devices*, March 1977, pp. 241–253.

Faggin, F. and Klein, T., "Silicon Gate Technology," *Solid-State Electronics*, 13, 1974, pp. 1125–1144.

Fang, F. F. and Rupprecht, R. S., "High Performance MOS Integrated Circuit Using the Ion-Implantation Technique," *IEEE J. Solid-State Circuits*, SC-10, 1975, pp. 205–211.

Fisher, M. and Young, A., "CMOS Technology Can Do It All in Digital and Linear Systems," *EDN*, June 1979, pp. 106–114.

Frohman-Bentchkowsky, D. and Grove, A. S., "Conductance of MOS Transistors in Saturation," *IEEE Transactions on Electron Devices*, Vol. ED-16, No. 1, January 1969, pp. 108–113.

Glaser, A. B. and Subak-Sharpe, G. E., *Integrated Circuit Engineering: Design, Fabrication and Applications*. Reading, MA: Addison-Wesley, 1977.

Goetzberger, A., "Ideal MOS Curves for Silicon," *Bell System Technical Journal*, Vol. 45, 1966, pp. 1097–1122.

Gray, P. R., Hodges, D. A. and Brodersen, R. W. (Eds.), "Analog MOS Integrated Circuits," *IEEE Press*, 1980.

Grove, A. S., *Physics and Technology of Semiconductor Devices*, New York: Wiley, 1967.

Herman, F. P. and Hofstein, S. R., "Metal Oxide Semiconductor Field Effect Transistors," *Electronics*, November 1964, pp. 50–62.

Hofstein, S. R. and Heiman, F. P. E., "The Silicon Insulated-Gate Field-Effect Transistor," *Proc. Inst. Elec. Electron Eng.*, 51, 1963, pp. 1190–1202.

Hon, B. and Sequin, C. H. *Guide to LSI Implementation*, 2nd Edition, Palo Alto, CA: Xerox PARC, January 1980.

IEEE Journal of Solid State Circuits.

IEEE Proceedings of the Custom Integrated Circuit Conference.

Lyon, R. F., "Simplified Design Rules for VLSI Layouts," *Lambda*, Vol. 2, No. 1, 1981, pp. 54–59.

Masuda H. et al., "Characteristics and Limitations of Scaled-Down MOSFETs Due to Two-Dimensional Field Effect," *IEEE Transactions on Electron Devices*, Vol. ED-26, No. 5, June 1979.

Mavor, J., *MOST Integrated-Circuit Engineering*, London: Peter Peregrinus Ltd., 1973.

Mead, C. and Conway, L., *Introduction to VLSI Systems*, Reading, MA: Addison-Wesley, 1980.

Motamdei, M. E., Tam, K. Y. and Steckl, A. J., "Design and Evaluation of Ion-Implanted CMOS Structures," *IEEE Transactions Electron Devices*, Vol. ED-27, No. 3, March 1980, pp. 578, 583.

Motorola Semiconductor Products Inc., *McMOS Handbook*, 1974.

Muroga, S., *VLSI System Design — When and How to Design Very Large Scale Integrated Circuits*, New York: Wiley, 1982.

Nathan, I. N. (Ed.), "MOS Integrated Circuits and Their Applications," *Philips Application Book*, Mullard Limited, 1970.

Penny, W. M. and Lau, L., *MOS Integrated Circuits: Theory, Fabrication, Design and Systems Applications of MOS LSI*, New York: Van Nostrand, Reinhold Co., 1973.

Prince, J. L., "VLSI Device Fundamentals," *Very Large Scale Integration: Fundamentals and Applications*, Barbe, D. F. (Ed.), Berlin, New York: Springer-Verlag, 1982.

Proceedings of the European Solid State Circuits Conference (ESSCIRC).

Proceedings of the IEEE Custom Integrated Circuits Conference.

Proceedings of the IEEE Design Automation Conference.

Proceedings of the IEEE Solid State Circuits Conference (ISSCC).

RCA Solid State, *COS/MOS Integrated Circuits Manual*, CMS-272, 1979.

Richman, P., *MOS Field-Effect Transistors and Integrated Circuits*, New York: Wiley, 1973.

Richman, P., *Characteristics and Operation of MOS Field-Effect Devices*, New York: McGraw-Hill, 1967, pp. 63–73.

Sarace, J. C., Kerwin, R. E., Klein, D. L. and Edwards, R., "Metal-Nitride-Oxide Silicon Field Effect Transistors with Self-aligned Gates," *Solid State Electron.*, No. 11, 1968, pp. 653–660.

Sequin, C. H., "Generalized IC Layout Rules and Layout Representation," *VLSI 81*, J. P. Gray (Ed.), Academic Press, 1981.

Streetman, B. G., *Solid State Electronic Devices*, Englewood Cliffs, NJ: Prentice-Hall, 1980.

Till, C. W. and Luxon, J. T., *Integrated Circuits: Materials, Devices and Fabrication*, Englewood Cliffs, NJ: Prentice-Hall, 1982.

Toshiba Inc., "C^2MOS Integrated Circuits — Technical Data," AT-101D-1, First Ed., June 1981.

Wang, R., Dunkley, J., DeMassa, T. and Jelsma, L., "Threshold Voltage Variations with Temperature in MOS Transistors," *IEEE Transactions on Electron Devices*, Vol. ED-18, No. 6, 1971, pp. 386–388.

INDEX